Methods in Enzymology

Volume 130
ENZYME STRUCTURE
Part K

METHODS IN ENZYMOLOGY

EDITORS-IN-CHIEF

Sidney P. Colowick   Nathan O. Kaplan

Methods in Enzymology

Volume 130

Enzyme Structure

Part K

EDITED BY

C. H. W. Hirs

DEPARTMENT OF BIOCHEMISTRY, BIOPHYSICS, AND GENETICS
UNIVERSITY OF COLORADO MEDICAL CENTER
DENVER, COLORADO

Serge N. Timasheff

GRADUATE DEPARTMENT OF BIOCHEMISTRY
BRANDEIS UNIVERSITY
WALTHAM, MASSACHUSETTS

1986

ACADEMIC PRESS, INC.
Harcourt Brace Jovanovich, Publishers
Orlando   San Diego   New York   Austin
Boston   London   Sydney   Tokyo   Toronto

O 83321

ACADEMIC PRESS, INC.
Orlando, Florida 32887

United Kingdom Edition published by
ACADEMIC PRESS INC. (LONDON) LTD.
24–28 Oval Road, London NW1 7DX

LIBRARY OF CONGRESS CATALOG CARD NUMBER: 54-9110

ISBN 0–12–182030–0

PRINTED IN THE UNITED STATES OF AMERICA

86 87 88 89   9 8 7 6 5 4 3 2 1

Table of Contents

Section I. Macromolecular Self-Associations and Structural Assemblies

Section II. Macromolecular Conformation: Spectroscopy

v

Section III. Macromolecular Conformational Stability and Transitions

Contributors to Volume 130

Article numbers are in parentheses following the names of contributors.
Affiliations listed are current.

GARY K. ACKERS (9), *Department of Biology, The Johns Hopkins University, Baltimore, Maryland 21218*

JOSE MANUEL ANDREU (5), *Centro de Investigaciones Biologicas, C.S.I.C., 28006 Madrid, Spain*

PATRICK ARGOS (10), *European Molecular Biology Laboratory, D-6900 Heidelberg, Federal Republic of Germany*

DALILA BENDEDOUCH (7), *Nuclear Engineering Department, Massachusetts Institute of Technology, Cambridge, Massachusetts 02139*

SHERMAN BEYCHOK[1] (22), *Departments of Biological Sciences and Chemistry, Columbia University, New York, New York 10027*

RODNEY L. BILTONEN (23), *Department of Pharmacology and Interdisciplinary Program in Biophysics, University of Virginia, Charlottesville, Virginia 22908*

STEVEN G. BOXER (20), *Department of Chemistry, Stanford University, Stanford, California 94305*

MICHAEL BRENOWITZ (9), *Department of Biology, The Johns Hopkins University, Baltimore, Maryland 21218*

RODNEY R. BUCKS (20), *Department of Chemistry, Stanford University, Stanford, California 94305*

D. MICHAEL BYLER (13), *Eastern Regional Research Center, Agricultural Research Service, United States Department of Agriculture, Philadelphia, Pennsylvania 19118*

JOHN R. CANN (3), *Department of Biochemistry/Biophysics/Genetics, University of Colorado Health Sciences Center, Denver, Colorado 80262*

SOW-HSIN CHEN (7), *Nuclear Engineering Department, Massachusetts Institute of Technology, Cambridge, Massachusetts 02139*

E. L. ELSON (19), *Division of Biological Chemistry, Washington University Medical School, St. Louis, Missouri 63110*

FRANK R. N. GURD (17, 18), *Department of Chemistry, Indiana University, Bloomington, Indiana 47405*

HANS-JÜRGEN HINZ (6), *Institut für Biophysik und Physikalische Biochemie, Universität Regensburg, D-8400 Regensburg, Federal Republic of Germany*

BARTON HOLMQUIST (12), *Center for Biochemical and Biophysical Sciences and Medicine, Harvard Medical School, Boston, Massachusetts 02115*

BRUCE HUDSON (15), *Department of Chemistry, University of Oregon, Eugene, Oregon 97403-1210*

MICHAEL L. JOHNSON (23), *Department of Pharmacology and Interdisciplinary Program in Biophysics, University of Virginia, Charlottesville, Virginia 22908*

JEREMY LUCHINS (22), *Department of Biological Sciences, Columbia University, New York, New York 10027*

HUGO M. MARTINEZ (11), *Department of Biochemistry and Biophysics, University of California, San Francisco, San Francisco, California 94143*

JAMES B. MATTHEW (17, 18), *Central Research & Development Department, E. I. du Pont de Nemours & Company, Experimental Station, Wilmington, Delaware 19898*

[1] Deceased.

LELAND MAYNE (15), *Department of Chemistry, University of Oregon, Eugene, Oregon 97403-1210*

J. K. MOHANARAO (10), *Department of Biological Sciences, Purdue University, West Lafayette, Indiana 47097*

LAWRENCE W. NICHOL (2), *Office of the Vice-Chancellor, University of New England, Armidale, New South Wales 2351, Australia*

ALEJANDRO A. PALADINI, JR. (21), *Instituto de Ingenieria Genética y Biología Molecular (INGEBI)–CONICET and Facultad de Ciencias Exactas (UBA), 1427 Buenos Aires, Argentina*

N. O. PETERSEN (19), *Department of Chemistry, The University of Western Ontario, London, Ontario, Canada N6A 5B7*

MOLLIE PFLUMM (22), *Department of Biological Sciences, Columbia University, New York, New York 10027*

V. PRAKASH (1), *Biophysical Chemistry Section, Department of Food Chemistry, Central Food Technological Research Institute, Mysore-570 013, India*

DONALD F. SENEAR (9), *Department of Biology, The Johns Hopkins University, Baltimore, Maryland 21218*

PHILIP SERWER (8), *Department of Biochemistry, The University of Texas Health Science Center, San Antonio, Texas 78284-7760*

MADELINE A. SHEA (9), *Department of Biology, The Johns Hopkins University, Baltimore, Maryland 21218*

HEINO SUSI (13), *Eastern Regional Research Center, Agricultural Research Service, United States Department of Agriculture, Philadelphia, Pennsylvania 19118*

SERGE N. TIMASHEFF (1, 5), *Graduate Department of Biochemistry, Brandeis University, Waltham, Massachusetts 02154*

WILLIAM W. VAN OSDOL (23), *Department of Pharmacology and Interdisciplinary Program in Biophysics, University of Virginia, Charlottesville, Virginia 22908*

ROBERT W. WILLIAMS (14), *Department of Biochemistry, Uniformed Services University of the Health Sciences, Bethesda, Maryland 20814-4799*

ROBLEY C. WILLIAMS, JR. (4), *Department of Molecular Biology, Vanderbilt University, Nashville, Tennessee 37235*

DONALD J. WINZOR (2), *Department of Biochemistry, University of Queensland, St. Lucia, Queensland 4067, Australia*

CHUEN-SHANG C. WU (11), *Cardiovascular Research Institute, University of California, San Francisco, San Francisco, California 94143*

JEN TSI YANG (11), *Department of Biochemistry and Biophysics and Cardiovascular Research Institute, University of California, San Francisco, San Francisco, California 94143*

NAI-TENG YU (16), *School of Chemistry, Georgia Institute of Technology, Atlanta, Georgia 30332*

Preface

Enzyme Structure, Part J, Volume 117 of *Methods in Enzymology,* was devoted to physical methods. This volume and Volume 131 also deal in detail with physical methods. It is hoped that they present up-to-date coverage of techniques currently available for the study of enzyme conformation, interactions, and dynamics.

As in the past, these volumes present not only techniques that are currently widely available but some which are only beginning to make an impact and some for which no commercial standard equipment is as yet available. In the latter cases, an attempt has been made to guide the reader in assembling his own equipment from individual components and to help him find the necessary information in the research literature.

In the coverage of physical techniques, we have departed somewhat in scope from the traditional format of the series. Since, at the termination of an experiment, physical techniques frequently require much more interpretation than do organic ones, we consider that brief sections on the theoretical principles involved are highly desirable as are sections on theoretical and mathematical approaches to data evaluation and on assumptions and, consequently, limitations involved in the applications of the various methods.

The organization of the material is similar to that of the previous volumes, with Part K being devoted primarily to techniques related to molecular protein interactions and protein conformation.

We wish to acknowledge with pleasure and gratitude the generous cooperation of the contributors to this volume. Their suggestions during its planning and preparation have been particularly valuable. The staff of Academic Press has provided inestimable help in the assembly of this volume. We thank them for their many courtesies.

C. H. W. HIRS
SERGE N. TIMASHEFF

METHODS IN ENZYMOLOGY

EDITED BY

Sidney P. Colowick and Nathan O. Kaplan

VANDERBILT UNIVERSITY SCHOOL OF MEDICINE NASHVILLE, TENNESSEE

DEPARTMENT OF CHEMISTRY UNIVERSITY OF CALIFORNIA AT SAN DIEGO LA JOLLA, CALIFORNIA

METHODS IN ENZYMOLOGY

EDITORS-IN-CHIEF

Sidney P. Colowick and Nathan O. Kaplan

VOLUME XXXII. Biomembranes (Part B)
Edited by SIDNEY FLEISCHER AND LESTER PACKER

VOLUME XXXIII. Cumulative Subject Index Volumes I–XXX
Edited by MARTHA G. DENNIS AND EDWARD A. DENNIS

VOLUME XXXIV. Affinity Techniques (Enzyme Purification: Part B)
Edited by WILLIAM B. JAKOBY AND MEIR WILCHEK

VOLUME XXXV. Lipids (Part B)
Edited by JOHN M. LOWENSTEIN

VOLUME XXXVI. Hormone Action (Part A: Steroid Hormones)
Edited by BERT W. O'MALLEY AND JOEL G. HARDMAN

VOLUME XXXVII. Hormone Action (Part B: Peptide Hormones)
Edited by BERT W. O'MALLEY AND JOEL G. HARDMAN

VOLUME XXXVIII. Hormone Action (Part C: Cyclic Nucleotides)
Edited by JOEL G. HARDMAN AND BERT W. O'MALLEY

VOLUME XXXIX. Hormone Action (Part D: Isolated Cells, Tissues, and Organ Systems)
Edited by JOEL G. HARDMAN AND BERT W. O'MALLEY

VOLUME XL. Hormone Action (Part E: Nuclear Structure and Function)
Edited by BERT W. O'MALLEY AND JOEL G. HARDMAN

VOLUME XLI. Carbohydrate Metabolism (Part B)
Edited by W. A. WOOD

VOLUME XLII. Carbohydrate Metabolism (Part C)
Edited by W. A. WOOD

VOLUME XLIII. Antibiotics
Edited by JOHN H. HASH

VOLUME XLIV. Immobilized Enzymes
Edited by KLAUS MOSBACH

VOLUME XLV. Proteolytic Enzymes (Part B)
Edited by LASZLO LORAND

VOLUME 85. Structural and Contractile Proteins (Part B: The Contractile Apparatus and the Cytoskeleton)
Edited by DIXIE W. FREDERIKSEN AND LEON W. CUNNINGHAM

VOLUME 86. Prostaglandins and Arachidonate Metabolites
Edited by WILLIAM E. M. LANDS AND WILLIAM L. SMITH

VOLUME 87. Enzyme Kinetics and Mechanism (Part C: Intermediates, Stereochemistry, and Rate Studies)
Edited by DANIEL L. PURICH

VOLUME 88. Biomembranes (Part I: Visual Pigments and Purple Membranes, II)
Edited by LESTER PACKER

VOLUME 89. Carbohydrate Metabolism (Part D)
Edited by WILLIS A. WOOD

VOLUME 90. Carbohydrate Metabolism (Part E)
Edited by WILLIS A. WOOD

VOLUME 91. Enzyme Structure (Part I)
Edited by C. H. W. HIRS AND SERGE N. TIMASHEFF

VOLUME 92. Immunochemical Techniques (Part E: Monoclonal Antibodies and General Immunoassay Methods)
Edited by JOHN J. LANGONE AND HELEN VAN VUNAKIS

VOLUME 93. Immunochemical Techniques (Part F: Conventional Antibodies, Fc Receptors, and Cytotoxicity)
Edited by JOHN J. LANGONE AND HELEN VAN VUNAKIS

VOLUME 94. Polyamines
Edited by HERBERT TABOR AND CELIA WHITE TABOR

VOLUME 95. Cumulative Subject Index Volumes 61–74 and 76–80
Edited by EDWARD A. DENNIS AND MARTHA G. DENNIS

VOLUME 96. Biomembranes [Part J: Membrane Biogenesis: Assembly and Targeting (General Methods; Eukaryotes)]
Edited by SIDNEY FLEISCHER AND BECCA FLEISCHER

Section I

Macromolecular Self-Associations and Structural Assemblies

[1] Criteria for Distinguishing Self-Associations in Velocity Sedimentation

By V. Prakash and Serge N. Timasheff

Bimodality in a sedimentation velocity pattern or the derivative with respect to elution volume of a band front in chromatography can arise from several sources. The principal ones are (1) true heterogeneity of the protein, or the existence of the protein in more than one state of stable aggregation; (2) rapidly reequilibrating self-association to polymers greater than a dimer; and (3) rapidly reequilibrating self-association mediated by a strongly binding small ligand. The purpose of this chapter is to describe the criteria which may be applied to distinguish between these three cases.

True Heterogeneity. In this case, bimodality appears as soon as the boundary starts separating from the meniscus. As sedimentation progresses, the pattern gets resolved completely, and eventually the minimum between the two peaks comes down to the baseline of the schlieren pattern. This reflects the fact that the species of various molecular weights sediment each according to its size, shape, and molecular weight. In this case, the areas under the peaks correspond to the concentrations of the individual components, after correction for radial dilution and the Johnston–Ogston effect.

Rapidly Reequilibrating Self-Association. This is the system described by the Gilbert theory.[1] When the polymerization is cooperative to an end product which is a trimer or greater, the pattern resolves into a bimodal one. Here, resolution of the schlieren pattern sets in immediately after separation from the meniscus. As transport progresses, the two maxima become clearly resolved from each other. The minimum between the two peaks, however, never descends fully to the baseline, even in the late stages of the experiment. This is due to the fact that, in such a system, chemical equilibrium between the monomer and aggregate exists throughout the boundary. Therefore, the areas under the peaks do not correspond to the concentrations of any given molecular species and their measurement may not be used directly to determine the distribution of species.

Ligand-Mediated Self-Association. This is the system described by

---

[1] G. A. Gilbert, *Discuss. Faraday Soc.* **20,** 68 (1955); L. M. Gilbert and G. A. Gilbert, this series, Vol. 27, p. 273.

the Cann and Goad theory.[2] In this system the self-association of a protein to a dimer or higher species is linked to the binding of one or more molecules of a small ligand during the polymerization reaction. According to the Wyman theory of linked reactions,[3] the ligand must bind more strongly to the polymers than to the monomers if its binding enhances the self-association process. As discussed in detail elsewhere in this volume,[4] this creates a ligand pump from the zone centripetal to the boundary toward the bottom of the cell, generating a concentration gradient of unbound ligand across the boundary. In the early stages of sedimentation, this gradient is perturbed by back diffusion, and the self-association is reflected only by broadening of the boundary. As sedimentation progresses, the back diffusion can no longer balance the generated gradient and bimodality sets in. At this stage, the pattern resembles greatly that obtained from true heterogeneity in intermediate stages of sedimentation; it displays two distinct maxima, separated by a minimum which is raised above the baseline. In this system, however, there is never total separation of the two peaks and the minimum remains above the baseline to the very end of sedimentation. In this system again there is chemical equilibrium throughout the boundary and the areas under the peaks may not be used as a direct measure of the relative concentrations of the monomer and polymer present. When there is a large excess of ligand in the system, the boundary never becomes bimodal, unless it reduces to a Gilbert system, but it assumes an asymmetric shape, being skewed forward.

Examples

The three cases discussed in this chapter are well illustrated by our studies on the self-association of tubulin under various circumstances.[5] This is shown in Fig. 1. The schlieren patterns of Part A are for a ligand-mediated self-association, namely the isodesmic polymerization of tubulin in the presence of the anti-cancer drug, vincristine.[6] It is seen that in the early stages of sedimentation, there is no clear bimodality, resolution of the peak into 6 S and 9 S components becoming evident only after 29 min of sedimentation. The clearly defined minimum, however, never blends with the baseline. Part B is for a truly heterogeneous system. It corresponds to tubulin which had partially aggregated to a stable 9 S

---

[2] J. R. Cann and W. B. Goad, *Arch. Biochem. Biophys.* **153,** 603 (1972); J. R. Cann and W. B. Goad, this series, Vol. 27, p. 296.
[3] J. Wyman, Jr., *Adv. Protein Chem.* **19,** 223 (1964).
[4] G. C. Na and S. N. Timasheff, this series, Vol. 117, p. 459.
[5] V. Prakash and S. N. Timasheff, *Anal. Biochem.* **131,** 232 (1983).
[6] V. Prakash and S. N. Timasheff, *J. Biol. Chem.* **258,** 1689 (1983).

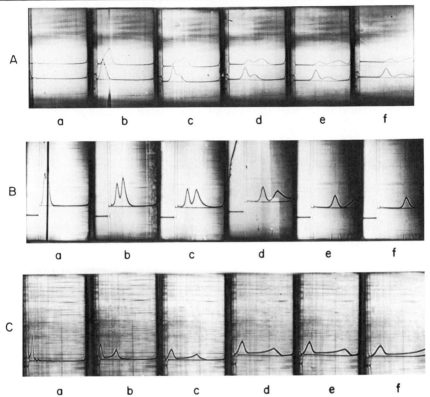

FIG. 1. Sedimentation velocity patterns of (A) tubulin in the presence of vincristine concentrations of 2.5×10^{-5} M (bottom) and 5×10^{-5} M (top) in PG buffer; (B) tubulin incubated in PG buffer at 4° for 47.5 hr; (C) tubulin in the presence of 13 mM MgCl$_2$ in PG buffer. All the sedimentation experiments were performed at 20° and at 60,000 rpm and photographs were obtained at a bar angle of 70° unless stated otherwise. The time of the photograph after reaching two-thirds full speed is indicated for each frame. (A) (a) 4 min, (b) 20 min, (c) 29 min, (d) 53 min, (e) 61 min, and (f) 84 min; (B) (a) 10 min, (b) 20 min, (c) 30 min, (d) 35 min, (e) 45 min, and (f) 60 min; (C) (a) 4 min, (b) 12 min, (c) 20 min, (d) 24 min (60° bar angle), (e) 32 min (60° bar angle), and (f) 36 min (60° bar angle). A protein concentration of 11.1 mg/ml was used in the vincristine and MgCl$_2$ experiments (i.e., A and C) and a protein concentration of 15.3 mg/ml was used in the aggregation experiment (i.e., B). PG buffer is a pH 7.0 Na phosphate buffer, containing 1 mM GTP.

species.[7] Here bimodality is clear early in the run (frame a, after 10 min of sedimentation), and the minimum between the peaks descends to the baseline late in the experiment. Part C is characteristic of a Gilbert system. This is tubulin in rapid equilibrium between the 6 S heterodimer and

[7] V. Prakash and S. N. Timasheff, *J. Mol. Biol.* **160,** 499 (1982).

the 43 S double ring.[8] Again resolution of the peaks sets in early (frame a, 4 min of sedimentation), but the minimum between the peaks never touches the baseline.

These examples clearly illustrate the application of the criteria for distinguishing between the three types of systems that give rise to bimodality in velocity sedimentation. These criteria can serve as powerful diagnostic tools to identify the type of aggregation which a given system is undergoing.

Acknowledgment

This work was supported in part by NIH Grants CA-16707 and GM-14603.

[8] R. P. Frigon and S. N. Timasheff, *Biochemistry* **14**, 4559 (1975); R. P. Frigon and S. N. Timasheff, *Biochemistry* **14**, 4567 (1975).

[2] Calculation of Asymptotic Boundary Shapes from Experimental Mass Migration Patterns

By LAWRENCE W. NICHOL and DONALD J. WINZOR

The discovery[1] that α-chymotrypsin exists as a mixture of polymers in rapidly established equilibrium was followed by similar findings for a series of proteins with widely differing biological functions. For proteins such as hemerythrin,[2] α-chymotrypsin,[3] neurophysin,[4] and α-amylase,[5] the self-association seems to be restricted to an equilibrium between monomer and a single higher polymer, but in others (e.g., hemoglobin[6] and β-lactoglobulin A[7]) monomer is considered to coexist with a range of polymers: indeed, several proteins, including glutamate dehydrogenase,[8]

[1] G. W. Schwert, *J. Biol. Chem.* **179**, 655 (1949).
[2] N. R. Langerman and I. M. Klotz, *Biochemistry* **8**, 4746 (1969).
[3] L. W. Nichol, W. J. H. Jackson, and D. J. Winzor, *Biochemistry* **11**, 585 (1972).
[4] P. Nicolas, M. Camier, P. Dessen, and P. Cohen, *J. Biol. Chem.* **251**, 3965 (1976).
[5] R. Tellam, D. J. Winzor, and L. W. Nichol, *Biochem. J.* **173**, 185 (1978).
[6] E. Chiancone, L. M. Gilbert, G. A. Gilbert, and G. L. Kellett, *J. Biol. Chem.* **243**, 1212 (1968).
[7] L. M. Gilbert and G. A. Gilbert, this series, Vol. 27, p. 273.
[8] H. Eisenberg and G. M. Tomkins, *J. Mol. Biol.* **31**, 37 (1968).

insulin,[9] and lysozyme,[10] are known to undergo indefinite self-association. In view of the widespread occurrence of self-association phenomena and of their biological implications,[11] it is of fundamental importance in the study of any protein in a given environment to establish methods for the detection of such interactions and their characterization in thermodynamic terms. It is conventional wisdom in such an investigation to subject the system to a mass migration process such as electrophoresis, sedimentation velocity, or chromatography in order to examine protein purity; and it is indeed in this type of experiment that the manifestation of self-association will likely first be seen. Thus, positive dependence of the weight-average sedimentation coefficient on total concentration,[1,12] or negative dependence of weight-average elution volume in gel chromatography,[13] is a readily available index of the operation of self-association. Certainly a variety of methods, including the equilibrium techniques of sedimentation equilibrium[14,15] and light scattering,[16,17] may then be used to characterize such interactions, but it is true that those analyses are materially assisted by probing further the migration results already available to the experimenter. It is indeed axiomatic that mass migration patterns must reflect, in the nature of the boundary forms, the operation of self-interactions; and the purpose of this chapter is to outline one type of approach which may be used to unravel their complexities.

Discussion is limited to consideration of reversibly polymerizing solutes for which rates of interconversion between monomeric and polymeric forms are sufficiently rapid for equilibrium to be established effectively at all positions within the boundary, despite the tendency of the species to separate. Such a situation applies to all of the above-mentioned systems, and may further be exemplified by the sedimentation of β-lactoglobulin A at low temperature,[18] where a reaction boundary, shown in Fig. 1, is observed in schlieren patterns recorded at different times. Quantitative description of the boundary shape requires consideration not only of the total flux of all solute species in equilibrium, due to mass migration,

[9] P. D. Jeffrey, B. K. Milthorpe, and L. W. Nichol, *Biochemistry* 15, 4660 (1976).
[10] P. R. Wills, L. W. Nichol, and R. J. Siezen, *Biophys. Chem.* 11, 71 (1980).
[11] L. W. Nichol, W. J. H. Jackson, and D. J. Winzor, *Biochemistry* 6, 2449 (1967).
[12] L. W. Cunningham, F. Tietze, N. M. Green, and H. Neurath, *Discuss. Faraday Soc.* 13, 58 (1953).
[13] D. J. Winzor and H. A. Scheraga, *Biochemistry* 2, 1263 (1963).
[14] D. C. Teller, this series, Vol. 27, p. 346.
[15] K. C. Aune, this series, Vol. 48, p. 163.
[16] R. F. Steiner, *Arch. Biochem. Biophys.* 39, 333 (1952).
[17] E. P. Pittz, J. C. Lee, B. Bablouzian, R. Townend, and S. N. Timasheff, this series, Vol. 27, p. 209.
[18] J. M. Armstrong and H. A. McKenzie, *Biochim. Biophys. Acta* 147, 93 (1967).

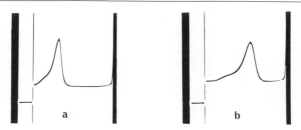

Fig. 1. Schlieren patterns of the reaction boundary formed during centrifugation of β-lactoglobulin A (15 g/liter) in acetate buffer (pH 4.6, $I = 0.1\ M$) for (a) 64 min and (b) 120 min at 59,780 rpm and 2.5°. Adapted from Armstrong and McKenzie, *Biochim. Biophys. Acta* **147**, 93.[18]

but also of the diffusional flows which arise to counter the gradients of chemical potential thereby created. Mention will be made later of approaches involving numerical integration which attempt such complete descriptions, but we are primarily concerned with an alternative approach which can be outlined in two steps. First, theory is presented which deliberately neglects the effects of diffusional flow, and in so doing permits simple calculations to be made of the diffusion-free or asymptotic shape of the migrating boundary, whose shape, therefore, is solely determined by the self-association effects and differential migration of the species.[19,20] Second, means are described whereby experimentally observed reaction boundaries may be analyzed to yield patterns free from diffusional effects, suitable for direct comparison with the theoretical asymptotic solutions.

Calculation of Asymptotic Boundary Shapes

Upon neglect of diffusional flow the continuity equation for self-associating systems in rapidly (instantaneously) established equilibrium may be expressed[21] as

$$u = d(\bar{v}\bar{c})/d\bar{c} \tag{1a}$$

$$\bar{c} = \sum_{i=1}^{i=n} K_i c_1{}^i; \qquad K_1 = 1 \tag{1b}$$

$$\bar{v} = \sum_{i=1}^{i=n} v_i K_i c_1{}^i \bigg/ \sum_{i=1}^{i=n} K_i c_1{}^i \tag{1c}$$

[19] G. A. Gilbert, *Discuss. Faraday Soc.* **20**, 68 (1955).
[20] G. A. Gilbert, *Proc. R. Soc. London Ser. A* **250**, 377 (1959).
[21] L. W. Nichol and A. G. Ogston, *Proc. R. Soc. London Ser. B* **163**, 343 (1965).

where u defines the velocity of a lamina of constant composition with total (constituent) concentration $\bar{c}$. Equation (1b) defines this constituent concentration in terms of monomer concentration, c_1, utilizing equilibrium constants, K_i (liter^{i-1} g^{1-i}), that are products of those governing stepwise additions of monomer. Equation (1c) describes the constituent velocity in similar fashion, v_i being the velocity of species i. From Eqs. (1a)–(1c) it is readily shown that

$$u = \sum_{i=1}^{i=n} iv_iK_ic_1^{i-1} \bigg/ \sum_{i=1}^{i=n} iK_ic_1^{i-1} = Q/P \qquad (2a)$$

which on differentiation with respect to $\bar{c}$ and subsequent inversion yields

$$d\bar{c}/du = P^3/(PQ' - QP') \qquad (2b)$$

where $P' = dP/dc_1$ and $Q' = dQ/dc_1$. A plot of $\bar{c}$ versus u represents a time-normalized concentration distribution, whereas a plot of $d\bar{c}/du$ versus u is the corresponding time-normalized schlieren profile. It is evident that systems undergoing indefinite self-association, e.g., lysozyme[10] and glutamate dehydrogenase,[8] require, in principle, an infinite number of terms to be included in the summations contained in the expressions for $\bar{c}$, P, Q, P', and Q', which are required for the calculation of theoretical migration patterns. In practice, the concentrations of successive polymers decrease exponentially, and hence the summations may be truncated at a value, n, of i such that Eq. (1b) approximates $\bar{c}$.

In order to illustrate the use of Eqs. (1b), (2a), and (2b), we shall calculate asymptotic sedimentation velocity patterns for β-lactoglobulin A, a system in which monomer (M_r 36,000) is considered[7,18] to undergo tetramerization via dimers and trimers as intermediate polymers. For such a system the expressions required for calculation of time-normalized patterns become

$$\bar{c} = c_1 + K_2c_1^2 + K_3c_1^3 + K_4c_1^4 \qquad (3a)$$
$$P = 1 + 2K_2c_1 + 3K_3c_1^2 + 4K_4c_1^3 \qquad (3b)$$
$$Q = v_1 + 2v_2K_2c_1 + 3v_3K_3c_1^2 + 4v_4K_4c_1^3 \qquad (3c)$$
$$P' = 2K_2 + 6K_3c_1 + 12K_4c_1^2 \qquad (3d)$$
$$Q' = 2v_2K_2 + 6v_3K_3c_1 + 12v_4K_4c_1^2 \qquad (3e)$$

and hence $\bar{c}$, u, and $d\bar{c}/du$ may be calculated for any selected value of c_1 provided that values are assigned to the velocity (v_i) and equilibrium constant (K_i) pertaining to each species. Figure 2 summarizes the results of such calculations with v_1 taken as the sedimentation coefficient ($s_{2.5,b}$) of monomer, 1.68 S, and the other v_i obtained on the basis of spherical

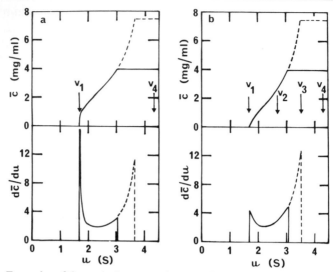

FIG. 2. Examples of theoretical asymptotic mass migration patterns for a solute undergoing rapid, reversible tetramerization. Parameters used for the calculation of these patterns from Eqs. (1)–(3) were as follows: $v_1 = 1.68$ S, $v_n = v_1 n^{2/3}$; and (a) $K_2 = K_3 \simeq 0$, $K_4 = 0.0086$ liter^3 g^{-3}; (b) $K_2 = 0.12$ liter g^{-1}, $K_3 = 0.0084$ liter^2 g^{-2}, $K_4 = 0.021$ liter^3 g^{-3}. Upper patterns represent the time-normalized concentration distributions and lower patterns the corresponding schlieren profiles.

geometry ($s_n = s_1 n^{2/3}$). The first situation simulated (Fig. 2a) is a monomer ⇌ tetramer system ($K_4 = 0.0086$ liter^3 g^{-3}, $K_2 = K_3 \simeq 0$) in experiments with total concentrations, $\bar{c}$, of 4 g/liter (solid line) and 7.5 g/liter (dashed line), the patterns being coincident in the concentration range 0–4 g/liter. Figure 2b presents the corresponding simulated patterns for a system with $K_2 = 0.12$ liter g^{-1}, $K_3 = 0.0084$ liter^2 g^{-2}, and $K_4 = 0.021$ liter^3 g^{-3}, this being the model considered[7,22] to provide the best description of the self-association of β-lactoglobulin A under the conditions (pH 4.6, 2.5°) to which Fig. 1 refers.

Use of Migration Patterns for the Quantitative Definition of Self-Association Equilibria

Calculations of asymptotic migration patterns such as those presented in Fig. 2 have served a useful purpose by providing valuable insight into basic features of reaction boundaries for self-associating systems. Their calculation has shown that even self-associating systems with instanta-

[22] D. J. Winzor, R. Tellam, and L. W. Nichol, *Arch. Biochem. Biophys.* **178,** 327 (1977).

neously established equilibrium between species can yield bimodal reaction boundaries provided that dimer is not the sole polymer present. Their calculation has also illustrated the experimental observation that the size of the slower migrating boundary is insensitive to the total concentration of solute in the range of $\bar{c}$ for which a particular solute exhibits a bimodal reaction boundary in migration experiments. However, quantitative comparisons of theoretical migration patterns, such as those in Fig. 2, with experimental patterns, such as those for β-lactoglobulin A in Fig. 1, are clearly rendered very difficult by the unrealistic sharpness of the asymptotic patterns due to the neglect of diffusional effects in their computation. This dilemma has led to the development of three approaches to the use of mass migration patterns for the quantitative definition of self-association equilibria.

1. The simplest approach is to ignore altogether the detailed shape of the reaction boundary, except in the definition of its median bisector, the rate of movement of which is related unequivocally to the composition of the solution in the plateau region ahead of the reaction boundary. Such use of weight-average migration parameters to define solution composition and hence the self-association characteristics of a solute has been described in detail elsewhere.[23]

2. A second approach employs computer simulation to incorporate the effects of diffusional spreading into the theoretical patterns. Reaction boundaries are thereby generated which bear greater resemblance to their experimental counterparts. Of the several methods that have been used[24-27] for numerical solution of the full continuity equation (inclusive of diffusional terms), that involving the use of weight-average sedimentation and diffusion coefficients in conjunction with alternation of the sedimentation and diffusion processes[26,28] has proved to be the most popular. This approach, together with its application to analysis of the β-lactoglobulin A system, has been described in an earlier volume of this series.[7]

3. The aim of the present chapter is to consider in detail the third approach, namely, the determination of asymptotic boundary shapes by eliminating the effects of diffusion from experimental mass migration patterns.

[23] D. J. Winzor, in "Protein-Protein Interactions" (C. Frieden and L. W. Nichol, eds.), Chap. 3. Wiley, New York, 1981.
[24] J. R. Cann and W. B. Goad, J. Biol. Chem. **240**, 148 (1965).
[25] M. Dishon, G. H. Weiss, and D. A. Yphantis, Biopolymers **4**, 449 (1966).
[26] D. J. Cox, Arch. Biochem. Biophys. **119**, 230 (1967).
[27] J.-M. Claverie, H. Dreux, and R. Cohen, Biopolymers **14**, 1685 (1975).
[28] D. J. Cox and R. S. Dale, in "Protein-Protein Interactions" (C. Frieden and L. W. Nichol, eds.), Chap. 4. Wiley, New York, 1981.

Experimental Determination of Asymptotic Boundary Shapes

Research in this area has been along two lines.

Extrapolation to Eliminate Diffusional Effects in Sedimentation Velocity

This method[22] for the determination of asymptotic boundary shapes from sedimentation velocity experiments for self-associating solutes is an adaptation of procedures developed for elimination of diffusional effects in the analysis of heterogeneous, noninteracting systems.[29,30] Its underlying principle is that boundary spreading arising from differences between the migration rates of monomeric and polymeric species is proportional to time, t, of centrifugation, whereas diffusional spreading is proportional to $\sqrt{t}$. At infinite time, therefore, diffusional effects tend to be negligible in relation to those emanating from sedimentation and chemical reaction.[20,31]

Experimental Protocol. The procedure used to obtain asymptotic boundary shapes by this method entails the following steps.[22]

1. The mean boundary position, $\bar{r}$, is determined as the square root of the second moment of the reaction boundary[32,33] in each of several schlieren patterns recorded during a sedimentation velocity experiment. Specifically,

$$\bar{r}^2 = \int r^2 (d\bar{c}/dr)dr \bigg/ \int (d\bar{c}/dr)dr \tag{4}$$

in which the limits of integration are the solvent and solution plateaux.

2. The linear plot of log $\bar{r}$ versus time is then extrapolated to the meniscus (log r_m) to find the effective time, t, of centrifugation at constant angular velocity ω.

3. Trapezoidal integration is then used to obtain values of $\bar{c}(r)$, the total weight-concentration of solute at radial distance r, from each schlieren pattern. In more detail, the ratio of the area under the peak at a particular r to the total peak area is multiplied by the plateau concentration, the latter being obtained from the original concentration by multiplying by $(r_m/\bar{r})^2$ to account for the effect of radial dilution. This procedure yields the actual concentration at each selected r.

[29] R. L. Baldwin, *J. Am. Chem. Soc.* **76,** 402 (1954).
[30] R. L. Baldwin, *J. Phys. Chem.* **63,** 1570 (1959).
[31] G. A. Gilbert, *Proc. R. Soc. London Ser. A* **276,** 354 (1963).
[32] R. J. Goldberg, *J. Phys. Chem.* **57,** 194 (1953).
[33] R. Trautman and V. Schumaker, *J. Chem. Phys.* **22,** 551 (1954).

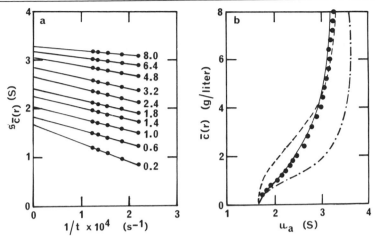

FIG. 3. Experimental determination of the asymptotic sedimentation velocity pattern for β-lactoglobulin A: (a) the extrapolation procedure used to eliminate effects of diffusion; (b) the asymptotic profile (●) so obtained, and theoretical patterns for three suggested modes of self-association. Adapted from Winzor et al., Arch. Biochem. Biophys. 178, 327.[22]

4. The points $[\bar{c}(r), r]$ thus obtained are then transformed to $[\bar{c}(r), \bar{s}]$ values by use of the integrated form of the definition of a sedimentation coefficient[30]:

$$\bar{s} = [\ln(r/r_m)]/\omega^2 t \qquad (5)$$

5. Plots of $\bar{c}(r)$ versus $\bar{s}$ are constructed for each t; and values of $(\bar{s}_{\bar{c}(r)}, t)$, found by interpolation for a particular value of $\bar{c}(r)$, are then used to plot $\bar{s}_{\bar{c}(r)}$ versus $1/t$. The extrapolated value (at infinite time; $1/t = 0$) may be regarded as the sedimentation coefficient, u_a, corresponding to the chosen $\bar{c}(r)$. Nevertheless, u_a must be considered to be an *apparent* velocity (per unit field) of the lamina of constant composition with constituent concentration $\bar{c}(r)$, since its evaluation on the basis of Eq. (1a) entails assumptions that sedimentation proceeds in a cell of uniform cross-sectional area under the influence of a homogeneous field, the "rectangular" approximation. Similar treatment of other $\bar{c}(r)$ values then leads to a plot of $\bar{c}(r)$ versus u_a, which is suitable for comparison with asymptotic profiles [$\bar{c}(r)$ versus u] calculated for any selected model of the migrating self-associating system, using Eqs. (2a) and (3a)–(3c).

Application to β-Lactoglobulin A. This procedure for experimental determination of diffusion-free boundary shapes is illustrated in Fig. 3, which presents the results of such analysis[22] of the sedimentation velocity experiment to which Fig. 1 refers.[18] Figure 3a shows the extrapolations required to obtain u_a for a range of $\bar{c}(r)$ values (0.2–8.0 g/liter) from five exposures covering the time interval $t = 72$ to $t = 128$ min. Values of u_a, the ordinate intercepts obtained by linear least-squares calculations, are then used to construct the experimentally determined asymptotic sedimentation velocity pattern shown in Fig. 3b. Also shown are theoretical profiles calculated on the basis of Eqs. (1b) and (2a) with the sedimentation coefficient of each species, s_i ($\equiv v_i$) related to its limiting value, s_i^0, by the expression $s_i = s_i^0(1 - 0.007\bar{c})$, and the values of s_i^0 used for the construction of Fig. 2. Theoretical patterns are presented for simple monomer–tetramer systems with $K_4 = 0.0086$ (dashed line) and 0.086 (dot-dashed line) liter^3 g^{-3}, models considered initially[18,34,35] to provide adequate descriptions of molecular weight and optical rotation measurements on β-lactoglobulin A under these conditions. The solid line represents the theoretical profile for the monomer–dimer–trimer–tetramer system (Fig. 2b) invoked by Gilbert and Gilbert[7] as a better description of the system. Their conclusion, based on numerical simulation of sedimentation velocity patterns, is clearly supported by the present analysis.[22]

A general warning must be sounded concerning the application of this procedure. It is entirely possible that in the theoretical calculation a concentration region may be encountered where $du/d\bar{c}(r)$ assumes negative values as illustrated by one curve (dot-dashed line) in Fig. 3b, an effect also exhibited by the solid curve at concentrations greater than 8 g/liter. In such regions the present theory based on the concept of time-independent velocities is inapplicable because description would require development of the concept that a region of the boundary (theoretically hypersharp[31]) migrates with an ever-changing velocity because of the decreasing plateau concentration arising from radial dilution in the sector-shaped cell.

Other Migration Methods. In view of the general nature of Eqs. (1a) and (2a) it is logical to enquire whether results obtained by other mass migration procedures might also be subjected to similar analysis. The first point to note is that the theoretical requirement for the retention of a plateau of original composition restricts such consideration to two methods, namely, moving boundary electrophoresis and frontal gel chromatography. From the theoretical viewpoint the assumed uniformity of cell cross-sectional area and applied field in the present analysis is a much better approximation in moving boundary electrophoresis than in sedimentation velocity, and hence there is every reason to believe that it

[34] R. Townend and S. N. Timasheff, *J. Am. Chem. Soc.* **82**, 3168 (1960).
[35] H. A. McKenzie, W. H. Sawyer, and M. B. Smith, *Biochim. Biophys. Acta* **147**, 73 (1967).

could be applied with even greater validity to electrophoresis, particularly a spread descending pattern, which is free from effects of conductivity and pH changes[36]: in such an experiment the ascending boundary is hypersharp due to these effects and chemical reaction. However, very few self-associating systems, β-lactoglobulin A being a notable exception,[37] could be studied by this technique because of the additional requirement that the electrophoretic mobilities of monomeric and polymeric species differ sufficiently. Only in the event of either a lack of charge conservation or a pronounced change in shape on polymerization is this situation likely to pertain.[38]

In principle, the present method is directly applicable to frontal gel chromatography provided that an optical scanning device[39] is used to obtain the required concentration–distance profiles at fixed times. Unfortunately, despite the use of precision-bore tubing for chromatography columns, the assumptions of uniform cross-sectional area and uniform field are invalidated by the nonuniform packing of the gel column, a phenomenon clearly evident from optical scans of frontal gel chromatographic experiments.[39,40] In instances where the experimental record is an elution profile, that is, the concentration of solute in the mobile phase as a function of time taken to migrate a fixed distance (the column length), it is reasonable to consider that the requirements of uniformity with respect to cross-sectional area and field are met, since all molecules encounter the same variations in the degree of column packing as they traverse the total column length. In such experiments the analog of time is the length of the gel column,[41] and accordingly the experimental results required for the analysis would in principle be obtained by recording elution profiles from a series of experiments conducted with gel columns of different lengths. However, no experiments of this type have been reported, presumably a reflection of the inherent difficulty in attempting to ensure that a similar average degree of packing applies to gel beds with markedly different lengths. Instead, a chromatographic method amenable to asymptotic interpretation of elution profiles has been developed[6,42] in which nonuniform packing of the gel column is of no concern.

[36] L. G. Longsworth, in "Electrophoresis: Theory, Methods, and Applications" (M. Bier, ed.), Chap. 3. Academic Press, New York, 1959.
[37] M. P. Tombs, Biochem. J. 67, 517 (1957).
[38] J. M. Creeth and L. W. Nichol, Biochem. J. 77, 230 (1960).
[39] E. E. Brumbaugh and G. K. Ackers, J. Biol. Chem. 243, 6315 (1968).
[40] J. K. Zimmerman and G. K. Ackers, Anal. Biochem. 57, 578 (1974).
[41] G. E. Hibberd, A. G. Ogston, and D. J. Winzor, J. Chromatogr. 48, 393 (1970).
[42] G. A. Gilbert and G. L. Kellett, J. Biol. Chem. 246, 6079 (1971).

The Difference Boundary Method

A second approach[6,42] to the problem of determining the asymptotic shape of a reaction boundary from experimental mass migration patterns involves the performance of a series of layering experiments[43,44] in which the migration of the difference boundary between two solutions with slightly different concentrations of solute is observed. From considerations of mass conservation it may be shown that the velocity of a difference boundary with concentration difference $\Delta \bar{c}$ is given by

$$v_\Delta = \Delta(\bar{v}\bar{c})/\Delta\bar{c} \tag{6}$$

As the size of the difference boundary becomes vanishingly small ($\Delta\bar{c} \rightarrow$ 0), the right-hand side of Eq. (6) becomes identical with that of Eq. (1a), and hence $v_\Delta \rightarrow u$, the velocity of a lamina of constant composition with constituent concentration $\bar{c}$. Because of the requirement that $\Delta\bar{c}$ be small in comparison with $\bar{c}$, this method is likely to be of very restricted use in sedimentation velocity and moving boundary electrophoresis studies, where a relatively large concentration increment (>0.5 g/liter) must be used for the formation of a difference boundary that is detectable in experiments with conventional cells and optical systems. The difference boundary procedure is, however, ideally suited to evaluating asymptotic elution profiles in frontal gel chromatography.[6,42] Application of Eq. (6) to elution profiles merely requires the direct substitution of elution volumes for velocities,[45-47] and hence a plot of V_Δ, the elution volume of the difference boundary, versus $\bar{c}$, the mean of the plateau concentrations used for its formation, yields the diffusion-free elution profile.

Results obtained[6] for human oxyhemoglobin in a series of difference boundary experiments with $\Delta\bar{c} = 0.1\bar{c}$ are shown in Fig. 4, which also contains the theoretical asymptotic profile calculated on the basis of Eqs. (3a)–(3c) with $K_2 = 4.8 \times 10^3$ liter g^{-1}, $K_3 = 8.6 \times 10^2$ liter^2 g^{-2}, $K_4 = 4.1 \times 10^8$ liter^3 g^{-3}, and the elution volumes V_i ($\equiv v_i$) as indicated in the diagram. Clearly, this mode of self-association[6] provides a very good description of the asymptotic elution profile for human oxyhemoglobin over the range of concentration ($\bar{c}$) shown in Fig. 4, and, indeed, of the frontal gel chromatographic behavior of the protein over a much wider range of $\bar{c}$ ($0 < \bar{c} < 35$ g/liter).[6] Simplicity in terms of the treatment of the experimental record and freedom from empiricism are clearly attributes

[43] R. Hersh and H. K. Schachman, *J. Am. Chem. Soc.* **77**, 5228 (1955).
[44] G. A. Gilbert, *Nature (London)* **212**, 296 (1966).
[45] G. K. Ackers and T. E. Thompson, *Proc. Natl. Acad. Sci. U.S.A.* **53**, 342 (1965).
[46] G. A. Gilbert, *Nature (London)* **210**, 299 (1966).
[47] L. W. Nichol, A. G. Ogston, and D. J. Winzor, *J. Phys. Chem.* **71**, 726 (1967).

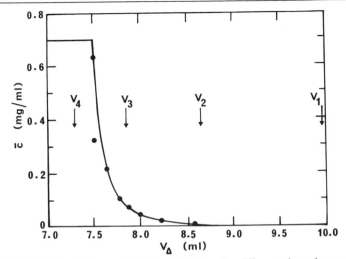

FIG. 4. Asymptotic elution profile (●) obtained by the difference boundary method in a gel chromatographic study of the self-association of human oxyhemoglobin (0.7 g/liter) on Sephadex G-100. The solid curve is the theoretical relationship calculated from Eqs. (3a)–(3c) with $V_\Delta = v_u$ for a monomer–tetramer system with $K_2 = 4.8 \times 10^3$ liter g^{-1}, $K_3 = 8.6 \times 10^2$ liter^2 g^{-2}, $K_4 = 4.1 \times 10^8$ liter^3 g^{-3}, and species elution volumes V_i ($\equiv v_i$) denoted by the arrows. Adapted from Chiancone et al., J. Biol. Chem. **243**, 1212.[6]

which make this procedure for determining asymptotic boundary shapes preferable to that based on extrapolation of data to infinite time; but its application is clearly restricted to chromatography.

Concluding Remarks

Figures 3 and 4 serve to demonstrate the feasibility of using the asymptotic shapes of reaction boundaries to assess the relative merits of various self-association patterns as descriptions of the mass migration behavior of polymerizing solutes. However, although comparisons of experimental profiles with theoretical predictions for various models of the self-association process may thus be made either by means of the experimental determination of the asymptotic boundary shape or by numerical simulations of theoretical patterns inclusive of diffusional spreading, the value of such exercises must be examined fairly closely. Selection of a particular mode of self-association in preference to other self-association patterns on the degree of correspondence between the shapes of experimental and theoretical reaction boundaries does, of course, imply correctness of the velocities (elution volumes) assigned to the postulated polymeric species; and it must be remembered that these values are only estimates which are empirically based on geometrical considerations. Under those circumstances, is perseverance with such endeavors worthwhile when there are other methods which do not require assignment of characteristic parameters for all species in order to obtain thermodynamic characterization of the interaction pattern? Indeed, analytical procedures are now available which permit determination, from experimental results obtained by sedimentation equilibrium,[48] light-scattering,[17] and partition chromatography of special design,[49] of the dependence of monomeric concentration on total concentration, independent of the operative self-association pattern: in such analyses only the characteristic parameter (molecular weight or partition coefficient) for monomer need be determined—by extrapolation of results to infinite dilution.

In answer to that question it seems only fair to point out that these less subjective equilibrium approaches become more tractable when some idea of the self-association pattern is available. For example, although sedimentation equilibrium has possibly the greatest potential for comparisons of different postulated modes of self-association based on the concentration dependence of monomer weight-fraction, the extent to which that potential may be realized is very dependent on basic matters such as selection of the appropriate angular velocity for the experiment. Estimates of association constants obtained from mass migration studies may be used to simulate the solute distribution in a sedimentation equilibrium experiment at any given angular velocity,[50] and hence to aid in the selection of the optimal speed for refinement of the self-association pattern by sedimentation equilibrium studies. It also seems fair to note that for very large protein structures such as the pyruvate dehydrogenase complex,[51] sedimentation equilibrium becomes a less attractive method, not only because of the much longer time period required for attainment of equilibrium but also because of difficulty in maintaining accurate speed control at the very low angular velocity required for analysis of a system with a molecular weight of 6 million. Under those circumstances analysis of the boundary shape obtained in sedimentation velocity experiments takes on the key role of providing the only reliable quantitative assessment of the system, be it in terms of homogeneity, as exemplified by studies with the pyruvate dehydrogenase complex,[52] or in terms of the mode of self-association.

[48] B. K. Milthorpe, P. D. Jeffrey, and L. W. Nichol, *Biophys. Chem.* **3,** 169 (1975).
[49] L. W. Nichol, R. J. Siezen, and D. J. Winzor, *Biophys. Chem.* **9,** 47 (1978).
[50] G. J. Howlett, P. D. Jeffrey, and L. W. Nichol, *J. Phys. Chem.* **74,** 3607 (1970).
[51] M. J. Danson, G. Hale, P. Johnson, R. N. Perham, J. Smith, and P. Spragg, *J. Mol. Biol.* **129,** 603 (1979).
[52] G. A. Gilbert and L. M. Gilbert, *J. Mol. Biol.* **144,** 405 (1980).

[3] Effects of Microheterogeneity on Sedimentation Patterns of Interacting Proteins and the Sedimentation Behavior of Systems Involving Two Ligands

By JOHN R. CANN

Because the interaction of proteins and other macromolecules with each other and with small ligand molecules is central to current biological thought, it is essential to delineate and define as precisely as possible the various factors which may govern the sedimentation behavior of such systems. Two of these factors, microheterogeneity and interaction with two different ligands, are discussed below.

Microheterogeneity

It is common knowledge that certain highly purified proteins exhibit molecular heterogeneity when scrutinized by high-resolution immuno-chemical, biochemical, or biophysical techniques. This property is termed microheterogeneity. Biologically the most interesting source of microheterogeneity is polymorphism, i.e., genetic variations in primary structure as, for example, in the case of isozymes,[1] polyclonal antibodies,[2,3] and β-lactoglobulins.[4,5] Molecular heterogeneity can also be due to possible differences in glycosylation or phosphorylation; *in vivo* deamidation of a single asparagine residue, Asn 258, in rabbit muscle aldolase leading to a random distribution of tetramers containing zero to four deamidated subunits[5a]; intramolecular disulfide interchange in plasma albumin[6]; hypothesized enzymatic nicking prior, perhaps, to isolation of

---

[1] C. L. Markert, ed., "Isozymes I Molecular Structure." Academic Press, New York, 1975.

[2] E. A. Kabat, *Adv. Protein Chem.* **32**, 1 (1978).

[3] H. N. Eisen, "Immunology. An Introduction to Molecular and Cellular Principles of the Immune Responses," 2nd ed., Chaps. 16 and 19. Harper, New York, 1980.

[4] W. G. Gordon, J. J. Basch, and E. B. Kalan, *J. Biol. Chem.* **236**, 2908 (1961).

[5] E. B. Kalan, R. Greenberg, M. Walter, and W. G. Gordon, *Biochem. Biophys. Res. Commun.* **16**, 199 (1964).

[5a] B. L. Horecker, *in* "Isozymes I Molecular Structure" (C. L. Markert, ed.), p. 11. Academic Press, New York, 1975.

[6] M. Sogami, H. A. Petersen, and J. F. Foster, *Biochemistry* **8**, 49 (1969).

concanavalin A[7,8]; and adventitious modification (e.g., deamidation and limited proteolytic attack) during purification, subsequent manipulation, and/or storage. In any case, microheterogeneity can have pronounced effects on the electrophoretic and sedimentation behavior of interacting systems.

Thus, microheterogeneity accounts for the electrophoretic resolution of the N and F forms of plasma albumin undergoing rapid isomerization near pH 4[6,9-11]; heterogeneity of pyruvate dehydrogenase multienzyme complex with respect to molecular weight causes excessive spreading of its sedimenting boundary[12]; the nicked subunits present in some preparations of concanavalin A have a much decreased competency to self-association as determined by sedimentation equilibrium[13]; and the genetic variants of β-lactoglobulin differ strikingly with respect to association constant for low-temperature tetramerization at pH 4.4–4.7 as reflected in their electrophoretic and analytical sedimentation patterns.[14-17] The effects of microheterogeneity on the sedimentation patterns of the α-hemocyanin of *Helix pomatia* under associating–dissociating conditions are discussed by Kegeles and Cann in Vol. 48 [12] of this series. Recently, Tai and Kegeles[17a] have computer simulated these effects with a model which assumes extensive microheterogeneity of the protein. Nor is heterogeneity restricted to proteins. For example, the trimodal zonal sedimentation patterns of *Escherichia coli* ribosomes have been explained[18] as being due to heterogeneity of formation constants of 70 S particles from 30 and 50 S subunits (for a discussion, see this series, Vol. 48 [12]).

These examples along with poorly understood but suggestive observations in the literature prompted a recent extension of our theoretical in-

[7] J. L. Wang, B. A. Cunningham, and G. M. Edelman, *Proc. Natl. Acad. Sci. U.S.A.* **68,** 1130 (1971).

[8] J. L. Wang, B. A. Cunningham, M. J. Waxdal, and G. M. Edelman, *J. Biol. Chem.* **250,** 1490 (1975).

[9] M. Sogami and J. F. Foster, *J. Biol. Chem.* **238,** 2245 (1963).

[10] H. A. Peterson and J. F. Foster, *J. Biol. Chem.* **240,** 2503 (1965).

[11] H. A. Petersen and J. F. Foster, *J. Biol. Chem.* **240,** 3858 (1965).

[12] G. A. Gilbert and L. M. Gilbert, *J. Mol. Biol.* **144,** 405 (1980).

[13] D. F. Senear and D. C. Teller, *Biochemistry* **20,** 3076 (1981).

[14] R. A. Brown and S. N. Timasheff, *in* "Electrophoresis Theory, Methods and Practice" (M. Bier, ed.), Chap. 8. Academic Press, New York, 1959.

[15] R. Townend, R. J. Winterbottom, and S. N. Timasheff, *J. Am. Chem. Soc.* **82,** 3161 (1960).

[16] S. N. Timasheff and R. Townend, *J. Am. Chem. Soc.* **83,** 464 (1961).

[17] T. F. Kumosinski and S. N. Timasheff, *J. Am. Chem. Soc.* **88,** 5635 (1966).

[17a] M.-S. Tai and G. Kegeles, *Biophys. Chem.* **19,** 113 (1984).

[18] J. B. Chaires and G. Kegeles, *Biophys. Chem.* **7,** 173 (1977).

vestigations into the mass transport of interacting systems (this series, Vols. 25 [11], 27 [12], 48 [11,12], 61 [10]; and Cann[19]) to include the effects of microheterogeneity on sedimentation behavior. Two classes of interaction were considered: self-association of proteins with particular reference to rapidly equilibrating, tetramerization reactions[20] and complex formation as exemplified by the specific univalent antigen–antibody and bivalent hapten–antibody reactions.[21]

Self-Association. The starting point of this investigation was the tetramerization of a homogeneous protein M schemetized by the rapid equilibrium

$$4M \rightleftarrows M_4, \quad K \tag{1}$$

which assumes negligible concentrations of intermediate dimer and trimer. Gilbert[22,23] was the first to describe the salient features of the sedimentation behavior of such a system. At sufficiently low protein concentration, the analytical sedimentation pattern shows a single sedimenting peak of monomer, which grows in area as the concentration is increased to a critical value determined by the value of the equilibrium constant K. Upon increasing the concentration still further, a second more rapidly sedimenting peak appears and grows in area, while the area of the slower peak now remains unchanged. In general, this behavior is predicted for $nM \rightleftarrows M_n$ with $n \geq 3$ and critical concentration dependent upon n and K. A number of self-associating proteins show the features predicted by the Gilbert theory, e.g., the low-temperature tetramerization of β-lactoglobulin A at pH 4.65[14–17] and the formation of skeletal muscle myosin filaments.[24] The approach used in our study simulated experimental procedure in that sedimentation patterns were computed for a range of protein concentrations by numerical solution of the set of transport equations and mass action expressions appropriate for the particular model system under consideration. As shown in Fig. 1A, the computed patterns for tetramerization of a homogeneous protein (Reaction 1) agree qualitatively with the predictions of the Gilbert theory.

This classical perception of the sedimentation behavior of self-associating proteins is modified significantly by microheterogeneity with respect to association constant, other molecular parameters held constant. Three

[19] J. R. Cann, in "Electrokinetic Separation Methods" (P. G. Righetti, C. J. van Oss, and J. W. Vanderhoff, eds.), p. 369. Elsevier, Amsterdam, 1979.
[20] J. R. Cann and N. H. Fink, *Biophys. Chem.* **17**, 29 (1983).
[21] J. R. Cann, *Mol. Immunol.* **19**, 505 (1982).
[22] G. A. Gilbert, *Discuss. Faraday Soc.* **20**, 68 (1955).
[23] G. A. Gilbert, *Proc. R. Soc. Ser. A* **250**, 377 (1959).
[24] R. Josephs and W. F. Harrington, *Biochemistry* **5**, 3474 (1966).

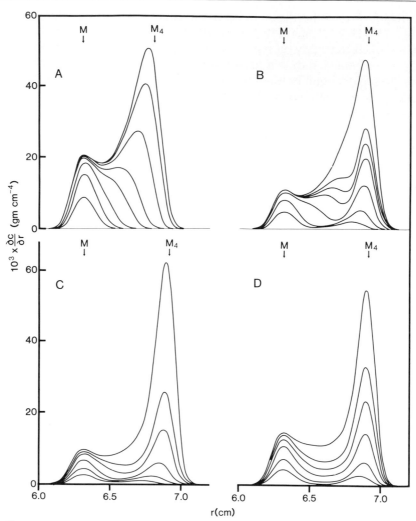

FIG. 1. Theoretical sedimentation patterns for rapidly equilibrating monomer–tetramer systems computed for a range of initial constituent protein concentrations. (A) Homogeneous protein, reaction (1) with $K = 7.407 \times 10^{10} \, M^{-3}$. Reading from bottom to top the initial concentrations are 1.75, 3.50, 5.25, 7.875, 10.5, 15.75, 21.0, and 24.5 mg ml^{-1}. (B) Microheterogeneous protein, reaction (2) with $K(1) = 3.0720 \times 10^{15} \, M^{-3}$ and $K(2) = 5.9259 \times 10^{11}$ M^{-3}: 1.75, 3.50, 7.00, 10.5, 12.25, 14.0, and 21.0 mg ml^{-1}; trimodal patterns obtained for a broad range of K values. (C) Microheterogeneous protein, reaction set (3): 0.875, 1.75, 3.50, 7.00, 10.5, and 21.0 mg ml^{-1}. (D) Microheterogeneous protein, reactions (4) with following distribution of monomer: 50% of class 1 with $K(1) = 4.736 \times 10^{16} \, M^{-3}$ and 12.5% each of the other four classes with $K(2) = 7.324 \times 10^{14}$, $K(3) = 7.407 \times 10^{13}$, $K(4) = 8.468 \times 10^{12}$, and $K(5) = 1.446 \times 10^{12} \, M^{-3}$. Initial concentrations: 1.75, 3.50, 7.00, 10.5, 14.0, and 21.0 mg

different models of microheterogeneity were considered. The first model assumes that the protein is a 1 : 1 mixture of either two polymorphic monomers or, possibly, native and modified forms of a monomorphic protein, designated as $M(1)$ and $M(2)$. The two have the same sedimentation and diffusion coefficients, but they tetramerize independently with different equilibrium constants according to the reactions

$$4M(1) \rightleftarrows M(1)_4, \quad K(1) \tag{2a}$$
$$4M(2) \rightleftarrows M(2)_4, \quad K(2) \tag{2b}$$

In striking contrast to the homogeneous protein, the theoretical sedimentation patterns for this model of microheterogeneity can show three peaks depending upon the protein concentration (Fig. 1B). This trimodality is a direct result of superimposing the individual patterns of the two admixed systems. Thus, the slowest sedimenting peak is a mixture of the two monomers. To a first approximation, the leading peak corresponds to the faster of the two peaks which the more strongly associating system would show, if it were to be isolated and examined alone, while the intermediate peak belongs to the weaker system. The monomer peak at first grows with concentration even though the leading peak has already emerged, but its area remains constant once the intermediate peak appears.

The shape of the patterns is quite different when, in accordance with our second model, the two monomers interact to form hybrid tetramers as schematized by the reaction set

$$4M(1) \rightleftarrows M(1)_4, \quad\quad K(1) \tag{3a}$$
$$3M(1) + M(2) \rightleftarrows M(1)_3M(2), \quad K(1)^{3/4}K(2)^{1/4} \tag{3b}$$
$$2M(1) + 2M(2) \rightleftarrows M(1)_2M(2)_2, \quad K(1)^{1/2}K(2)^{1/2} \tag{3c}$$
$$M(1) + 3M(2) \rightleftarrows M(1)M(2)_3, \quad K(1)^{1/4}K(2)^{3/4} \tag{3d}$$
$$4M(2) \rightleftarrows M(2)_4, \quad\quad K(2) \tag{3e}$$

In this case, the patterns show only two peaks at all concentrations (Fig. 1C) and there is no critical concentration beyond which the area of the monomer peak remains constant. In fact, the area apparently does not approach a limiting value, probably due to the change in distribution of material among the various tetramers with increasing constituent concentration.

In practice, it would be difficult to distinguish between hybrid forma-

ml[-1]. For plots of area of monomer peak vs constituent concentration see Cann and Fink.[20] Arrows at the top of this and following figures indicate where the peaks in the patterns would have been located had sedimentation been carried out on a mixture of noninteracting proteins having the same sedimentation coefficients as the designated species. Sedimentation proceeds to the right. From Cann and Fink, *Biophys. Chem.* **17**, 29[20]; reprinted with permission of Elsevier Biomedical Press.

tion and our third model, which is for rather extensive microheterogeneity with five classes of monomer associating independently of one another,

$$4M(i) \rightleftarrows M(i)_4, \qquad K(i) \qquad (4)$$

where $i = 1, 2 \ldots 5$. Comparison of Fig. 1C and D points out the similarity between the shape of the sedimentation patterns predicted by the two models. There is, however, a qualitative difference: in the case of rather extensive microheterogeneity the area of the monomer peak does approach a limiting value, where the gradient of each of the five components of the mixture is bimodally distributed along the centrifuge cell.

Although the foregoing results are for tetramerization reactions, the qualitative conclusions also apply to high-order associations even when the equilibrium constants are quite large (see footnote on p. 33 of Cann and Fink[20]). Thus, they may bear on the *in vitro* formation of smooth muscle myosin filaments.[25] In contrast to rabbit skeletal muscle myosin, the sedimentation behavior of calf aorta smooth muscle myosin under self-associating conditions cannot be analyzed in terms of the Gilbert theory. Rather, the shape of the patterns and the monomer area–concentration profile are strikingly similar to those predicted for rather extensive microheterogeneity, and the SDS–gel electrophoretic pattern of the preparation shows heterogeneity in the region of the heavy chains. Microheterogeneity due to a difference in phosphorylation of the light chains could be at least in part responsible[26] for the aberrant sedimentation behavior; but, in view of the high concentration of plasminogen activator in vascular structures[27] as compared to its absence in rabbit muscle,[28] extra- or intracellular proteolytic nicking cannot be dismissed. Chemical modification such as the phosphorylation of aorta myosin[26] is one way of ascertaining possible sources of microheterogeneity and assessing its effects on the electrophoretic and sedimentation behavior of interacting systems. Another, especially powerful approach is fractionation accompanied by biochemical and biophysical analyses as practiced in the case of plasma albumin,[29,30] concanavalin A,[7,8,13] β-lactoglobulin,[14,17] and α-hemocyanin of *Helix pomatia* (this series, Vol. 48 [12]). This approach can be applied to complex formation between different proteins as well as to isomerization and self-association.

[25] J. Megerman and S. Lowey, *Biochemistry* **20**, 2099 (1981).

[26] K. M. Trybus and S. Lowey, *Biophys. J.* **41**, 299a (1983).

[27] E. W. Davie and O. D. Ratnoff, *in* "The Proteins Composition, Structure, and Function" (H. Neurath, ed.), 2nd ed., Vol. 3, Chap. 16. Academic Press, New York, 1965.

[28] O. K. Albrechtsen, *Acta Physiol. Scand.* **47** (Suppl. 165), 1 (1959).

[29] V. R. Zurawski, Jr., W. J. Kohn, and J. F. Foster, *Biochemistry* **14**, 5579 (1975).

[30] K.-P. Wang and J. F. Foster, *Biochemistry* **8**, 4096 (1969).

Complex Formation. Reversible complex formation between dissimilar macromolecules is one of the most important of biological interactions, and the several methods of mass transport are particularly well suited for their characterization.[31,32] Here we are concerned with the effect of microheterogeneity on the shape of their analytical sedimentation patterns as exemplified by the antigen–antibody reaction. The prominence of immunology in cellular biology (e.g., cell surface antigen; monoclonal antibodies) and medicine (autoimmune disease; organ transplantation) warrants a fresh evaluation of the factors governing the sedimentation behavior of soluble antigen–antibody complexes, particularly those formed between a univalent antigen (Ag) and its bivalent antibody (Ab) in accordance with the only two possible reactions

$$Ag + Ab \rightleftarrows AgAb, \qquad 2k^0 \qquad\qquad (5a)$$
$$Ag + AgAb \rightleftarrows Ag_2Ab, \qquad k^0/2 \qquad\qquad (5b)$$

where k^0 is the intrinsic association constant. The classical system studied by Pepe and Singer[33] is BSA-S-R_1 : anti-R antibody where BSA-SH represents bovine serum mercaptalbumin, and

$$R_1 = CH_2CONH \bigotimes - AsO_3H_2 \text{ and } R = \bigotimes - AsO_3H_2$$

The experimental sedimentation pattern of a mixture of BSA-S-R_1 and anti-R under conditions of molar excess of antigen shows three peaks (inset to Fig. 2A). The slower sedimenting boundary corresponds to pure antigen, while the two faster peaks evidently constitute a bimodal reaction boundary across which the consecutive reactions (5a) and (5b) reequilibrate during differential sedimentation of the interacting species. But, theoretical sedimentation patterns computed for a homogeneous antibody assuming rapid reaction rates show only two peaks (pattern a in Fig. 2A), a boundary of pure antigen and a unimodal reaction boundary, thereby failing to simulate experiment. This discrepancy would not seem to be due to slow reaction rates,[21,34] but it occurred to us that it might be due to the well known heterogeneity of polyclonal anti-hapten antibodies with respect to binding affinity.[3]

The heterogeneous antibody population is usually described by either

[31] J. R. Cann, "Interacting Macromolecules; The Theory and Practice of Their Electrophoresis, Ultracentrifugation, and Chromatography," pp. 133–149 and 173–176. Academic Press, New York, 1970.

[32] D. J. Winzor, *in* "Protein-Protein Interactions" (C. Frieden and L. W. Nichol, eds.), p. 155. Wiley, New York, 1981.

[33] F. A. Pepe and S. J. Singer, *J. Am. Chem. Soc.* **81,** 3878 (1959).

[34] S. J. Singer, *in* "The Proteins Composition, Structure and Function" (H. Neurath, ed.), 2nd ed., Vol. 3, Chap. 15. Academic Press, New York, 1965.

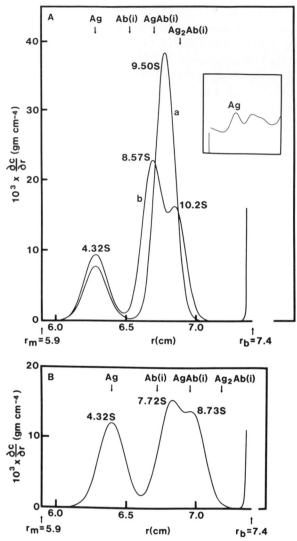

FIG. 2. Theoretical sedimentation patterns for univalent antigen–antibody reactions. (A) Computed pattern a is for a homogeneous antibody, reaction set (5) with $k^0 = 9.2836 \times 10^4$ M^{-1}, $\overline{Ag} = 7.8359 \times 10^{-5}\ M$, and $\overline{Ab} = 3.28 \times 10^{-5}\ M$. Pattern b is for a heterogeneous antibody, reaction set (6) with $\langle k^0 \rangle = 9.2836 \times 10^4\ M^{-1}$, $\overline{Ag} = 7.8359 \times 10^{-5}\ M$, $\overline{Ab}(1) = 2.18$ $\times\ 10^{-4}\ M$, $k_1^0 = 2.5 \times 10^4\ M^{-1}$, $\overline{Ab}(2) = 1.10 \times 10^{-5}\ M$, $k_2^0 = 1.25 \times 10^6\ M^{-1}$. Inset, experimental pattern of BSA-S-R$_1$: anti-R system (Adapted with permission from Pepe and Singer.[33] Copyright 1959 American Chemical Society.); $\overline{Ag}/\overline{Ab} = 2.22$. (B) Pattern computed for heterogeneous antibody, reaction set (6) with $\langle k^0 \rangle = 1 \times 10^4\ M^{-1}$, $\overline{Ag} = 7.8359 \times 10^{-5}\ M$, $\overline{Ab}(1) = 1.968 \times 10^{-5}\ M$, $k_1^0 = 5 \times 10^3\ M^{-1}$, $\overline{Ab}(2) = 1.312 \times 10^{-5}\ M$, $k_2^0 = 2.8284 \times 10^4\ M^{-1}$. Reprinted with permission from *Molecular Immunology* **19**, J. R. Cann, "Theory of Sedimentation for Antigen–Antibody Reactions: Effect of Antibody Heterogeneity on the Shape of the Pattern," copyright (1982), Pergamon Press, Ltd.

a Gaussian distribution of binding free energy or a Sips distribution. In some cases, however, the distribution in antiserum from a single animal may be discrete, consisting of relatively few molecular species. For example, there is a report that hapten-binding data for a particular antiserum could be analyzed in terms of only two binding constants (for references see Cann[21]).

With these considerations in mind we have computed theoretical sedimentation patterns for reaction of a univalent antigen with an antibody heterogeneous with respect to binding affinity but homogeneous in sedimentation and diffusion coefficients. The model assumes a discrete, bimodal distribution of intrinsic association constants and is schematized by the set of reactions

$$\text{Ag} + \text{Ab}(i) \rightleftarrows \text{AgAb}(i), \qquad 2k_i^0 \qquad\qquad (6a)$$
$$\text{Ag} + \text{AgAb}(i) \rightleftarrows \text{Ag}_2\text{Ab}(i), \qquad k_i^0/2 \qquad\qquad (6b)$$

where Ab(1) is the antibody of lower affinity, i.e., the one possessing the small intrinsic association constant, k_1^0, and Ab(2) is the antibody of higher affinity, k_2^0. Representative sedimentation patterns computed for this model are displayed in Fig. 2A, pattern b, and Fig. 2B. It is immediately apparent that with judicious choice of parameters the assumed binding heterogeneity of the antibody can give a bimodal reaction boundary. In fact, the theoretical pattern shown in Fig. 2B for the experimental values of the average binding constant $\langle k^0 \rangle = 1 \times 10^4 \, M^{-1}$ and the molar ratio of antigen to antibody $\overline{\text{Ag/Ab}} = 2.4$ is a rather good qualitative simulation of the shape of the observed pattern for the BSA-S-R$_1$: anti-R system (inset to Fig. 2A).

The shape of the computed pattern can be understood as follows. Because the slower sedimenting antigen is in molar excess, a boundary of pure antigen is left behind as the other species migrate differentially down the centrifuge cell to generate the bimodal reaction boundary across which the four reactions comprising reaction set (6) are continually reequilibrating. To a good approximation, the bimodality can be viewed as the result of two overlapping unimodal reaction boundaries, one for each class of antibody. Because of the difference between k_1^0 and k_2^0, the composition and, thus, the average sedimentation velocity of these virtual, unimodal boundaries also differ. The situation is in fact more subtle than this due to coupling between the reactions of Ab(1) and Ab(2) through the agency of a common antigen. Accordingly, the two peaks actually constitute a single reaction boundary. Concordantly, the two peaks do not correspond to separated antigen–antibody complexes.

These results are for a discrete bimodal distribution of binding affinities. In order to approach a continuous distribution, calculations have also been made for a discrete trimodal distribution using the simplified set of reactions

$$Ag + Ab(i) \rightleftarrows AgAb(i), \qquad 2k_i^0 \qquad i = 1, 2, 3 \qquad (7)$$

The conclusions reached are that a continuous distribution skewed toward low affinity or a bell-shaped distribution would not give bimodal reaction boundaries, but that a distribution skewed toward high affinity or a continuous bimodal distribution would. A distribution skewed toward high affinity makes biological sense, and a continuous bimodal distribution is not unreasonable. (For reference to the immunochemical literature see Cann.[21]) As to whether interpretation of the sedimentation patterns shown by the BSA-S-R_1 : anti-R system in terms of binding heterogeneity is realistic, Pepe and Singer[33] found that the value of the intrinsic association constant derived from electrophoretic patterns decreased with increasing concentration. As they pointed out, a possible explanation for this behavior is antibody heterogeneity.

Antibody heterogeneity also bears on the interpretation of the sedimentation behavior of mixtures of the bivalent T-hapten

$$R''-\!\!\!\bigcirc\!\!\!-R'', \ R''\!\!=\!\!CONH-\!\!\!\bigcirc\!\!\!-AsO_3H_2$$

with anti-R antibody.[35] Reaction of the two results in formation of linear polymers, i.e., essentially the dimer, trimer, etc. of the antibody molecule. The theory of sedimentation for ligand-mediated association (this series, Vol. 27 [12] and Vol. 48 [11]), of which this is a specific case, would predict bimodal sedimentation patterns for the T : anti-R system, were the antibody to be homogeneous; but the observed patterns for a molar ratio of hapten to antibody $\overline{Hap}/\overline{Ab} = 1.06$ are actually trimodal (Fig. 3A). The possibility that antibody heterogeneity is responsible for this trimodality was explored with the heuristic system

$$2Ab(1) + Hap \rightleftarrows Ab(1)_2Hap, \qquad K_1 \qquad (8a)$$
$$3Ab(2) + 2Hap \rightleftarrows Ab(2)_3Hap_2, \qquad K_2 \qquad (8b)$$

where the antibody of lower affinity, Ab(1), is assumed to form dimers with negligible amounts of high-order polymers, and the one of higher affinity, Ab(2), forms trimers with negligible amounts of both intermediate dimer and high-order polymers. Further simplification of the calculation was achieved by neglecting complexes such as AbHap, $AbHap_2$, Ab_2Hap_2, etc. The shape of the theoretical sedimentation pattern computed for this system (Fig. 3B) simulates the shape of the experimental

[35] S. I. Epstein and S. J. Singer, *J. Am. Chem. Soc.* **80**, 1274 (1958).

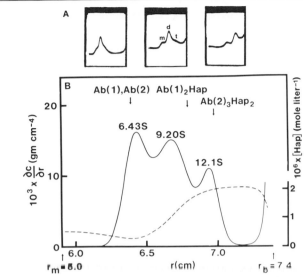

FIG. 3. Sedimentation patterns for bivalent hapten–antibody reaction. (A) Experimental patterns of T : anti-R system (Adapted with permission from Epstein and Singer.[35] Copyright 1958 American Chemical Society.): $\overline{Hap}/\overline{Ab}$ = 1.06, patterns obtained at different times of sedimentation are presented chronologically from left to right. (B) Theoretical pattern computed for the heuristic reaction set (8): (———) protein gradient curve; (–––) unbound hapten concentration. $\overline{Hap}$ = 2.475 × 10^{-5} M, $\overline{Ab(1)}$ = 5.5 × 10^{-5} M with K_1 = 9.0910 × 10^9 M^{-2}, $\overline{Ab(2)}$ = 5.5 × 10^{-5} M with K_2 = 3.3333 × 10^{23} M^{-4}. Reprinted with permission from *Molecular Immunology* **19**, J. R. Cann, "Theory of Sedimentation for Antigen–Antibody Reactions: Effect of Antibody Heterogeneity on the Shape of the Pattern," copyright (1982), Pergamon Press, Ltd.

pattern in showing three peaks. The peaks do not correspond to separated macromolecular species but to different equilibrium compositions each, in turn, rich in antibody monomer, dimer, or trimer. Resolution of this trimodal reaction boundary is dependent upon production and maintenance of gradients of unbound hapten along the centrifuge cell by reequilibration during differential transport of monomer, dimer, and trimer. The gradients are sufficiently strong (Fig. 3B) that the nonlinear coupling of transport with the hapten–antibody reactions causes the macromolecular composition to change markedly along the cell, thereby giving rise to resolution. In a sense, the trimodality results from superposition of two ligand-mediated self-associations, but this view is only an approximation since the reactions of Ab(1) and Ab(2) are coupled through a common ligand. That heterogeneity of the antibody with respect to binding affinity is a reasonable explanation of the observed trimodality finds support in light-scattering measurements on the T : anti-R system, which indicate that the association constant decreases with increasing total concentration.[36]

While we have dwelled on the antigen–antibody reaction, the described results have important implications for fundamental studies on other systems such as subunit proteins and protein assemblies, in the context of disassembly as well as reassembly experiments. Unrecognized microheterogeneity of one or more of the reactants with respect to association constant could give rise to multimodal sedimentation patterns which might prove conceptually misleading. Such microheterogeneity could be an inherent property of the biological system or merely adventitious due, for example, to deamidation or intramolecular disulfide interchange during isolation of the reactants. We noted early on that the microheterogeneity of plasma albumin resulting from intramolecular disulfide interchange has a profound effect on its electrophoretic patterns under isomerizing conditions. This effect is due to the fact that the various species of albumin undergo the N–F transition at slightly different characteristic pH values.[6] An example of the untoward effects of adventitious modification on interacting systems is afforded by the neuropeptide substance P (SP, Arg^1-Pro^2-Lys^3-Pro^4-Gln^5-Gln^6-Phe^7-Phe^8-Gly^9-Leu^{10}-Met^{11}-NH_2). The C-terminal residue Met^{11} undergoes slow oxidation to the sulfoxide,[37] so that some preparations of SP may contain small amounts of SP-sulfoxide. Whereas SP binds reversibly to tubulin and in so doing inhibits reassembly into microtubules, SP-sulfoxide does not inhibit reassembly.[38] Biologically active peptides and their analogs can be synthesized by classical and solid-phase methods and purified by conventional techniques. In the case of proteins, fractionation along with biochemical and biophysical characterization of the separated components is the established procedure for dealing with the effects of microheterogeneity on interactions. In the particular case of the antigen–antibody reaction, the techniques are now available for preparing monoclonal anti-R antibodies with differing binding affinities to test the ideas set forth above.

Systems Involving Two Ligands

The heuristic modeling of the T : anti-R reaction which we have just considered illustrates how microheterogeneity can give rise to multimodal

[36] S. I. Epstein, P. Doty, and W. C. Boyd, *J. Am. Chem. Soc.* **78,** 3306 (1956).

[37] J. M. Stewart, *in* "Substance P-Dublin 1983" (P. Skrabanek and D. Powell, eds.), p. 6. Boole Press, Dublin, 1983.

[38] J. R. Cann, I. Rahim, A. Vatter, and J. M. Stewart, *in* "Substance P-Dublin 1983" (P. Skrabanek and D. Powell, eds.), p. 24. Boole Press, Dublin, 1983.

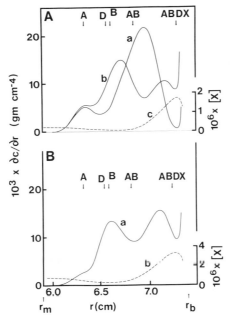

FIG. 4. Theoretical sedimentation patterns for ligand-mediated complex formation between dissimilar proteins. (A) Pattern a, control without ligand mediation: $K_1 = 2 \times 10^4\,M^{-1}$, $K_2 = 4 \times 10^4\,M^{-1}$; $\bar{A} = 1 \times 10^{-4}\,M$, $\bar{B} = 7.5 \times 10^{-5}\,M$, $\bar{D} = 5 \times 10^{-5}\,M$. Pattern b, computed for complex formation mediated in part by a small ligand molecule, reaction set (9): $K_1 = 2 \times 10^4\,M^{-1}$, $K_2 = 2 \times 10^{10}\,M^{-2}$, and $\bar{X} = 2.7 \times 10^{-5}\,M$. Curve c, concentration profile of unbound ligand corresponding to pattern b. (B) Pattern a, computed for reaction set (9): $\bar{A} = \bar{B} = \bar{D} = 7.5 \times 10^{-5}\,M$, $\bar{X} = 4 \times 10^{-5}\,M$. Curve b, concentration profile of unbound ligand. From Cann, *Biophys. Chem.* **16**, 41[39]; reprinted with permission of Elsevier Biomedical Press.

sedimentation patterns in the case of ligand-mediated self-association. Recently,[39] it was found that other, mechanistic factors can also dictate multimodal patterns for rapidly equilibrating interactions even when the reactants are homogeneous. Consider, for example, complex formation between three dissimilar protein molecules A, B, and D mediated in part by a small ligand molecule X as schematized by the reaction set

$$A + B \rightleftarrows AB, \qquad\qquad K_1 \qquad\qquad (9a)$$
$$AB + D + X \rightleftarrows ABDX, \qquad K_2 \qquad\qquad (9b)$$

Sedimentation patterns computed for this reaction set are displayed in Fig. 4. The contrasting behavior of systems with and without ligand mediation (with, reaction set 9; without, the sequence $A + B \rightleftarrows AB$, AB +

[39] J. R. Cann, *Biophys. Chem.* **16**, 41 (1982).

D $\rightleftarrows$ ABD) is illustrated in Fig. 4A. Whereas patterns for complex formation without ligand mediation shows a boundary of A and a unimodal reaction boundary sedimenting at a velocity between that of AB and ABD, pattern b for ligand mediation shows three peaks: a boundary of A and two more rapidly sedimenting peaks constituting a bimodal reaction boundary. The faster of the two peaks corresponds to an equilibrium mixture rich to ABDX and, thus, sediments between AB and ABDX; the slower one being rich in AB sediments between B and AB. Resolution of the bimodal reaction boundary is dependent, through the agency of mass action, upon the generation of a concentration gradient of ligand along the centrifuge cell (compare pattern b with the ligand concentration profile c). These results are for nonstoichiometric proportions of A, B, and D; but, as shown in Fig. 4B, qualitatively similar patterns were obtained for stoichiometric proportions.

There are several examples of the structural role of metal ions in stabilizing subunit proteins and protein assemblies, each evidently involving a single kind of ligand (for references, see Cann[39]), but it is conceivable that the stability of some protein assemblies might depend upon two different ligands. Accordingly, a model system was examined in which two dissimilar proteins A and B assemble into a complex with the mediation of two different ligands X and Y acting in a stepwise fashion,

$$A + B + X \rightleftarrows ABX, \qquad K_1 \qquad (10a)$$
$$2ABX + Y \rightleftarrows (ABX)_2Y, \qquad K_2 \qquad (10b)$$

in which X is obligatory for complex formation between A and B, and Y is obligatory for dimerization of the complex ABX. It was found that this model can give sedimentation patterns exhibiting four peaks: a boundary of A and a trimodal reaction boundary [Fig. 5A(b)]. Resolution of the reaction boundary is dependent upon generation of concentration gradients of both ligands along the centrifuge cell, and each of its three peaks corresponds to a different equilibrium composition and not to an individual protein species. There are two limiting cases which give sedimentation patterns showing only three peaks. (1) For sufficiently high concentration of Y relative to the concentration of X the two slowest peaks are boundaries of pure A and essentially pure B, which are virtually uncoupled due to depletion of X in the upper half of the centrifuge cell, while the fast third peak is a unimodal reaction boundary rich in $(ABX)_2Y$. (2) For vanishingly low concentration of Y reaction set (10) collapses to A + B + X $\rightleftarrows$ ABX, and the pattern shows a boundary of A and a leading, bimodal reaction boundary, the faster peak of the reaction boundary being rich in ABX and the slower one rich in B.

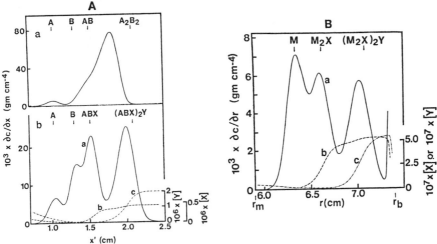

FIG. 5. Theoretical sedimentation patterns for complex formation or self-association mediated by two different ligands acting in a stepwise fashion. (A) (a) Control without ligand mediation [i.e., analogous to reaction set (10) but without participation of X and Y]; $K_1 = 4 \times 10^4\ M^{-1}$, $K_2 = 5 \times 10^3\ M^{-1}$, $\bar{A} = \bar{B} = 2.5 \times 10^{-4}\ M$. (b) Pattern a, unbound ligand X-concentration profile b, and unbound ligand Y-concentration profile c were computed for mediated complex formation, reaction set (10): $K_1 = 8 \times 10^{10}\ M^{-2}$, $K_2 = 5 \times 10^9\ M^{-2}$, $\bar{A} = \bar{B} = 1.38 \times 10^{-4}\ M$, $\bar{X} = 1.005 \times 10^{-4}\ M$, $\bar{Y} = 2.7 \times 10^{-5}\ M$. (B) Pattern a, ligand X-concentration profile b and ligand Y-concentration profile c were computed for mediated self-association, reaction set (11) with $K_1 = 2 \times 10^{10}\ M^{-2}$, $K_2 = 4 \times 10^{10}\ M^{-2}$, $\bar{M} = 1.5 \times 10^{-4}\ M$, $\bar{X} = 5.05 \times 10^{-5}\ M$, and $\bar{Y} = 1.3 \times 10^{-5}\ M$. From Cann, Biophys. Chem. 16, 41[39]; reprinted with permission of Elsevier Biomedical Press.

Finally, we return to self-association, this time mediated by two different ligands acting in a stepwise fashion. The model is for the tetramerization schema

$$2M + X \rightleftharpoons M_2X, \qquad K_1 \qquad (11a)$$
$$2M_2X + Y \rightleftharpoons (M_2X)_2Y, \qquad K_2 \qquad (11b)$$

As illustrated in Fig. 5B this schema can give sedimentation patterns exhibiting a well resolved trimodal reaction boundary. The slowest peak in the reaction boundary sediments slightly faster than monomer, the central peak, slower than dimer, and the fastest peak, slower than tetramer. As for the mechanism of resolution, the accompanying ligand concentration profiles speak for themselves. There are three situations for which the patterns are only bimodal. (1) For sufficiently high concentration of ligand Y relative to ligand X self-association is driven toward tetramer so that the central peak, which is rich in dimer, disappears. (2) In the limit of vanishingly low concentration of Y, reaction set (11) collapses to ligand-mediated dimerization [reaction (11a)] so that the fastest peak disappears (this series, Vol. 27 [12] and Vol. 48 [11]). (3) Decreasing the strength of the interaction occasions a bimodal reaction boundary in which the slow peak sediments about midway between monomer and dimer and the fast peak between dimer and tetramer.

These findings extend the guidelines arrived at previously (this series, Vol. 27 [12] and Vol. 48 [11]) for interpretation of sedimentation patterns in the case of ligand-mediated dimerization. Thus, in general, the sedimentation velocities of peaks comprising a reaction boundary are not characteristic parameters of products or reactants, and their areas cannot be used to calculate association constants. On the other hand, significant mechanistic insights can be obtained by systematically varying the proportions of protein reactants and the concentration of ligand(s) which promotes their association.

Recapitulation

The results described above for various model interactions broaden the theoretical base for biophysical investigations into the architectural and regulatory roles played by protein association *in vivo*. A rather provocative result is that microheterogeneity with respect to association constant can give sedimentation patterns exhibiting three peaks, but other mechanisms can also give rise to trimodal patterns. These include protein associations mediated by two ligands, the system, monomer $\rightleftarrows$ trimer $\rightleftarrows$ nonamer,[40] and slow reaction rates (this series, Vol. 48 [12]). It is essential, therefore, that appeal be made to additional biophysical methods, including fractionation, in order to establish the exact nature of an associating system and to characterize it as precisely as possible in terms of thermodynamic and other parameters.

In regard to conventional ultracentrifugal analysis, many of the patterns displayed herein bear a strong resemblance to patterns shown by mixtures of noninteracting proteins. This points up once again that proof of inherent heterogeneity depends upon isolation of the various components. In a related vein, the results for self-association mediated by two ligands could conceivably have import for active enzyme sedimentation[41] in solutions containing allosteric affectors, cofactors, or inhibitors, which

[40] J. L. Bethune and P. J. Grillo, *Biochemistry* 6, 796 (1967).
[41] D. J. Winzor, *in* "Protein-Protein Interactions" (C. Frieden and L. W. Nichol, eds.), p. 154. Wiley, New York, 1981.

might promote association of the enzyme as in the case of carbamoyl-phosphate synthetase from *E. coli*.[42]

Acknowledgment

Supported in part by Research Grant GM 28793 from the National Institute of General Medical Sciences, National Institutes of Health, U.S. Public Health Service.

[42] P. M. Anderson and S. V. Marvin, *Biochemistry* **9,** 171 (1970).

[4] Measurement of Linear Polymer Formation by Small-Angle Light Scattering

By ROBLEY C. WILLIAMS, JR.

Introduction

The problem of measuring quantitatively the assembly of linear, helical, or rod-shaped polymers from their constituent monomers has been addressed by measurement of a number of properties, including sedimentation, viscosity, quasielastic light scattering, and turbidity. Methods for making these measurements have been well treated in this series and elsewhere.[1-6] The use of light scattering to measure polymer formation has a number of advantages. It requires relatively little sample, it is effectively instantaneous, and it can easily be carried out over a wide range of solution conditions. The use of turbidity as a measure of light scattering in a solution of rodlike polymers has been widespread because only a spectrophotometer is needed to make measurements. Turbidity is related to the number of molecules of protein that have entered the polymer via two assumptions. First, it is assumed that the only quantitatively important species in solution are the monomer and large, rodlike polymers. Second, it is assumed that the intensity of light scattering produced by the monomers is negligible and that the intensity produced by the large polymers is proportional only to their mass, and not also to their molecular weight. The first assumption will be justified for many kinds of cooperative assem-

[1] J. A. Cooper and T. D. Pollard, this series, Vol. 85, p. 182.
[2] F. Gaskin, this series, Vol. 85, p. 433.
[3] D. L. Purich, T. L. Karr, and D. Kristofferson, this series, Vol. 85, p. 439.
[4] C. Montague and F. D. Carlson, this series, Vol. 85, p. 562.
[5] C. E. Smith and M. A. Lauffer, *Biochemistry* **6,** 2457 (1967).
[6] R. L. Nagel and H. Chang, this series, Vol. 76, p. 760.

bly processes. The second assumption is valid[7,8] for rods of length greater than about one wavelength of the incident light. (In the case of microtubules, such a length is attained by a tubule of about 1000 subunits.) There will thus be many situations in which these assumptions will hold. In cases where they cannot be assumed to hold, however, such as polymerization in the presence of many nuclei, or noncooperative polymerization, or polymerization at very low concentration, or polymerization under circumstances where exact quantitation of the number of intermonomer bonds formed is an important objective, a better founded technique is desirable. Small-angle light scattering offers most of the advantages of turbidity measurement, and is also somewhat easier to interpret in theoretically meaningful terms.

The scattering technique described in this chapter is well suited to the measurement of polymerization which results in formation of linear or rodlike structures of length not greatly exceeding 2 μm. The scattering from a solute in the course of polymerization can be described by the well-known expression

$$Kc/R_\theta = (1/M_w + 2Bc + \ldots)\,[1/P(\theta)] \tag{1}$$

where $K = 2\pi^2 n_0^2 (dn/dc)^2 (1 + \cos^2\theta)/N\lambda^4$

$$R_\theta = I_\theta r^2/(I_0 V) \tag{2}$$

Here, M_w is the weight-average molecular weight of the solute, B is the second virial coefficient, c is the concentration in grams per cm^3, n_0 is the refractive index of the solvent, (dn/dc) is the specific refractive index increment of the solute, θ is the scattering angle, I_θ is the scattered intensity, I_0 is the intensity incident on the scattering volume, and V is the magnitude of the scattering volume in cm^3. $P(\theta)$ is the form factor, which relates the observed scattering to the scattering that would be observed if the solute were composed of small Rayleigh scatterers. When $P(\theta)$ is near 1, the measured values of Kc/R_θ can be interpreted to give the weight-average molecular weight. For particles of arbitrary shape and dimensions not too much greater than the wavelength of light, it is possible to write

$$1/P(\theta) \cong 1 + (16\pi^2/3\lambda^2)R_G^2 \cdot \sin^2(\theta/2) \tag{3}$$

where R_G is the radius of gyration of a solute particle. Alternatively, for rigid isotropic rods it is well known that

$$P(\theta) = \frac{2}{qL}\int_0^{qL}\frac{\sin W}{W}\,dW - \left(\frac{2}{qL}\sin\frac{qL}{2}\right)^2 \tag{4}$$

---

[7] P. Doty and R. F. Steiner, *J. Chem. Phys.* **18**, 1211 (1950).
[8] B. J. Berne, *J. Mol. Biol.* **89**, 755 (1974).

where $q = (4\pi/\lambda)\sin(\theta/2)$ and L is the length of the rod. In the case of either Eq. (3) or (4), as the scattering angle, θ, becomes small, the value of $P(\theta)$ approaches 1, with a consequent simplification of the interpretation of the observed value of Kc/R_θ. Thus, at small angles, the scattered intensity is nearly proportional to the molecular weight of the observed particles to a good approximation. Taking as an example a microtubule-shaped particle of diameter 30 nm and length 100 nm (one which would contain 160 tubulin dimers) and a scattering angle of 5°, one would estimate by Eq. (3) or (4) that $P(\theta) = 0.996$. For a similar particle of length 1000 nm (1600 dimers) Eq. (3) or (4) gives $P(\theta) = 0.964$. By the time 2000 nm is reached, $P(\theta)$ begins to fall off, and a value of 0.87 is obtained.

It might be said that helical polymerization reactions are particularly well suited to study by small-angle light scattering. The reason for this is that one can usually contrive to introduce the solution to be studied into the scattering cell in a monomeric state. This strategy allows a very fine filter to be used to clarify the solution, so that "dust," which ordinarily plagues light scattering experiments, is excluded. The particles to be studied, of course, are polymers much larger than the monomers that were introduced, and consequently also larger than the largest "dust" particles that pass the filter. The objects of study in this system thus become much stronger scatterers than the "dust." This happy situation contrasts with the difficulties encountered in studying particles of fixed size, where a filter that passes the object of study must also pass "dust" particles of the same size.

Description of the Instrument

A practical small-angle light scattering photometer built according to a modification of a design of Kaye[9-11] is available as the Chromatix KMX-6.[12] This instrument operates over the angular range 2 to 7°, making use of a small He-Ne laser as a light source. Its major features are shown in Fig. 1. The laser beam is directed into the sample cell via a series of calibrated attenuators, as in most light scattering photometers. A chopper and phase-sensitive amplifier provide stability of response. The sample is contained within a cell consisting of a small hole in a block of plastic or metal that is "sandwiched" between two silica windows of thickness 5 cm. Scattered light leaving the cell passes through an annulus that defines the scattering angle, then through a relay lens of large aperture, and then through a small field stop located in the image plane of the lens. (The

[9] W. Kaye and A. J. Havlik, *Appl. Opt.* **12,** 541 (1973).
[10] W. Kaye, *Anal. Chem.* **45,** 221A (1973).
[11] W. Kaye and J. B. McDaniel, *Appl. Opt.* **13,** 1934 (1974).
[12] Available from LDC/Milton Roy Co., Riviera Beach, Florida.

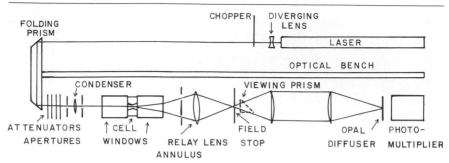

FIG. 1. Schematic diagram of important optical elements of the Chromatix KMX-6 small-angle light scattering photometer. A number of elements have been omitted for clarity.

annulus can be momentarily replaced by an aperture and attenuator to allow measurement of the intensity of the incident light beam.) This combination serves to isolate a small (approximately 0.02 μl) effective scattering volume and to overcome the otherwise severe problem of intense forward scattering from small imperfections in the windows of the cell. The large relay lens has a small depth of field, so that scattered light originating at a moderate distance from its object plane will not pass through the hole in the field stop. The thick windows have their outer surfaces far removed from the object plane of the relay lens, so that scattering from these surfaces does not contribute to the measured intensity. The inner surfaces of the windows are nearer to the object plane and do contribute marginally to the observed intensity, but because their refractive index is near to that of water, the contribution is small in comparison to the scattered intensity that arises at a silica–air interface. The effectiveness of this design is such that the Rayleigh ratio for water can easily be measured.[9] The size of the small scattering volume, V in Eq. (2), in this instrument is defined geometrically, so that no external calibration is required. Absolute, rather than relative, measurement of molecular weight is obtained. A viewing prism can be introduced into the beam to allow inspection of the scattering volume via an external ocular. Such observation facilitates alignment and allows one to see dust particles and other foreign materials that may be present in the scattering volume. The relatively small, thermostatted sample cell (200 μl or less) is filled through tubing in a "flow-through" manner.

The KMX-6 has been used with satisfactory results in a number of studies of aqueous solutions.[13–16]

[13] S. Hershenson and M. L. Ernst-Fonberg, *Biochim. Biophys. Acta* **751,** 412 (1983).

Cleaning the Cell and the Flow System

Principle. As is the case in all light scattering systems,[4,17,18] the presence of "dust" (by which is meant adventitious small particles, hereafter written without quotation marks) is a nontrivial practical problem to be overcome before meaningful measurements can be made with the KMX-6. The problem is particularly severe at small angles where, because of the immense forward scattering exhibited by large particles, a single piece of dust that finds its way into the narrowly focused scattering beam can completely overwhelm the less intense light scattered from the structures of interest. The problem is compounded when aqueous solutions are being studied, because of the difficulty of removing all small particles from water. The limited range of conditions to which proteins and other biological macromolecules can be subjected also limits the manipulations that can be carried out on a solution to remove dust. Ordinary clarification techniques, such as centrifugation of the sample, or filtering it into a separate vessel and then filling the cell with the filtered solution, are wholly inadequate to prepare samples for small-angle scattering observations.[19] These difficulties notwithstanding, the "flow-through" design of the cell of the KMX-6 allows one to exclude most dust by placing a filter in the input line. With careful use, the filter seldom needs to be replaced, and once the level of dust within the cell and its associated tubing is reduced to an acceptable level, only minor rinsing operations are required to achieve satisfactory results over several cycles of introduction of sample and buffer. It is necessary, however, to avoid opening the connection between the filter and the scattering cell. Exposure of the input solution to the atmosphere, even for a brief time, leads to the introduction of unsatisfactorily large amounts of dust into the scattering volume.[10] The technique presented below is one that minimizes the number of changes of filter and that has been found to work consistently in several investiga-

---

[14] K. J. Angelides, S. K. Akiyama, and G. G. Hammes, *Proc. Natl. Acad. Sci., U.S.A.* **76,** 3279 (1979).

[15] K. Adachi, T. Asakura, and M. L. McConnell, *Biochim. Biophys. Acta* **580,** 405 (1979).

[16] J. P. Weir, Doctoral dissertation, Vanderbilt University, 1980.

[17] S. N. Timasheff and R. Townend, *in* "Physical Principles and Techniques of Protein Chemistry, Part B" (S. J. Leach, ed.), p. 147. Academic Press, New York, 1970.

[18] E. P. Pittz, J. C. Lee, B. Balblouzian, R. Townend, and S. N. Timasheff, this series, Vol. 27, Part D, p. 209.

[19] There are two general approaches that can be taken to deal with the presence of dust: one can attempt to compensate for its presence, or one can attempt to remove it. An example of the former course of action can be found in K. J. Stelzer, D. F. Hastings, and M. A. Gordon, *Anal. Biochem.* **136,** 251 (1984). The author has chosen the latter approach.

tions in which the KMX-6 was employed to observe self-assembling systems.

Cleaning the Cell Windows. The cell windows may be cleaned with soap and water, applied *carefully* either with the clean fingertips or with a clean lens tissue. The background scattering level is quite sensitive to the presence of scratches on the inner surfaces of the windows, and care should be exercised (e.g., wash with filtered soap solution, rinse with distilled water) to avoid scratching them. It is neither necessary nor desirable to dry the inner faces of the windows before the cell is reassembled. The outer faces can be dried with lens tissue after reassembly. The manufacturer's cleaning recommendations are not satisfactory for use with aqueous solutions.

Description of the Flow System. The newly cleaned cell and associated tubing are freed of dust by passing rinsing solutions (see below) through it by means of a pump. Figure 2A shows a satisfactory arrangement for doing this. Rinsing solution (filtered through a 0.45-μm Millipore filter to prevent premature clogging of the further filters, and partially degassed under vacuum to prevent formation of air bubbles within the system) is drawn from the reservoir (a) by a piston pump (b) and is passed through filters (e and g) to the cell (h). The pressure gauge (c) shows by a rise in indicated pressure whether the filters are becoming clogged. The first filter (e) is a prefilter, which functions to prevent the main filter (g) from becoming prematurely clogged. It is changed when necessary. The three-way stopcock (d) is employed to remove air bubbles from the tubing. Both it and the second stopcock (f) also facilitate changing the prefilter. Solution exiting the cell passes upward to a final stopcock (i) which is closed when the pump is turned off for long periods of time to prevent retrograde flow of the solvent. Maintaining the cell as the lowest point in the system helps to prevent air bubbles from forming or lodging there.

Filters. Any of a number of types of membrane filters may be used. Generally an effective pore size of 0.45 μm or smaller is required. The author has found both the Millex-GV filters of 0.22 μm pore size (Millipore Corporation) and the Nucleopore filters of 0.2 μm pore size (Nucleopore Corporation) to be satisfactory, although the latter membranes appear to filter a smaller volume of a given solution before becoming clogged. It is a good idea to test any new type of filter before employing it in actual experiments. Some types appear to decompose partially, slowly shedding small particles into the buffer stream. It is also advisable to rinse a new prefilter with 20–50 ml of buffer before attaching it to the system. Omitting this step shortens the working life of the main filter, which is to be changed as seldom as possible. No stopcock is used between the main

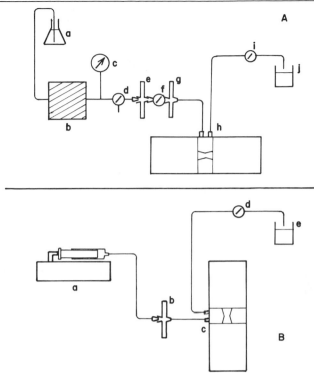

Fig. 2. Arrangement of flow components for cleaning and filling the light scattering cell. Components are described in the text. (A) Arrangement for rinsing the cell. (B) Arrangement for filling the cell. Luer taper fittings are used to make connections throughout the scheme. Tubing with fitted connectors, as well as stopcocks and other compatible fittings, can be obtained from the Hamilton Co., Reno, NV. The piston pump (item b in panel A) is a Milton-Roy Minipump in our apparatus, and the syringe pump (item a in panel B) is a Sage syringe pump. The pressure gauge (item c in panel A) has a pressure range of 0–50 psi.

filter and the cell because the operation of a stopcock appears to introduce small particles into the flowing stream.

Rinsing the System. The pump is equipped with a cyclic timer that turns it on for 4 min and then off for 4 min. The justification for this intermittent flow is as follows. First, it is observed that dust particles pass rapidly through the beam during flow, giving rise to large "spikes" in the intensity of scattered light. These spikes occupy only a small fraction of the total time of observation. Second, it is observed that when flow is stopped, dust particles soon appear in the beam. Under these static conditions, however, they remain in the beam for a substantial fraction of the time of observation. Third, when the flow is turned off for a time of several minutes and is then turned on again, it is observed that a "swarm"

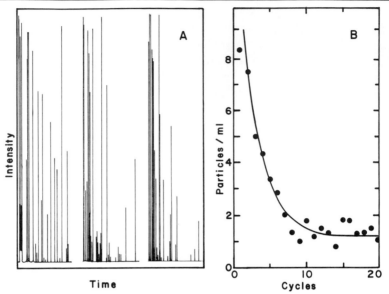

FIG. 3. Rinsing of the scattering cell under intermittent flow. (A) Dust "spikes" recorded during three successive 4-min intervals of solvent flow early in the rinsing process (direct tracings). The flow was interrupted for 4 min after each interval. Conditions of observation: flow rate, 2 ml/min; scattering angle, 6–7°; field stop, 0.2 mm; no attenuator in the beam; photomultiplier voltage, 600 V. (B) Number of spikes, of height greater than 1% of full scale, detected per ml of rinsing buffer during the rinsing process following a disassembly of the cell and changing of the filter. Each cycle represents 4 min of flow followed by 4 min of nonflow. Conditions were the same as in A.

of dust spikes appears immediately after resumption of flow, but that the number of spikes per unit time (or per unit volume) then diminishes rapidly as flow continues. If flow is interrupted and restarted for a second time, a similar initial swarm of spikes is seen again, as shown in Fig. 3A. A reasonable interpretation of these three observations is to suppose that during flow dust particles become trapped in eddies that occur in recesses in the filters, connectors, valves, and cell. When flow ceases, some of the dust particles are moved by Brownian motion and convection out of these recesses. When flow resumes, they are swept along, to appear in the beam as a swarm of scattering spikes. According to these notions, intermittent flow should gradually empty the recesses, while steady flow should remove dust inefficiently. In practice, intermittent flow works very well. Figure 3B indicates the rate of removal of dust that is seen to occur under intermittent flow. In contrast, it requires in excess of 12 hr of steady flow at the same rate to produce a situation in which the cell will be reliably dust free for 1 hr after flow is stopped. The use of intermittent flow in this application is strongly recommended.

The periods of time during which the pump is on and off, the rate of flow, and the solvents to be used in rinsing the system have not been optimized. The following represent a workable set of conditions for removing the dust that is present after the cell has been dismantled for cleaning of the windows; other solvents, flow rates, and times may work as well or better under a given set of circumstances. Solvents are pumped at a rate of 1–2 ml/min while the pump is on. About 40 ml each of filtered and degassed water, 5% acetic acid, 0.5 M KCl, and buffer are run under continuous flow through the prefilter and the main filter before they are connected to the cell in order to precondition them. (This can be done with a second pump at any convenient time.) The filters are then connected to the cell, and buffer is pumped through. The number of dust spikes of a significant amplitude is monitored under the conditions described in the legend to Fig. 3. When this number is reduced below 2–3 per ml, the system is ready for use.

Loading the Cell

Flow System. A flow system for introducing protein solution into the cell is schematized in Fig. 2B. A syringe pump (a), which in our system is contained within a small refrigerator mounted close to the scattering apparatus, causes degassed solution to flow through narrow-bore Teflon tubing directly to the main filter (b), and from there into the cell. The prefilter and stopcocks (d, e, and f in Fig. 2A) have been removed in this application to minimize the holdup volume that must be filled before the sample reaches the cell. (One may wish to include those two elements in a system where several milliliters of solution is available.) The cell is placed on end for filling, with the input line entering at the bottom and the exit line at the top. This orientation takes advantage of the fact that the density of the incoming solution is greater than the density of the buffer that it displaces. The orderly flow produced by having dense material below thus greatly reduces the amount of solution required to fill the cell. In practice, with the narrow tubing described, a single Millex filter unit, and the 200-μl scattering cell, the concentration within the scattering volume can be brought to >95% of its final concentration after approximately 0.9 ml has been injected with the syringe pump. As an example, we have routinely filled the cell with the use of a 3-ml syringe containing 1.2 ml of degassed solution. This liquid is pumped at the rate of 0.75 ml/min, so that about 2 min is required to fill the cell. (The limits to the flow rate are set by the resistance of the protein to denaturation due to shear in the filter. Faster flow could be achieved in principle.) The cell is then reinstalled in the scattering apparatus in another 15 sec, and measurements of scattering can begin.

Preparing for Another Sample. The cell is simply rinsed with buffer to prepare it for another sample. The syringe pump is disconnected from the main filter, the rinsing pump with its prefilter is reconnected, and approximately 40 ml of buffer is pumped through under intermittent flow as described above. Clearance of the cell is most rapidly achieved if it is removed from the instrument and placed vertically on the bench, with its exit port at the bottom.

Making Measurements

Figure 4A shows the scattered intensity from a solution of bovine serum albumin, recorded as a function of time at various values of the scattering angle. It is evident that as the scattering angle decreases the magnitude of the experimental "noise" increases. Figure 4B shows scattering intensity recorded as a function of time at various values of the diameter of the field stop. The smaller stops bring to the photomultiplier light from smaller volumes within the cell, and it is likely that fluctuations in the number of minute dust scatterers occupying this effective scattering volume contribute to the increasing "noise" that is observed with smaller stops.[20,21] The smaller amount of light falling on the photomultiplier must also contribute to the "noise." Figure 4C shows the end result obtained with bovine serum albumin. The molecular weight obtained, 69,400, is in good agreement with a value of $71,000 \pm 2000$ measured for the same sample by sedimentation equilibrium and probably reflects the presence of a small amount of dimer in the sample. The second virial coefficient estimated from the data, $2.2 \times 10^{-4} \text{ mol} \cdot \text{ml g}^{-2}$, is in satisfactory agreement with a value of $1.9 \times 10^{-4} \text{ mol} \cdot \text{ml g}^{-2}$ that can be calculated from previous measurements[22,23] made by light scattering techniques under similar but not identical conditions. The lack of variation of apparent molecular weight with scattering angle is in accord with expectations and indicates that the instrument is well calibrated geometrically.

Polymerization of Tubulin. Data from the temperature-induced polymerization of tubulin to make microtubules are shown in Fig. 5. A direct

[20] D. W. Shaefer and B. J. Berne, *Phys. Rev. Lett.* **28**, 475 (1972).

[21] D. W. Shaefer, *Science* **180**, 1293 (1973).

[22] J. T. Edsall, H. Edelhoch, R. Lontie, and P. R. Morison, *J. Am. Chem. Soc.* **72**, 4641 (1950).

[23] C. Tanford, S. A. Swanson, and W. S. Shore, *J. Am. Chem. Soc.* **77**, 6414 (1955).

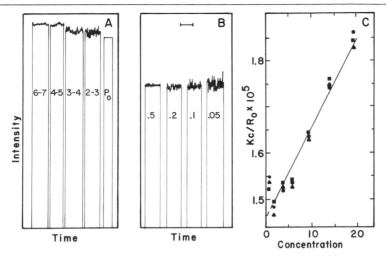

Fig. 4. Measures of the performance of the light scattering system with aqueous solutions. (A) Direct tracings of the scattered intensity observed at different scattering angles. The angles are as shown. The tracing labeled P_0 is a recording of the attenuated incident beam. Sample: bovine serum albumin at 5 mg/ml. Conditions: 0.2 mm field stop; 4-fold attenuator in the beam; photomultiplier voltage, 640 V; time constant, 0.3 sec. (B) Direct tracings of the scattered intensity observed with field stops of different diameter. Diameters are shown in mm. The photomultiplier voltage was adjusted to keep the recorded intensity approximately constant. Conditions: scattering angle, 6–7°; no attenuator in the beam; photomultiplier voltage, 589–643 V. The scale mark at the top indicates a time of 1 min and applies also to A. (C) Reciprocal of the apparent molecular weight of a sample of bovine serum albumin as a function of its concentration in 0.1 M NaCl, 0.1 M sodium phosphate, 1 mM dithioerythritol, pH 7.0, at 24°. Conditions: scattering angles, (●) 6–7°, (▲) 4.5–5.5°, (■) 3–4°; field stop, 0.2 mm; photomultiplier voltage, 600 V; attenuators in beam as appropriate.

recorder tracing of a reaction induced by a change in temperature at zero time is shown in Fig. 5A. The sensitivity is excellent over three orders of magnitude. It can be seen that the amplitudes of the minor fluctuations in observed intensity remain in rough proportion to the magnitude of the intensity itself. These slow fluctuations are not seen in observations of small macromolecules (e.g., bovine serum albumin or tubulin dimer), and they are probably not of instrumental origin. They appear to arise from the fact that the microtubules being observed are very large structures that diffuse very slowly.[10] The numbers of microtubules present at the concentrations employed here are expected to be large enough (order of 10^9 per μl) that fluctuations in the number of particles present in the scattering volume (occupation number) would be expected to be a small fraction of the total.[20,21] Correlation times of seconds to minutes, however, are present in the quasielastic scattering from particles of the size of

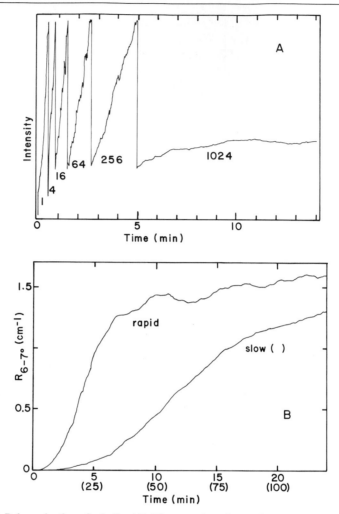

FIG. 5. Polymerization of tubulin. (A) Direct tracing of recorder output, showing assembly of sea urchin egg tubulin [H. W. Detrich III and L. Wilson, *Biochemistry* **22**, 2453 (1983)] at 24°, at a concentration of 0.76 mg/ml in 0.1 *M* PIPES buffer, 2 m*M* MgSO₄, 1 m*M* EGTA, 1 m*M* dithioerythritol, 0.1 m*M* GTP, pH 6.9. Conditions: scattering angle, 6–7°, field stop, 0.2 mm; time constant, 1 sec. Each vertical line represents the insertion of a measuring attenuator into the incident light beam. Each number represents the attenuation factor pertaining to the trace immediately above it. (B) Calculated values of the Rayleigh ratio, $R_{6-7°}$, through the course of two polymerizations of sea urchin egg tubulin. The curve labeled "rapid" is that shown in A, and corresponds to the times not in parentheses on the horizontal axis. The curve labeled "slow" represents sea urchin egg tubulin at a concentration of 1.5 mg/ml, polymerized at 18°, and corresponds to the times in parentheses. Other conditions as in A.

microtubules (dimensions of the order of a micron), especially at small scattering angle,[24,25] so that random thermal motion is the most likely source of these fluctuations. The courses of two representative polymerizations are shown in Fig. 5B. The data are of fairly good quality. It can be seen that the reaction carried out more rapidly shows much larger fluctuations than does the one carried out more slowly. The tubulin used in these studies has none of the associated proteins that ordinarily promote nucleation of microtubule formation. Rather, nucleation occurs during the time course of the reaction. Consequently, the slower polymerization has more and smaller microtubules than does the more rapid one, and therefore it would be expected to show less pronounced fluctuations.

Interpretation of observed values of R_θ in terms of molecular weights of the assembled particles formed is beyond the scope of this chapter. For rods that do not become too long, and for particles of other shapes that do not become too big, the approximation that $P(\theta)$ is near to 1.0 at the angles employed in the KMX-6 will be a valid one. This property of the data makes interpretation relatively simple. The technique, while not convenient enough for truly routine measurements, is nonetheless well suited to the detailed investigation[26] of helical or linear polymerization reactions of proteins.

[24] B. J. Berne and R. Pecora, "Dynamic Light Scattering with Application to Chemistry, Biology and Physics." Wiley (Interscience), New York, 1976.
[25] V. A. Bloomfield and T. K. Lim, this series, Vol. 48, Part F, p. 415. .
[26] H. W. Detrich, M. A. Jordan, L. Wilson, and R. C. Williams, Jr., *J. Biol. Chem.* **260**, 9479 (1985).

[5] The Measurement of Cooperative Protein Self-Assembly by Turbidity and Other Techniques

By JOSE MANUEL ANDREU and SERGE N. TIMASHEFF

The measurement of assembly processes of proteins into large asymmetric structures is complicated by the fact that their size precludes the use of standard macromolecular techniques, such as sedimentation velocity, sedimentation equilibrium, and classical light scattering. Methods which have been used with success are turbidity, rapid centrifugation, and small angle light scattering. While the last technique is discussed in Chapter [4] of this volume, the first two will form the subject of this chapter, with a particular emphasis on their limitations and criteria of validity.

METHODS IN ENZYMOLOGY, VOL. 130

Cooperative Polymerization

Assemblies into large structures usually proceed in cooperative manner, via a nucleated condensation polymerization mechanism. While the kinetic pathways of assembly in nucleated polymerization can be diverse and complicated, basically, for equilibrium thermodynamic purposes, the process can be reduced to two steps: (1) linear assembly with the formation of a polymerization nucleus, and (2) growth of the polymer in cooperative fashion. Frequently, examination of the process takes place when the system is in a state of pseudo-equilibrium and the most amenable parameter is the apparent equilibrium constant for the addition of each protein subunit to the growing polymer in the cooperative part of the pathway. The theoretical aspects have been exposed recently elsewhere.[1]

The two step nucleated polymerization reaction can be summarized as follows:

Step I: Nucleation:

$$P_1 + P_1 \rightleftharpoons P_2$$
$$P_2 + P_1 \rightleftharpoons P_3$$
$$\vdots \qquad \vdots$$
$$P_{n-1} + P_1 \rightleftharpoons P_n \tag{1}$$

$$K_n = \frac{[P_i]}{[P_{i-1}][P_1]}$$

where n is the number of subunits in the nucleus. This is a linear polymerization reaction in which addition of consecutive subunits involves the formation of only one intersubunit contact. The values of the equilibrium constants for all steps, K_n, are identical and small.

Step II: Growth:

$$P_n + P_1 \rightleftharpoons P_{n+1}$$
$$\vdots \qquad \vdots$$
$$P_{n+j} + P_1 \rightleftharpoons P_{n+j+1} \tag{2}$$
$$\vdots \qquad \vdots$$

$$K_g = \frac{[P_{n+j+1}]}{[P_{n+j}][P_1]}$$

Here each subunit is added in cooperative fashion, i.e., its addition involves the formation of contacts with two or more subunits of the already

[1] S. N. Timasheff, in "Protein-Protein Interactions" (C. Frieden and L. W. Nichol, eds.), p. 315. Wiley, New York, 1981.

assembled structure. The values of the equilibrium constants for all growth steps, K_g, are identical and large.

A necessary condition for a self-assembly to belong to the nucleated polymerization class is that $K_g \gg K_n$. This gives rise necessarily and trivially to the observation of a critical concentration, C_r, in such polymerizing systems. The critical concentration is a unique concentration value below which essentially all of the protein exists in monomeric subunit form and above which all of the protein exists as a mixture of monomers, present at a concentration $\simeq C_r$, and large polymers, present at a concentration of $C_t - C_r$, where C_t is total protein concentration.

Oosawa and co-workers[2] have examined the theory of nucleated polymerization, and have shown that, within a very close approximation, $C_r = K_g^{-1}$. Thus, in order to measure the apparent equilibrium constant for the growth reaction it is necessary simply to determine the critical concentration. This can be determined most easily by turbidity or by rapid centrifugation, if certain criteria are satisfied by the system.

Measurement of the Mass of Polymer Formed by Turbidity

Turbidity, i.e., measurement of the total light scattered by attenuation of the incident beam, has been used extensively to follow the assembly of microtubules and actin filaments. The theoretical validity of this approach is based on the theoretical analysis of Berne[3] of the scattering of light by long thin rods (see also Ref. 1). Berne[3] has shown that turbidity is a direct measure of the mass of protein polymerized if the dispersed system falls within the Rayleigh–Gans approximation and if it satisfies four criteria. These are (1) the assembled structure must have the geometry of rigid rods composed of optically isotropic monomeric subunits, (2) the rods must be randomly oriented in solution (thermodynamic ideality), (3) they must be monodisperse, and (4) their thickness must be small relative to the wavelength of the light, λ, and to the length of the rod, L. When all of these criteria are fulfilled, the turbidity becomes proportional to $(L/\lambda)^3$ at rod lengths of $(L/\lambda) > 3.5$, which leads to the result that turbidity becomes a linear function of the mass of protein polymerized,[1] i.e.,

$$\Delta\tau = \alpha C_h \qquad (3)$$

where $\Delta\tau$ is the turbidity increase on assembly, C_h is the mass of protein polymerized, and α is a proportionality constant characteristic of the nature of the polymer.

[2] F. Oosawa and M. Kasai, in "Subunits in Biological Systems, Part A" (S. N. Timasheff and G. D. Fasman, eds.), p. 261. Dekker, New York, 1971.
[3] B. J. Berne, J. Mol. Biol. 89, 755 (1974).

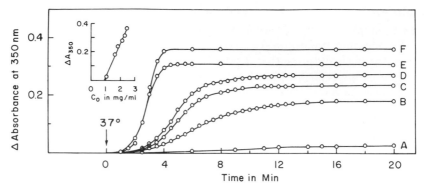

FIG. 1. Turbidimetric determination of the critical concentration of tubulin self-assembly into microtubules at 37° in a pH 7.0, 0.1 M phosphate buffer, containing 1.6×10^{-2} M MgCl$_2$, 10^{-4} M GTP, 3.4 M glycerol, and 10^{-3} M EGTA. [Reprinted by permission of the publisher from Thermodynamic Examination of the Self-Association of Brain Tubulin to Microtubules and Other Structures, by S. N. Timasheff, *in* "Physical Aspects of Protein Interactions" (N. Catsimpoolas, ed.), pp. 219–273. Copyright 1978 by Elsevier Science Publishing Co., Inc.

    In a typical experiment, the protein is permitted to polymerize at a set of total concentrations. At steady state, or pseudo-equilibrium, the turbidity attains a constant (plateau) value. A plot of the plateau values of turbidity as a function of total protein concentration is then extrapolated to $\Delta\tau = 0$, the intercept on the concentration coordinate being the critical concentration. A typical experiment of this type is shown as an example in Fig. 1. This is the heat-induced polymerization of tubulin into microtubules.[4] It is evident that an increase in tubulin concentration results in a gradual increase in the value of the plateau turbidity. In the inset of this figure, extrapolation of the plateau turbidity values yields a value of the protein critical concentration $C_r = 0.9$ mg/ml.

    The assembly reaction may be started in a variety of ways which perturb the state of the system from one in which monomers are favored to one in which polymers predominate. Frequently it is started by a temperature jump. This can be done with two water baths set at the two extreme temperatures and three way stopcocks connected to a jacketed cell placed in a spectrophotometer. Assembly can also be started by the addition of a necessary component (e.g., ATP or GTP) or by releasing an inhibition (e.g., in the case of microtubules, reducing the free Ca^{2+} concentration with a chelator). The turbidity is recorded as a function of time at a wavelength where the solution components do not absorb light. Once a plateau is reached, the reversibility can be tested by cooling the sample

[4] J. C. Lee and S. N. Timasheff, *Biochemistry* **16,** 1754 (1977).

to its initial temperature, or adding an inhibitor and recording the decrease of the turbidity. Thermodynamic reversibility can also be tested by consecutive polymerization cycles. These reversibility controls are necessary in order to ascertain that the polymerization process studied can be treated as an equilibrium (or a pseudo-equilibrium) and to correct the turbidity increments for any small contributions resulting from the formation of nonspecific protein aggregates that do not disappear upon cooling, e.g., denatured protein. It is also desirable to obtain electron micrographs at several points during the time course of the turbidity, as well as at different protein concentrations in order to identify the types of polymers formed, since these may contribute differently to the turbidity.

The time course of turbidity furnishes important qualitative information. If the polymerization proceeds by a nucleated condensation mechanism, the time course of the mass of polymer formed will consist of a lag period, followed by a sigmoidal rise of the turbidity which ends in a plateau.[1,5] During the approach to the plateau, the predominant reaction is that of elongation of the polymer. This generates an exponential time course which is characterized by an apparent rate constant, k_{app}, that is proportional to the concentration of nuclei. The reciprocal of the lag time and the apparent elongation rate constant are strongly dependent on protein concentration. The slope of log k_{app} versus log [protein] is a kinetic index of cooperativity which is related theoretically to the number of protomers within the nucleus.[5,6]

The quantitative use of the time course of turbidity requires, first of all, that the light scattering by the growing polymers conform to the Berne theory in the four criteria listed above over the entire range of the experiment. In the case of the assembly of long rod-like structures, such as microtubules or actin filaments, this is not difficult to accomplish near the plateau region. Strong deviations may be encountered, however, in the initial stages of polymerization due to departures from the long rod (length $>3.5\lambda$) geometry required, as is the case for short microtubules (i.e., microtubules shorter than 1.2 μm when 350 nm light is used). Departure from the correct geometry can be reflected also in the observation of turbidity overshoots before the plateau region. In the case of microtubules, this is observed frequently when polymerization is carried out at high protein concentration or in particular types of buffers.[7] While polymerization overshoots due to the formation of large metastable aggregates

[5] F. Oosawa and S. Asakura, in "Thermodynamics of the Polymerization of Protein." Academic Press, London, 1975.

[6] M. F. Carlier and D. Pantaloni, Biochemistry 17, 1908 (1978).

[7] J. M. Andreu and S. N. Timasheff, Arch. Biochem. Biophys. 211, 151 (1981).

have been reported for the tobacco mosaic virus protein,[8] in the microtubule system formation of open sheets is observed frequently. These obviously scatter light differently from microtubules and could account for the turbidity overshoots in some cases.

In general, the quantitative use of the time course of turbidity development to obtain kinetic information has to be approached with extreme caution. The presence of sheets or any other polymers with shapes different from long thin rods can render the entire turbidity analysis invalid and, in particular, any kinetic analysis. Furthermore, conformity to the rod-like geometry per se is not sufficient to validate the use of turbidity for quantitative conclusions about the equilibria, and, in particularly, about the kinetics of the polymerization reaction. The reason for this is that the Berne theory assumes ideal behavior of the polymer rods formed, in that they must be randomly oriented in solution, i.e., there must be no interactions between them. In fact, long rods would tend to align parallel to each other leading to long-range ordering effects,[9] resulting in external interference of the scattered light. This interference can be described in terms of virial coefficients.[10,11] For an identical total mass of protein polymerized, long rods may have more freedom of motion than short rods, because of tighter side-by-side packing for the latter.[9] As a result, attenuation of scattering at a given finite concentration may be less for long rods than short ones. It is known that the nucleated polymerization process is complex, the system reaching equilibrium with respect to the mass of protein polymerized prior to equilibrium with respect to the length of the rods. In fact, due to the slowness of the nucleation process, the initial structural state attained should be that of very long rods. As additional nuclei form, the number of rods increases, but the total mass of protein polymerized remains invariant. The length of the rods decreases with the final attainment of a steady state consisting of a larger number of shorter rods. As a result, the time course of the turbidity may display a maximum (or overshoot) prior to attainment of the steady state plateau, which reflects neither a change in mass of material polymerized nor a change in the overall structural geometry of the polymerized species.

The plateau turbidity can give important information with respect to the characteristics of the polymers formed and the interactions involved in polymerization. The nucleated condensation polymerization mechanism predicts the formation of negligible amounts of large polymers below

[8] T. M. Schuster, R. B. Scheele, and J. H. Khairallah, *J. Mol. Biol.* **127**, 461 (1979).
[9] L. Onsager, *Ann. N.Y. Acad. Sci.* **51**, 627 (1949).
[10] G. Fournet, *Acta Crystallogr.* **4**, 293 (1951).
[11] S. N. Timasheff and B. D. Coleman, *Arch. Biochem. Biophys.* **87**, 63 (1960).

the critical concentration, C_r, and the incorporation of essentially all of the protein above C_r into large polymers. If these conform to the Berne criteria, a plot of the plateau turbidity increment, $\Delta\tau$, versus total protein concentration, C_t, gives a straight line with a positive slope and an intercept of the abscissa at C_r. This is obtained experimentally for microtubule assembly *in vitro* at concentrations not far from C_r.[12,13] At higher protein concentrations, $\Delta\tau$ is frequently no longer proportional to $C_t - C_r$, but a significant downward curvature sets in, reflecting the external interference due to nonrandom orientation of the rods. This can be expressed in terms of a virial expansion of the turbidity. Carlier and Pantaloni[14] have proposed the following empirical expression:

$$C_h/\Delta\tau = 1/\alpha + \beta C_h \tag{4}$$

where $C_h = C_t - C_r$ and α^{-1} and β are the slope and intercept of a straight line plot of $1/\Delta\tau$ versus $1/C_h$. This expression is formally equivalent to a virial expansion of the turbidity, keeping only the first and second virial coefficients and setting all higher terms equal to zero.[1] The external interference effect may be so strong that turbidity decreases with polymer concentration. In fact, microtubule pellets are nearly transparent gels.[6] Rearrangement of Eq. (4) gives for $\Delta\tau'$, the turbidity corrected for concentration effects:

$$\Delta\tau' = \alpha C_h = \Delta\tau/(1 - \beta\Delta\tau) \tag{5}$$

Once straight line plots of turbidity vs protein concentration are obtained, their intercept may be found to vary with changes in the solution variables, since these may affect thermodynamically the critical concentration. The slope, however, must remain invariant as long as all the protein participates in the assembly and the polymers formed are of the same geometry, since the parameter α [Eqs. (3), (4), and (5)] is a property of the polymers. Participation of only a part of the protein in the assembly reaction (caused, e.g., by denaturation) would result in a decrease of slope, whereas a marked increase in slope indicates the formation of polymers different from long thin rods.

The second effect is illustrated in Fig. 2A, which shows $\Delta\tau$ vs C_t plots for the assembly of purified tubulin into microtubules (filled circles) and into polymers formed by tubulin liganded to the antimitotic drug colchicine (open circles) under identical solution conditions. The tubulin–colchicine polymers have been shown to be abnormal aggregates different

[12] F. Gaskin, C. R. Cantor, and M. L. Shelanski, *J. Mol. Biol.* **89,** 737 (1974).
[13] J. C. Lee and S. N. Timasheff, *Biochemistry* **14,** 5183 (1975).
[14] M. F. Carlier and D. Pantaloni, *Biochemistry* **20,** 1918 (1981).

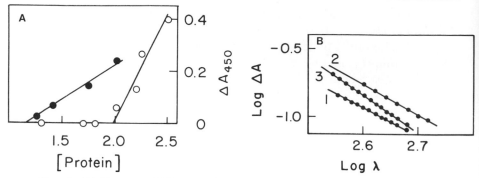

FIG. 2. (A) Plateau turbidity as a function of total protein concentration during the assembly of microtubules from purified tubulin (filled circles) and the polymerization of the tubulin–colchicine complex (open circles). Both reactions were performed in assembly buffer containing 10 mM sodium phosphate, 16 mM MgCl$_2$, 1 mM [ethylenebis(oxyethyl-enenitrilo)]tetraacetic acid, 0.1 mM GTP, 3.4 M glycerol at pH 7.0 and 37°. Reprinted with permission from Andreu *et al.* (1983).[15] Copyright 1983 American Chemical Society. (B) Wavelength dependence of the plateau turbidity generated by the polymerization of the tubulin–colchicine complex and of purified tubulin. Line 1: tubulin–colchicine polymer in assembly buffer (see A) without glycerol. Line 2: tubulin–colchicine polymer in assembly buffer. Line 3: microtubules in assembly buffer. (Taken from Andreu and Timasheff.[16])

from microtubules.[15] According to the Berne theory (see above), a plot of log Δτ vs log λ is expected to give a straight line with a slope of −3 for long thin rods. This has been checked experimentally for microtubules, and a value of −3.3 was found.[12] Small deviations from the theoretical value may be caused by instrumental imperfections.[1] When this is true, and provided that the electron microscope appearance of the polymer is as expected, the Berne approach can be applied safely and turbidity can be used as a quantity directly proportional to the mass of protein polymer-ized. Any strong deviation from the correct geometry is likely to be de-tected in the log Δτ vs log λ plots. As an example, such plots are shown in Fig. 2B for structures assembled from the tubulin–colchicine complex (lines 1 and 2, slope −2.1)[16] and from purified tubulin (line 3, slope −2.8).

Even when the Berne wavelength dependence criterion is satisfied, it is always desirable to calibrate Δτ in terms of the mass of protein polymer-ized, which can be determined by a direct procedure, such as centrifuga-tion (see below). This permits an empirical check of the proper propor-tionality relationship, as well as the determination of the turbidity per unit concentration of polymer, which is equal to the parameter α in Eqs. (3)

[15] J. M. Andreu, T. Wagenknecht, and S. N. Timasheff, *Biochemistry* **22**, 1556 (1983).
[16] J. M. Andreu and S. N. Timasheff, *Proc. Natl. Acad. Sci. U.S.A.* **79**, 6753 (1982).

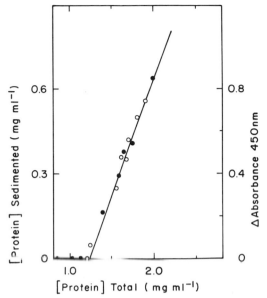

FIG. 3. Dependence of tubulin–colchicine polymer concentration and turbidity on total protein concentration. The tubulin–colchicine solutions were heated at 37° for 45 min in 10 mM sodium phosphate, 16 mM MgCl₂ 0.1 mM GTP, pH 7.0 buffer, centrifuged at 30,000 g at 37 ± 1° for 15 min, and the amount of protein sedimented was measured (solid circles). In another set of samples the plateau turbidity was measured under identical conditions (open circles). (Taken from Andreu and Timasheff.[16])

and (4). Typical results are shown in Fig. 3 for the tubulin–colchicine polymer, where $\Delta\tau$ and the mass of polymer sedimented were determined in parallel under identical conditions as a function of total protein concentration. Although the polymers formed did not satisfy the Berne criteria, a linear relationship was obtained, nevertheless, between $\Delta\tau$ and polymer mass, yielding an absorbance of 1.3 per mg/ml of polymer at 450 nm. This correlation has permitted to use turbidity as a strictly empirical tool for following polymerization in this case.

Sedimentation and Other Techniques

The amount of protein assembled can be measured directly by differential centrifugation in which only high polymers sediment. This procedure requires[1] (1) that the separation be performed under external conditions close to those of polymerization and in a time short enough so that no significant depolymerization takes place due to perturbation of the equilibrium, and (2) that the assembly equilibrium not be disturbed significantly by the hydrostatic pressure generated at the bottom of the tubes during high speed centrifugation, since assembly reactions are characterized frequently by sizable changes of the partial specific volume. The first requirement can be met by centrifuging the sample at the assembly temperature[12] (see also Fig. 3), employing as medium a buffer that stabilizes assembled polymers.[17] Alternately, one may use short centrifugation times in a table-top ultracentrifuge (where, e.g., microtubules are sedimented in 2 min at $1.6 \times 10^5 \, g$).[14] Perturbation of the polymerization equilibrium by the separation procedure and pressure can be ascertained by examining the protein in the supernatant for ability to assemble at all at atmospheric pressure and by checking that its concentration is equal to the critical concentration. Controls for nonspecific protein aggregation must also be run. Once these limitations are overcome, the sedimentation assays and, in particular, the fast sedimentation procedure[14] have the advantage that the mass of protein incorporated into large aggregates can be determined unequivocally for a large number of samples in a short time.

Electron microscopy has to be employed together with the turbidity and sedimentation assays to verify the morphology of the assembled structures. Electron microscopy can be used for two other important purposes, namely the measurement of the length distribution of polymers[18] and the monitoring of the rates of polymerization and dissociation at the two ends of the polymer.[19]

Transport techniques, such as viscosity and flow birefringence, have also been applied to monitor the formation of large protein assemblies. These techniques can give information on the critical concentration and on the general pattern of the assembly reaction, although more quantitative use is limited by the nonlinear relationships of the properties measured to the mass of protein polymerized and their marked dependence on the exact shape, flexibility, and cross-linking of the polymers.[1]

In the case of microtubule assembly, fast filtration procedures have been developed to measure assembly.[20,21] These are based on the stabilization of microtubules in glycerol-containing buffers and their retention in glass fiber filters. These procedures are useful to measure the incorpora-

[17] R. L. Margolis and L. Wilson, *Cell* **13,** 1 (1978).
[18] D. Kristofferson, T. L. Karr, T. R. Malefyt, and D. L. Purich, *Methods Cell Biol.* **24A,** 133 (1982).
[19] G. G. Borisy and L. G. Bergen, *Methods Cell Biol.* **24A,** 171 (1982).
[20] R. B. Maccioni and N. W. Seeds, *Arch. Biochem. Biophys.* **185,** 262 (1978).
[21] L. Wilson, K. B. Snyder, W. C. Thompson, and R. Margolis, *Methods Cell Biol.* **24A,** 159 (1982).

tion of labeled protein into polymers and its release in small samples. Their usefulness to measure the mass of protein polymerized depends on the ability to control carefully the stability of the polymers during the separation procedure, the retention efficiency, and nonspecific adsorption of protein onto the filters.

Another technique which has been used in studies of actin polymerization is that of fluorescence photobleaching recovery.[22,23] In this method the protein is labeled with a fluorophore. The assembled structure is then bleached with a short pulse of light of appropriate wavelength and the time course of the recovery of fluorescence is followed. This method can give information on the degree of polymerization, the relative size of the filaments formed, and the formation of networks. In order to apply this technique, controls must be performed to show that modification of the protein with the fluorophore does not affect the assembly process and that the irradiation does not damage the protein.

Interpretation of C_r

Once a nucleated condensation polymerization mechanism has been established, the most useful piece of information that can be obtained from the measurement of any parameter related to the mass of polymer formed as a function of protein concentration is the critical concentration, C_r. Since its measurement involves extrapolation to zero polymer concentration, it should not entail large errors. C_r is, in good approximation, the apparent dissociation equilibrium constant of the growth reaction. Its dependence on the experimental variables permits the determination of the apparent thermodynamic parameters of the growth reaction. As an example, Fig. 4 shows the apparent enthalpy and entropy contributions to the apparent standard free energy changes for the growth of the tubulin–colchicine polymers[16] and of microtubules.[4] These results, which indicate a close thermodynamic resemblance between the two polymerization reactions, were obtained from simple turbidimetric measurements of C_r at different temperatures and nonlinear integrated van't Hoff fitting of the data that resulted in apparent heat capacity changes in the vicinity of -1500 cal deg^{-1} mol^{-1} in both cases.[4,16] This apparent value of ΔC_p for microtubule assembly has been verified by direct microcalorimetric measurements.[24]

[22] J. F. Tait and C. Frieden, *Biochemistry* **21**, 3666 (1982).

[23] A. Mozo-Villarias and B. R. Ware, *J. Biol. Chem.* **259**, 5549 (1984).

[24] H.-J. Hinz, M. J. Gorbunoff, B. Price, and S. N. Timasheff, *Biochemistry* **18**, 3084 (1979).

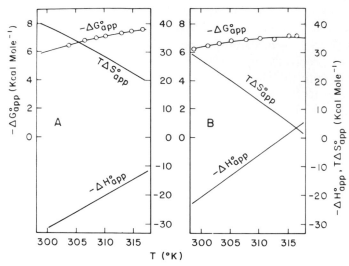

FIG. 4. (A) Enthalpy and entropy contributions to the apparent standard free energy change of polymerization of tubulin–colchicine. (B) Enthalpy and entropy contributions to the apparent standard free energy change of polymerization of tubulin during microtubule assembly.[4] Reprinted with permission from Andreu et al. (1983).[15] Copyright 1983 American Chemical Society.

Application of Time-Resolved X-Ray Scattering to Study Protein Self-Assembly

High intensity synchrotron X-ray sources and adequate detection systems have been applied recently to obtain solution X-ray scattering patterns in seconds. As a result, it is possible now to follow the pathway of relatively slow assembly reactions.[25,26] X-Ray scattering furnishes information on the average degree of polymerization, the shape of the aggregates, and the arrangement and conformation of the monomers. In the application of this technique to the microtubule system, the scattering patterns of purified tubulin heterodimers, microtubule protein assembled in the cold (with the formation of rings) and microtubule protein assembled at 37° (with the formation of microtubules) were obtained and analyzed. The patterns obtained during slow temperature scans of microtubule protein were interpreted in terms of these data. This has enabled the authors[26] to identify prenucleation events, microtubule nucleation, micro-

[25] E. M. Mandelkow, A. Harmsen, E. Mandelkow, and J. Bordas, Nature (London) 287, 595 (1980).
[26] J. Bordas, E. M. Mandelkow, and E. Mandelkow, J. Mol. Biol. 164, 89 (1983).

tubule growth, and postassembly events. This powerful technique has led to important conclusions,[26] some of which are (1) rings do not participate directly in microtubule assembly, but break down into smaller oligomers and tubulin heterodimers, (2) the nucleation stage is characterized by an increase in the average degree of polymerization without the formation of microtubules; during a few seconds a scattering pattern appears that can be modeled by a few associated protofilaments, structures which are good candidates for nucleation centers,[27] and (3) the bond length between α and β tubulin subunits is 4 nm in microtubules, while it is approximately 4.3 nm in rings (which are probably coiled protofilaments)[28] and longer in soluble $\alpha\beta$ heterodimers, indicating monomer structural changes between the three aggregation states.

[27] H. P. Erickson and D. Pantaloni, *Biophys. J.* **34,** 293 (1981).
[28] E. Mandelkow, E. M. Mandelkow, and J. Bordas, *J. Mol. Biol.* **167,** 179 (1983).

[6] Thermodynamic Parameters for Protein–Protein and Protein–Ligand Interaction by Differential Scanning Microcalorimetry

By HANS-JÜRGEN HINZ

Introduction

The usefulness of noninvasive methods to characterize protein–protein interactions has long been appreciated because of the inherent ambiguity of probe techniques which do not monitor the intrinsic properties of the system but those imparted by the probe. Among these noninvasive methods microcalorimetry holds a prominent place as a result of the practically universal linkage of chemical and biological processes to heat effects. Precise determination of reaction enthalpies has been recognized to be of potential value in giving insight into molecular details and the nature of interaction forces operative within systems when properly combined with nonthermodynamic information. There is no doubt that the preferred method of obtaining reliable enthalpy parameters is by direct calorimetric measurements. Although, so far, in the majority of cases, these measurements have been performed isothermally at various temperatures,[1-13] there are a number of endothermic biological reactions in which scanning microcalorimetry could provide information difficult to obtain otherwise.

Calorimetrically determined energy parameters, in contrast to those derived from the dependence on temperature of the equilibrium constant, the van't Hoff method, are independent of the model assumed for the reaction and lack the indeterminacy with respect to the size of the mole inherent in all van't Hoff ΔH_{vH} values. Another important property of calorimetric techniques, in general, and scanning methods, in particular, is the fact that calorimetric measurements involve the time parameter, so that in principle they can provide more information than the time-independent data obtained from equilibrium studies.

However, universality also has its price. Due to the ubiquitous occurrence of enthalpy changes, calorimetry may fall victim to misinterpretation more easily than other more specific methods. It is, therefore, of vital importance in biochemical applications to work with highly purified and well-defined systems.

Instrumentation

This discussion will be restricted to scanning microcalorimetry. There are some excellent treatments in the literature, which the interested reader can consult to obtain deeper insight into the method[14-16] and fields of application.[17-23] Only some salient features of the techniques will be

[1] R. L. Biltonen and N. Langerman, this series, Vol. 61, p. 287.
[2] N. Langerman and R. L. Biltonen, this series, Vol. 61, p. 261.
[3] C. H. Spink and I. Wadso, *Methods Biochem. Anal.* **23**, 2 (1976).
[4] C. H. Spink, *CRC Crit. Rev. Anal. Chem.* **9**, 1 (1980).
[5] I. Wadso, *Q. Rev. Biophys.* **3**, 383 (1970).
[6] I. Wadso, *in* "New Techniques in Biophysics and Cell Biology" (R. H. Pain and B. J. Smith, eds.), Vol. 2. p. 85. Wiley, New York, 1975.
[7] I. Wadso, *Pure Appl. Chem.* **52**, 465 (1980).
[8] H.-J. Hinz, *Annu. Rev. Biophys. Bioeng.* **12**, 285 (1983).
[9] H.-J. Hinz, *in* "Topics in Molecular Pharmacology" (A. S. V. Burgen and G. C. K. Roberts, eds.), p. 71. Elsevier, Amsterdam, 1983.
[10] H. D. Brown, ed., "Biochemical Microcalorimetry." Academic Press, New York, 1969.
[11] N. Langerman, this series, Vol. 57, p. 540.
[12] G. Rialdi and R. L. Biltonen, *MTP Int. Rev. Sci. Phys. Chem. Ser. 2* **10**, 147 (1975).
[13] I. Wadso, *MTP Int. Rev. Sci. Phys. Chem. Ser. 2* **10**, 1 (1972).
[14] E. Gmelin, *Thermochim. Acta* **29**, 1 (1979).
[15] K. S. Krishnan and J. F. Brandts, this series, Vol. 49, p. 1.
[16] P. L. Privalov, *Pure Appl. Chem.* **52**, 479 (1980).
[17] S. Mabrey and J. M. Sturtevant, *Methods Membr. Biol.* **9**, 237 (1978).
[18] B. G. Barisas and S. J. Gill, *Annu. Rev. Phys. Chem.* **29**, 141 (1978).

briefly mentioned here to provide a basis for understanding the advantages and problems of its application in extracting thermodynamic interaction parameters of proteins or proteins and other ligands.

The table gives a survey of some of the commercially available differential thermal analysis instruments (DTA) and scanning calorimeters (DSC). The compilation does not claim complete coverage, but it probably lists the majority of instruments which might be applied.

The distinction between DTA and DSC is maintained according to nomenclature suggestions of IUPAC.[24] The basic difference between these two categories of instruments resides in the measuring principle during a temperature scan. DTA apparatus registers temperature differences between the reference and sample cells resulting from a reaction in the sample cell which are proportional to the enthalpy change of the reaction under investigation, the heat capacity of the reactants, and the total thermal resistance to heat flow of the cell arrangement. Measurements using samples of known thermodynamic properties are required for calibration. DSC apparatus employs the temperature difference between the sample and reference cell only as a signal to activate electronically regulated heaters to maintain a "zero" temperature difference between the cells by extra supply and/or reduction of Joule heating of either cell. Thus the additional energy required to nullify the "out of balance signal" can be monitored directly. Depending on further construction details such as the use of removable or fixed cells, etc. (see e.g., Refs. 15, 16), calibration is preferentially performed electrically or by means of a sample of known thermodynamic characteristics. A common principle of all high sensitivity scanning calorimeters is the differential arrangement of physically identical cells. Particularly when measuring in highly diluted solutions of biopolymers, where usually more than 99% of the heat capacity is that of the solvent and where changes in the residual heat capacity are to be measured with an accuracy of better than 10%, compensation of the solvent heat capacity by an identical filling of a reference cell is the only way to provide the required sensitivity. However, sensitivity alone is not a sufficient prerequisite for high precision and accuracy. Of preponderant importance for determination of differences in heat capacity is the repro-

[19] J.-J. Hinz, in "Biochemical Thermodynamics" (M. N. Jones, ed.), p. 116. Elsevier, Amsterdam, 1979.
[20] J. M. Sturtevant, in "Physical Methods of Chemistry" (A. Weissberger and B. W. Rossiter, eds.), Vol. 1, Pt. 5, p. 347. Wiley (Interscience), New York, 1971.
[21] J. M. Sturtevant, this series, Vol. 26, p. 227.
[22] J. M. Sturtevant, Annu. Rev. Biophys. Bioeng. 3, 35 (1974).
[23] J. L. McNaughton and C. T. Mortimer, IRS Phys. Chem. Ser. 2 10 (1975).
[24] "Recommendations for Nomenclature of Thermal Analysis." IUPAC, London, 1972.

CALORIMETRIC PROPERTIES OF COMMERCIAL DSC AND DTA APPARATUS[a]

Property (unit)	Volume of cell (ml)	Temperature range (K)	Heating rates = r (K/min)	Noise level[b] [electrical] (value/ml) r = 1 K/min [μW(μW/ml)]	Baseline reproducibility with the same filling (value/ml) r = 1 K/min [μW(μW/ml)]	Baseline reproducibility with different fillings of the same sample (value/ml) r = 1 K/min [μW(μW/ml)]	Relative error of heat capacity determination (enthalpy) (%)
Daini Seikoshi SSC-50	0.07	120–400	0.01–5.0	1.3 (19)	2.5 (36)		0.5
DASM-1M Acad. Sci. USSR	1	278–378	0.1–2.0	0.5 (0.5)	2.0 (2.0)	5 (5)	0.01
DASM-4 Acad. Sci. USSR	0.6	278–408	0.125–2.0	0.5[15[c]] (0.8)	2.0 (3.0)	3 (5)	0.005
Dupont 910-DSC	0.05	100–1000	0.1–100	5.0 (100)	20		1
Mettler DSC-30	0.04–0.16	100–870	0–100	50 (310–1250)			2
Microcal	1.3	253–383	0.17–1.5	(0.2)	(1)	(1.7)	0.002
Netsch DSC444	0.04–0.12	130–800	0.1–20	1.5[50[d]] (12.5–37.5)			3(1)
Perkin Elmer DSC 7	0.03–0.075	100–1000	0.1–500	2.0 (27–67)	5.0 (67–167)		0.8
Setaram Bio DSC	1.2	263–373	0.002–10	0.2–2 (0.17–1.79)			0.005

[a] Data based on information given by the manufacturers. The instruments by Dupont, Mettler, Netsch, and Setaram must be classified DTA apparatus and heat flux calorimeter, respectively, according to Ref. 24.

[b] These numbers alone are not very good criteria for the sensitivity of the instruments as long as the frequency of the noise is not specified, which is unfortunately the case in all manuals. High frequency noise can be taken care of easily by proper filtering; that necessitates, however, long time constants. The usefulness of a DSC or DTA instrument for solving certain problems may depend, however, more on long-term drifts or changes in the shape of the baseline on refilling the cells, since such changes introduce more serious errors into the determination of enthalpies and heat capacities. This is particularly the case if the processes to be studied extend over a wide temperature range. This is illustrated by the following considerations. The magnitude of the heat effect depends on the amount of substance in the sample cell. According to Eq. (1) the detectability of the heat, i.e., the power signal dH/dT, is a function of the heating rate.

Example: 1 mg of a sample of molar mass 10,000 g/mol; molar enthalpy: $\Delta H = 100$ kJ/mol; heating rate: $r = 1$ K/min $= 1$ K/60 sec; 1 mg $= 10^{-7}$ mol; overall heat effect: $\Delta Q = 10^{-7}$ mol $\cdot 10^5$ J/mol $= 10^{-2}$ J $= 10$ mWs.

If one considers for reasons of simplicity a square-shaped heat absorption peak, which extends over 20 K, it will take 1200 sec to record the peak. Thus the observable power signal dH/dt relative to the baseline can be calculated from the relation: $dH/dt \cdot 1200$ sec $= 10$ mWs.

$\rightarrow dH/dt\ [\mu W] = 10,000/1200\ [\mu Ws/\text{sec}] = 8.3\ [\mu W].$

This power signal is close to the detection limit of several instruments. Another criterion for choosing an instrument useful for biochemical studies is the question of whether highly diluted solutions have to be studied. If one needs, as illustrated in the above example, 1 mg of the sample to get a detectable signal, the sample will yield a 0.1% solution in a sample cell of volume 1 ml, but a 3.3% solution in a sample cell of volume 30 μl. Thus the result of the measurements may be affected by interactions between the molecules. If interactions do not play a role nor do slow kinetics limit the heating rate, the signal can be improved by increase of the heating rate. In such a case smaller sample cells are more practical, because they are not subject to temperature inhomogeneities as much as larger cells. In general, before tackling a specific problem, it is advisable to perform test measurements with various instruments.

[c] Thermal noise.

[d] Lowest detectable power signal.

ducibility of the baseline at repeated fillings. A baseline is defined as the temperature scan of the difference signal of identically (e.g., solvent) filled cells.

For enthalpy determinations this requirement is less stringent, since the shape of the baseline rather than its exact position is of significance. More detailed discussions of these problems, which concern DTA and DSC instruments somewhat differently, can be found in the references.[16,25,26]

Quantities Determined by DSC

Apparent Specific Heat Capacity

The quantity primarily determined is the variation with temperature of the apparent heat capacity of the sample cell relative to that of the reference cell.

Equation (1) shows a transformation of the definition of isobaric heat capacity which illustrates the dependence of the recorded power signal, (dH/dt) [W], on the three variables: (1) mass m [g] of the sample in the sample cell, (2) its apparent specific isobaric heat capacity, c_p [J g^{-1} K^{-1}], and (3) the heating rate employed, $r = (dT/dt)$ [K sec^{-1}].

$$\text{apparent heat capacity} = mc_p = \left(\frac{dH}{dT}\right)_p = \left(\frac{dH}{dt}\right)_p \left(\frac{dt}{dT}\right)_p$$

$$= \left(\frac{dH}{dt}\right)_p \frac{1}{r} \rightarrow \left(\frac{dH}{dt}\right)_p = mc_p r \qquad (1)$$

Thus increase of the measured signal for a given heat capacity of a sample can be achieved by increasing the mass of the sample and/or increasing the heating rate of the scan. In the majority of the DTA instruments heat transfer problems limit the size of the cells, which are usually small removable pans with or without lids, to volumes below 100 μl. Therefore sensitivity is increased by using heating rates up to 500 K min^{-1}. Heating rates of this magnitude are permissible for small amounts of pure substances, when comparative measurements with a reference sample are performed under identical conditions. However, they cannot be applied to solutions of biopolymers. In addition to formation of temperature gradients within the cell, slow conformational changes could be measured

[25] R. N. Goldberg and E. Prosen, *Thermochim. Acta* **6**, 1 (1973).
[26] B. Wunderlich, *in* "Techniques of Chemistry" (A. Weissberger and B. W. Rossiter, eds.), Vol. 1, p. 427. Wiley (Interscience), New York, 1971.

under nonequilibrium conditions. Though in such a case variation of the measured parameter with heating rate could provide interesting information on the kinetics of the reaction, the equilibrium parameter could only be obtained by extrapolation to zero heating rate. Thus, in summary, one may state that heating rates ≤ 5 K min^{-1} and often ≤ 1 K min^{-1} are advisable for biochemical problems. In any case influence of heating rate on the thermodynamic property measured should be checked by preferentially going to as low a heating rate as possible.

Figure 1 shows the baseline, including an electrical calibration peak, and a sample curve, which would be typical for denaturation of a small globular protein. The heat capacity of the sample cell is lower than that of the reference cell before and after the transition, because part of the cell volume is occupied by protein, which has in both its native and denatured state a lower heat capacity than the equivalent volume of solvent. The heat capacity of the sample apparently increases only in the transition range due to the energy consumption of the reaction. Thus the temperature-dependent difference in heat capacity, Δc_p^{app} c_p(reference cell) c_p(sample cell), is given by the following equation[27]:

$$\Delta c_p^{app}(T) = c_{p,P}(T)m_P - c_{p,S}(T)\Delta m_S(T) \tag{2}$$

$c_{p,P}$ refers to the apparent specific heat capacity of the sample P, m_P to its mass, $c_{p,S}$ is the specific heat capacity of the solvent, and $\Delta m_S(T)$ is the mass of solvent displaced by the sample at temperature T. $\Delta m_S(T)$ can be calculated from the partial specific volumes of the sample, v_P, and solvent, v_S, according to the equation

$$\Delta m_S(T) = m_P v_P(T)/v_S(T) \tag{3}$$

Inserting Eq. (3) into Eq. (2) and rearranging, the temperature dependence of the apparent specific heat capacity of the sample is given by Eq. (4).

$$c_{p,P}(T) = c_{p,S}(T)\frac{v_P(T)}{v_S(T)} - \frac{\Delta c_p^{app}(T)}{m_P} \tag{4}$$

Enthalpies

Many applications in biochemistry do not require absolute values of the temperature dependence of the apparent specific heat capacity of the sample but rather the enthalpies and entropies of the conformational transitions or reactions induced by a temperature increase. According to the

[27] P. L. Privalov and N. N. Khechinashvili, *J. Mol. Biol.* **86**, 665 (1974).

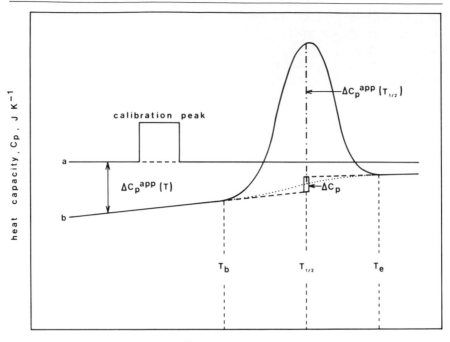

Temperature

FIG. 1. Schematic diagram of the measurements required for determination of heat capacity and enthalpy data from differential scanning microcalorimetric experiments. (a) So-called "baseline" obtained with sample and reference cell filled with solvent. (b) Variation with temperature of the apparent heat capacity, $C_{p,P}(T)$, of the sample. (---) Linearly extrapolated heat capacities of the pre- and posttransitional states. ($\cdots$) Extrapolation of the pre- and posttransitional heat capacities in proportion to the degree of completion of the transition. These extrapolations are referred to as C_p(baseline) in Eq. (6). ΔC_p, heat capacity change at the transition temperature $T_{1/2}$ assuming a two-state transition. (-·-·-) Apparent excess heat capacity at the transition temperature $T_{1/2}$ required to calculate ΔH_{vH} according to Eq. (9). $\Delta C_p^{app}(T)$, temperature-dependent heat capacity difference between the heat capacity curve (b) of the sample and the baseline (a) required to calculate the apparent specific heat capacity of the sample P according to Eq. (4). T_b and T_e, temperature at the beginning and the end of the transition, respectively.

van't Hoff equation any process involving a positive enthalpy change, ΔH_{vH}, will be favored by a temperature rise

$$d \ln K/dT = \Delta H_{vH}/RT^2 \qquad (5)$$

where K is the equilibrium constant, R the gas constant, T absolute temperature, and ΔH_{vH} the so-called van't Hoff enthalpy. Thus in a scanning calorimetric experiment the problem comes down to determining the enthalpy of the process from the heat capacity curve of the sample. The *molar* reaction enthalpy, ΔH_r, is given by the integral

$$\Delta H_r = M \int_{T_b}^{T_e} [c_{p,P}(T) - c_p(\text{baseline})]dT \tag{6}$$

in which T_b and T_e refer to the temperatures at the beginning and end of the reaction, respectively. $c_{p,P}(T)$ is the experimental apparent specific heat capacity of the sample P, $c_p(\text{baseline})$ the assumed temperature course of the heat capacity in the absence of the reaction, which connects the $c_{p,P}$ values corresponding to T_b and T_e, respectively, and M the molar mass of the sample [g mol^{-1}]. It is obvious that for narrow, single, symmetric transition peaks having linear variation with temperature of the heat capacity before and after the transition no serious problems arise as to how to draw the baseline. A straight line between $c_{p,P}(T_b)$ and $c_{p,P}(T_e)$ usually constitutes a good approximation for calculation of the enthalpy. However, wider transition curves showing more than one maximum as well as large differences between the apparent specific heat capacity before and after the transition pose a "baseline problem," as occurs in many other types of measurements for which integration of peaks is required. Though no correct a priori procedure exists, a reasonable approach is the following. The temperature course of the pre- and posttransitional heat capacity is extrapolated to the transition region, and a baseline is calculated by successive approximations as changing from the pretransition line to the posttransition line in proportion to the increasing area under the curve. In cases in which valid molecular models can be assumed for the reaction under investigation on the basis of nonthermodynamic evidence, other baselines may be drawn. However, usually the baseline problem is not a serious source of inaccuracy. Awareness of this inherent source of uncertainty will increase with the advent of constantly improved microcalorimeters and more sophisticated curve analysis.

Examples of Studies on Protein–Ligand Interaction by
    Scanning Microcalorimetry

Reactions

For obtaining insight into intramolecular interactions between proteins and ligands high sensitivity scanning microcalorimetry is a useful tool with a high information content. Complex formation between the S-peptide and S-protein of ribonuclease to give ribonuclease S' has been studied by Tsong *et al.*[28] and Hearn *et al.*[29] by both isothermal and scanning heat capacity microcalorimetry. Interpretation of the data obtained by scanning calorimetry has been based predominantly on the application of the additivity principle of enthalpies as expressed by Hess' law. If interaction between reacting macromolecules such as S-peptide and S-protein occurs and both molecules can undergo reversible thermal unfolding, then the sum of the reaction enthalpies of the reactants can be compared to the ΔH value of the complex. With interaction the sum of the unfolding enthalpies of the components will deviate from the unfolding enthalpy of the complex. A reasonable comparison of the ΔH values is of course possible only if all data refer to the same temperature. This is particularly important in view of the large heat capacity changes which are usually observed with protein unfolding reactions.[27,30]

Complex formation is often associated with an increase in thermal stability of the complex relative to its constituents. Therefore it is essential to know the temperature dependence of the unfolding enthalpies of the individual reacting species. This value is given by the so-called denaturational heat capacity change, ΔC_p, which can be directly determined from the difference between the pre- and posttransitional heat capacities extrapolated to the transition temperature, $T_{1/2}$, at which half of the transition has been completed. Remembering the "baseline problem" such extrapolation is uniquely possible only for strictly two-state processes. Another type of heat capacity change must be formally differentiated from the denaturational heat capacity change discussed so far, although the quantities may be interconnected, as has been shown for the reaction of S-peptide and S-protein of ribonuclease.[29] That is the temperature dependence of the reaction enthalpy, which is best determined by isothermal calorimetric measurements of the molar reaction enthalpy, ΔH_R, at various temperatures. The molar temperature derivative, $\Delta C_{p,R}$, of the reaction enthalpy, averaged over the temperature interval $T_2 - T_1$, is then given by the ratio

$$\frac{\Delta H_R(T_2) - \Delta H_R(T_1)}{T_2 - T_1} = \Delta C_{p,R} \tag{7}$$

Although in the majority of the reactions studied so far,[8] $\Delta C_{p,R}$ within the accuracy of the measurements is temperature independent, for associa-

[28] T. Y. Tsong, R. P. Hearn, D. P. Wrathall, and D. M. Sturtevant, *Biochemistry* **9**, 2666 (1970).

[29] R. P. Hearn, F. M. Richards, J. M. Sturtevant, and G. D. Watt, *Biochemistry* **10**, 806 (1971).

[30] W. Pfeil, *Mol. Cell. Biochem.* **40**, 3 (1981).

tion of ribonuclease it has been found to vary strongly with temperature.[29] This apparent temperature dependence of $\Delta C_{p,R}$ is largely the result of the conformational transitions undergone by S-protein and RNase S' as the temperature is raised.

Protein–nucleic acid interaction has been studied using the coat protein, various RNAs, and RNA–protein complexes of alfalfa mosaic virus (AMV).[31] For calculation of an apparent specific heat capacity of the protein–nucleic acid complex Eq. (4) must be modified by introducing a value for the partial specific volume of the complex, $v_p(\text{Mix})$, instead of v_p. That can be accomplished with the reasonable assumption that the partial specific volumes do not change considerably on complex formation. Then $v_p(\text{Mix})$ is given by the term

$$v_p(\text{Mix}) = \frac{v_{\text{Pro}}m_{\text{Pro}} + v_{\text{RNA}}m_{\text{RNA}}}{m_{\text{Pro}} + m_{\text{RNA}}} \tag{8}$$

in which v_{Pro}, m_{Pro} and v_{RNA}, m_{RNA} are the partial specific volumes and masses of protein and nucleic acid, respectively. Characteristic differences in the energetics of protein–nucleic acid interaction have been found to occur for complex formation of the coat protein with RNA 1 and RNA 4, a genomic and subgenomic RNA of AMV.

Interaction enthalpies per mole of protein estimated from a comparison of the sum of the unfolding enthalpies of RNAs and proteins with the unfolding enthalpy of the various RNA–protein complexes are surprisingly all positive. Furthermore there is a pronounced quantitative difference between the interaction enthalpies when only a few protein molecules bind as compared to complexes with a high protein to RNA ratio. The larger interaction enthalpies for complexes having a small protein to RNA ratio (6 or 26) can be interpreted as indicating that major structural changes are induced in the RNA by binding of the first coat protein molecules. The magnitude of the interaction enthalpies is suggestive of melting of base pairs on complex formation.

It is typical of calorimetric information that large enthalpy changes can often be interpreted as being indicative of structural changes; however, the existence of only small ΔH values is no valid criterion for the absence of significant molecule-wide conformational changes.

An illustrative example has been provided by Takahashi *et al.*[32] in a study on the binding of D-glucose to yeast hexokinase. According to the results of X-ray crystallography[33] the hexokinase monomer contains two

[31] J.-J. Hinz, S. Srinivasan, and E. M. J. Jaspars, *Eur. J. Biochem.* **95,** 107 (1979).
[32] K. Takahashi, J. L. Casey, and J. M. Sturtevant, *Biochemistry* **20,** 4693 (1981).
[33] W. S. Bennett and T. A. Steitz, *Proc. Natl. Acad. Sci. U.S.A.* **75,** 4848 (1978).

domains separated by a cleft. Binding of glucose into the cleft causes one of the domains to rotate 12° relative to the other, resulting in a movement of 8 Å in the polypeptide backbone and closing of the cleft between the two lobes. The transition curves of the enzyme in the presence and absence of glucose as well as under conditions of different ionic strength revealed some intriguing features. The pretransitional baselines in the absence of glucose had slopes corresponding to unusually large temperature dependences of the apparent specific heat capacity of the protein, amounting to 0.0146–0.0251 J K^{-2} g^{-1}, whereas the usual value for small globular proteins is approximately 0.00837 J K^{-2} g^{-1}.[34]

At low ionic strength unfolding of hexokinase in the absence of glucose is associated with a double-peaked endotherm. Addition of glucose changes this to a single-peaked endotherm having a slightly higher transition temperature. This result suggests that the two peaks in the absence of glucose result from independent unfolding of the two domains of the native enzyme, which appear in the X-ray structure and that binding of glucose induces a conformational change in the enzyme which brings the two lobes into close proximity. It is a striking result of these studies, which cannot be rationalized yet, that the net change in interaction enthalpy on glucose binding is very small despite the occurrence of a molecule-wide conformational change.

Subunit Interaction

Though thermodynamic quantities such as enthalpies or heat capacities cannot give information on the details of structural changes, they often provide good evidence for the very existence of these changes. An enlightening study has been published by Vickers *et al.*[35] on the mutual interactions between catalytic and regulatory subunits in aspartate transcarbamylase (ATCase) using a DTA instrument. ATCase from *E. coli* is composed of two catalytic trimers, which are linked noncovalently by three regulatory dimers. Since the subunits are readily obtained by dissociation of the native enzyme with mercurials, it is possible to study the thermal stability of the individual subunits and to examine changes in this characteristic parameter resulting from their incorporation into the intact enzyme and from ligand binding.

A significant stabilization of both regulatory and catalytic subunits relative to their isolated state is indicated by the transition temperatures obtained from the thermograms. While isolated regulatory and catalytic subunits show transition temperatures of 55 and 80°, respectively, the

[34] P. L. Privalov, *Adv. Protein Chem.* **33**, 167 (1979).
[35] L. P. Vickers, J. M. Donovan, and H. K. Schachman, *J. Biol. Chem.* **253**, 8493 (1978).

intact ATCase exhibits a double peaked thermogram, with a small endotherm near 73° and a major endotherm around 82°. The small endotherm could be ascribed to denaturation of the regulatory subunits, while the large endotherm at 82° was associated with unfolding of the catalytic subunits. Since transition temperatures are proportional to the Gibbs energy of stabilization, ΔG, they reflect energetic as well as entropic alterations of the native enzyme relative to the isolated subunits. Energetic changes become evident in the denaturational enthalpies. The results obtained for ATCase and its subunits fully support the conclusions drawn on the basis of the transition temperatures. Both isolated regulatory and catalytic subunits of ATCase were characterized by a surprisingly low specific enthalpy of denaturation (regulatory subunits: 8 J g^{-1}; catalytic subunits: 16.5 J g^{-1}; typical globular proteins: 16.7 to 25 J g^{-1} [34]), whereas in the native enzyme their specific enthalpies of denaturation are increased by 125 and 38%, respectively (regulatory subunits: 18 J g^{-1}; catalytic subunits 22.7 J g^{-1}). It is likely that these profound changes in the transition enthalpies are associated with considerable changes in the secondary and tertiary structure of the proteins, a view which is supported by other results. It is found that reactivity of the sulfhydryl groups of the regulatory subunits is drastically reduced on incorporation into ATCase and that the circular dichroism spectrum is significantly different from the sum of the spectra of the isolated subunits.

Binding of ligands need not necessarily result in an increase of the enthalpy of denaturation of a protein. An example is provided by binding of the bisubstrate analog PALA [N-(phosphonacetyl)-L-aspartate].

Binding of PALA stabilizes the free catalytic subunits and those within ATCase, but it causes a decrease in the enthalpy of denaturation of the regulatory subunits within ATCase. An interpretation consistent with the finding of a 6-fold increase in the reactivity of the sulfhydryl groups of the regulatory subunits within ATCase on PALA binding invokes a "loosening" of the structure as a result of the interaction.

Stability of Protein–Ligand Complexes

The influence of the binding of ligands on the stability of proteins has been probed by scanning calorimetry for various systems. Relimpio et al.[36] performed an extensive study on the thermal stability of cytoplasmic aspartate transaminase in the absence and presence of coenzyme and substrate derivatives. The thermodynamic transition parameters revealed the contributions of various groups in the coenzyme and the substrates to

[36] A. Relimpio, A. Iriarte, J. F. Chlebowski, and M. Martinez-Carrion, *J. Biol. Chem.* **256,** 4478 (1981).

thermal stability of the dimeric protein. However, no comparison has been made with isothermal binding enthalpies to ascertain whether the stabilizing enthalpic effects can be rationalized on the basis of these quantities, taking their variation with temperature into account.

An interesting contribution to this problem of how to interpret the frequently observed increase in thermal stability by ligand binding is the study by Fukada et al.[37] on the thermodynamics of the binding of L-arabinose and D-galactose to the L-arabinose binding protein of E. coli. The protein undergoes reversible unfolding on being heated, with an increase in enthalpy at 53.5° of 635 ± 4.6 kJ mol^{-1}, and a heat capacity change of 13.2 ± 0.3 kJ K^{-1} mol^{-1}.

In the presence of arabinose the unfolding enthalpy is increased to 840 ± 75 kJ mol^{-1} at 59.0°. This increase could be shown by parallel isothermal calorimetric binding studies to be solely due to the enthalpy of dissociation of the ligand which amounts to 130 kJ mol^{-1} at the unfolding temperature. Displacement of the unfolding-dissociation equilibrium by the added ligand was calculated to be the source of the increase in the unfolding temperature.

When the ligand bound to a protein is itself a protein, the various energetic and structural rearrangements on complex formation can be, at least in principle, more equally distributed among the reactants than for complex formation between a protein and a small ligand. Examples for that have been provided by studies on thermal denaturation of subtilisin BPN' with and without its inhibitor (SSI).[38] Unfolding of the inhibitor in the absence of the protease is reversible, while in the complex it is irreversible. This result must be rationalized by assuming that the inhibitor is not dissociated from the protein in the course of the denaturation, for if it were it would be expected to refold on cooling as it does in the absence of the enzyme.

An interesting kinetic analysis of the irreversible thermal denaturation of the association complexes of bovine β-trypsin with soybean trypsin inhibitor or ovomucoid using a DTA instrument was reported by Donovan and Beardslee.[39] Similar studies have been performed on the avidin–biotin complex,[40] on bovine or porcine β-trypsin with chicken ovoinhibitor,[41] and on complexes of chymotrypsin, subtilisin, and trypsin with chicken ovoinhibitor and with lima beam protease inhibitor.[42] It was typical in

[37] H. Fukada, J. M. Sturtevant, and F. A. Quiocho, J. Biol. Chem. 258, 13193 (1983).
[38] K. Takahashi and J. M. Sturtevant, Biochemistry 20, 6185 (1981).
[39] J. W. Donovan and R. A. Beardslee, J. Biol. Chem. 250, 1966 (1975).
[40] J. W. Donovan and K. D. Ross, Biochemistry 12, 512 (1973).
[41] J. C. Zahnley, J. Biol. Chem. 254, 9721 (1979).
[42] J. C. Zahnley, Biochim. Biophys. Acta 613, 178 (1980).

these experiments to employ large heating rates (10 K min^{-1}) and high protein concentrations (6–8%). Denaturation temperatures are defined operationally and vary with heating rates, since they are kinetically determined parameters. Due to the nonequilibrium situation different pathways of thermal denaturation have to be taken into account. The rate constant obtained from the measurements can be analyzed in terms of Arrhenius plots and activation energies. In cases where the parameters can be demonstrated to be heating rate independent they can also be used for thermodynamic analysis. Studies of this kind can potentially provide interesting information, complementary to that of equilibrium unfolding, on subtle changes in the conformation of the proteins produced by association, by way of proper analysis of the activation parameters.

Polymerization Reactions

Endothermic polymerization reactions of proteins can be successfully examined by scanning microcalorimetry. Three biologically important assembly processes have been investigated, and the thermodynamic studies have been supplemented by structure-specific optical studies to facilitate the correlation between structural changes and thermodynamic parameters.

In vitro reconstitution of microtubules from purified tubulin was studied in an adiabatic differential heat capacity microcalorimeter using different protein concentrations and various heating rates.[43] A complex heat uptake pattern was observed, which involved exothermic subreactions, indicating that the shape of the thermogram should depend on both the kinetic and the equilibrium properties of the assembly system. Although these calorimetric data were not amenable to exact kinetic analysis, they could be used to gain qualitative insight into the pathway of the polymerization reaction. A careful comparison of the heat uptake pattern with the light scattering curve, which is susceptible to the occurrence of large rodlike structures, led to the suggestion of a reaction sequence consistent with recent time-dependent X-ray studies.[44]

Due to the kinetic nature of the exothermic process visible in the thermograms the reaction enthalpy has not been calculated. However, the heat capacity change on polymerization was found to be independent of the heating rate and therefore can be assumed to reflect equilibrium properties. This conclusion is supported by the excellent agreement between the value obtained from the calorimetric studies (-6694 ± 2092 J mol^{-1} K^{-1}; $M = 110,000$) and that derived from a van't Hoff analysis of the

[43] H.-J. Hinz, M. J. Gorbunoff, B. Price, and S. N. Timasheff, *Biochemistry* **18,** 3084 (1979).
[44] E. Mandelkow, personal communication.

assembly reaction (-6272 J mol^{-1} K^{-1}). The heat capacity change is suggestive of a structural change which the tubulin undergoes during the polymerization process.

Similar studies involving turbidimetry and sedimentation velocity measurements as structure-specific control experiments have been performed on the endothermic polymerization of coat protein of tobacco mosaic virus (TMV-P) by Sturtevant et al.[45] The polymerization could be studied close to equilibrium employing a low heating rate of 0.5 K min^{-1}. Since a large number of structural studies has been available for a detailed molecular interpretation of the thermodynamic parameters, the transition enthalpies and the variation with temperature of the apparent specific heat capacity could be used to validate or invalidate molecular models of polymerization. It was a surprising result of the study that the isodesmic model, using identical polymerization enthalpies and equilibrium constants for each step, fitted the experimental heat capacity data best. It is worth noting that for TMV-P polymerization a negative change in heat capacity is associated with assembly at pH 6.5–6.8. This change is, however, pH dependent and becomes positive at pH 7.0 [pH 6.5–6.8: $\Delta C_p(T_m)$ $= -1464$ J K^{-1} (mol of monomer)$^{-1}$ (17,500 molecular weight); pH 7.5: $\Delta C_p(T_m) = +628$ J K^{-1} (mol of monomer)$^{-1}$]. The enthalpy of polymerization at low protein concentration varies from 52 kJ mol^{-1} under conditions where the product is largely a mixture of short helical rods to 25 kJ mol^{-1} for the formation of double disks containing 34 monomer units. These results constitute evidence that these two types of polymerization involve intersubunit bonds of quite different chemical character.

The third biologically interesting protein whose polymerization behavior has been studied by scanning and isothermal microcalorimetry is flagellin.[44-48] The most prominent result of the isothermal studies was the observation of a very large negative heat capacity change associated with polymerization of $\Delta C_{p,R} = -12.6$ kJ mol^{-1} ($M = 40,000$), rendering the enthalpy of polymerization positive at 10° ($+56$ kJ mol^{-1}) and strongly negative at 30° (-195 kJ mol^{-1}). Thus obviously the intersubunit interactions operative in the polymerization reaction are associated with considerable structural changes in the proteins, as is also indicated by the changes in the ellipticity at 220 nm.

The scanning calorimetric study[47] of the polymerization process is not

[45] J. M. Sturtevant, G. Velicelebi, R. Jaenicke, and M. A. Lauffer, Biochemistry 20, 3792 (1981).
[46] W. Bode, H.-J. Hinz, R. Jaenicke, and A. Blume, Biophys. Struct. Mech. 1, 55 (1974).
[47] W. Bode and A. Blume, FEBS Lett. 36, 318 (1973).
[48] O. V. Fedorov, N. N. Khechinashvili, R. Kamiya, and S. Asakura, in preparation (1986).

quite consistent with the isothermal measurements, since no proper division of the depolymerization and the denaturation steps of the reaction could be made. In a more recent investigation[48] these two processes were resolved. A fair agreement with the results from the isothermal study was found by taking into account the different molecular weight used in the latter study for calculating the molar quantities.

Interactions between Protein Domains

An extremely important application of high sensitivity scanning microcalorimetry in the study of protein–protein interactions is shape analysis of the transition curves. The high information content of this type of thermodynamic analysis results from the fact that both quantitative energy parameters and structural characteristics can be extracted from the calorimetric measurements. Although the calorimetrically determined domains within proteins are primarily energetically defined, in several cases where X-ray information is available, these energetically defined domains could be nicely correlated with analogous structural domains.[49] However, even proteins too large to be studied by present X-ray methods can be analyzed and their domain structure can be quantitatively characterized, which challenges future structural studies.

A powerful thermodynamic criterion for the inter- and intramolecular cooperativity of a system is the ratio of the van't Hoff, ΔH_{vH}, and the calorimetric enthalpy, ΔH_{cal}.[15–17,34,50–53] It is a unique property of scanning microcalorimetric measurements that both ΔH values can be obtained from the same transition curve. For a process of the type A ↔ B the van't Hoff enthalpy can be calculated from the equation

$$\Delta H_{vH} = 4R(T_{1/2})^2 \frac{\Delta C_{p,P}(T_{1/2})}{\Delta H_{cal}} \tag{9}$$

while for a process of the type A ↔ B + C the numerical factor in Eq. (9) is 6 instead of 4.[29] $T_{1/2}$ is the temperature of half-completion of the reaction, R the gas constant, $\Delta C_{p,P}(T_{1/2})$ the molar excess heat capacity of the sample at $T_{1/2}$, and ΔH_{cal} the molar transition enthalpy. It should be mentioned that these latter two quantities need not be expressed in molar units, but that any consistent units serve the purpose, as long as the ratio

[49] P. L. Privalov, Adv. Prot. Chem. 35, 1 (1982).
[50] H.-J. Hinz and J. M. Sturtevant, J. Biol. Chem. 247, 6071 (1972).
[51] R. Lumry, R. Biltonen, and J. F. Brandts, Biopolymers 4, 917 (1966).
[52] C. Tanford, Adv. Protein Chem. 23, 121 (1968).
[53] C. Tanford, Adv. Protein Chem. 24, 1 (1970).

$\Delta C_{p,P}(T_{1/2})/\Delta H_{cal}$ has the dimension of reciprocal temperature. ΔH_{cal} is proportional to the peak area and is determined according to Eq. (6).

If the reaction studied is a two-state process with a negligible percentage of thermodynamically relevant intermediate states, the ratio $\Delta H_{vH}/\Delta H_{cal}$ will be close to 1.0. Such a situation of molecule-wide cooperation on unfolding has been observed for many small globular proteins.[30,34,54] On the other hand, if via intermolecular interaction, as, e.g., in phospholipid systems, large clusters of molecules change their structure cooperatively, the ratio $\Delta H_{vH}/\Delta H_{cal}$ can approach values of several hundred.[17] Ratios smaller than 1.0 can be encountered if a single polypeptide chain is folded into several domains which are of similar thermal stability and therefore undergo unfolding in the same temperature range. Although visibly the single-peaked transition curve of such a molecule is not different from the heat capacity curve of a fully cooperative system, the thermodynamic analysis will reveal the existence of these domains. One example is furnished by papain,[55] and several others are summarized in Ref. 49. The simple approach of inspecting the ratio $\Delta H_{vH}/\Delta H_{cal}$ is no longer practical when interdependent reactions occur on thermal unfolding. For a sequential reaction scheme of the type

$$I_0 \overset{\Delta h_1}{\longleftrightarrow} I_1 \overset{\Delta h_2}{\longleftrightarrow} I_2 \ldots I_{n-1} \overset{\Delta h_n}{\longleftrightarrow} I_n \tag{10}$$

Freire and Biltonen[56,57] provided a quantitative theoretical framework which permits quantitative deconvolution of the experimental transition curve in a unique manner. They noted that the experimentally observable $\langle \Delta H \rangle$

$$\langle \Delta H \rangle = \int_{T_b}^{T} [c_{p,P}(T) - c_p(\text{baseline})]dT \tag{11}$$

allows calculation of the partition function $Q(T)$ according to the relationship

$$\ln Q(T) = \int_{T_b}^{T} \frac{\langle \Delta H \rangle}{RT^2} dT \tag{12}$$

where T_b, in a rigorous sense, should be absolute zero, since no finite temperature exists in which a single state can account for 100% of the population. However, practically the smallest finite temperature at which

[54] E. Moses and H.-J. Hinz, *J. Mol. Biol.* **170**, 765 (1983).
[55] E. I. Tiktopulo and P. L. Privalov, *FEBS Lett.* **91**, 57 (1978).
[56] E. Freire and R. Biltonen, *Biopolymers* **17**, 463 (1978).
[57] R. Biltonen and E. Freire, *Crit. Rev. Biochem.* **5**, 85 (1978).

a physically detectable change in the system can be observed can be used without introducing significant errors. This temperature is T_b, the beginning of the heat absorption of the transition. On the basis of the partition function Q molecular averages and distributions can be calculated. Noting that the fraction of molecules in the initial state, $F_0 = I_0/\Sigma_i I_i$, is

$$F_0 = \frac{1}{Q} = \exp\left(-\int_{T_b}^{T} \frac{\langle \Delta H \rangle}{RT^2} \, dT\right) \tag{13}$$

the fraction of all molecules in the residual states is $1 - F_0$. It was shown[56] that the quantity $\langle \Delta H \rangle/(1 - F_0)$ is an S-shaped curve whose lower limit is equal to the enthalpy difference, Δh_1, between I_1 and I_0. By the use of recursion relations[56,57] complete thermodynamic information of all intermediate states as well as their population can be obtained.

Some modifications to reduce error propagation in such an analysis have been introduced by Potekhin and Privalov.[58] It is obvious that intramolecular interactions of domains can be conveniently analyzed in the framework of the above theory. Stabilizing and destabilizing interactions can occur between the domains and examples for both have been observed. Tsalkova and Privalov[59] studied thermal unfolding of troponin C and its fragments in the presence of Ca^{2+} and Mg^{2+}. It was found that the separated fragment that includes domains 3 and 4 has a cooperative structure significantly more stable than within the intact molecule. This result suggests destabilizing interactions between the domains of this sequence and the other domains in the native protein (Ref. 49, p. 14, unpublished results). Stabilizing interactions have been found to occur in the E fragment of fibrinogen.[60]

A generalized macromolecular partition function has been formulated by Gill et al.[61] using the formalism of the chemical potential of the macromolecule (μ_m) as developed by Wyman.[62,63] μ_m is usually not only dependent on temperature (T) but also on pressure (p) and the chemical potentials, $\mu_{x,i}$, of various ligands X_i as can be described by the Gibbs–Duhem equation

$$d\mu_m = -S \, dT + V \, dp + \sum_i X_i \, d\mu_{X,i} \tag{14}$$

[58] S. A. Potekhin and P. L. Privalov, *J. Mol. Biol.* **159**, 519 (1982).
[59] T. N. Tsalkova and P. L. Privalov, *Biochim. Biophys. Acta* **624**, 196 (1980).
[60] P. L. Privalov and L. V. Medved, *J. Mol. Biol.* **159**, 665 (1982).
[61] S. J. Gill, B. Richey, G. Bishop, and J. Wyman, *Biophys. Chem.* **21**, 1 (1985).
[62] J. Wyman, *J. Mol. Biol.* **11**, 631 (1965).
[63] J. Wyman, *Proc. Natl. Acad. Sci. U.S.A.* **72**, 1464 (1975).

The entropy (S), volume (V), and amount of chemical ligand $i(X_i)$ have been normalized to the moles of macromolecule. The partition function is given by

$$Q^* = \sum_{i=0}^{q} \sum_{j=0}^{r} [(K_{i,j})_0 \cdot e^{-(\Delta H_{i,j}/R)(\tau - \tau_0)}$$

$$\cdot e^{-(\Delta V_{i,j}/RT)(p - p_0)} \cdot e^{-j(\ln a_x - \ln a_{x,0})}] \tag{15}$$

i and j refer to macromolecular conformation i having j binding sites for ligands. τ_0, p_0, and $\ln a_{x,0}$ are reference temperature, pressure, and ligand concentration, respectively, to which the equilibrium constants $(K_{i,j})_0$ refer. Partial differentiation of $\ln Q^*$ with respect to the appropriate intensive variable, τ, p, or $\ln a_x$, permits calculation of ΔH, ΔV, or X.

The generalized partition function Q^* can be employed for analysis of a sequential transition scheme as shown in Eq. (10) as well as for analysis of independent two-state transitions. The two models have been shown to provide identical answers for tRNA if the population of the states is taken into account in the independent transition model.

Considering only temperature effects the pertinent equations for analyzing scanning calorimetric heat capacity curves are the following.[61]

Independent two state transition model:

$$C_{p,P}(T) - C_p(\text{baseline}) = \frac{\tau^2}{R} \sum_{i=1}^{n} \frac{(\Delta H_i)^2 \cdot e^{-\Delta H_i(\tau - \tau_i^m)/R}}{(1 + e^{-\Delta H_i(\tau - \tau_i^m)/R})^2} \tag{16}$$

$C_{p,P}(T) - C_p(\text{baseline})$ is the experimental molar excess heat capacity and τ is the reciprocal absolute temperature $(\tau = 1/T)$. τ_i^m is the reciprocal absolute temperature at the midpoint of the ith transition, R the gas constant, and ΔH_i the molar enthalpy of the ith transition.

Sequential allosteric transitions:

$$C_{p,P}(T) - C_p(\text{baseline}) = \frac{\tau^2}{R} \left[\frac{\Sigma(\Delta H_i)^2 \cdot L_i^0 \cdot e^{-\Delta H_i(\tau - \tau_0)/R}}{\Sigma L_i^0 \cdot e^{-\Delta H_i(\tau - \tau_0)/R}} \right.$$

$$\left. - \frac{(\Sigma \Delta H_i \cdot L_i^0 \cdot e^{-\Delta H_i(\tau - \tau_0)/R})^2}{(L_i^0 \cdot e^{-\Delta H_i(\tau - \tau_0)/R})^2} \right] \tag{17}$$

Equation (17) reflects the heat capacity effects which are associated with alterations in the distribution of allosteric form resulting from a change in temperature. $C_{p,P}(T) - C_p(\text{baseline})$, τ, and R have the same meaning as defined above; L_i^0 is the standard state $(\tau = \tau_0)$ equilibrium constant, ΔH_i, the change in enthalpy for the reaction between the ith and zeroth forms. τ_0 can be identified with $1/T_b$, T_b being defined as in Eq. (11). Both L_i^0 and

ΔH_i depend in general on chemical ligand activity and that contribution can in principle be separated from the overall quantities.

Concluding Remarks

The present survey was intended to demonstrate by some typical examples the potential of the thermodynamic approach based on high sensitivity scanning calorimetry measurements to provide quantitative energy parameters for protein–ligand interactions. These interactions play a crucial role in virtually all biological processes, since the interplay between energetic and entropic changes associated with complex formation determines the conformational adaptations of the macromolecules which are essential for their function. A prerequisite for good biophysics is good biochemistry. This motto is particularly important for calorimetric studies, since practically all processes involve either heat release or absorption which renders these measurements strongly susceptible to side reactions. However, when properly executed, scanning microcalorimetry is a powerful method for providing information on energetic details and even structural aspects of protein–ligand interaction. It is obvious that the "structural" interpretation of thermodynamic data gains reliability in combination with other nonthermodynamic evidence.

[7] Structure and Interactions of Proteins in Solution Studied by Small-Angle Neutron Scattering

By SOW-HSIN CHEN and DALILA BENDEDOUCH

Introduction

The use of the small-angle neutron scattering (SANS) technique as an analytical tool for the study of global structures of macromolecules in solution follows naturally from the earlier development of the small-angle X-ray scattering (SAXS) technique.[1,1a,2]

It is well known that a detailed knowledge of the three-dimensional structure of a macromolecule is the key to the understanding of its biological function. At the present time, X-ray and neutron crystallography are

---

[1] A. Guinier and G. Fournet, "Small-Angle Scattering of X-Rays." Wiley, New York, 1955.

[1a] O. Kratky, *Prog. Biophys.* **13**, 105 (1963).

[2] O. Glatter and O. Kratky, eds., "Small-Angle X-Ray Scattering." Academic Press, New York, 1982.

METHODS IN ENZYMOLOGY, VOL. 130

the only methods by which detailed structural information may be obtained at the level of atomic resolution.[3,4] But the method is time consuming and aside from its enormous complexity has two major limitations. First, for many proteins it is a serious problem to grow crystals required for the diffraction experiment. Second, it is not certain whether protein molecules retain their conformation when crystals are dissolved in water. This leaves room for other methods, characterized by a lower resolution such as small-angle scattering, for investigation of the coarse structural properties of macromolecules in solution. The greatest advantage of SANS and SAXS lies in the possibility of performing the measurements in any desired solvent simulating a given biological environment, and in the ability to follow changes of the particle structure which may occur on changing the external conditions. However, it must be immediately qualified that while the diffraction techniques give a spatial resolution of the order of 1 Å, the small-angle scattering technique generally gives at best a picture with a resolution of the order of 10 Å. In some special cases the resolution can be improved to the 5-Å level by specific isotopic substitution of various functional groups.

In this chapter we shall present the method of SANS as applied to soluble globular proteins in solution. Its advantages over the SAXS technique are also brought out in the discussion. In contrast to other review articles[5,6] our emphasis is put on analysis of scattering data from systems at both low and moderately high concentrations, which better simulate physiological conditions. This means that not only the structure of the macromolecules, but also their mutual interactions have to be taken into account explicitly in the analysis. Owing to the considerable number of charges on proteins in solution at pH 7, there exists a long range electrostatic interaction between the molecules even at the 1% (g/dl) concentration level. It should be emphasized that the dilute regime where interactions can be neglected is an impractical regime in which to perform experiments because of the severe loss in scattered intensity for solutions with less than 0.1% protein. Furthermore, for globular proteins with known conformation, experiments performed at high concentrations provide additional information such as the effective surface charge, the amount of bound water, and the local spatial ordering of the molecules in solution.

[3] T. L. Blundell and L. N. Johnson, "Protein Crystallography." Academic Press, London, 1976.
[4] D. M. Engleman and P. B. More, *Annu. Rev. Biophys. Bioeng.* **4**, 219 (1975).
[5] H. B. Stuhrmann and A. Miller, *J. Appl. Crystallogr.* **11**, 325 (1978).
[6] B. Jacrot, *Rep. Prog. Phys.* **39**, 911 (1976).

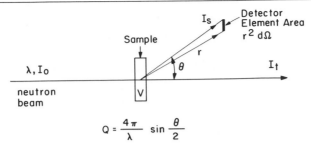

$$Q = \frac{4\pi}{\lambda} \sin\frac{\theta}{2}$$

FIG. 1. Scattering geometry. I_0, incident neutron intensity; I_t, transmitted neutron intensity; I_s, scattered neutron intensity; λ, wavelength of the neutron beam; θ, scattering angle; Q, Bragg wavenumber; $d\Omega$, solid angle subtended by a detector element; V, volume of the scattering sample.

The Technique

Geometry of a SANS Experiment

In a scattering experiment (see Fig. 1), a parallel beam of subthermal neutrons (wavelength λ between 4 and 12 Å) of intensity I_0 (number of neutrons per cm^2 per second), incident on a flat sample cell of volume V containing N particles in solution, is scattered in a small cone around the forward direction. The measurement of the scattered neutron intensity I_s at an angle θ is carried out with a detector subtending an area $r^2 d\Omega$, defined by the solid angle $d\Omega$, at a distance r away from the sample. The basic quantity measured in the experiment is the ratio

$$d\sigma(Q) = \frac{I_s r^2 d\Omega}{I_0} \qquad [\text{cm}^2]$$

which has the dimension of an area and is called the differential cross-section of the scattering specimen. In practice, a multicell two-dimensional position-sensitive detector is used to simultaneously detect the scattered intensity at all angles θ of interest.

The basic geometrical arrangement in the scattering of waves demands that the fundamental variable in the experiment is the so-called Bragg-wave vector

$$\mathbf{Q} = \mathbf{k}_i - \mathbf{k}_s$$

where $\mathbf{k}_i$ and $\mathbf{k}_s$ are, respectively, the incident and scattered wave vectors of the neutron beam. In a diffraction experiment from an isotropic macromolecular solution one measures predominantly the elastic scattering for which $|\mathbf{k}_i| = |\mathbf{k}_s| = 2\pi/\lambda$. Consequently, the $\mathbf{Q}$-vector has a magnitude

$Q = (4\pi/\lambda) \sin(\theta/2)$, and the scattering is axially symmetric in the forward direction with an intensity depending on the scalar variable Q only.

A fundamental theorem in the theory of scattering of waves by an extended object connects the real-space density distribution to the Q-space scattered intensity distribution in terms of a Fourier transform relationship. It follows that the characteristic size in real-space, R, is reciprocally related to the characteristic width of the intensity distribution in Q-space. Therefore for a typical macromolecular size of interest, say R $\sim$50 Å, one is interested in an intensity distribution over a typical Q-range of $(2\pi/R) = 0.13$ Å^{-1}. This implies a scattering angle θ which is less than 6° if $\lambda = 5$ Å. Thus SANS is of immediate relevance to investigation of the structure of macromolecules.

Theoretical Considerations

In this chapter we shall present all the data in terms of a quantity defined as the scattering cross-section per unit volume, i.e.,

$$\frac{d\Sigma}{d\Omega}(Q) = \frac{1}{V}\frac{d\sigma}{d\Omega} \quad [\text{cm}^{-1}]$$

This is "the probability of a neutron being scattered in the θ-direction per unit solid angle around that direction, when it traverses a unit length of the sample." This quantity, often named scattered intensity $I(Q)$, can be shown to be equal to[7]

$$I(Q) \equiv \frac{d\Sigma}{d\Omega}(Q) = nP(Q)\bar{S}(Q) \quad [\text{cm}^{-1}] \tag{1}$$

where $n = N/V$ is the number density of the particles in the sample. $P(Q)$ is called the particle structure factor and $\bar{S}(Q)$ is the effective interparticle structure factor.

The standard theory of SANS gives[6-8]

$$P(Q) = \langle|F(\mathbf{Q})|^2\rangle \tag{2}$$

and

$$\bar{S}(Q) = 1 + \frac{|\langle F(\mathbf{Q})\rangle|^2}{\langle|F(Q)|^2\rangle}[S(Q) - 1] \tag{3}$$

where $S(Q)$ is the interparticle structure factor describing the positional correlations among particles as a result of their mutual interactions. We

[7] D. Bendedouch and S.-H. Chen, *J. Phys. Chem.* **87**, 1473 (1983).
[8] M. Kotlarchyk and S.-H. Chen, *J. Chem. Phys.* **Sept.** (1983).

TABLE I
SCATTERING PROPERTIES OF SOME ELEMENTS[a]

Element	$b_{coh}[10^{-12}$ cm$]^b$	$\sigma_{inc}[10^{-24}$ cm$^2]^c$	$f_x(0)[10^{-12}$ cm$]^d$
Hydrogen	−0.3742	80.0	0.28
Deuterium	0.6671	2.0	0.28
Lithium	−0.214	0.6	0.84
Carbon	0.6651	0.0	1.69
Nitrogen	0.919	0.3	1.97
Oxygen	0.5804	0.0	2.25
Sodium	0.351	1.8	3.09
Magnesium	0.516	0.1	3.38
Phosphorus	0.52	0.3	4.23
Sulfur	0.2847	0.2	4.5
Chlorine	0.9584	3.4	4.8
Calcium	0.49	0.4	5.6

[a] G. E. Bacon, "Neutron Diffraction," 3rd Ed. Oxford Univ. Press, 1980.
[b] Coherent neutron scattering length.
[c] Incoherent neutron scattering refers to that part of the scattering of neutrons by individual nuclei in a sample which does not result in a constructive or destructive interference effect. Therefore the incoherent scattering is essentially isotropic in all directions and can be regarded as a Q-independent background in the forward direction.
[d] X-Ray scattering amplitude in the forward direction.

shall discuss the computation of $S(Q)$ in a later section. $F(Q)$ is called the form factor of the particle in solution and is given by the Fourier transform of the excess coherent scattering length density (csld) distribution of the particle.

$$F(\mathbf{Q}) = \int d^3r[\rho(\mathbf{r}) - \rho_s]\exp(i\mathbf{Q} \cdot \mathbf{r}) \qquad (4)$$

The integral is over the entire volume v of the solvated particle which is not accessible to the bulk solvent. ρ_s is the csld of the solvent, in this case being treated as a continuum because of the low spatial resolution of the scattering probe. The formal definition of the csld of the particle $\rho(\mathbf{r})$ is

$$\rho(\mathbf{r}) = \sum_i b_i\delta(\mathbf{r} - \mathbf{r}_i)$$

where b_i is the bound coherent scattering length of a nucleus i situated at a position $\mathbf{r}_i$ in the particle, δ is the Dirac delta function, and b is a measure of the "strength" of the neutron–nucleus interaction. The values of b for various atoms are tabulated in Table I.

Inspection of Table I shows that the coherent neutron scattering lengths are different for atoms with similar atomic numbers, whereas the X-ray scattering lengths are proportional to the atomic number. The dif-

TABLE II
MEAN NEUTRON SCATTERING LENGTH
DENSITIES OF VARIOUS COMPOUNDS

Compound	10^{10} cm^{-2}
H_2O	-0.56
D_2O	6.34
CH_2	-0.31
CH_3	-0.85
$C_5H_{13}NPO_4$ (lipid)	1.1
Hydrated protein in H_2O	1.1
Hydrated protein in D_2O	2.9

ference is most significant for hydrogen and its isotope deuterium. This fact is used with great advantage by SANS methods in studying biological systems rich in hydrogen atoms.

In practical terms, the estimation of a csld involves the computation of the sum of the scattering length of a group of atoms occupying a volume v_g. For example, to compute the mean csld of an alkane $CH_3(CH_2)_mCH_3$ we may write $\Sigma b_i = (m + 2)b_c + (2m + 6)b_H$, $v_g = mv_{CH_2} + 2v_{CH_3}$ and the csld $= \Sigma b_i/v_g$. The csld of various compounds are listed in Table II. Note that H_2O has a negative csld and D_2O a positive csld.

The angular brackets in Eqs. (2) and (3) denote averages over either the random orientation of nonspherical particles in solution, when they can be considered monodispersed, or over their sizes when they are poly-dispersed. Specific calculations can be found in Ref. 8. In the simplest case of a monodispersed system of homogeneous particles with a spheri-cal shape of radius R

$$F(Q) = v(\bar{\rho} - \rho_s)\frac{3j_1(QR)}{QR} \tag{5}$$

where $\bar{\rho}$ is the mean csld of the particle defined as

$$\bar{\rho} = \frac{1}{v}\int d^3r\rho(\mathbf{r})$$

and $j_1(x)$ is the first-order spherical Bessel function

$$j_1(x) = \frac{\sin x - x\cos x}{x^2}$$

It should be noted that in this particularly simple case

$$[|\langle F(Q)\rangle|^2/\langle|F(Q)|^2\rangle] = 1$$

so that $\bar{S}(Q) = S(Q)$. In the case of an ellipsoidal particle of semi-major and semi-minor axis a and b

$$F(Q, \mu) = (\bar{\rho} - \rho_s)v3j_1(u)/u \tag{6}$$

with

$$u = Q[a^2\mu^2 + b^2(1 - \mu^2)]^{1/2} \tag{7}$$

μ is the cosine of the angle between the directions of the vectors $\mathbf{a}$ and $\mathbf{Q}$. In an isotropic solution, one should average over the orientations of the particles by integrating $F(Q, \mu)$ with respect to μ from 0 to 1. In this way, the form factor becomes a function of the magnitude of $\mathbf{Q}$ only.

One can generalize this formalism to more complicated cases. For instance, we consider a spherical particle with an internal core of radius R_1 and a csld ρ_1, surrounded by a shell with an outer radius R_2 and a csld ρ_2. The form factor for this composite particle computed with the help of Eq. (4) is

$$F(Q) = \frac{4\pi}{3} R_1^3(\rho_1 - \rho_2) \frac{3j_1(QR_1)}{QR_1} + \frac{4\pi}{3} R_2^3(\rho_2 - \rho_s) \frac{3j_1(QR_2)}{QR_2} \tag{8}$$

For a special case when $\rho_1 = \rho_s$, i.e., a structure similar to that of a spherical vesicle, the expression simplifies to

$$F(Q) = \frac{4\pi}{3} (\rho_2 - \rho_s) \left[R_2^3 \frac{3j_1(QR_2)}{QR_2} - R_1^3 \frac{3j_1(QR_1)}{QR_1} \right] \tag{9}$$

The particle structure factors for the cases where $y = R_1/R_2 = 0.9$ and $y = 0$ (solid sphere of radius R_2) as a function of $x = QR_2$ are plotted in Fig. 2. It is interesting to note that while for a solid sphere, $P(Q)$ decays monotonically like a Gaussian function in Q, for a hollow sphere $P(Q)$ decays to zero faster than in the former case and shows a more pronounced second peak. This feature can be used to determine the outer-layer thickness $R_2 - R_1$.

The simplest way to extract a size parameter from a measurement of $P(Q)$ is as follows. For a homogeneous sphere of radius R, a small-Q expansion leads to

$$P(Q) \simeq (\bar{\rho} - \rho_s)^2 v^2 \left[1 - \frac{1}{5} Q^2R^2 + \cdots \right]$$

$$\simeq (\bar{\rho} - \rho_s)^2 v^2 \exp\left(-\frac{Q^2R^2}{5} \right) \tag{10}$$

Therefore a plot of $\ln P(Q)$ versus Q^2 would yield a straight line with a negative slope equal to $-R^2/5$. This type of analysis which uses the small-

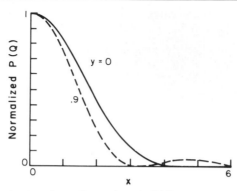

FIG. 2. Particle structure factor for a sphere (solid line) and a spherical shell (dotted line). $x = QR_2$ and $y = R_1/R_2$, where R_1 and R_2 are the inner and outer radii, respectively.

Q data (i.e., $QR < 2$) to obtain the equivalent spherical size parameter can be rigorously extended to the case of nonspherical particles, and is called the Guinier approximation.[1]

An important quantity obtainable from a SANS experiment is the scattered intensity extrapolated to $Q = 0$, i.e., the scattered intensity at zero angle. We shall explain in a later section the technical reasons why this quantity is easily deduced from the experiment. For a monodispersed system of quasi-spherical particles

$$I(0) = n(\bar{\rho} - \rho_s)^2 v^2 S(0) \tag{11}$$

The factor $n(\bar{\rho} - \rho_s)^2 v^2$ can be related to the molecular weight and to the solvent excluded volume of the hydrated protein in solution, as will be demonstrated. $S(0)$ relates to the osmotic compressibility at constant temperature.[9] This is an important thermodynamic quantity characterizing the nonideality of solutions. We shall discuss the use of these parameters in later sections.

Special Features of SANS

Although in many ways SANS is similar to the well-known SAXS technique, it has nevertheless three unique features which distinguish it from SAXS and which can be used with great advantage in the study of biological macromolecules. They are (1) the relative ease with which one can measure the absolute cross-section of the scattering [Eq. (1)]; (2) the possibility of changing the csld of the solvent by large amounts (by simply

[9] J. P. Hansen and I. R. McDonald, "Theory of Simple Liquids." Academic Press, New York, 1976.

changing the D_2O content of the water) to vary the solvent/solute contrast; and (3) the possibility for selective deuteration of particular functional groups in a macromolecule to enhance the spatial resolution. We shall elaborate on these points in later sections.

An immediate consequence of Eq. (11) is that one can determine the molecular weight of a macromolecule in a dilute aqueous solution from one simple measurement. We describe here a simple method of molecular weight determination for soluble proteins in H_2O solvent as recommended by Jacrot and Zaccai,[10] but with some modifications to account for the hydrated state of the protein molecules in solution.

At a protein concentration of 0.5% or less with 0.1 M salt, the electrostatic interaction between charged protein molecules is effectively screened out, and the structure factor $S(Q)$ at small Q is to a good approximation unity. Then, using the Guinier approximation[1] and plotting $\ln I(Q)$ vs Q^2 the extrapolated intensity at zero-angle is obtained as

$$I(0) \sim nv^2(\bar{\rho} - \rho_s)^2 \tag{12}$$

independent of the shape and internal structure of the protein. One can rewrite Eq. (12) in a more suggestive form by noting that

$$n = cN_A/M$$

$$\bar{\rho} = \sum_i b_i/v$$

$$v = \bar{v}_h M_h/N_A$$

with c the protein concentration in g/cm³, M the molecular weight of the dry protein, N_A Avogadro's number, M_h and $\bar{v}_h$ the molecular weight and specific volume of the protein in the hydrated state, respectively, then,

$$I(0) = c \frac{M_h^2}{M} N_A \left[\frac{\sum b_i}{M_h} - \rho_s \frac{\bar{v}_h}{N_A} \right]^2 \tag{13}$$

It has been noted by Jacrot and Zaccai[10] that for nonhydrated proteins the mean value for $\sum_i b_i/M \simeq 2.27 \times 10^{-14}$ cm/Da and the partial specific volume $\bar{v} \simeq 0.74$ cm³/g. Since we require the corresponding two quantities for the hydrated state, we here give their estimates

$$\sum_i b_i/M_h = 1.66 \times 10^{-14} \text{ cm/Da} \tag{14}$$

and

$$\bar{v}_h = \bar{v} + w\bar{v}^0 = 1.09 \text{ cm}^3/\text{g} \tag{15}$$

[10] B. Jacrot and G. Zaccai, *Biopolymers* **20**, 2413 (1981).

with w taken to be 0.35 g of bound H_2O per gram of dry protein, an average value for many globular proteins.[11] $\bar{v}^0 = 1$ cm³/g is the specific volume of bulk H_2O. Using Eqs. (14) and (15) we finally arrive at an empirical formula

$$I(0)[\text{cm}^{-1}] \simeq 4.32 \times 10^{-4}\, cM_h \tag{16}$$

valid for dilute solutions of proteins in aqueous buffer.

As an illustration, we determined $I(0) = 0.100$ cm^{-1} for a solution in H_2O of essentially fatty acid-free bovine serum albumin (BSA) in acetate buffer, pH 5.1, 0.03 M salt with $c = 0.003$ g/cm³. Equation (16) leads to $M_h = 77,160$. From the amino acid sequence of BSA[12] we know $M = 66,114$ and therefore $M_h = 1.35 \times M = 89,254$, assuming again $w = 0.35$ g/g. The empirical formula in this case predicts a value of M_h which is off by about 10%. In general, an experiment with H_2O as solvent is time consuming especially at low concentration because of the high level of flat incoherent background due to the protons. Nevertheless, when the detector is well calibrated a simple measurement is able to give M_h with 10% accuracy. This compares favorably with the standard method of sodium dodecyl sulfate–gel electrophoresis which gives M_h with 5 to 10% accuracy.[11]

Since either method immediately gives M_h and thus M, we will henceforth assume that the molecular weight of a protein is a known parameter and use it to derive other relevant properties from the experimental data.

SANS Instrument and Experimental Procedures

The Spectrometer

A modern small-angle scattering spectrometer is characterized by a well-calibrated two-dimensional multidetector controlled by a dedicated computer which performs the on-line data acquisition and storage. A complete software package for data processing is usually available.

Figure 3 depicts a schematic arrangement of the SANS apparatus at the National Center for Small-Angle Scattering Research (NCSASR), Oak Ridge National Laboratory (ORNL), aside from its computer system. The neutron source is located at the High Flux Reactor core from which a thermal neutron beam emerges. The beam passes through a beryl-

[11] C. R. Cantor and P. R. Schimmel, "Biophysical Chemistry," Part II. Freeman, San Francisco, 1980.

[12] J. R. Brown, in "Albumin Structure, Functions and Uses" (W. M. Resenoer, M. Oratz, and M. A. Rothschild, eds.). Pergamon, New York, 1977.

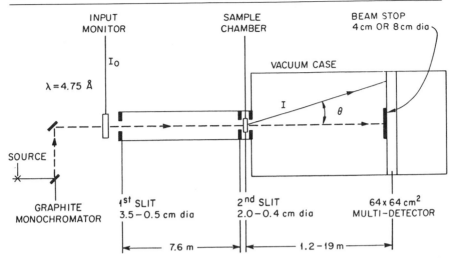

FIG. 3. Schematic layout of the 30-m SANS instrument at the National Center of Small-Angle Scattering Research at the Oak Ridge National Laboratory.

lium filter which is followed by a couple of graphite monocromator crystal sets which select a wavelength $\lambda = 4.75$ Å with a spread $\Delta\lambda/\lambda = 6\%$. A nuclear reactor provides a neutron beam with a large cross-sectional area, thus specimen of large cross-sections can be used to increase the scattered intensity. The incident beam flux I_0 is continuously monitored by a low efficiency neutron proportional counter before entering a double-slit vacuumed flight path of length $L_1 = 7.6$ m. The entrance slit S_1 has a circular diameter ranging from 0.5 to 3.5 cm, and the sample slit S_2 has a diameter which can be varied from 0.4 to 2.0 cm. The two-slit geometry defines a spot of the transmitted beam I_t at the center of the detector which is covered by the beam stop. Neutrons which are scattered at an angle θ are detected by the area detector elements situated on a ring at a distance d from the center of the detector. The multidetector consists of 4096 detector elements covering an area 64×64 cm^2. For a large sample-to-detector distance L_2, the Bragg wavenumber can be written approximately as

$$Q \simeq \frac{2\pi}{\lambda} \tan \theta = \frac{2\pi}{\lambda} \frac{d}{L_2} \qquad (17)$$

L_2 can be varied from 1.2 to 19 m in order to span a Q-range from 0.005 to 0.3 Å^{-1}.

Figure 4 shows another small-angle neutron scattering spectrometer located at the high flux reactor of the Brookhaven National Laboratory,

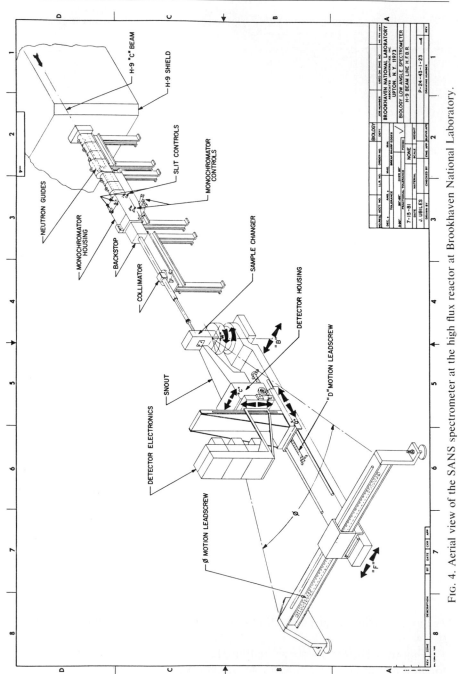

FIG. 4. Aerial view of the SANS spectrometer at the high flux reactor at Brookhaven National Laboratory.

Upton, New York. This spectrometer is situated at the exit of a beam port equipped with a cold moderator and thus has a supply of neutrons in the wavelength range of 4.0 to 8 Å with $\Delta\lambda/\lambda \simeq 10\%$. The corresponding neutron flux is also higher (about 10^6 neutrons per cm^2 per second) than that of the spectrometer at NCSASR. This spectrometer is designed primarily for biological applications, and at present is mostly used for solution studies. The spectrometer has a smaller two-dimensional detector area, 17×17 cm^2, with an individual pixel area 1.6×1.6 mm^2. Since the detector dimensions are smaller, the spectrometer has correspondingly smaller L_1 and L_2. The sample cells can be mounted in a temperature-controlled automatic sample changer containing 8 sample positions. The spectrometer is controlled by a minicomputer (PDP11/34) which also handles on-line data acquisition, storage, and processing.

Experimental Procedures

Choice of the Scattering Geometry. At constant S_2, the flux at the sample is approximately proportional to the area S_1. However, a large S_1 can be used only in combination with a small L_2 because of the finite size of the beam stop. Furthermore, the Q-resolution $\Delta Q/Q$ has to be kept at a reasonably small value (say 5%). This means Q needs to be kept at a relatively large value. The net result is that it is always preferable to work at the largest possible Q consistent with the size of the particle, namely $0.5 < QR < 5$. In this way L_2 can be kept at the minimum and S_1 at the maximum value.

In general, the scattering geometry should be set up to suit the requirements that the transmitted beam is completely intercepted by the beam stop and that the size of the sample slit S_2 is approximately half that of the neutron source split S_1.

Typical Data Collection and Correction Sequence. Any SANS measurement from a sample solution requires the evaluation also of the scattered intensity contributions from the solvent (I_{so}), the empty container (I_{emp}), and the ambient background when a piece of solid cadmium (a strong neutron absorber) is blocking the beam. The transmission (I_t/I_o) for each sample should also be determined to correct for the attenuation of the neutrons in traversing the sample. The transmission of the protein solution T should be kept above 70% to minimize multiple scattering effects[13] by a judicious choice of the concentration and sample cell path length. The scattering from the sample cell I_{emp} is in practice hardly different from background (BKG). The sample cells are usually flat quartz cells with 1-mm-thick windows and path lengths of 1 to 10 mm. The empty cells

[13] J. Schelten and W. Schmartz, *J. Appl. Crystallogr.* **13**, 385 (1980).

should have a very high transmission ($\geq 95\%$) meaning that their absorption and scattering cross-sections are small. The measured intensity from the sample I_{sam} is corrected for these various contributions according to

$$
I_s = \frac{1}{D_{sam}} \left(\frac{I_{sam} - BKG}{T_{sam}} - \frac{I_{emp} - BKG}{T_{emp}} \right)
$$

$$
- (1 - \eta) \frac{1}{D_{sol}} \left(\frac{I_{sol} - BKG}{T_{sol}} - \frac{I_{emp} - BKG}{T_{emp}} \right) \tag{18}
$$

where η is the volume fraction of the solute, D_{sam} and D_{sol} the path lengths, and T_{sam} and T_{sol} the transmission factors of sample cells and solvent cells, respectively. For dilute solutions $\eta \ll 1$, and if the sample cell and the solvent cell have the same path length D, then the transmission factors would be nearly equal ($=T$) and so Eq. (18) simplifies to

$$
I_s = \frac{I_{samp} - I_{sol}}{DT} \tag{19}
$$

In this case two measurements have to be made of the sample cell and the solvent cell only. Because of the nonuniformity of the multidetector, these corrections, plus the sensitivity correction as described in the next section, have to be applied to the two-dimensional array of the intensity data before the latter can be submitted to a circular averaging reducing them to one-dimensional data $I(Q)$.

It should be noted that because of their short wavelengths, neutrons are insensitive to the condition of the surface of the quartz cells, to bubbles of the order of a micron, and to a small amount of dust particles in the sample solution. These are the reasons why $I(0)$ is easily obtained by interpolation of the SANS data.

Relative Sensitivity Correction of the Detector Elements. Since the detector consists of many elements, each of which has a slightly different neutron detection efficiency, their relative efficiency has to be calibrated using a sample which has an angular-independent cross-section in the small-angle region. Two well-characterized candidates for this purpose are H_2O at room temperature and a vanadium plate of about 1 mm thickness. The measured scattered intensity I_i at a detector cell i is proportional to the efficiency ε_i of the detector element. Hence, the efficiency factor

$$
I_i \bigg/ \frac{\Sigma I_i}{N_d} = \frac{\varepsilon_i}{\bar{\varepsilon}}
$$

should divide, cell by cell, the counts detected at each detector element to correct for their sensitivity. N_d is the total number of detector cells. The sensitivity correction of the raw data is usually made before all other corrections.

Absolute Intensity Calibration. By definition,

$$\frac{d\Sigma}{d\Omega}(Q) = \frac{1}{V}\frac{J_s}{\varepsilon T \Delta a}\frac{L_2^2}{I_0} \tag{20}$$

The first factor $1/V$ takes into account the fact that one is interested in the cross-section per unit volume of the scattering sample. The second factor is the scattered neutron intensity in the direction of θ. J_s is the measured and sensitivity-corrected integrated number of counts on a detector element with surface area Δa, over a time period of a unit input monitor count. T is the transmission factor of the sample which divides the expression to correct for the attenuation of the beam in the sample of finite thickness t. The third factor contains in the numerator the sample-to-detector distance squared, and in the denominator the incident flux at the sample in a time interval corresponding to a unit monitor count. This expression is a practical form of Eq. (1). Since all the detector elements have the same area Δa and the average efficiency factor of the detector elements $\bar{\varepsilon}$ is constant, it is a standard practice to define

$$K_N = I_0 \bar{\varepsilon} \Delta a \tag{21}$$

which is proportional to the incident flux at the sample. At constant reactor power, K_N is a function of the size of the entrance pinhole S_1, only. For example, the spectrometer at NSCSAR has the following K_N for various S_1:

S_1(cm)	3.5	3.0	1.0	0.5
K_N	1660 ± 177	1220 ± 130	160 ± 18	41 ± 4

The largest S_1 size, 3.5 cm, can be used for $L_2 = 2.1$ m while $S_1 = 1.0$ cm is necessary for $L_2 = 14.1$ m. We can see a tremendous loss in intensity as a result of demanding a very small Q measurement at large sample-to-detector distance.

The absolute calibration of the detector is equivalent to the determination of K_N. This is normally done by using a standard sample such as a well-calibrated polymer or an aluminum single crystal with a known void distribution. When carefully done K_N can be determined to a few percent accuracy.

Weakly Interacting Systems

From the considerations of signal-to-background ratio, in a typical SANS experiment, the lowest practical protein concentration to use is

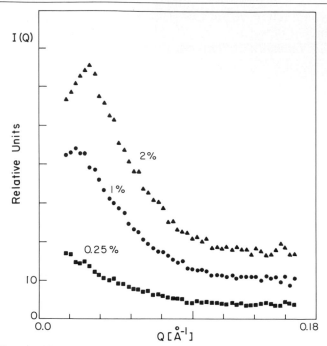

FIG. 5. Scattered intensity distributions of BSA solution in D_2O: 0.25% (triangles), 1% (dots), and 2% (squares).

about 0.2% or 2 mg/ml. A more convenient concentration range is usually around 1%. Depending on the chemical properties of a solution such as the pH or ionic strength, a 1% protein solution can be considered as either a weakly or a strongly interacting system. As an example, we show in Fig. 5 the scattered intensity distributions $I(Q)$ for 0.25, 1, and 2% BSA solutions in pure D_2O (99.8 atom% deuterium) at 37°. It can be seen that while for the 0.25% case $I(Q)$ is essentially given by the particle structure factor $P(Q)$ of a randomly oriented ellipsoid, for the 1 and 2% cases the effect of the interparticle structure factor $S(Q)$ is evident. For a strongly repulsive system of particles, $S(Q)$ tend to be depressed in the small-Q region. This behavior results in a curving down toward the origin of $I(Q)$ [e.g., Eq. (22)]. We shall call the peak (or the shoulder) thus formed from the product of $P(Q)$ by $S(Q)$ the interaction peak. It is evident from Fig. 5 that at the 1% concentration level, the interaction peak is already present. On the other hand, Fig. 6 shows $I(Q)$ and $S(Q)$ for 1% BSA solution in D_2O with 0.3 M LiCl added, at 35°. The intensity distribution function is nearly identical to the $P(Q)$ function since $S(Q)$ in this case is rather flat in the low-Q region, and is nearly unity elsewhere. We shall call a solution of this type a weakly interacting system.

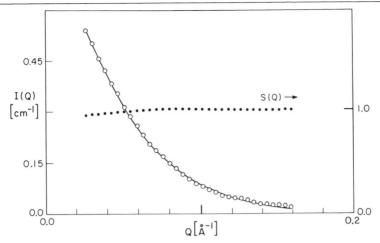

FIG. 6. One percent BSA, 0.3 M LiCl in D_2O solvent. The circles are the data points and the line represents theoretically calculated $I(Q)$. $S(Q)$ computed by using $z = -20$ is also shown as the dotted line. $S(0) = 0.94$ deviates appreciably from the ideal solution value of unity. (From Bendedouch and Chen, *J. Phys. Chem.* **87**, 1473.[7])

It is the purpose of this section to discuss the method of data analysis for this latter type of system. A rule of thumb is that at 1% protein concentration level, if sufficient amount of salt is present, the method of data analysis presented below should apply.

Debye Plot

For a monodisperse system of weakly interacting ellipsoidal particles, Eq. (1) reduces to

$$I(Q) = nP(Q)S(Q) \tag{22}$$

because in this case $\bar{S}(Q) \simeq S(Q)$. This fortunate situation arises from the following reasons: in the low-Q region where $S(Q)$ is different from unity the ratio $|\langle F(Q)\rangle|^2/\langle |F(Q)|^2\rangle$ is about one, and in the large-Q region where this ratio is significantly different from unity, $S(Q)$ is hardly different from one.

Furthermore, as can be seen from Fig. 6 $S(Q)$ in the low-Q region is rather constant as compared to the variation of $P(Q)$. Thus, it is a good approximation to write $S(Q) \simeq S(0)$ in Eq. (22) for the small-Q region. A thermodynamic relation gives[9]

$$S(0) = k_B T/[\partial\pi/\partial n]_T \simeq \frac{1}{1 + 2B_2 n} \tag{23}$$

where $n[\partial\pi/\partial n]_T$ is the inverse osmotic compressibility of the solution. The virial expansion of the osmotic pressure π leads to the expression on the right-hand side where B_2 is the second virial coefficient. The Guinier approximation[1] which is valid in the small-Q region can be expressed as $P(Q) \simeq P(0) \exp[-Q^2Rg^2/3]$, where Rg is the radius of gyration of the particle defined as

$$Rg^2 = \frac{\int r^2[\rho(\mathbf{r}) - \rho_s]d^3r}{\int[\rho(\mathbf{r}) - \rho_s]d^3r} \tag{24}$$

The integrals are over the solvent excluded volume of the particle v. The value for Rg can be given explicitly in terms of the geometrical parameters in a few simple cases. For example for a sphere with radius R

$$Rg^2 = 3R^2/5$$

for an ellipsoid of revolution with semi-axis, a, b, c

$$Rg^2 = \frac{1}{5}(a^2 + b^2 + c^2)$$

for a circular cylinder with radius R and height h

$$Rg^2 = \frac{1}{4}\left(2R^2 + \frac{h^2}{3}\right)$$

and for a spherical shell with inner and outer radius R_1 and R_2

$$Rg^2 = \frac{3}{5}\left[\frac{R_2^5 - R_1^5}{R_2^3 - R_1^3}\right]$$

As a consequence of the approximations discussed above, a way of reducing useful quantities from the low-Q data is to generate the so-called Debye plot.[14,15] We can write

$$I(Q) = I(0) \exp[-Q^2Rg^2/3] \tag{25}$$

where

$$I(0) = nP(0)/(1 + 2B_2n) \tag{26}$$

and

$$P(0) = (\bar{\rho} - \rho_s)^2v^2 \tag{27}$$

Therefore, Guinier plots [$I(Q)$ vs Q^2] for a concentration series of protein solutions would give Rg and $I(0)$ (as a function of n) directly. We shall

[14] P. Debye, *J. Appl. Phys.* **15**, 338 (1944).
[15] B. J. Berne and R. Pecora, "Dynamic Light Scattering." Wiley, New York, 1976.

discuss the interpretation of Rg in the next section, but we wish to point out immediately that $I(0)$ contains valuable information. Assuming that the molecular weight M of the protein under investigation is known, or determined as described previously, it is useful to plot $n[I(0)]^{-1}$ vs n which is the Debye plot by definition. Since

$$n[I(0)]^{-1} = (1 + 2B_2 n)/P(0) \tag{28}$$

$P(0)$ and B_2 are deduced from the zero intercept and the slope of the straight line, respectively. $2B_2$ is the total volume inaccessible to the center of the surrounding particles because of the presence of a particle with an equivalent hard sphere diameter of σ^*,[16]

$$2B_2 = \frac{4\pi}{3} \sigma^{*3} \tag{29}$$

where

$$\sigma^* = \sigma(1 + 3\delta)^{1/3} \tag{30}$$

$$\delta = \int_1^\infty [1 - \exp(\ V_R(x)/k_B T)]x^2 dx \tag{31}$$

k_B is the Boltzmann constant and T the absolute temperature.

Here we regard the pair-potential of interaction as consisting of a hard-core potential with a diameter σ, plus an electrostatic repulsive potential $V_R(x)$ valid for $x > 1(x = r/\sigma)$. When σ can be independently determined by other means (such as the dry volume of the protein molecule) then δ, as deduced experimentally from Eqs. (29) and (30), can be used in conjunction with Eq. (31) to reduce for example the surface charge of the protein from $V_R(x)$.[16]

Contrast Variation

One of the distinguishing features of SANS technique is the relative ease by which one can vary the solvent/solute contrast. The csld of the aqueous solvent can be varied by large amounts from -0.56×10^{10} to 6.34×10^{10} cm^{-2} by simply changing the volume fraction of D_2O in water β, then

$$\rho_s = [-0.56 + 6.90\beta] \times 10^{10} \ \text{cm}^{-2} \tag{32}$$

The variation of ρ_s is large enough to cover all the csld of both protonated and deuterated biological molecules of interest, as can be seen from Table II.

[16] D. Bendedouch and S.-H. Chen, *J. Phys. Chem.* **87**, 1653 (1983).

The so-called contrast variation method consists of measurements of a series of scattered intensity spectra from solutions in various H_2O/D_2O mixtures. In the literature,[6] the term contrast is defined as

$$\hat{\rho} = \bar{\rho} - \rho_s \tag{33}$$

where $\bar{\rho}$ is the volume average csld of the macromolecule in solution. In practice, a particle may have a nonuniform csld distribution, so in order to treat this general case, we write

$$\rho(\mathbf{r}) = \bar{\rho}[1 + \Delta(\mathbf{r})] \tag{34}$$

Then Eqs. (2) and (4) lead to

$$[P(0)]^{1/2} = |\bar{\rho} - \rho_s|v \tag{35}$$

for a monodisperse system of particles in solution.

The conventional method of contrast variation consists of a series of measurements of $I(Q)$ as a function of β in the small-Q region to determine both the radius of gyration Rg and the scattered intensity at zero angle. A good example is given in Fig. 7 for the case of myoglobin.[17] Figures 7a and b show the SANS spectra obtained with aqueous solvents containing 0 to 97% D_2O. Note that the intensity goes through a minimum for β between 38.7 and 43.6%. Figure 7c presents a plot of the square root of $I(0)$ versus β. The straight line goes through zero for $\beta_0 = 40\%$ which corresponds to $\rho_s = \bar{\rho} = 2.2 \times 10^{10}$ cm^{-2}, assuming that $S(0) = 1$.

Our prescription is that $P(0)$ be computed from the following equation

$$nP(0) = I(0)[1 + 2B_2n] \tag{36}$$

in accordance with the arguments presented in the previous section, for each H_2O/D_2O mixture. To correct effectively for the interparticle interaction effect, the protein concentration chosen to do the contrast variation series must be similar to the largest concentration of the series used to get the Debye plot, as described previously.

After correction for interparticle interference effects according to Eq. (36), a plot on an absolute scale of $[P(0)]^{1/2}$ versus ρ_s would yield both $\bar{\rho}$ and v in principle [Eq. (35)]. The volume v obtained in this way would be the dry volume of the particle which is independent of β. ρ_s is given by

$$\rho_s(\beta) = \beta\rho_{D_2O} + (1 - \beta)\rho_{H_2O} \tag{37}$$

However, for proteins the matter is complicated by the fact that $\bar{\rho}$ is also a linear function of the volume fraction of D_2O in water β. The reasons are the following. On the one hand, proteins are known to exchange their

[17] K. Ibel and H. B. Sturhmann, *J. Mol. Biol.* **93**, 255 (1975).

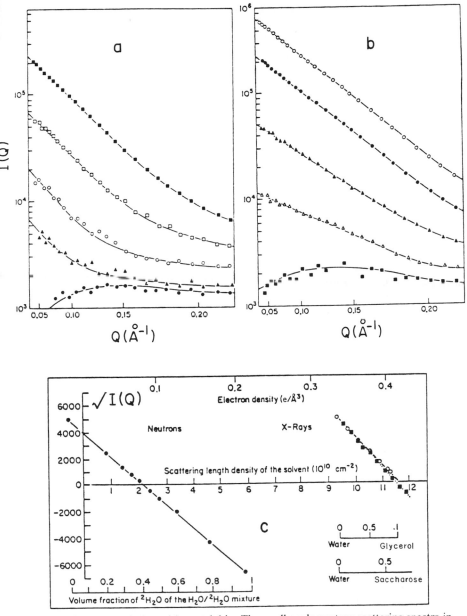

FIG. 7. Contrast variation with myoglobin. The small-angle neutron scattering spectra in various H_2O/D_2O mixtures are shown: (a) 0% D_2O (■), 29% D_2O (○), 38.7% D_2O (●), 19% D_2O (□), 33.8% D_2O (▲); (b) 97% D_2O (○), 57.7% D_2O (▲), 43.6% D_2O (■), 77.3% D_2O (●), 48.5% D_2O (△). The square root of the zero-angle scattering of the myoglobin solutions is a linear function of the scattering length density of the solvent: (c) neutron scattering (●); X-ray scattering (water/glycerol mixtures) (○); X-ray scattering (sugar solutions) (■). (From Ibel and Stuhrmann, *J. Mol. Biol.* **93**, 255.[17])

TABLE III
SCATTERING LENGTHS OF AMINO ACID RESIDUES CALCULATED FROM
THEIR CHEMICAL COMPOSITIONS AT pH 7[a]

Amino acid residue	Chemical composition	M	N_e	$\Sigma b(H_2O)$ $[10^{-12}$ cm]	Partial volume (Å^3)
Alanine	C_3NOH_5	71	1	1.645	91.5
Arginine	$C_6H_4OH_{13}$	157	6	3.466	180.8
Asparagine	$C_4N_2O_2H_6$	114	3	3.456	135.2
Aspartic acid	$C_4NO_3H_4$	114	1	3.845	113.6
Cysteine	C_3NOSH_5	103.1	2	1.930	105.6
Glutamic acid	$C_5NO_3H_6$	128	1	3.762	140.6
Glutamine	$C_5N_2O_2H_8$	128	3	3.373	161.1
Glycine	C_2NOH_3	57	1	1.728	66.4
Histidine	$C_6N_3OH_{6.5}$	136.5	1.5	4.959	167.3
Isoleucine	C_6NOH_{11}	113	1	1.396	168.8
Leucine	C_6NOH_{11}	113	1	1.396	167.9
Lysine	$C_6N_2OH_{13}$	129	4	1.586	176.2
Methionine	C_6NOSH_9	131	1	1.764	170.8
Phenylalanine	C_9NOH_9	147	1	4.139	203.4
Proline	C_5NOH_7	97	0	2.227	129.3
Serine	$C_3NO_2H_5$	87	2	2.225	99.1
Theonine	$C_4NO_2H_7$	101	2	2.142	122.1
Tryptophan	$C_{11}N_2OH_{10}$	186	2	6.035	237.6
Tyrosine	$C_9NO_2H_9$	163	2	4.719	203.6
Valine	C_5NOH_9	99	1	1.479	141.7

[a] Adapted from Jacrot and Zaccai.[10]

labile protons with deuterium, and the extent of this exchange is proportional to β. On the other hand, proteins are heavily hydrated in solution and this bound water has a scattering length which depends also on β. Thus, in an aqueous solvent with D_2O volume fraction β, the mean csld of a protein is given by

$$\bar{\rho}(\beta) = \left[\sum_i b_i + N_e(b_D - b_H)\beta + mb_w\right]/lv \qquad (38)$$

where $\Sigma_i b_i$ = sum of the scattering lengths of all the atoms in the native protein, N_e = number of exchangeable protons, m = number of water molecules bound to one protein molecule, and

$$b_w = \beta b_{D_2O} + (1 - \beta)b_{H_2O} \qquad (39)$$

is the mean scattering length of a water molecule in any H_2O/D_2O mixture.

Since the number of labile protons in a given amino acid residue is known (see Table III), the maximum number of exchangeable protons can be calculated for a protein with a determined amino acid sequence. For instance, this number is 1018 for BSA. But in pure D_2O, it is usually estimated that only 80% of the labile protons are accessible to the solvent and hence exchange with deuterium atoms. Thus for BSA, N_e can be estimated to be 814.

Using Eqs. (37) to (39), one can rewrite Eq. (35) as

$$[P(0)]^{1/2} = \left|\left|\sum_i b_i + mb_{H_2O} - v\rho_{H_2O}\right| - \beta[v(\rho_{D_2O} - \rho_{H_2O})\right.$$
$$\left. - N_e(b_D - b_H) - m(b_{D_2O} - b_{H_2O})]\right| \tag{40}$$

Thus a plot of $[P(0)]^{1/2}$ versus β would yield a straight line giving two independent quantities:

(1) the zero β intercept: $\sum_i b_i + mb_{H_2O} - v\rho_{H_2O}$
(2) the slope: $V(\rho_{D_2O} - \rho_{H_2O}) - N_e(b_D - b_H) - m(b_{D_2O} - b_{H_2O})$

Thus one obtains two equations relating four unknown quantities $\sum_i b_i$, v, N_e, and m, and one needs to determine independently two of these four parameters. It is most convenient to determine v by fitting the finite-Q data of $P(Q)$. In practice this can easily be done by postulating a model for the shape of the protein molecules in solution, as we shall demonstrate below.

We analyzed the scattered intensity spectrum $I(Q)$ obtained from a 0.3% BSA solution in acetate buffer, pH 5.1, 0.03 M salt in D_2O. Since $S(Q)$ is essentially unity in this condition, we simply had to assume a model for $P(Q)$. A successful model representing the BSA molecule in solution was found to be a randomly oriented homogeneous prolate ellipsoid with semi-major axis a and b, respectively. Thus

$$I(Q) \simeq nP(Q) = n(\bar{\rho} - \rho_s)^2 \left[\frac{4\pi}{3} ab^2\right]^2 \int_0^1 d\mu \left[\frac{3j_1(u)}{u}\right]^2 \tag{41}$$

with u given by Eq. (7). The resulting fit is excellent as can be seen from Fig. 8 when we choose $a = 70 \pm 2$ Å and $b = 20 \pm 1$ Å giving the hydrated volume of BSA inaccessible to bulk solvent $v = 117,300$ Å^3. A normalization to the absolute intensity scale also gives $\bar{\rho}(\beta = 1) = 3.89 \times 10^{10}$ cm^{-2}.

A contrast variation experiment was carried out for a 1% BSA solution containing 0.3 M LiCl. Figure 9 shows that the contrast match point occurs for $\beta_0 = 0.405$ corresponding to $\bar{\rho}(\beta_0) = 2.24 \times 10^{10}$ cm^{-2}. Since for BSA we know the amino acid sequence,[12] it can be calculated that $\sum_i b_i = 1485 \times 10^{-12}$ cm and $N_e = 814$. Using these parameters and setting $\beta = \beta_0$

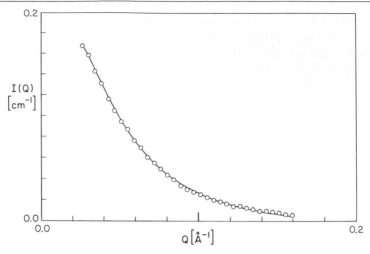

FIG. 8. Intensity spectrum (circles) and theoretical fit (line) for a solution containing 0.3% BSA in acetate buffer, pH 5.1, 0.03 M. The fit is obtained by assuming a prolate ellipsoidal shape for the solvated BSA ($a = 70$ Å and $b = 20$ Å) and $z = -1$. $S(Q)$ computed as explained in the text turned out to be nearly unity, rendering this system almost ideal. (From Bendedouch and Chen, *J. Phys. Chem.* **87,** 1473.[7])

in Eq. (38), it is found that $m = 1180$ which corresponds to 0.34 g of bound H_2O per gram of BSA.

An alternative way of obtaining the weight fraction of the bound water (w) is to note the relation

$$w = \bar{v}_b \left(\frac{N_A v}{M} - \bar{v} \right)$$

Therefore by using a known value of $\bar{v} = 0.734$ cm³/g for BSA[11] one can compute $w = 0.334$ g/g assuming that the specific volume of the bound water, $\bar{v}_b$, can be taken as unity. Agreement of this number with that derived previously from the contrast variation data indicates that the assumption of 80% H/D exchange was valid. The often quoted literature value of w is about 0.4 g/g.[11] By setting $\beta = 1$ in Eq. (38) we obtain a consistent value for $\bar{\rho}(\beta = 1)$ with that reduced from the finite-Q fitting of $I(Q)$.

Besides the zero-intercept $I(0)$, the Guinier plots also yield the radius of gyration Rg defined by Eq. (24). Using Eq. (34) one can rewrite Rg as

$$Rg^2 = R_s^2 \left[1 + \frac{\bar{\rho}}{\hat{\rho}} \alpha \right] \tag{42}$$

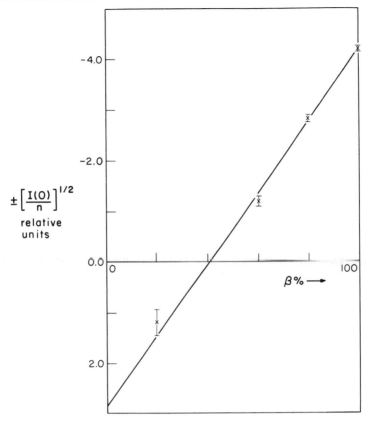

FIG. 9. The square root of the extrapolated forward scattering intensity of various BSA samples is plotted vs the D_2O volume fraction of the solvent β. The zero intercept occurs for $\beta = 40.5\%$ for a solution containing 1% BSA and 0.3 M LiCl. (From Bendedouch and Chen, *J. Phys. Chem.* **87**, 1473.[7])

where

$$R_s^2 = \frac{1}{v} \int d^3r \, r^2 \tag{43}$$

and

$$\alpha = \frac{1}{R_s^2 v} \int d^3r \Delta(r) r^2 \tag{44}$$

$$= \frac{1}{R_s^2 v} \int_0^R dr 4\pi r^2 \Delta(r) r^2 \tag{45}$$

TABLE IV

MEAN NEUTRON SCATTERING LENGTH DENSITIES OF
A FEW PROTEINS AND DETERGENTS

Compound	$\bar{\rho}[10^{10}\ cm^{-2}]$	Fraction of exchanged protons
Acetylcholine receptor	2.09	0
	2.27	1
Bovine serum albumin	1.83	0
	3.89	0.8
Triton X-100	0.63	
Lithium dodecyl sulfate	0.96	
Acetylcholine–Triton X-100 complex	1.65	

R_s is the radius of gyration associated with the shape of the particle and α given by Eq. (45) is for the case where the center of the csld distribution coincides with the geometrical center of a spherical particle of radius R. From this expression one can see that if the csld distribution were uniform then $\alpha = 0$. For soluble globular proteins with a hydrophobic core surrounded by a hydrophilic region, $\Delta(r)$ is negative in the core and positive in the outer-layer. Since the integration is weighted by r^4 the net value of α tends to be positive in this case. Therefore a plot of Rg^2 vs $\bar{\rho}^{-1}$, the reciprocal contrast, would give a line with a positive slope. This in fact is the case for myoglobin as shown by Ibel and Stuhrmann.[17]

In general, the csld of native unhydrated proteins is around 1.8×10^{10} cm^{-2} and is contrast matched in 40% D_2O. The average csld of a protein is positive because N and O contribute positively to the csld in spite of the fact that there is a large number of hydrogen atoms present. This is not the case of pure hydrocarbons which have negative csld.

Insoluble proteins can be studied indirectly by investigating the structure of the soluble complexes they form with detergents. In Table IV are listed the measured csld of two proteins and two detergents. Because of the sufficient differences in csld between proteins and detergents, it is possible to match out selectively the protein or the detergent component of the complex by chosing a suitable H_2O/D_2O mixture. This technique has been used by Wise et al.[18] to investigate the acetylcholine receptor from the electric tissue of *Torpedo californica* in Triton X-100 solutions. The receptor is a large integral membrane protein insoluble in water. The measurements on the receptor-detergent complex in solution yield a ra-

[18] D. S. Wise, A. Karlin, and B. P. Schoenborn, *Biophys. J.* **28,** 473 (1979).

dius of gyration for the receptor monomer of 46 ± 1 Å and a molecular volume of 305,000 Å^3. The neutron scattering data yield in addition the molecular weight of the receptor protein and the extent of Triton X-100 bound (~0.4 g/g protein). The shape of the receptor-detergent complex is determined to be an oblate ellipsoid of axial ratio 1/4.

Differential Method

This technique makes use of the possibility of a selective deuteration of specific functional groups. Since deuteration is not expected to lead to appreciable changes of the conformation of the molecular groups, a differential measurement would give specific information on their spatial distribution. This serves as a way of visualizing the specific groups of interest and to enhance the spatial resolution which would otherwise be obtained in an ordinary small-angle scattering experiment.

The technique consists in the determination of the scattering curves of two identically prepared samples of labeled and unlabeled material, $I^*(Q)$ and $I(Q)$, respectively. The data are then processed to obtain the absolute value of the scattering amplitude of the csld distribution $\rho_g(\mathbf{r})$ of the labels

$$A(Q) = |\sqrt{I(Q)/S(Q)} - \sqrt{I^*(Q)/S(Q)}| \tag{46}$$
$$= n^{1/2}|F(Q) - F^*(Q)| \tag{47}$$

In Eq. (46) one can use an identical $S(Q)$ to correct $I(Q)$ and $I^*(Q)$ for the interparticle interference effect because the deuteration does not alter the interparticle interaction. If one writes

$$\rho(\mathbf{r}) = \rho_0(\mathbf{r}) + \rho_g(\mathbf{r})$$

and   (48)

$$\rho^*(\mathbf{r}) = \rho_0(\mathbf{r}) + \rho_g^*(\mathbf{r})$$

where $\rho_0(\mathbf{r})$ represents the spatial distribution of the undeuterated part of the molecule, then

$$A(Q) = n^{1/2} \left| \int d^3r(\rho_g(\mathbf{r}) - \rho_g^*(\mathbf{r})) \exp(i\mathbf{Q} \cdot \mathbf{r}) \right| \tag{49}$$

For a given molecular group with a scattering length b_g and its associated number density distribution $n_g(\mathbf{r})$,

$$\rho_g(\mathbf{r}) = b_g n_g(\mathbf{r}) \tag{50}$$

therefore Eq. (49) can be rewritten as

$$A(Q) = n^{1/2}|b_g - b_g^*| \left| \int d^3r \, n_g(\mathbf{r}) \exp(i\mathbf{Q} \cdot \mathbf{r}) \right|, \tag{51}$$

b_g^* is the scattering length of the deuterated group. The integral in $A(Q)$ is over the volume v of the molecule. Aside from the known quantity $A = n^{1/2}|b_g - b_g^*|$, the differential data $A(Q)$ are essentially a Fourier integral of the number density distribution of the deuterated groups. In the case of a spherically symmetric distribution Eq. (51) reduces to

$$A(Q) = A \left| \int_0^R dr 4\pi r^2 \frac{\sin Qr}{Qr} n_g(r) \right| \qquad (52)$$

with R being the maximum radial extent of the distribution.

A good illustration of this technique is provided by an experiment by Laggner et al.[19] on selectively deuterated human plasma low-density lipoproteins (LDL). Their SANS results indicate that LDL is a spherical particle with the phospholipids forming a spherical monolayer shell in the range between 75 and 103 Å around a core of cholesteryl esters and triglycerides. The protein moiety is located, on average, 5 to 10 Å from the polar phospholipid headgroups and forms the outer layer of this three-layer particle. The spatial distribution of the polar phospholipid headgroups was deduced from the SANS experiment by selectively deuterating the $N(CH_3)_3$ groups in the LDL. The intensity distributions for native and deuterated LDS, and the resulting differential curve $A(Q)$ vs Q are shown in Fig. 10. The characteristic oscillations in $A(Q)$ can be understood as follows. Assuming that the deuterated phospholipid headgroup layer of thickness ΔR and uniform number density n_g is located at a mean distance R_1 from the center of the particle with $\Delta R \ll R_1$, then Eq. (52) can be integrated to yield

$$A(Q) = A n_g 4\pi R_1^2 \Delta R \left| \frac{\sin Qr_1}{QR_1} \right| \qquad (53)$$

Thus the differential curve plotted against Q will oscillate like $|\sin x/x|$ with the zero-crossings occurring at $QR_1 = \pi, 2\pi$, etc. The dotted line in Fig. 10 is essentially such a curve with $R_1 = 100$ Å. The actual differential curve shows some damping in the oscillation because this layer is not infinitely thin and R_1 spreads over a distance from 75 to 103 Å.

Strongly Interacting Systems

Proteins often exist in substantial concentrations in physiological environments. For example, the concentration of hemoglobin in human blood

[19] P. L. Laggner, G. M. Kostner, U. Rakusch, and D. Worcester, J. Biol. Chem. 256, 11832 (1981).

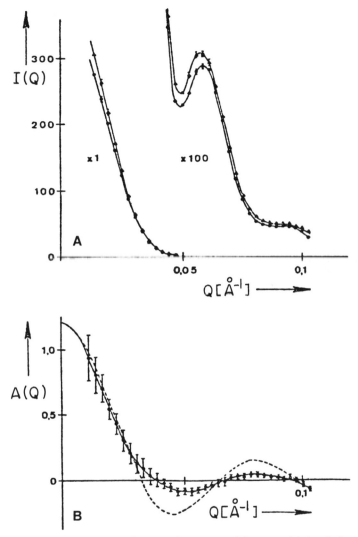

FIG. 10. (A) Neutron small-angle scattering curves of deuterated (triangles) and native LDL (circles) measured in 100% D_2O buffers at sample-to-detector distance of 2.603 m. Some typical statistical errors are shown by vertical bars. (B) Amplitude difference, $A(Q)$, between deuterated and native LDL obtained from the data in (A) with statistical error bars. Broken line, theoretical amplitude curve for an infinitely thin, spherical shell of radius 100 Å. (From Laggner *et al. J. Biol. Chem.* **256,** 11832.[19])

is about 14.5% by weight and that of serum albumin is 4.2%.[20] Furthermore, at the physiological plasma pH (7.0) and salt concentration (0.15 M) protein molecules are usually heavily charged. Therefore, electrostatic interaction between protein molecules is strong and cannot be neglected. A direct consequence of this interaction is that even for 1% solutions with a low salt concentration the interparticle structure factor $S(Q)$ becomes significantly depressed from unity in the small-Q region. Thus the conventional Guinier analysis would give smaller apparent radius of gyration Rg and molecular weight for the protein molecule. At higher protein concentration, the Guinier plot would have a positive slope giving a seemingly unphysical negative Rg^2. In such a case, it is mandatory that a theoretical model taking into account a realistic protein–protein interaction be used to analyze the data properly. We shall show in this section that a SANS experiment for a concentrated protein solution is useful when combined with such an analysis. Besides the results that we have been discussing in the previous section, added information can be gained on such quantities as the surface potential of the protein molecule and the interparticle ordering, especially in highly concentrated solutions.

Model for the Interparticle Structure Factor

$S(Q)$ can be calculated according to the standard theory of liquid structure if a plausible form of the pair potential between molecules is assumed.[9]

Since globular proteins are rigid particles in their native conformation in aqueous solution because of strong hydrophobic interactions, it is appropriate to approximate the interparticle repulsive potential by a superposition of an effective hard core and a screened Coulomb potential. The effective hard core diameter is determined from an excluded volume consideration. For example, for a globular protein of semi-major and semi-minor axis a and b, we may simply take

$$\sigma^3 = 8ab^2 \tag{54}$$

But when the axial ratio a/b is larger than 3, a more accurate prescription is

$$\sigma^3 = 2(f + 1)ab^2 \tag{55}$$

$$f = \frac{3}{4}\left(1 + \frac{\sin^{-1}(p)}{p(1 - p^2)^{1/2}}\right)\left(1 + \frac{1 - p^2}{2p^2}\ln\frac{1 + p}{1 - p}\right) \tag{56}$$

$$p^2 = (a^2 - b^2)/a^2 \tag{57}$$

---

[20] T. Peters, *in* "The Plasma Proteins" (F. W. Putnam, ed.), Vol. 1, p. 133. Academic Press, New York, 1975.

when $a \rightarrow b, f \rightarrow 3$, and Eq. (55) becomes identical to Eq. (54). In the case of BSA taking $a = 70$ Å and $b = 20$ Å, Eq. (54) predicts $\sigma = 60.7$ Å and Eq. (55) $\sigma = 69.3$ Å.

The screened Coulomb potential is of the form

$$V(x)\Big/ k_B T = \gamma \exp(-k(x - 1))/x, \; x \geq 1 \qquad (58)$$

where γ is the surface potential of the protein molecule in units of $k_B T$ and is related to the surface charge by

$$\gamma = \frac{4z^2 e^2}{\varepsilon\sigma(2 + k)^2 k_B T} \qquad (59)$$

with

$$k = \kappa\sigma \qquad (60)$$

and

$$\kappa^2 = \frac{8\pi N_A e^2 I}{\varepsilon k_B T \, 10^3} \qquad (61)$$

where ε is the dielectric constant of the medium, e is the electronic charge, and I is the ionic strength of the solution in equivalent^2/mol · liter. This form of the potential is valid for $k < 6$. When $k > 6$, i.e., $I > 0.2$ for particles of colloidal sizes, the repulsive potential can be replaced by an effective hard sphere potential with an increased diameter σ^*.[16]

When a solution is at its isoelectric pH, the effective charge of the solute protein molecule is zero but there remains the van der Waals attractive interaction. This attraction can be taken into account phenomenologically by setting the surface potential γ negative.

The method of obtaining $S(Q)$ is as follows: the pair-correlation function $g(r)$ is calculated according to the standard liquid state theory[9] which relates $g(r)$ to the pair-potential of interaction. A convenient computer program exists for computing $g(r)$ using the mean spherical approximation (MSA).[21-23] The structure factor $S(Q)$ is then deduced from $g(r)$ by a Fourier transform relation

$$S(Q) = 1 + n \int_0^\infty dr 4r^2(g(r) - 1) \frac{\sin Qr}{Qr} \qquad (62)$$

[21] J. B. Hayter and J. Penfold, J. Mol. Phys. **42**, 109 (1981).

[22] J.-P. Hansen and J. B. Hayter, Mol. Phys. **46**, 561 (1982).

[23] J. B. Hayter and J.-P. Hansen, "The Structure Factor of Charged Colloidal Dispersions at Any Density." Institut Laue–Langevin, Report No. 82 HA14T, 1982.

The pair-correlation function $g(r)$ has the simple physical meaning that

$$N(r)dr = 4\pi r^2 ng(r)dr \qquad (63)$$

is the number of neighboring particles located in a spherical shell defined by $[r, r + dr]$ from a given central particle. Thus a plot of $g(r)$ as a function of r gives a visual picture of the spatial correlation between the particles in solution. For example, if the liquid consists of point particles randomly distributed in space, then the number of particles in the spherical shell is simply $n4\pi r^2 dr$ ($=$ number density $\times$ volume of the spherical shell). Equation (63) would then imply that $g(r) = 1$. This corresponds to the ideal gas limit of an infinite dilution. As the volume fraction of the particles in solution increases, the effect of the hard core of the particles becomes significant. Since no particle can approach a given central particle within a distance of one hard core diameter σ, $g(r)$ will be identically zero for $r \leq \sigma$. From this point on $g(r)$ will gradually rise above unity and peak at the first-neighbor position. It then oscillates around and rapidly approaches unity. As the distance r becomes large a given particle at position r has vanishing correlation with the central particle, and the same argument as above shows that $g(r) = 1$.

Model for the Particle Structure Factor

A reasonable model for the shape of a globular soluble protein molecule is to take its average hydrodynamic shape, which is usually a prolate or an oblate ellipsoid with semi-major and -minor axes a and b, respectively. Since the hydrophobic interior of a protein is rather uniform, and the outer layer containing primarily the hydrated hydrophilic amino acid residues is thin compared to the dimensions of the molecule, the particle structure factor can be modeled, to a good approximation, as a uniform particle with a mean csld $\bar{\rho} = 3 \times 10^{10}$ cm^{-2} in D$_2$O. Thus

$$P(Q) = (\bar{\rho} - \rho_s)^2 v^2 \int_0^1 d\mu \left[\frac{3j_1(u)}{u} \right]^2 \qquad (64)$$

In the case of a prolate ellipsoid

$$u = Q[a^2\mu^2 + b^2(1 - \mu^2)]^{1/2} \qquad (65)$$

and

$$v = 4\pi ab^2/3 \qquad (66)$$

But in the case of an oblate ellipsoid

$$u = Q[b^2\mu^2 + a^2(1 - \mu^2)]^{1/2} \qquad (67)$$

and

$$v = 4\pi a^2 b/3 \qquad (68)$$

According to these formulas, *when a/b is less than about 3, one cannot distinguish a prolate ellipsoid from an oblate ellipsoid with an equal volume*. However, when a/b is larger than 3, the $P(Q)$ for these two cases are sufficiently different even for particles with identical volumes.

For example, we were able to confirm that BSA is a prolate ellipsoid with an axial ratio a/b equal to 3.5. This has been described already in the former section and illustrated in Fig. 8 for the case of 0.3% BSA in pH 5.1 buffer solution in D_2O with 0.03 M salt.

In general model fitting on an absolute intensity scale taking into account both $S(Q)$ and $P(Q)$ would give a unique set of values for $\bar{\rho}$ and v in the case of monodispersed systems of globular proteins. These two parameters alone would give the amount of bound water to the protein molecule, as explained earlier.

Some Examples

When the concentration of BSA is increased to 1% in D_2O at neutral pH with 0.3 M LiCl added, the surface charge is about $-20e$ and the interparticle interaction cannot be neglected.[7] Therefore, the analysis is made by taking into account the model $S(Q)$ described previously. As can be seen from Fig. 6 the model fits the data very well giving a surface potential $\gamma = 1.1$ in units of $k_B T$ and a range for the interaction of 5.6 Å. This latter value is the Debye–Hückel screening length κ^{-1} equivalent to the thickness of the electric double-layer surrounding the particles and depends on the ionic strength of the solution. The dotted line in the figure indicates $S(Q)$ which is slightly depressed from unity in the small-Q region. A Guinier analysis of $I(Q)$ in this case would yield a smaller Rg which would not be consistent with the values of a and b deduced from the analysis of Fig. 8.

As the protein concentration is increased to 4% and above, at neutral pH with no salt added, the electrostatic interactions between protein molecules become so strong that $S(Q)$ is significantly depressed in the forward direction. This is manifested in $I(Q)$ by a well-defined interaction peak with the curve plunging down steeply in the small-Q region. It is obvious that the Guinier analysis in this case would give a completely unphysical result with an imaginary Rg. In the literature, it is sometimes mistakenly taken to be the result of the formation of dimeres, but our analysis clearly shows that the interaction peak reflects the existence of a screened Coulomb potential (i.e., electrical double layer repulsion) be-

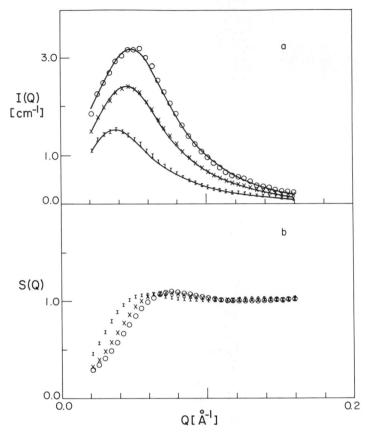

FIG. 11. (a) Intensity distribution for 4% (I), 8% (×), and 12% (○) BSA in D_2O solutions. The lines are theoretical fits. The sharpening of the interaction peak as the concentration increases is due to the building up of Coulombic repulsions between charged BSA molecules. (b) Corresponding $S(Q)$ calculated using $z = -8$ for all three concentrations. (From Bendedouch and Chen, *J. Phys. Chem.* **87,** 1473.[7])

tween charged particles in solution. Figure 11 shows the results of such an analysis for 4, 8, and 12% BSA in pure D_2O, which gave a surface charge of $-8e$ and a surface potential of 3.4, 2.6, and 2.2 in units of k_BT, respectively. The agreement between the theory and the experiment is excellent except for one reservation. The scattered intensity $I(Q)$ does not approach zero for large-Q as it should for a uniform rigid particle. We had to subtract about 5% of the peak-height intensity from each spectrum as a flat background. This background intensity is much more than what can be accounted for by the incoherent scattering from the protons in the molecules. This excess background seems to be flat and proportional to the concentration in all cases studied and its origin is at present uncertain. We will however remark that in a comparable study of a micellar system[24] such a background does not exist. This suggests that it may be due to inelastic scattering from large short-wavelength density fluctuations in the protein interior. The origin of the background is probably related to the density variations on the scale of the tertiary structure of the protein[25] as can be inferred from the Q-range in which it is observed. Because of the absence of a "tertiary" structure in the micellar core, the density fluctuations would have a shorter wavelength and would be smaller in amplitude.

In Fig. 12 we present the structure factor $S(Q)$ for the highest BSA concentration (12%) in the extended Q-range, and the corresponding pair-correlation function $g(r)$,

$$g(r) = 1 + \frac{1}{2\pi n} \int_0^\infty [S(Q) - 1]Q^2 \frac{\sin Qr}{Qr} \, dQ \qquad (69)$$

It is noted that $g(r)$ is indeed equal to zero up to about 61 Å (hard core diameter σ), gradually rises above unity, and shows a first neighbor peak centered at about $r_0 = 110$ Å. After that, $f(r)$ becomes flat and equal to unity showing that no significant local order exists beyond the first neighbor shell. It should be remarked here that the position of the first diffraction peak Q_0 of $S(Q)$ is not related to the position of the first-neighbor peak r_0 of $g(r)$ by a Bragg relation $Q_0 = 2\pi r_0$. This relation is valid only when the system of particles displays a perfect crystalline order. In this latter case, $g(r)$ would have more successive peaks than just the first neighbor peak. Take our data as an example, $Q_0 = 0.075$ Å^{-1} and $r_0 = 110$ Å; therefore $2\pi/r_0 = 0.057$ which is smaller than Q_0. In general, $Q_0 \geq 2\pi r_0$ for a homogeneous liquid-like particle distribution.

In the case of a BSA solution at its isoelectric pH, the electrostatic repulsive interaction is expected to be much smaller than the van der Waals attraction because the mean surface charge is zero. An approximate analysis can be made by using the same potential function Eq. (58) with a negative γ. However, in this case the interaction range cannot be specified by the Debye–Hückel length κ^{-1} and is chosen from a physical consideration. It seems reasonable to choose it to be of the order of the van der Waals diameter of a water molecule, i.e., 2.9 Å. With the effective hard core diameter σ taken as before to be 60.7 Å, the screening parameter $k = \kappa\sigma$ is equal to 20.9. $P(Q)$ is still taken to be that for a monomer BSA molecule. The resulting fit is shown in Fig. 13. The solid line is the

[24] D. Bendedouch, S.-H. Chen, and W. C. Koehler, J. Phys. Chem. 87, 2621 (1983).
[25] M. YU. Pavlov and B. A. Fedorov, FEBS Lett. 88, 114 (1978).

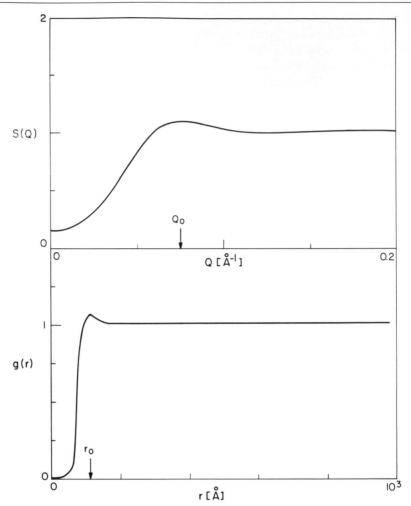

FIG. 12. Structure of a protein solution where the dominant interaction is strongly repulsive. The experimentally extracted structure factor $S(Q)$ and the pair correlation function $g(r)$ are shown for a 12% BSA solution in D_2O (neutral pH). The particles are distributed in a disordered liquid state. For this type of distribution the position of the first diffraction peak Q_0 is generally less than $2\pi/r_0$, where r_0 is the mean position of the first-neighbor shell.

computed curve and is seen to agree very well with the experimental data points. The effective attractive surface potential turned out to be $\gamma = -8$ in units of k_BT. This number is very reasonable and can be shown to correspond to a Hamaker constant $A_H = 15.3 \ k_BT$.[26] The dotted line

[26] D. Bendedouch and S.-H. Chen, *J. Phys. Chem.* **88**, 648 (1984).

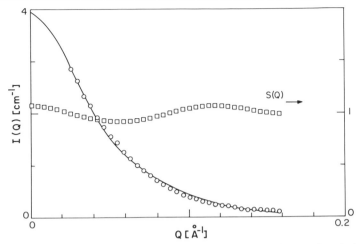

FIG 13 Scattered intensity distribution and structure factor of a protein solution where the net interaction is attractive. This is the case of a 5% BSA in acetate buffer, pH 5.1, 0.03 M (nearly isoelectric conditions) in D_2O.

represents $S(Q)$ which has been extracted from the experimental data. It is interesting to note that in the small-Q region, $S(Q)$ is larger than unity signifying that the interaction is predominately attractive. We remember from the previous figures that $S(Q \sim 0)$ is generally less than unity for a predominantly repulsive interaction. It should be remarked here that the SANS data show no appreciable fraction of dimers in the solution, even under the condition where the attractive van der Waals force dominates. Instead, the effective attractive force acting mutually between the particles serves only to increase the concentration fluctuation of the system. This effect manifests itself as an increase of $S(Q)$ above unity for $Q \to 0$.

Prospect

We have described a practical procedure by which a SANS spectrum from a globular protein solution can be analyzed to give useful information. This procedure can be extended straightforwardly to other macromolecules. We wish to emphasize that working exclusively at the infinite dilution limit is neither profitable nor necessary from the point of view of understanding the structures and functions of protein molecules in physiological environments. It has been explicitly demonstrated that by taking properly into account the intermolecular interactions using standard liquid theory,[9] experiments at moderate concentrations (say from 1 to 10%) and with no substantial amount of salt added can be carried out rapidly and analyzed simply. Among the parameters which can be recovered from data analysis are (1) the molecular weight, (2) the size and shape parameters Rg, a, and b, (3) the solvent excluded volume of the hydrated particle v, (4) the mean csld $\bar{\rho}$, (5) the amount of bound water, (6) the number of exchangeable protons, (7) the surface potential and charge of the molecule, and, finally, (8) the nature of the interaction and the correlation.

This procedure is applicable to soluble globular proteins with fairly homogeneous interiors. In the case of composite particles such as serum lipoproteins, the selective deuteration and contrast variation techniques can reveal the structure of the particles with a resolution of a few Å.

By being able to analyze the scattering data at high concentration, a wide range of possibilities is open for studying ordered structures in liquid-crystalline-like phases. As far as simple globular proteins are concerned, a study of similar scope as that of BSA (described in this chapter) can be achieved with either the BNL or the NCSASR spectrometer in less than 1 weeks time. Therefore the SANS technique can be now considered as another part of the repertoire of standard biochemical techniques such as SAXS or light scattering. The procedures we describe in this chapter do not demand an excessive expertise with the technique, and thus should become very popular among researchers in biochemistry in the years to come.

Acknowledgments

We are grateful to Dr. Benno Schoenborn for providing SANS spectrometer time, to Dr. Dieter Schneider for technical assistance, and also to the Biology Department at BNL for hospitality. This research is supported by grants from the Sloan Fund for Basic Research at M.I.T. and from the National Science Foundation.

[8] Use of Gel Electrophoresis to Characterize Multimolecular Aggregates

By PHILIP SERWER

Introduction

The functions of macromolecules are sometimes performed while part of a multimolecular complex. Such cellular functions include protein synthesis, DNA replication, transcription, respiration, transport, and purposeful motion.[1] Viruses are multimolecular complexes whose function is to protect a nucleic acid genome and to deliver the genome to the interior of a host cell. Inside the host the nucleic acid directs the synthesis of viral components; these components assemble to form progeny.[2] To understand their mechanisms of assembly and function, attempts are made to isolate and characterize multimolecular complexes. One of the procedures for identifying, isolating, characterizing, and quantitating multimolecular complexes is electrophoresis in gels. Electrophoresis in multisample slab gels can be used to detect and quantitate the particles in hundreds of samples simultaneously subjected to electrophoresis in a single gel bed (see details below). Agarose gels appear to have the best handling characteristics and range of pore sizes for the electrophoresis of multimolecular complexes with radii above 5–10 nm (i.e., almost all viruses).[3] To characterize such complexes, the techniques presented below are designed for determining electrophoretic mobility (μ) in agarose gels as a function of agarose percentage (A) in the gel. From these data are calculated (1) μ in the absence of a gel (μ_0; directly proportional to the average electrical surface charge density[4]), and (2) outer radius (R) of spherical particles (or effective R of nonspherical particles). Although agarose gels are used with the procedures described below, some of these procedures have potential for use (possibly with modification) with other types of gels.

Horizontal Slab Gels

Because agarose gels deform less when used for electrophoresis in a horizontal orientation than they do when used in a vertical orientation,[5,6] horizontal slab gels are used in the procedures described below. To further avoid damage to the less concentrated gels (A < 0.5%) and to avoid entry of sample into gels after loading and prior to electrophoresis, gels are submerged beneath electrophoresis buffer before loading of samples and during electrophoresis. The electrophoresis apparatus used in the procedures described below are modified versions of an apparatus originally designed by F. W. Studier and marketed by Aquebogue Machine and Repair Shop, P. O. Box 205, Aquebogue, New York 11931 (Model

[1] B. Alberts, D. Bray, J. Lewis, M. Raff, K. Roberts, and J. D. Watson, "Molecular Biology of the Cell." Garland, New York, 1983.
[2] S. E. Luria, J. E. Darnell, Jr., D. Baltimore, and A. Campbell, "General Virology." Wiley, New York, 1978.
[3] P. Serwer, *Electrophoresis* **4**, 375 (1983).
[4] D. J. Shaw, "Electrophoresis." Academic Press, London, 1969.
[5] P. Serwer and M. E. Pichler, *J. Virol.* **28**, 917 (1978).
[6] P. Serwer, *Anal. Biochem.* **112**, 351 (1981).

850). These modified apparatus and the other equipment described below (template, water bath for control of temperature) are also available commercially from Aquebogue Machine and Repair Shop. Apparatus related to the Aquebogue Model 850 are sold by several companies. Unless otherwise indicated, all of the equipment described here has been made in the Department of Instrumentation, the University of Texas Health Science Center, San Antonio.

Electrophoresis Apparatus: Pouring an Agarose Slab Gel

To form an agarose gel, agarose is dissolved by boiling in glass-distilled water (we usually use a microwave oven for boiling). The amount of water lost during boiling is determined by weighing before and after boiling; water lost is replaced. The agarose solution (tightly sealed to prevent evaporation of water) is equilibrated at 50–60° and the appropriate amount of buffer at the temperature of the agarose is added (we usually use a 10× dilution from a stock buffer). This mixture is poured (usually from a flask) into an electrophoresis apparatus to form the gel used for electrophoresis. Spontaneous gelation occurs as the agarose cools.[7,8] When Isogel, a galactomannan-containing agarose,[9] is used the agarose solutions are equilibrated at 70–75° before pouring because of the comparatively high gelling temperature of Isogel. Isogel (obtained from the Marine Colloids Division of the FMC Corporation, Rockland, Maine 04841) was developed for isoelectric focusing, but can also be used in the procedures described below.

The electrophoresis apparatus (all Plexiglas) is a rectangular, horizontal gel bed (Fig. 1A) with buffer tanks (Fig. 1B) attached at two opposing sides and walls (Fig. 1C) at the other two sides. To pour a uniform gel in the bed (nonuniform gels are described below), access to the buffer tanks is obstructed with two plates (Fig. 1D) sealed for leaks with agarose. The plates are held in place by (1) grooves (Fig. 1B,1) at either end of the gel bed (the plates can slide in the grooves), and (2) a détente (Fig. 1B,2) placed beneath the plate. After the pouring of the agarose solution and subsequent solidification of the agarose gel (45–60 min later), the détentes are removed and the plates are slid below the level of the gel bed. Either just before or after lowering the plates of Fig. 1D, the buffer tanks are

---

[7] S. Arnott, A. Fulmer, W. E. Scott, I. C. M. Dea, R. Moorhouse, and D. A. Rees, *J. Mol. Biol.* **90,** 269 (1974).

[8] D. A. Rees, E. R. Morris, D. Thom, and J. K. Madden, *in* "The Polysaccharides" (G. O. Aspinall, ed.), Vol. 1, p. 196. Academic Press, New York, 1982.

[9] R. B. Cook and H. J. Witt, U.S. Patent 4,290,911 (1981).

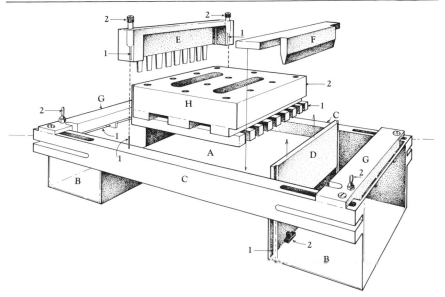

FIG. 1. The electrophoresis apparatus, template for forming multigels, and a sample well-forming comb that fits the template. The electrophoresis apparatus discussed in the text is shown (A–D,G). The apparatus is designed like an Aquebogue Model 850 with the following modifications. (1) To prevent warping-induced changes in the distance between guide pins, the sides (C) and 21 cm long gel bed (A) are made of 7/16 in. Plexiglas (instead of 1/8 in. Plexiglas). (2) To bring the electrodes (G,1) as close as possible to the gel during electrophoresis, the electrodes are mounted on a sliding Plexiglas bar (G). During loading of samples, this Plexiglas bar is moved away from the gel so that it does not interfere with loading. The template for forming multigels (H) and a comb (E) whose fingers are aligned with the Teflon template bars (H,1) also are shown. The comb and template fit snugly together so that the template bars are centered on the comb fingers and are parallel to the walls of the electrophoresis apparatus. Further description of the template is in Serwer.[6] To pour several gels in series, the gel bed can be lengthened, separators (F) can be placed in the gel bed to form gaps between the gels, and a separate comb placed in the gel bed for each gel. Gel beds as long as 72 cm have been constructed. To prevent warping of an electrophoresis apparatus this long, Plexiglas braces are placed between the gel bed and buffer tanks.

filled with buffer, and additional buffer is added until the buffer level is 2–4 mm above the surface of the gel.

For the loading of samples, wells are formed in the slab gel by placing a comb (Fig. 1E) in the gel bed before pouring of the gel. Several different combs are used; the comb illustrated in Fig. 1E is designed for a use described below. The teeth of the comb extend to 1–2 mm above the bottom of the gel. If the teeth touch the gel bed, sample will usually leak out of wells during loading. After covering the gel with buffer (above), the comb is slowly withdrawn from the gel 10–20 min later. Premature removal of the comb results in the formation of a raised "bump" of agarose at the bottom of the sample wells.

The dimensions of comb teeth and the intertooth distance vary with the use of the comb. For 0.7% and more concentrated gels of underivatized agarose, we have used a 30-tooth comb (the teeth are 2.7 × 2.7 mm in cross section) with 1.1 mm between teeth (purchased on custom order from Aquebogue Machine and Repair Shop).[10] However, as **A** decreases, the chances of interwell septum breakage increases unless the distance between teeth is increased. For 0.15–0.50% agarose gels, a comb with 13 teeth (2.5 mm deep × 5.9 mm in cross-section) and 3 mm between teeth can be safely used.

To increase the number of samples that can be simultaneously subjected to electrophoresis in one gel bed, several gels with combs have been simultaneously poured in series in one bed (series gels). To prevent samples from migrating out of one gel into another, and to facilitate gel removal after electrophoresis, separators (Fig. 1F) are placed in the gel bed before gel pouring; the separators produce a gap between gels. These separators are tapered by ~20° at the bottom to facilitate removal and are attached at the top to a flat bar that lies on the sides of the gel bed to prevent falling of the separator during pouring of the gels. Electrophoresis apparatus with gel beds up to 72 cm long have been constructed here for electrophoresis of series gels. As the length of gel beds increases, warping of the gel bed after construction increases. To prevent warping, some parts of the electrophoresis apparatus have been thickened and bracing has been added (see legend to Fig. 1). The number of electrophoresis power supplies, number of recirculating pumps (see below), and the amount of electrophoresis buffer needed are less by a factor of $1/n$ when n series gels are used than they are when n gels in separate apparatus are used. The labor required to prepare the series gels is also considerably less than the labor required to prepare the same number of separate gels. The sharpness of bands in series gels is not detectably different from the sharpness of the bands in gels not in series (the gel in Fig. 4 is one of two series gels).

Sample Preparation. Samples are prepared for electrophoresis by diluting them in a buffer such that after dilution: (1) the conductivity of the sample is comparable to or lower than the conductivity of the electrophoresis buffer (ideally the sample would be in the electrophoresis buffer, but this is often not practical and is not necessary); (2) the density of the sample is greater than the density of the electrophoresis buffer so that the sample will not float out of the sample well during loading in the gel (a

[10] P. Serwer, R. H. Watson, S. J. Hayes, and J. L. Allen, *J. Mol. Biol.* **170**, 447 (1983).

nonionic compound such as a dextran, sucrose, Ficoll, or metrizamide is usually used to control sample density). A tracking dye is also usually (but not necessarily) added to the sample at this stage. If the conductivity of the sample after loading is too high, the sample will initially not move when the voltage gradient is applied. The portion of sample closest to the edge of the gel will achieve a low enough conductivity to move before portions further from the edge do, and will begin to separate from sample further from the edge of the gel. As this process (detectable by observing the behavior of the tracking dye) proceeds, the sample will be spread and banding patterns will become diffused or lost.

Electrophoresis. The voltage gradient is applied to samples through platinum electrodes (Fig. 1G,1) mounted on sliding Plexiglas bars (Fig. 1G). The electrodes are connected by a banana plug (Fig. 1G,2) to the electrophoresis power supply. Sliding the electrodes toward the gel after loading of samples reduces the voltage drop between the electrodes and the gel and, therefore, improves control of voltage gradient.[11] To obtain a uniform voltage gradient in the gel, either the surface on which the electrophoresis apparatus rests is leveled or the electrophoresis apparatus is leveled with self-contained leveling screws. If during electrophoresis a particle migrates "downhill," the voltage gradient will be a decreasing function of distance migrated. In theory, downhill migration should, therefore, have a band-sharpening effect. Conversely, uphill migration should have a band-broadening effect. To test this possibility, two slab gels were poured on separate glass plates, and the two plates were placed in series in the bed of an electrophoresis apparatus such that samples in one gel ran uphill by 20° and samples in the other gel ran downhill by 20°. Using double-stranded DNA as the sample, it was found that downhill migration produced bands sharper than bands produced by uphill migration. If the apparatus is not level, it is, therefore, desirable to run downhill rather than uphill. This procedure is not deliberately used to sharpen bands because of the risk of the sample migrating electrophoretically out of the gel during downhill migration.

To prevent the formation of pH gradients within the slab gel during electrophoresis, buffer is pumped from one buffer tank to the other. Buffer returns to the first buffer tank over the surface of the gel. The pumping rate required to prevent the formation of a pH gradient increases with decreasing buffering capacity, increasing buffer conductivity, and increasing buffer height above the gel. Pumping rates needed to control pH are most easily determined empirically (the pumping rate was 50 ml/min in Figs. 2–4). To prevent possible washing of sample from sample

[11] P. Serwer, *Electrophoresis* **4**, 227 (1983).

wells, circulation is delayed for a time sufficient for migration of particles into the gel (45–60 min delay in Figs. 2–4). The pump used must be reliable for pumping over hundreds of hours. Most of the pumps we have used are single head chemical feed pumps obtained from Cole-Parmer (p. 545 of the 1985 catalogue).

Staining. After electrophoresis, particles in the gel are detected by either staining or autoradiography. Staining of nucleic acids can be performed by soaking the gel in ethidium bromide and observing the fluorescence enhancement of nucleic acid-bound ethidium bromide during illumination with ultraviolet light.[12–14] The detection limit for virus-associated double-stranded DNA during ethidium bromide staining was 2 ng in our previous studies.[14] However, more sensitive detection is probably achievable by adjustment of staining conditions, ultraviolet light wavelength spectrum, and filters used for photography.[12,13]

The intensity with which ethidium bromide stains double-stranded DNA packaged in mature bacteriophages is 5- to 6-fold less than the intensity with which ethidium bromide stains DNA released from the bacteriophages.[14] The ethidium bromide staining characteristics of a viral capsid with packaged DNA can, therefore, be used to help determine the length of DNA in the capsid.[15] Ethidium bromide stains RNA and DNA, but not the DNA-free protein capsids of any bacteriophage used in our studies (T7, T3, T4, P22). Ethidium bromide appears, therefore, not to stain proteins.

To stain proteins in agarose gels without staining nucleic acids, Coomassie Blue can be used. Bands formed by 0.5 μg or more of a protein-containing particle can be detected by staining in 0.05% Coomassie Blue, 10% acetic acid at room temperature for 1.5 hr, followed by diffusion destaining in 10% acetic acid. The 1.5 hr staining time appears to be an upper limit for effective destaining of the gel. Longer staining times result in backgrounds that are either difficult or impossible to completely destain. Staining with either Coomassie Blue (in 10% acetic acid) or ethidium bromide does not change the dimensions of the gel. However, staining with Coomassie Blue in 10% acetic acid does weaken agarose gels. When stained with Coomassie Blue, 0.15–0.20% agarose gels are the most dilute gels that we have successfully stained without breakage.[16] Gels as

[12] C. F. Brunk and L. Simpson, *Anal. Biochem.* **82,** 455 (1977).
[13] E. M. Southern, this series, Vol. 68, p. 220.
[14] P. Serwer and S. J. Hayes, *Electrophoresis* **3,** 76 (1982).
[15] P. Serwer and R. Gope, *J. Virol.* **49,** 293 (1984).
[16] P. Serwer, *Anal. Biochem.* **101,** 154 (1980).

dilute as 0.035% agarose can be stained with ethidium bromide without breakage.[6]

To detect proteins in agarose gels at a sensitivity as low as 20 ng, silver staining has been used.[17] However, this procedure (and also autoradiography procedures that require exposures longer than a few hours) requires drying of the gel. Because of possible alterations of dimensions of the gel during drying, accuracy in μ measurements may be reduced by drying. Therefore, when accuracy in μ is needed (as in the procedures described below), the staining procedures that are known not to alter gel dimensions are favored over silver staining and autoradiography.

Control of Fluctuations in Voltage Gradient and Temperature

Because the observed μ during agarose gel electrophoresis is inversely proportional to buffer viscosity (for example, see Ref. 18) and directly proportional to voltage gradient, accuracy of μ measurements is limited by accuracy in voltage gradient and temperature control. To control the temperature of gels during electrophoresis, the electrophoresis apparatus of Fig. 1 has been immersed in a water bath thermally connected through a heat exchanger to a circulating constant temperature bath. A temperature probe placed on the gel surface regulates the circulating constant temperature bath so that the gel temperature is controlled ±0.3°. Details of the construction and use of these temperature-controlling devices have been previously described[11] and are not repeated here. To achieve full temperature control, recirculation of buffer during electrophoresis is necessary. We monitor both the voltage gradient and temperature continuously on a strip chart recorder.

Control of temperature improves control of voltage gradient. Use of the sliding electrodes (Fig. 1G) further improves voltage gradient control. Fluctuations in voltage gradient are 1–2% during electrophoresis at 1 V/cm.

When measuring μ as a function of **A**, for accuracy, it is desirable that the same residual fluctuations in voltage gradient and temperature are experienced by all of the gels of different **A**. To accomplish this, the gels of different **A** (running gels) have been embedded within a single frame of agarose (frame gel; the composite gel is referred to as a multigel[6,16]). After layering the same sample at the origins of running gels, the multigel is subjected to electrophoresis with buffer recirculation, voltage gradient control, and temperature control, as described above. The pore sizes of

[17] E. W. Willoughby and A. Lambert, *Anal. Biochem.* **130,** 353 (1983).
[18] P. Serwer and J. L. Allen, *Biochemistry* **23,** 922 (1984).

the agarose gels used (0.04–2.0% agarose) are apparently large enough[3,19,20] so that the electrical resistance of the running gels does not vary detectably with the gel concentration.[16] This is also concluded from the absence in running gels of detectable band distortion caused by either the frame gel or the buffer above the running gels (see gel profiles in Refs. 6 and 16 and Fig. 2).

Casting Multigels. To cast a multigel, the frame gel is poured with the following in the gel bed: (1) a template (Fig. 1H) with nine Teflon bars (Fig. 1H,1) that exclude the frame gel from the eventual locations of the running gels, and (2) the sample well-forming comb (made of Plexiglas). After solidification of the frame gel, the template and comb are removed from the gel bed. The trenches formed by the template bars are cleaned, the comb is replaced in the gel, and the agarose solutions for the running gels are poured in the trenches. Procedures for preparing and pouring agarose for the frame and running gels are those described above. To prevent warping of the template bars, the template is mounted with stainless-steel screws to a Plexiglas block (Fig. 1H,2) (for further detail see Ref. 6).

During pouring of the frame gel, the template bars are aligned with the teeth of the sample well-forming comb and are kept parallel to the sides of the gel bed by holding comb and template in a predetermined position as follows. (1) Guide pins (Fig. 1C,1) on both sides of the electrophoresis apparatus fit snugly in matching holes in the comb and hold the comb in place. (2) The template and comb fit together snugly to hold the template in place. If multigels are poured as described above, the sample wells are at the ends of the running gels. If during pouring of running gels a second set of guide pins (to the right of the guide pins in Fig. 1) are used, the sample wells are formed within the running gels (Fig. 2).

Damage to the multigel can occur during withdrawing of either the template or the comb. Damage is caused by either sticking to the gel, development of a partial vacuum, or, in the case of the more dilute running gels ($0.15 \le$ **A**), tearing caused by uneven motion. Teflon bars[6] stick to agarose gels considerably less than the Plexiglas bars originally used.[16] However, after multiple use of Teflon bars, increased sticking to the frame gel has sometimes occurred. This sticking is either reduced or eliminated by soaking the Teflon bars overnight in 10% acetic acid. Therefore, the sticking is presumed to be the result of a film (invisible) of unidentified material on the Teflon bars. Damage to gels from production

[19] P. G. Righetti, B. C. W. Brost, and R. S. Snyder, *J. Biochem. Biophys. Methods* **4,** 347 (1982).
[20] P. Serwer and J. L. Allen, *Electrophoresis* **4,** 273 (1983).

of a partial vacuum during removal of the comb and bars has been reduced by tapering the bars and the comb fingers by 1.0–1.5° so that the vacuum is released during removal.[6] The guide pins help reduce gel damage from uneven motion of the comb during removal from completed multigels.[6] Our more recent combs also have a bolt (Fig. 1E,2) threaded into an extension of the hole in Fig. 1E,1. To remove the comb, the bolts are turned so that they thread downward and gradually lift the comb when in contact with the guide pins.

The running gels with $A \leq 0.075$ can also be damaged by moving of the electrophoresis apparatus for photography after electrophoresis. To prevent this damage, we now have cameras mounted above the electrophoresis apparatus, so that the electrophoresis apparatus is not moved for photography.

Interpretation of Data

In the absence of adsorption to gel fibers, the μ of a particle in a gel is determined by μ_0, the electroosmosis[3,21,22] of the gel (described by the parameter, μ_E[22]; see below) and the extent to which particles are retarded by gel sieving. It has been assumed that the sieving-induced retardation is proportional to the probability that a particle with a random orientation and position intersects a randomly placed and oriented gel fiber.[23] If so and if the gel fiber length is either much greater than or much less than the size of the particle, then $\ln|\mu|$ is in theory a linear function of A.[23,24] Empirically, plots of $\ln|\mu|$ vs A, made with virus-sized (13.3–41.9 nm) spherical particles using the multigel procedure described above, are linear for $A \leq 0.9$. At higher A values, $|\mu|$ decreases with increasing A more rapidly than it would if the plot were linear (i.e., the plot has convex curvature).[16,25,26] In the region of linearity, it is found that[22]

$$\mu = (\mu_0 + \mu_E)e^{-K_R A} \tag{1}$$

with K_R (also known as the retardation coefficient) equal to the empirically derived slope of a $\ln|\mu|$ vs A plot (in the region of linearity). By extrapolating μ to an A of 0 and subtracting μ_E, μ_0 is obtained.

The data for spherical particles fit the theoretically predicted relationship[23]

[21] R. Quast, *J. Chromatogr.* **54**, 405 (1971).

[22] P. Serwer and S. J. Hayes, *Electrophoresis* **3**, 80 (1982).

[23] D. Rodbard and A. Chrambach, *Proc. Natl. Acad. Sci. U.S.A.* **65**, 970 (1970).

[24] A. G. Ogston, *Trans. Faraday Soc.* **54**, 1754 (1958).

[25] P. Serwer and S. J. Hayes, in "Electrophoresis '81" (R. C. Allen and P. Arnaud, eds.), p. 237. De Gruyter, Berlin, 1981.

[26] P. Serwer, S. J. Hayes, and J. L. Allen, *Electrophoresis* **4**, 232 (1983).

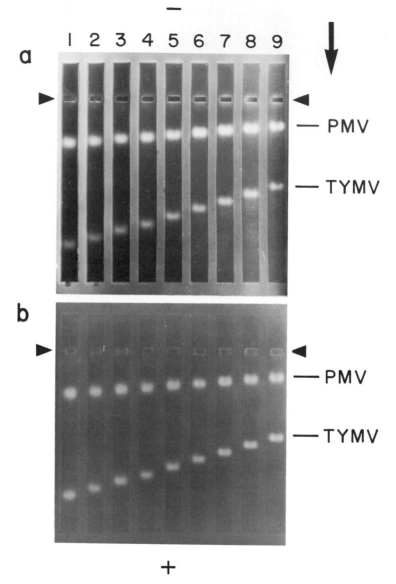

FIG. 2. μ vs **A** for PMV and TYMV. A sample of PMV was received from Drs. Philip Berger and Robert W. Toler, Texas A & M University. A sample of TYMV was received from Dr. R. Virudachalam, Purdue University. A mixture containing 0.5 mg/ml of PMV and 0.2 mg/ml TYMV was made by dilution into 0.05 M sodium phosphate, pH 7.4, 0.001 M MgCl$_2$ (electrophoresis buffer). To 160 μl of this mixture was added 310 μl of 5% sucrose, 400 μg/ml bromophenol blue in electrophoresis buffer. Fifty microliters of this latter mixture was layered at the origin of nine running gels of SeaPlaque agarose (Marine Colloids)

$$K_R{}^{1/n} = c(r + R) \tag{2}$$

with n, c, and r empirically determined constants describing the sieving characteristics of the gel. Equation (2) accurately describes the data for spherical particles with n assumed to have any value between 2 and 3.[25] However, for reasons previously described,[25] it is usually assumed that $n = 2$. This assumption is equivalent to assuming that the gel fibers are cylinders with radius, r, and that the fiber length is much greater than the particle R. By plotting $K_r{}^{1/2}$ vs R and extrapolating to a K_R of 0, values of r and c are obtained.[23,25–27] When measured values of r and c are used in Eq. (2), the K_R of a spherical particle of unknown R can be used to determine $R \pm 8\%$ in the presence of a standard of known R.[25]

In a 6.5% hydroxyethylated agarose (SeaPlaque, purchased from Marine Colloids), plots of μ vs $\mathbf{A}$ were more linear than semilogarithmic plots of the same data ($0.4 \le \mathbf{A} \le 1.6$). When μ vs $\mathbf{A}$ plots were extrapolated to a μ of 0 ($\mathbf{A}_0$), values of $\mathbf{A}_0$ were found to be a more accurate measure of R than is K_R. From $\mathbf{A}_0$, R is calculated $\pm 4\%$ with the following relationship:

$$\frac{R}{R^s} = \left(\frac{A}{\mathbf{A}_0^s}\right)^{-1.71} \tag{3}$$

R^s is the radius of a standard particle (of known R) and $\mathbf{A}_0^s$ is the $\mathbf{A}_0$ of the standard particle determined in the same multigel as the $\mathbf{A}_0$ of the particle of unknown R. There is (to the author's knowledge) no theoretical explanation for Eq. (3). Determination of μ as a function of $\mathbf{A}$ in SeaPlaque agarose is made for turnip yellow mosaic virus (TYMV)[28] and the major component (see below) of panicum mosaic virus (PMV)[29,30] in the multigel of Fig. 2. Because values of $\mathbf{A}$ in successive running gels in Fig. 2 differ by the same amount, the linearity of a μ vs $\mathbf{A}$ plot is apparent to the eye for both TYMV and PMV. Unlike previously presented[6,10,16,22] multigels, the

[27] A. Chrambach and D. Rodbard, *Science* **172**, 440 (1971).
[28] R. Koenig and D.-E. Lesemann, in "Handbook of Plant Virus Infections and Comparative Diagnosis" (E. Kurstak, ed.), p. 33. Elsevier, Amsterdam, 1981.
[29] C. L. Niblett and A. Q. Paulsen, *Phytopathology* **65**, 1157 (1975).
[30] P. H. Berger and R. W. Toler, *Phytopathology* **73**, 185 (1983).

embedded in a frame gel of Seakem ME agarose (Marine Colloids). Electrophoresis was performed at 0.71 V/cm, $25.0 \pm 0.3°$ for 22 hr. The gel was stained for 3 hr with (a) ethidium bromide in electrophoresis buffer with the $MgCl_2$ replaced with 0.001 M EDTA (this substitution increases the staining intensity of both viruses) and (b) Coomassie Blue. The $\mathbf{A}$'s of the running gels were (1) 0.40, (2) 0.55, (3) 0.70, (4) 0.85, (5) 1.00, (6) 1.15, (7) 1.30, (8) 1.45, and (9) 1.60. The origins of electrophoresis are indicated by the arrowheads; the direction of electrophoresis is indicated by the arrow. In (a), photographic dodging was used to increase the uniformity of backgrounds in running gels.

sample wells are not at the ends of the running gels in Fig. 2. The reason for this change is that the μ_0 of PMV is smaller in magnitude than the smallest μ_0 known to result in measured μ's independent of the μ_E of the frame gel.[22] Thus, to avoid possible effects of the frame gel μ_E on μ's, the sample wells were moved away from the frame gel.

Additional Uses of Gel Electrophoresis

Use in Combination with Other Fractionation Procedures. Because hundreds of samples can be analyzed simultaneously on agarose slab gels, agarose gel electrophoresis is a cost- and time-efficient way to assay for particles that form bands. For instance, the contents of sucrose gradients after velocity sedimentation can be assayed by agarose gel electrophoresis. In Fig. 3 is the agarose gel profile of a preparation of PMV after sedimentation through a sucrose gradient. PMV consists of two components that cosediment, but have different μ values (PMV[1] and PMV[2]; in Fig. 2 the sample was too dilute to detect PMV[2]). Using the procedures described above and in Refs. 20, 22, and 26, it has been shown that PMV(1) and PMV(2) differ in μ_0. Although PMV(1) migrates toward the cathode, PMV(1) has a negative μ_0. The cathode-directed migration of PMV(1) is caused by a μ_E greater in magnitude than μ_0. In some preparations of agarose, PMV(1) does not migrate perceptibly (i.e., $\mu_E = -\mu_0$). When first subjecting particles of unknown μ_0 to agarose gel electrophoresis, it is, therefore, advisable to use at least two different agarose preparations of different μ_E.

Also present in the gradient of Fig. 3 is a previously described[29] satellite particle (PMVS) that sediments more slowly than PMV. The satellite also has two components different in μ_0 (PMVS[1] and PMVS[2]). The finding of two electrophoretically distinguishable components of PMV and PMVS has not previously been reported.

Agarose gel electrophoresis can also be used to monitor the contents of fractions obtained from purification procedures other than velocity sedimentation. Agarose gel electrophoresis has been thus used in studies of Refs. 10 and 15.

Monitoring of Antigen–Antibody Binding. Aggregation of bacteriophages by specific antibodies has been detected by agarose gel electrophoresis.[31] Antibody-joined multimers form bands closer to the origin of electrophoresis than the band formed by the monomeric particle. The multimers have, however, the same μ_0 as the monomer.[31] Nonimmune sera do not aggregate bacteriophages[31] (see Fig. 4).

---

[31] P. Serwer and S. J. Hayes, *Electrophoresis* **3**, 315 (1982).

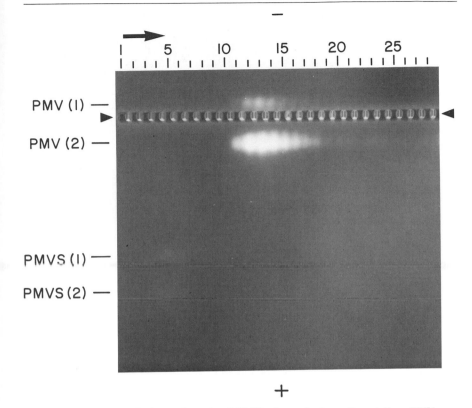

FIG. 3. Agarose gel electrophoresis of PMV after velocity sedimentation. PMV was diluted in the electrophoresis buffer of Fig. 2 to 1 mg/ml, and 200 μl of this mixture was layered over a 4.4-ml linear 5–25% sucrose gradient in the electrophoresis buffer of Fig. 2. The sucrose gradient had been poured over a 0.6-ml layer of metrizamide in electrophoresis buffer (density = 1.3 g/ml), contained in a centrifuge tube for an SW 50.1 rotor. The preparation of PMV then was centrifuged at 28,000 rpm, 18° for 60 min and was fractionated by pipeting from the top. Twenty microliters of fractions 3–28 was layered in the sample wells of a 1.5% agarose (Seakem ME) gel in electrophoresis buffer. Twenty microliters of fractions 1 and 2 also was layered after increasing density by adding a 1/10 dilution of 25% sucrose in electrophoresis buffer. These samples were subjected to electrophoresis at 0.96 V/cm, room temperature (25 ± 3°) for 16 hr, as described in the text. The gel was stained with ethidium bromide as described in the legend to Fig. 2 and photographed. The origins of electrophoresis are indicated by the arrowheads; the direction of sedimentation is indicated by the arrow. The fraction numbers of the sucrose gradient are indicated.

To determine antigenic relationships between two particles of different μ, a mixture of the two particles can be incubated with increasing amounts of specific antiserum and aggregation of the two particles can be separately monitored by agarose gel electrophoresis. Incubating a mix-

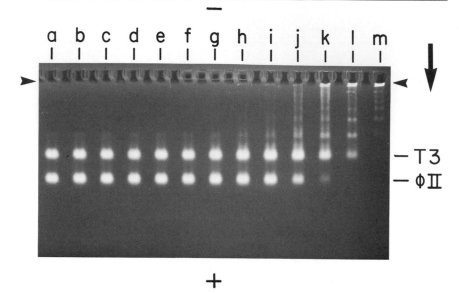

FIG. 4. Serotyping by agarose gel electrophoresis. A mixture containing 64 μg/ml each of the related bacteriophages, ϕII and T3, was made in 0.025 M sodium phosphate, pH 7.4, 0.001 M MgCl$_2$, 100 μg/ml bovine serum albumin (serum buffer). To separate 10-μl portions of the mixture was added 5 μl of rabbit antiserum to a third related bacteriophage, T7. Before addition to the bacteriophages, the antiserum was diluted in serum buffer to obtain the final dilutions indicated below (the first order rate constant of the undiluted serum for killing T7 was 9.1 $\times$ 10^4 min^{-1}). The bacteriophages were incubated with the sera at 30° for 5 hr. Similar incubations were performed using a nonimmune rabbit serum and using serum buffer without any serum. Subsequently, 10 μl of 8% sucrose, 400 μg/ml bromophenol blue in electrophoresis buffer was added to each mixture. Two-thirds of each mixture was subjected to agarose gel electrophoresis at 1.2 V/cm, room temperature for 16 hr, as described in the text. The gel was stained with ethidium bromide after bursting bacteriophages with 10% acetic acid, as previously described.[14] The electrophoresis buffer was 0.05 M sodium phosphate, pH 7.4, 0.001 M MgCl$_2$. (a) Nonimmune serum, 1/15; (b) serum buffer. In (c)–(m) anti-bacteriophage T7 serum was used at the following dilutions: (c) 1/375,000, (d) 1/150,000, (e) 1/75,000, (f) 1/37,500, (g) 1/15,000, (h) 1/7500, (i) 1/3750, (j) 1/1500, (k) 1/750, (l) 1/375, and (m) 1/150. The origins of electrophoresis are indicated by the arrowheads. The bands formed by unaggregated bacteriophages T3 and ϕII are indicated.

ture of the related bacteriophages, T3 and ϕII[10,32] (64 μg/ml each), with increasing amounts of antiserum to a third related bacteriophage, T7,[10,32] results in progressive loss of the bands formed by monomeric T3 and ϕII, and the appearance of bands formed by aggregates (Fig. 4, lanes h–m). Particles of ϕII were aggregated at roughly one-half the serum concentra-

[32] F. W. Studier, *Virology* **95,** 70 (1979).

tion necessary to comparably aggregate particles of T3. This indicates that the ϕII surface is serologically more closely related than the T3 surface to the T7 surface. ϕII is also genetically more closely related than T3 to T7.[32] Further development of agarose gel electrophoresis for the analysis of antibody-aggregated complexes of virus-sized antigens is in progress.

Discussion

With the procedures described here, agarose gel electrophoresis is used to determine the R of a spherical particle, and also the μ_0 and surface antigens of any particle that can be subjected to electrophoresis through agarose gels. Thus far, agarose gel sieving has not been used to determine either the shape of a particle or the size parameters of a nonspherical particle. The shape of virus-sized particles is usually determined by electron microscopy. However, particle shrinkage (probably from dehydration) and flattening that occur during preparation of specimens for electron microscopy can cause errors in particle size 2–7× greater than the errors of the gel sieving procedures described here.[33,34] Thus, at present, electron microscopy is combined with gel sieving to characterize virus-sized particles. An additional advantage of using gel sieving for determining particle size is that particles in crude extracts can be sorted during characterization. Sorting and then size determination of bacteriophage P22 capsids by agarose gel sieving of crude extracts has been previously performed.[15]

The procedures developed here for use with viruses should also be useful for characterizing other multimolecular complexes. Ribosomes form sharp bands during electrophoresis in either agarose or agarose–polyacrylamide gels[3,35,36]; nucleosomes and polynucleosomes form bands during electrophoresis in agarose–polyacrylamide gels.[37]

Acknowledgments

For construction of the apparatus described here, I thank Harry D. Stokes and co-workers of the Department of Instrumentation, the University of Texas Health Science Center at San Antonio. For drawing Fig. 1, I thank Carolyn E. Wittlif. I thank Elena T.

[33] R. C. Williams, *Cold Spring Harbor Symp. Quant. Biol.* **18,** 185 (1953).
[34] P. Serwer, *J. Ultrastruct. Res.* **58,** 235 (1977).
[35] A. E. Dahlberg, W. C. Dingman, and A. C. Peacock, *J. Mol. Biol.* **41,** 139 (1969).
[36] R. Brimacombe, J. Morgan, D. G. Oakley, and R. A. Cox, *Nature (London) New Biol.* **231,** 209 (1971).
[37] R. D. Todd and W. T. Garrard, *J. Biol. Chem.* **252,** 4729 (1977).

Moreno and Shirley J. Hayes for performing the experiments in Figs. 2–4; Philip Berger, Robert W. Toler, and R. Virudachalam for generous amounts of plant viruses; and Donna Scoggins for secretarial assistance. Support was received from the National Institutes of Health (Grants AI-16117 and GM-24365).

[9] Quantitative DNase Footprint Titration: A Method for Studying Protein–DNA Interactions

By MICHAEL BRENOWITZ, DONALD F. SENEAR, MADELINE A. SHEA, and GARY K. ACKERS

Introduction

A central problem in understanding the molecular mechanism of any biological system is that of ascertaining the roles played by the local parts of the system in determining its function. In many cases the system of interest is an assembly of interacting macromolecular components (e.g., proteins and nucleic acids) that is responsible for a specific biological function or its regulation. The interactions between these components and the roles of their local parts (e.g., subunits, binding sites, interfaces) must be understood in structural and functional terms. An extensive array of genetic, biochemical, and structural information may be required in even the simplest system.

Once a qualitative understanding has been developed for the roles of the various molecular entities in generating biological function the system becomes a candidate for quantitative studies. Such studies can provide an understanding of the physical basis of the interactions, including the energetics and driving forces responsible for the structural changes of functional significance. A quantitative characterization in terms of physical–chemical principles affords unequaled precision in the conceptual formulation of a molecular mechanism and also makes it possible to predict the system's behavior.[1]

In many systems the molecular interactions of functional significance take the form of equilibrium binding processes in which one or more of the macromolecular species may be regarded as a "ligand" in binding at multiple sites on a second macromolecular species (the "ligand-acceptor"). Regulatory proteins in gene control systems provide common examples of such "ligands" in their interactions at specific sites of DNA templates. In order to understand and predict the behavior of these sys-

[1] M. A. Shea and G. K. Ackers, *J. Mol. Biol.* **181,** 211 (1985).

tems it is necessary to have methods for determining the energetics of protein–DNA interactions in all combinations of occupied binding sites.

Most techniques for studying binding reactions are incapable of distinguishing between reactions at individual sites in a multisite system. Such techniques (e.g., filterbinding) can only resolve "macroscopic" binding constants, which are composite averages that include the constants for binding to individual sites. We will refer to binding curves from these techniques as "classical binding isotherms."[2] When there is interaction between sites (cooperativity) the fractional saturation at any site is dependent not only on its local binding constant and the ligand concentration. It also depends on the fractional saturation of the neighboring sites with which it is interacting. One consequence of such interaction is that the ligand concentration at half-saturation does not provide a valid determination of the binding constant. The complexity of this situation, which is common among gene regulatory systems, poses problems that are unresolvable, even in principle, by classical binding techniques.

By contrast, a technique that can resolve individual-site binding isotherms that separately represent fractional saturation of each site provides much more experimental information. It can be used to determine not only the intrinsic, or local-site, binding constants but also the thermodynamic constants for cooperative interactions between the individual sites of a multisite system. Even in a noncooperative system, if the local sites have different binding constants (site-heterogeneity) it is not possible to resolve their differences uniquely by classical binding techniques. However, the intrinsic binding constants can be resolved using individual-site binding methods. The quantitative footprint titration technique to be discussed in this chapter has this capability. Figure 1 shows a footprint titration, and the corresponding resolved isotherms, for the three-site system of the bacteriophage lambda left operator. These results and the methods for obtaining them will be discussed in detail in later sections of this chapter.

In a footprint titration experiment an end-labeled double-stranded DNA fragment containing specific protein binding sites is incubated with the protein "ligand" at a series of known concentrations. After equilibration, each mixture contains a distribution of filled and vacant binding sites. The equilibrium mixture is exposed to DNase I (deoxyribonuclease I, EC 3.1.4.5), an endonuclease that will introduce single-stranded nicks on DNA unprotected by bound protein. Under experimental conditions such that the DNase nicks only a fraction of the fragments, the extent of nicking at a particular base pair is a measure of the number of DNA

[2] G. K. Ackers, M. A. Shea, and F. N. Smith, *J. Mol. Biol.* **170,** 223 (1983).

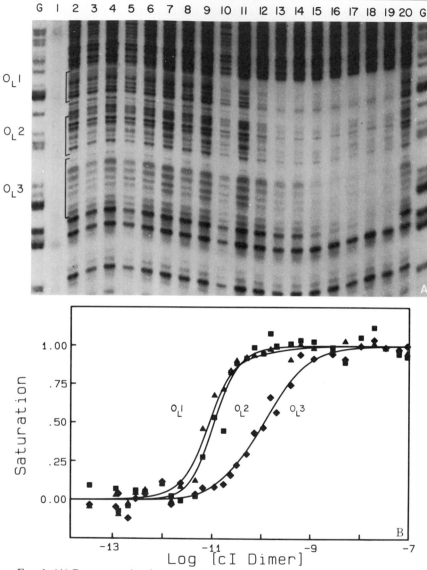

FIG. 1. (A) Representative footprint titration of repressor binding to wild-type O_L. The outside lanes (labeled G) show the guanine-specific reaction of Maxam and Gilbert[17] used to identify the three binding sites (outlined in brackets). Lane 1 is a control lane showing the degree of labeled strand nicking in the absence of exposure to DNase. Repressor dimer concentrations range from 0 (Lane 2) through 61 nM (Lane 20). (B) Resolved individual-site isotherms from footprint titration experiments. Extent of saturation for each of the three sites O_L1, O_L2, and O_L3 was determined (see text) from the data shown in (A) and data from a companion experiment. Triangles, O_L1; squares, O_L2; diamonds, O_L3. The data shown

fragments protected from nicking at that base pair by the bound protein molecules. The DNase cleavage products (shortened, labeled DNA fragments) are separated on DNA-sequencing gels and autoradiographed. Bands corresponding to fragments that terminate in binding sites are diminished in intensity because the bound protein will prevent DNase cleavage at those sites. The intensity of these bands can, in turn, be related to the fractional saturation of binding sites by the protein at each concentration to produce the individual-site binding isotherms. From these isotherms the equilibrium constants for binding and cooperativity can be resolved.

The theory of individual-site isotherms has recently been developed in detail.[2] As necessary, we will summarize the various aspects of the theory that are pertinent to the analysis of experimental results obtained by the footprint titration method. For a cooperatively interacting system the "pairwise cooperative interaction energy" between sites is defined by the fact that the total Gibbs free energy, ΔG_t for saturating all sites is different from the sum of the intrinsic free energies ΔG_1, ΔG_2, ..., ΔG_n. (The intrinsic energy is the energy of binding at one site in the absence of binding to any other sites.) The pair-wise cooperative interaction energy is defined in a two-site system by $\Delta G_{12} = \Delta G_t - (\Delta G_1 + \Delta G_2)$. The term ΔG_{12} represents the additional energy required to load either of the sites given that the other site has already been loaded. For each of these interactions the relationship between the free energy ΔG and the corresponding equilibrium constant K is given by the standard relationship $\Delta G = -RT \ln K$.

The determination of intrinsic binding free energies and the corresponding cooperative interaction terms may contribute to understanding DNA–protein interactions at two levels. (1) It can provide a basis for understanding the molecular origins of major effects in gene regulation. For example, in bacteriophage lambda, footprint titrations revealed the presence and magnitude of cooperative interactions between operator-bound cI repressor molecules. This cooperativity was found to dominate the regulation of the lysogenic-to-lytic switch.[1,3,4] (2) It can be used to probe the energetics of protein–DNA interactions at a molecular level.

[3] G. Ackers, A. Johnson, and M. Shea, *Proc. Natl. Acad. Sci. U.S.A.* **76,** 5061 (1982).
[4] M. Shea and G. Ackers, *in* "Mobility and Recognition in Cell Biology" (H. Sund and C. Veeger, eds.). De Gruyter, Berlin, 1983.

were fit to Eqs. (6) in the text to produce the solid curves through the points. The resolved free energies are listed in Table VII. Experimental conditions were 20°, pH 7.00, 0.20 M KCl (assay buffer).

The detailed molecular origins of site recognition and cooperativity may be understood through studies of both protein and DNA mutants (cf. Refs. 5 and 6 for examples from the lambda right and left operator systems). The availability of mutants through recent advances in DNA synthesis and site-specific mutagenesis facilitates site-specific modification studies of both the DNA templates and the protein ligand. The footprint titration technique is ideally suited for these studies because of its applicability over a wide range of precisely controlled experimental conditions of pH, salt concentration, and temperature. This should allow the determination of the enthalpies and entropies as well as the intrinsic binding and interaction free energies for multisite systems.

"Footprinting" (also called DNase protection mapping) was originally developed as a method for detection and localization of protein binding sites on DNA.[7] The method was subsequently extended for quantitative applications in studies of bacteriophage lambda cI repressor. These developments included (1) a titration method for detection of cooperativity from half-saturation points of the individual sites in wild-type and mutant operators[8] and (2) the development of analytical methods for determining values of the free energies of intrinsic binding and cooperativity.[3]

A diverse array of gene regulatory systems are well suited to study by footprint titration analysis. Examples of the variety of both prokaryotic and eukaryotic systems that have been studied include the transcription factor IIIA from *Xenopus laevis* oocytes,[9] the gal operon,[10] the lac repressor,[11,12] the heat shock transcription factor from *Drosophila*,[13] and the cellular transcription factor Sp1 from HELA cells.[14] The technique has also been used to study DNA-binding drugs.[15,16]

In this article we describe procedures for carrying out footprint titrations and review recent studies from our laboratory that place the method

---

[5] B. Meyer, R. Maurer, and M. Ptashe, *J. Mol. Biol.* **139**, 163 (1980).

[6] M. Hecht, H. Nelson, and R. Sauer, *Proc. Natl. Acad. Sci. U.S.A.* **80**, 2676 (1983).

[6a] C. M. H. Nelson and R. T. Sauer, *Cell* **42**, 549 (1985).

[7] D. Galas and A. Schmitz, *Nucleic Acids Res.* **5**, 3157 (1978).

[8] A. Johnson, B. Meyer, and M. Ptashne, *Proc. Natl. Acad. Sci. U.S.A.* **76**, 5061 (1979).

[9] D. R. Smith, I. J. Jackson, and D. D. Brown, *Cell* **37**, 645 (1984).

[10] M. H. Irani, Orosz, L., and S., Adhya, *Cell* **32**, 783.

[11] A. Schmitz and D. Galas, *Nucleic Acids Res.* **6**, 111 (1979).

[12] M. Becker and J. Wang, *Nature (London)* **309**, 682 (1984).

[13] J. Topol, D. M. Ruden, and C. S. Parker, *Cell* **42**, 527 (1985).

[14] D. Gidoni, J. T. Kadonaga, H. Barrera-Saldana, K. Takahasi, P. Chambon, and R. Tjian, *Science* **230**, 511 (1985).

[15] M. Van Dyke, R. Hertzberg, and P. Dervan, *Proc. Natl. Acad. Sci. U.S.A.* **79**, 5470 (1982).

[16] M. Lane, J. Dabrowiak, and J. Vournakis, *Proc. Natl. Acad. Sci. U.S.A.* **80**, 3260 (1983).

on a sound, thermodynamically valid foundation. We have analyzed the effects of reaction conditions and DNase exposure and the relationship between radioactivity and film density of the autoradiograms. We will show that the equilibrium population of DNA fragments is exposed to DNase I under conditions that do not alter the distribution of vacant and occupied sites.

The system of interactions of the bacteriophage lambda cI repressor with its operator sites has served as a model system for our methodological studies. The lambda operators O_R and O_L are three-site systems to which cI repressor binds cooperatively.[3,8,16a] We shall also discuss some of the considerations necessary to apply this technique to the study of other DNA-binding proteins.

Experimental Design and Protocol

Equipment Required

Much of the equipment necessary to conduct footprint titrations is similar to that required for DNA sequence analysis. The tools necessary to prepare and isolate DNA fragments and to conduct electrophoresis and autoradiography are well described by others.[17,18] However, there are two additional requirements specific to quantitative analysis of footprint titrations. (1) For digitization of the optical densities of an autoradiogram, it is necessary to use an optical film scanner capable of two-dimensional scanning. This instrument must have sufficient spatial resolution to allow the operator to distinguish individual bands within a lane. (2) The second specific requirement is for a computer capable of processing the digitized record of an autoradiogram. It must be able to communicate with the film scanner, whether by direct-link, magnetic tape, or disc, and have a graphics screen capable of displaying the autoradiogram image with either multiple gray levels or false-color contrast. The computer should also have sufficient memory to store 5–10% of an entire film scan to enable random and rapid access.

Experimental Methods

In this section, the procedures for the preparation of DNA fragments and binding protein, suitable for analysis by a footprint titration are pre-

[16a] A. D. Johnson, A. R. Poteete, G. Lauer, R. T. Sauer, G. K. Ackers, and M. Ptashne, *Nature (London)* **294,** 217 (1981).

[17] A. Maxam and W. Gilbert, this series, Vol. 65, p. 499.

[18] T. Maniatas, E. F. Fritsch, and J. Sambrook, "Molecular Cloning A-Laboratory Manual." Cold Spring Harbor Laboratory, Cold Spring Harbor, New York, 1982.

sented. A detailed protocol that outlines the procedures for the equilibration of protein and DNA, exposure to DNase, electrophoresis of the reaction products, and autoradiography is also presented. Then the procedures to convert the film densities on an autoradiogram to a binding isotherm are described.

Preparation of Materials. DNA preparation. The details of the procedures for the preparation, purification, and radiolabeling of the DNA restriction fragments necessary to conduct footprint titrations can be found in Maniatas *et al.*[18] An outline illustrating these procedures using *c*I lambda repressor binding at O_R and O_L as model systems is presented below.

Regions of the lambda genome containing the left and right operators have been cloned into plasmids. Strains of *Escherichia coli* harboring the appropriate plasmid are grown and amplified as described.[18] The plasmids are purified by the method of Birnboim and Doly[19] followed by two $CsCl_2$ gradient centrifugations.[18] The plasmid preparations are electrophoresed on 1% agarose gels in TBE buffer[19a] to assay for chromosomal DNA or RNA contamination. To check for contamination by proteins and RNA, the absorbance ratio A_{260}/A_{280} is measured. A ratio of 1.8 is a criterion for a suitable plasmid preparation. DNA concentrations are determined from the absorbance at 260 nm with 1.0 optical density unit corresponding to approximately 50 $\mu g/ml$.[18] Purified plasmids are stored in TE buffer[19b] at $-70°$. Plasmid pOR1 containing site O_R1 of the lambda right operator genome is described by Johnson[20] and was a gift from R. Sauer. Plasmid pKC30[21] contains the entire left operator of lambda and was a gift from R. McMacken.

Binding studies are conducted with fragments of the plasmids containing the appropriate binding site; they are excised from the plasmid by restriction endonucleases. (The use of restriction endonucleases is described in detail by Maniatis *et al.*[18]) For example, plasmid pOR1 is digested with *Eco*RI and the products electrophoresed on a 1% preparative agarose gel. The 570 base pair fragment containing the O_R1 site (shown in Fig. 2) is electroeluted from the gel[18] and purified by chromatography on a NACS Prepac column (Bethesda Research Laboratory) as described by the manufacturer. The concentration of the fragment is determined spectroscopically and the restriction fragment stored in TE buffer at $-70°$.

[19] H. Birnboim and J. Doly, *Nucleic Acids Res.* **7**, 1513 (1979).
[19a] TBE buffer: 0.089 M Tris–borate, 1 mM EDTA, pH 8.3.
[19b] TE buffer: 10 mM Tris–HCl, 1 mM EDTA, pH 8.0.
[20] A. Johnson, Dissertation, Harvard University, Cambridge, MA, 1980.
[21] H. Shimatake and M. Rosenberg, *Nature (London)* **292**, 126 (1981).

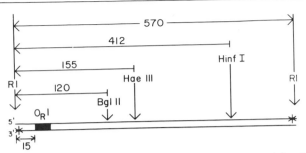

FIG. 2. Schematic representation of the 570 base pair O_R1-containing DNA fragment which was prepared by *Eco*RI digestion of plasmid pOR1. The asterisk indicates 3' recessed end suitable for incorporation of 4 ^{32}P-labeled nucleotides. Fragments labeled at only one end are prepared by digestion with restriction endonucleases as indicated. The length of the resulting fragments is indicated by an arrow.

Restriction fragments that will be used for many experiments are routinely prepared in microgram quantities. Digestion of 400 μg of plasmid pOR1 yields approximately 20 μg of the 570 base pair fragment.

The 570 base pair fragment generated by the *Eco*RI digest has a 3' recessed end 15 base pairs from the O_R1 site (Fig. 2). We label the DNA with α-^{32}P-deoxyribonucleotides (Amersham, 3000 Ci/mmol) using the large fragment of DNA polymerase (Klenow fragment, New England Biolabs) by the method described by Maniatas *et al.*[18] for preparing labeled DNA for the Maxam–Gilbert sequencing reactions. We substitute chromatography on NACS Prepac columns for the phenol extraction following termination of the labeling reaction. This labeling procedure is preferred because of the high incorporation of ^{32}P that can be achieved by completely filling in the restriction site with labeled nucleotides. Incorporation of 2×10^7 cpm/pmol 570 base pair fragment (80% of the theoretical maximum) has been achieved by filling in all four of the single-stranded positions available in the *Eco*RI cut. The addition of a high concentration of unlabeled nucleotides before termination of the labeling reaction (a cold-chase) ensures that all fragments are the same length, even if the efficiency of incorporation is low.

For the unambiguous analysis of the fragments generated by exposure to DNase the DNA fragments must be labeled at only one end.[17] The 570 base pair fragment (which is labeled at both ends) is cut with another enzyme, as shown in Fig. 2. The resulting end-labeled fragment containing the O_R1 site is purified by electrophoresis as described above. The labeled DNA is resuspended in TE buffer and stored in aliquots at $+4°$.

The fragment containing the O_L region of the lambda was prepared in an analogous manner.[21a]

For some DNA sequences of interest, it is possible to digest the fragment with two restriction endonucleases, only one of which produces a 3' recessed end. After purification, this results in a fragment that will incorporate ^{32}P-labeled deoxyribonucleotides at only one end during the labeling reaction. This avoids the necessity of a second electrophoretic isolation following labeling of the fragments.

Preparation of cI repressor. The cI repressor used in the studies discussed in this article was purified essentially as described by Johnson *et al.*[22] with modifications suggested by K. Hehir, H. Nelson, and R. Sauer (personal communication).[22a] Repressor prepared by this method is greater than 95% pure as estimated by electrophoresis on sodium dodecyl sulfate–polyacrylamide gels[23] and gives a ratio $A_{280}/A_{260} = 1.71$ (measured in preparation buffer (see Ref. 22a) plus 200 mM KCl). The yield of repressor is about 120 mg from 45 g of packed cells.

From stoichiometry experiments of the type described from Johnson *et al.*[22] and Sauer[24] we estimate that the cI repressor is 46% active. This is

---

[21a] The restriction fragment containing the left operator of the lambda genome is prepared by digesting plasmid pKC30 with *Cla*I and *Bgl*II. The 660 base pair fragment which results from this digestion is isolated on a 1.5% agarose gel and purified as described above. Following labeling the fragment is cleaved with *Hae*III and the 244 base pair fragment containing O_L purified on a 1.5% agarose gel.

[22] A. Johnson, C. Pabo, and R. Sauer, this series, Vol. 65, p. 839.

[22a] The lambda cI Repressor protein was isolated from *Escherichia coli* strain W3110I^q pEA300WT [E. Amnan, J. Brosius, and M. Ptashne, *Gene* **25**, 167 (1983); a gift from R. Sauer]. Eight liters of cells was grown to late-log phase ($A_{590} = 1.0$) and induced with 60 μg/ml isopropyl β-D-thiogalactopyranoside (Sigma Chemical Co.) for 4 hr. The supernatant from centrifugation of the crude lysate was made 0.2% in polyethyleneimine (Sigma Chemical Co.) by dropwise addition of a 10% w/v aqueous stock, and was stirred continuously at 4° for 5 min. Precipitate was removed by centrifugation at 13,200 g for 15 min. After precipitation with ammonium sulfate, the pellet was resuspended in 50 ml preparation buffer [10 mM Tris–HCl, pH 8.0, 2 mM CaCl$_2$, 0.1 mM EDTA, 0.1 mM dithiothreitol, and 5% (v/v) glycerol; identical to standard buffer of Johnson *et al.*[22]]. This was dialyzed versus 3 changes of preparation buffer plus 100 mM KCl. After centrifugation to remove a small amount of precipitate formed during dialysis, the dialyzed material was loaded onto a 2.5 by 12 cm column of Affigel Blue (Bio-Rad) equilibrated with preparation buffer plus 100 mM KCl at 4°. The column was washed with 5 volumes of preparation buffer plus 100 mM KCl before elution with a 1.5 liter linear gradient from preparation buffer plus 100 mM KCl to preparation buffer plus 750 mM KCl. Repressor eluted in the only major protein peak following the wash, at about 600 mM KCl. Pooled repressor containing fractions were concentrated to 7–8 mg/ml protein and loaded directly onto a 2.8 by 25 cm hydroxyapatite column (Bio-Rad). Chromatography was as described.[22]

[23] U. K. Laemmli, *Nature (London)* **227**, 680 (1970).

[24] R. Sauer, Dissertation, Harvard University, Cambridge, MA, 1979.

within the expected range.[22] All the protein concentrations have been corrected for this activity where the total protein concentration in monomer units is corrected prior to calculation of the dimer concentration.

A Protocol for Performing a Footprint Titration. This section gives a protocol for conducting a quantitative footprint titration. In subsequent sections, the justification for this particular set of experimental procedures will be presented.

Equilibration of cI Repressor and DNA. A series of 1.5-ml siliconized microfuge tubes containing 180 μl of a stock solution of assay buffer[24a] and the labeled DNA fragment containing the protein binding site(s) is prepared. Labeled DNA is added to the stock solution to yield an activity of 10,000–20,000 cpm per reaction mixture.

The repressor is serially diluted from its concentrated stock solution into assay buffer. The appropriate volumes of cI protein and buffer are added to each assay mixture to bring the final volume to 200 μl. One sample is prepared to serve as a control which is not exposed to DNase to assess background nicking of the DNA. The tubes are incubated in a regulated water bath for 30–40 min to ensure equilibration of the binding reaction.

Exposure to DNase. DNase[24b] is diluted to the appropriate concentration in assay buffer without bovine serum albumin or calf thymus DNA. To sample cI protein occupancy of binding sites, 5 μl DNase solution is added to each sample. The tubes are vortexed gently and replaced in the 20° bath. At the end of the reaction time (routinely 1 min or 30 sec), the DNase reaction is quenched with 700 μl of a solution of 92% ethanol, 0.57 M ammonium acetate, and 50 μg/ml tRNA (Sigma Chemical Co., type XX) prechilled in an ethanol–dry ice bath. The quenched samples are immediately placed in the ethanol–dry ice bath and the DNA precipitated for at least 15 min.

Preparation of the Samples for Electrophoresis. The DNA is pelleted in a microfuge. The supernatant is discarded, and the pellet is washed twice with 1 ml of 80% ethanol. The pellet is then dried in a Speed-Vac Concentrator (Savant). The DNA pellets are dissolved in 5 μl of a solution of 80% deionized formamide, TBE buffer, 0.2% bromophenol blue, and 0.2% xylene cyanole. The samples are electrophoresed immediately or

---

[24a] Assay buffer: 10 mM Bis-Tris, 200 mM KCl, 2.5 mM MgCl$_2$, 1 mM CaCl$_2$, 0.1 mM EDTA, 2 μg/ml sonicated calf-thymus DNA (PL-Biochemicals), and 100 μg/ml bovine serum albumin (Bethesda Research Labs ultrapure) titrated to pH 7.0 with HCl at 20°.

[24b] Deoxyribonuclease I; (EC 3.1.4.5), Worthington #2139, stored at a concentration of 2 mg/ml in 50 mM Tris–HCl, pH 7.2, 10 mM MgCl$_2$, 1 mM CaCl$_2$, 1 mM dithiothreitol, 50% glycerol at −70°.

stored at −70°. For electrophoresis, the samples are heated to 90° for 10 min, cooled rapidly in an ice bath, and loaded immediately onto a preelectrophoresed gel using a Hamilton syringe equipped with a 29-gauge blunt tip needle.

Gels and Autoradiography. Sequencing gels (38 cm × 33 cm × 0.4 mm) are cast and electrophoresis performed as described by Maxam and Gilbert.[17] A sample comb with 6-mm-wide lanes and 6-mm spacing is used. The larger plate is siliconized with 0.5% Surfa-sil (Pierce Chemical Co.) dissolved in chloroform and then cleaned with 95% ethanol. We routinely use 8% gels for the O_R1-containing fragments where the site is 15 base pairs from the labeled end and 6% gels for the O_L containing fragment where the site is 65 base pairs from the labeled end. For 6% gels, the plate which is not siliconized is coated with γ-aminopropyltriethoxysilane (Pierce Chemical Co.) to improve adherence of the gel.[25] Following electrophoresis, the gels are dried. Autoradiography is conducted at −70° with Kodak X-Omat film, preflashed to optical density of 0.15 above film base.[26] A single Dupont Cronex Lightening Plus Intensifying screen is used. With approximately 10,000 cpm loaded per lane, exposures of 1–5 days are common. We routinely make 2 or 3 exposures of each gel over a 2- to 3-fold range of optical density. The film is developed in Kodak GBX developer at 20° for 5 min. Controlled temperature and the replenishment of developer as described by the manufacturer are necessary for consistent and uniform development.

Quantitation of an Autoradiogram to Resolve Isotherms

In this section we will outline the procedures necessary for resolving an isotherm from a developed autoradiogram. These include digitization of the autoradiogram, correction of the measured optical density values for film background, and correction of integrated optical densities for variation in the amount of DNA loaded in each lane. The final procedure is the conversion of the protection of a binding site (i.e., the density of the bands in that region) to fractional saturation.

Digitization of Optical Density. The optical density of the film is read in two dimensions by an Optronix P-1700 rotating drum scanner and stored on magnetic tape. This film scanner is a single beam instrument that can discriminate 256 levels of optical density. Calibration of the scanner shows that the linear optical density range is 0 to 1.35 optical

[25] H. Garoff and W. Ansorge, *Anal. Biochem.* **115,** 450 (1981).
[26] R. Lasky and D. A. Mills, *FEBS Lett.* **82,** 314 (1977).

density units. The film base has an optical density of 0.2 and preflashing adds an additional 0.15. This leaves a useful range of 1.0 optical density units for the titration. The scanner has a variable picture element (pixel) size of 400 to 12.5 μm^2; positional accuracy is 2 μm. Scanning at a resolution of 200 μm^2 produces an 1150 × 1150 data point array for the 23 by 23 cm area that can be scanned.

The array of optical densities corresponding to 200 μm pixels on the gel is processed interactively using an HP1000 computing system. Analysis also requires a tape drive, graphics console, and printer. Our current generation of programs makes use of a Metheus 500 color graphics monitor. This hardware allows the display of either the full 23 × 23 cm scan or enlarged portions of a smaller region. The programs make conservative use of disc and memory space, both for program instructions and the storage of the scan data. The program allows random access to those data. The options within the analysis program coordinate the procedures necessary to integrate the optical density of any region of an autoradiogram and correct for the background density of the film. Analysis of autoradiograms can be accomplished in an efficient and timely manner (about 1 hr). These programs have been designed to be compatible with personal computers with graphics capability.

Integration of Optical Density. Under appropriate reaction conditions (as described above), bands or blocks (which are groups of contiguous bands) in each lane indicate the degree of protection afforded by a particular concentration of protein incubated with the DNA fragments before digestion. Thus, the integrated optical density of bands within binding sites (see Fig. 1A) may be used to generate an individual-site isotherm for each site. To measure the total optical density of a particular film region, the analysis program allows a quadrilateral to be drawn that defines the area of interest. (The variable shape of the contour accounts for nonuniformity of gel migration.) The program then calculates the integrated optical density within that contour as the sum of each individual pixel optical density contained therein. To establish a uniform criterion and ensure that the entire band or block is included in the analysis, the contour extends to the middle of the space between lanes (see Fig. 3).

Estimation of the Local Film Background Intensity. The raw integrated densities must first be corrected for the local background density of the film. To estimate the local background density in a lane, quadrilateral contours are drawn in the center of the spacer regions on either side of the lane. The width of these quadrilaterals is 800 μm (4 picture elements wide) and the length corresponds to the length of the block or band being integrated. In determining the background for individual bands in this man-

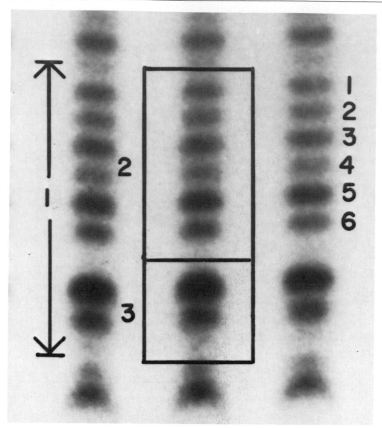

FIG. 3. Portion of the O_R1 footprint titration autoradiogram analyzed in Fig. 9 showing the bands and blocks used for analysis. Note that the contours are drawn extending to the middle of the space between adjacent lanes.

ner, it is frequently necessary to increase the length (in the direction of electrophoresis) of the quadrilateral over the length of the band, so that a statistically significant number of picture elements is included.

To define the background optical density, the distribution of individual picture element optical densities within each quadrilateral is plotted. The most probable value, i.e., the peak of the resulting histogram, is taken to be the correct background value.[27] The values on either side of a lane are averaged to give the background value for that lane. This value is subtracted from each individual pixel optical density.

[27] J. Bossinger, M. J. Miller, K. P. Vo, E. P. Geiduschek, and N. Xyong, *J. Biol. Chem.* **254,** 7986 (1979).

Correction for the Variation of Total DNA per Lane. Standard bands or blocks outside the regions protected by bound protein serve as measures of the amount of DNA loaded onto each lane. The degree of protection is given by the ratio of the integrated density of the "site" block to the integrated density of the standard block. This ratio is then referenced to a lane or lanes with zero repressor (lanes 2 and 20 in Fig. 1A).

A transformation to correct the integrated optical densities for the amount of DNA loaded and to convert protection to degree of saturation is given by Eq. (1):

$$\overline{Y}_{app} = 1 - \left\{\frac{D_{N,Site}}{D_{N,Std}}\right\} \bigg/ \left\{\frac{D_{R,Site}}{D_{R,Std}}\right\} \tag{1}$$

where $D_{N,Site}$ is the integrated optical density of a site minus the background for lane N and $D_{N,Std}$ is the integrated density of the standard minus background for lane N. D_R is integrated optical density of a lane to which no cI protein has been added (lanes 2 and 20 of the autoradiogram shown in Fig. 1A). The $\overline{Y}_{app}$ values for each site and their corresponding repressor concentrations (corrected for activity and dimerization constant) are then used to plot each individual-site isotherm. The resulting isotherms are analyzed by nonlinear least-squares fitting procedures to estimate the free energies of interactions and the 65% confidence interval parameters.[28,29]

The $\overline{Y}_{app}$ values frequently do not span the range, 0 to 1.0. One reason for this is that we do not account for the density due to nicking of the DNA prior to exposure to DNase (as discussed in the section on "Rationale for the Footprint Titration Protocol"). Therefore, in our model-dependent fits we treat the data as a transition curve and include the endpoints as adjustable parameters.

Range and Foundation of the Method

In the preceding section, we have described procedural aspects of the footprint titration method by succinctly presenting the experimental protocol in chronological order. However, many of the considerations that are critical to the selection of a biochemical system for study and to an understanding of our development of the general quantitative methodology were reserved for discussion. In this section, we will describe the motivation and justification for the experimental preparations and procedures required to resolve individual-site isotherm in an order that is roughly parallel to that of the previous section.

[28] M. Johnson, H. Halvorson, and G. Ackers, *Biochemistry* 15, 5363 (1976).
[29] B. Turner, D. Pettigrew, and G. Ackers, this series, Vol. 76, p. 596.

Selection of Protein–DNA System

There are several factors that determine whether a particular protein–DNA binding system is suitable for study by the footprint titration method. There is a minimal level of qualitative characterization that must be achieved before a system becomes a candidate for quantitative analysis. There are also both technical and theoretical limits on the technique's ability to measure binding. Some of these factors are reviewed below.

The most important preliminary information is that which thoroughly defines the state of association of the bound protein and the chemical identity of the binding sites in a DNA fragment. This fundamental definition of the system may pose great challenges. It is especially important to determine whether there are multiple classes of specific sites and whether the sites are nonoverlapping. Binding sites may be mapped using many methods including chemical modification of bases, UV-induced strand breakage, and sequence studies of mutants (see Refs. 20, 30, and 31 for a description of these approaches applied to the lambda O_R and O_L operators). The ratio of specific to nonspecific binding must also be considered.[7]

It is possible to estimate some limitations of the range of binding affinities that can be studied with our current protocol. Since the concentration of labeled DNA is too low to be measured spectroscopically, it is necessary that the concentration of DNA be at least two orders of magnitude less than the dissociation constants for protein so that free protein concentration is justifiably estimated as being equal to that of total protein. However, the DNA level must be high enough to allow exposure of an autoradiogram within a reasonable time. With our labeling procedures we have performed experiments in which the DNA concentration was as low as 0.5 pM. In the simplest case of a monomer protein binding to a single DNA site, calculation of the maximum affinity compatible with this criterion is approximately -15 kcal/mol at 20°. In practice, this limit can be extended in the case of a monomer–oligomer association where the oligomer binds the DNA.

There may be no theoretical lower limit to the magnitude of the intrinsic free energy of the weakest binding that can be determined accurately using the footprint titration technique. It is apparently not necessary that the exposure to DNase be extremely short relative to the half-life of the repressor–operator complex. Estimates of the half-life of the repressor–O_R1 complex, based either on the theory of Winter et al.[32] or on the

[30] Z. Humayan, A. Jeffrey, and M. Ptashne, *J. Mol. Biol.* **112**, 265 (1977).
[31] Z. Humayan, D. Kleid, and M. Ptashne, *Nucleic Acids Res.* **4**, 1595 (1977).
[32] R. Winter, O. Berg, and P. H. von Hippel, *Biochemistry* **20**, 6961 (1981).

measurements of Sauer,[24] indicate that under our experimental conditions, a 1-min period of exposure to DNase is long relative to the half-life
of the repressor–operator complex. Apparently, by minimizing the fraction of DNA molecules that are nicked by the DNase, our experiments
meet the more general criterion that the equilibrium binding of repressor
should not be perturbed by the assay technique. There may be limits set
by availability of materials or by interference due to nonspecific binding in
particular cases. In practice, we have measured free energies as low as −9
kcal/mol with the footprint titration technique (unpublished results).

A systematic error would occur in the correction for the amount of
DNA loaded in each lane if there were nonspecific binding outside the
specific binding sites. There is no evidence of protection outside the specific sites below 3 μM cI repressor monomer total, which is 10-fold higher
than the range of concentrations used in these experiments (unpublished
data).

Rationale for the Experimental Methods

Many experimental variables must be controlled carefully to obtain
reliable, thermodynamically valid footprint titration data. As in any other
quantitative method, the purity of materials, integrity of technical manipulations, choice of equipment, and calibration of instrumentation are all
important. Here we will first outline the criteria that govern acceptability
of protein and DNA samples and, in the subsequent sections, present
justification of the experimental procedures required to conduct a titration
per se.

Criteria for Acceptable Materials. The selection and preparation of a
labeled DNA fragment are preliminary but critical steps. When selecting
the restriction endonuclease sites to utilize for fragment excision, one
must consider the total length of the resulting fragment and the distance
between the edges of the protected region and the fragment ends.

We have conducted footprint titrations using fragments with the label
as close as 15 base pairs and as distant as 200 base pairs from the beginning of the first protein-binding site. The maximum distance between the
first site and the label is determined by the length of the gel apparatus
available to the investigator. The minimum distance is determined by the
inefficiency of DNase to nick near the end of a fragment[7,20] and the requirement that the integrity of the secondary structure of the binding site
be maintained. Although a clear footprint can be obtained with the binding
site 15 base pairs from the labeled end, we have observed a weakening of
the affinity of cI repressor for such fragments (unpublished data). This
observation will be considered in detail in a future article.

Throughout the section describing the rationale for the experimental
protocol, there will be a discussion of the implications of fragment length

measured as an absolute value, and one relative to the number of backbone positions protected by bound protein. We will only indicate the nature of those concerns here; they pertain to the selection, the purification, and the labeling of a fragment.

One requirement is that the protein be equilibrated with a solution containing a single species of site-containing DNA fragments. Another is that the final length of a labeled strand be an indicator of the backbone position nicks; thus all fragments must share a common origin. Incomplete "filling-in" of the 3' end following labeling will result in an altered distribution of DNA fragments after DNase exposure. We are able to use fragments with up to four ^{32}P-labeled nucleotides incorporated because DNase is inefficient at nicking the ends of fragments; thus, it will not create multiple labeled origins from which fragment length may be measured inconsistently.

High specific radioactivity is achieved by incorporating multiple α-^{32}P-deoxyribonucleotides in an individual fragment. Note that the choice of restriction enzyme cuts determines the availability of 4 nucleotide positions. This ensures a sufficient number of counts for rapid imaging of an autoradiogram but permits the DNA concentration in the titration assay mixture to remain as low as 0.5 pM.

Rationale for the Footprint Titration Protocol. We previously described the experimental protocol for a footprint titration. Many elements are crucial to the applicability of this method to quantitatively study cooperative, site-specific binding reactions. Although they were chosen on the basis of exploratory studies employing cI repressor, most are completely general. We present below the results of some of these studies that validate the approach and describe our rationale for many of the procedures.

Equilibration of binding protein and DNA. It is self-evident that the equilibrium between DNA and the binding form of the protein must be established. The association and dissociation rates for many DNA-binding proteins are highly dependent on the experimental conditions of salt, pH, and temperature, as well as the length of the DNA fragment. If no estimates of the rates are available, it may be sufficient to verify that identical results are obtained in experiments where the equilibration time is varied. It is equally important to ascertain that the binding form of the protein is in equilibrium with its subunits or monomer form and any effector ligands that may be included in the study.

Exposure to DNase. This enzyme makes single-stranded cuts in the phosphodiester backbone of double-stranded DNA with low sequence specificity. In order to have confidence in the use of DNase as a quantitative probe, experimental studies were conducted to explore the properties of the DNase nicking reaction, both in the absence and presence of a binding protein. They addressed several of the assumptions that allow us to relate the extent of nicking at a fragment position (as visualized by an autoradiogram) to the extent of site occupancy by a cI repressor dimer (as represented by the resolved isotherm). They will be reviewed here after formulating a simplified description of the nicking reaction in the absence of binding protein.

Characterization of DNase nicking in the absence of binding protein. We consider a DNA fragment of fixed length B backbone positions ($B + 1$ bases total) that receives an exposure to DNase of X, where X is equal to the product of the experimentally controlled variables of DNase specific activity (nicks/position/sec) and elapsed time (sec). The distribution of nicks introduced by DNase into a population of fragments will necessarily depend on the product XB. In all of the experiments described here, an amount of nicking is related to a concentration of DNase reported in ng/ml. Since the specific activity of DNase varies depending on the vendor, storage condition, and age, it must be calibrated.

If we assume that (1) all labeled DNA fragments in a reaction volume are independent targets, (2) they are all of uniform length throughout the reaction, and (3) all nicks at backbone positions within a fragment are introduced independently and with low probability, then the distribution of nicks should be described by Poisson statistics. The appropriate expression for the probability of k nicks in the labeled strand of a fragment is given by $p(k; XB) = e^{-XB}(XB)^k/k!$ Thus, the probability of finding no nicks (i.e., the fraction of intact fragments) is $p(0; XB) = e^{-XB}$ and the probability of finding one or more nicks is $p(k \neq 0; XB) = 1 - e^{-XB}$. The average number of nicks per fragment is given by the product XB. All of the nicked strands migrate in the gel according to their final length (the distance between the label and the closest nick); thus, it is impossible to distinguish between fragments nicked once and those nicked at several positions during exposure to DNase. To measure the distribution of nicks and determine that these expressions are applicable, we conducted experiments in which the concentrations and times of DNase exposure were varied (all other conditions were the same as those described in the section "Experimental Design and Protocol").

In the first series of experiments, the concentration of DNase was varied over a 20-fold range; however, the duration of exposure was held constant at 1 min. The nicked fragments were separated on a sequencing gel as previously described and the amount of intact material was determined. The fractions of original fragments that were intact following exposure to DNase are shown in Fig. 4. As predicted by Poisson statistics, an exponential function describes the relationship between the amount of intact material and DNase concentration. Calculations of the distribution of intact (uncut), singly nicked and multiply nicked fragments are shown in Table 1 for each of these DNase concentrations.

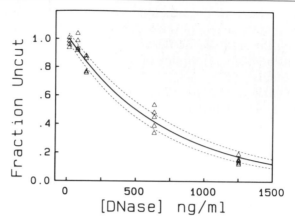

FIG. 4. Fraction of DNA fragments still intact after exposure to DNase. Aliquots of a stock solution of the 155 base pair O_R1 fragment were exposed to various concentrations of DNase for 1 min, in the absence of cI repressor, using the standard protocol. The reaction products were separated on an 8% sequencing gel that was cast with N_1N'-bisacryloyl-cystamine (Bio-Rad) instead of the N_1N'-methylene bisacrylamide cross-linker. The bands corresponding to uncut DNA were located by autoradiography of the wet gel. These bands were excised, dissolved in 150 μl 2-mercaptoethanol (J. Hansen, *Anal. Biochem.* **76**, 37, 1977) and counted in 20 ml of scintillation fluid. The solid curve is the best fit of the exponential function, $y = Ae^{kx}$, to the data. The dashed lines are the 65% confidence limits.

TABLE I

CALCULATED DISTRIBUTION OF DNase NICKS AMONG FRAGMENTS[a]

Uncut[b] (%)	Singly nicked[c] (%)	Multiply nicked[c] (%)	Average number of nicks per fragment[d]
92	7.7	0.3	0.08
85	14	1	0.16
42	37	21	0.87
19	32	49	1.66

[a] Before exposure to DNase, the O_R1-containing DNA fragments are 155 base pairs long.

[b] Experimentally measured percentage of fragments left intact after exposure to DNase. The O_R1-containing DNA fragments are 155 base pairs long.

[c] Percentages were calculated assuming a Poisson distribution of nicks; they are shown for each condition of DNase exposure in the experiments shown in Fig. 5.

[d] Calculated average number of nicks per labeled strand (i.e., the product XB).

Another prediction of the Poisson expression is that the product of DNase activity and the duration of digestion should uniquely determine the distribution of nicks among the DNA fragments. Experimental comparisons have demonstrated that this is true (data not shown). In one experiment, the 155 base pair DNA fragment was exposed to DNase concentrations which varied by a factor of two. The total exposure was maintained constant by proportional changes in the length of exposure. Changes in the distribution of fragment lengths were monitored by calculating the ratio of the number of fragments in two arbitrarily designated size classes, one containing shorter fragments than the other. The ratios of short to long fragments were 0.562 ($\pm$0.042 SD) and 0.561 ($\pm$0.023 SD) for the same total exposure, but 2-fold different DNase concentrations. At half the exposure, the ratio was 0.430 ($\pm$0.021 SD).

These studies showed that the observed behavior of DNase is indeed consistent with the assumptions listed above, although these assumptions are not entirely accurate as stated. In particular, DNase is sensitive to DNA sequence; thus, it does not nick at every position of an individual fragment with equal probability.[33,34] However, because each fragment has exactly the same sequence and length, the frequency of DNase nicking at a given backbone position is the same for all fragments exposed to DNase under identical conditions (i.e., the exposure could have been formulated in terms of populations of backbone positions, each with a characteristic nicking frequency). DNase is also sensitive to position relative to the ends of a fragment; it will not nick closer than approximately 10 base pairs with high frequency.[7,20] However, this effectively serves to shorten all of the fragments equivalently. These findings are obviously independent of the DNA-binding protein under investigation.

Characteristics of DNase nicking in the presence of binding protein. Additional studies are required to demonstrate that DNase samples (without perturbing) an equilibrium distribution of fragments with bound protein. It is necessary to determine that there is no direct physical interaction between DNase and the binding protein that would alter the concentration of bound or free protein. For cI repressor binding at O_R1 (in the 155 base pair fragment shown in Fig. 2), this was addressed by conducting titrations that were identical in all respects except that the factors determining DNase exposure were varied by a factor of five (i.e., [DNase] $\times$ 1 min and 1/5 [DNase] $\times$ 5 min). The titration conducted with a 5 min DNase exposure was fit to Eq. (2) (see section on "Analysis of an O_L Footprint Titration"); the resolved free energy (with 65% confidence in-

[33] G. Lomanosoff, P. Butler, and A. Klug, *J. Mol. Biol.* **149**, 745 (1981).
[34] R. Dickerson and H. Drew, *J. Mol. Biol.* **149**, 761 (1981).

terval) is $\Delta G_1 = -13.0 \pm 0.4$ kcal/mol. This is within the experimental error of a parallel titration conducted with a 1 min DNase exposure; $\Delta G_1 = -12.7 \pm 0.3$ kcal/mol. These results indicate that DNase is not participating in the processes governing the multiple equilibria between cI dimers and monomers, and cI dimers and operator sites.

Another constraint on the period of DNase exposure is the requirement that the apparent extent of cutting at all levels of binding protein be unambiguously related to the distribution of fragments obtained in the absence of DNA-bound ligand. If each DNA fragment that is cut by DNase is cut only once, each band observed on a gel unambiguously represents a single nicking event and this criterion will be fulfilled. A large fraction of the fragments must be intact to provide for a high ratio between those nicked once and those nicked more than once (see Table I). However, a much smaller fraction (approximately one-third) of the fragments must be intact to provide for an average of one nick per fragment. Both of these distributions have been described as "one-hit" or "single-hit" kinetics; our use of the term is restricted to the former case.

For several reasons, it has been thought generally that it is necessary to adhere rigorously to "one-hit" kinetics to resolve isotherms accurately. One concern is the effect of nicking on the protein–DNA equilibrium; multiple nicks may destroy or alter the population of sampled fragments. By contrast, when the number of single-stranded nicks is small and the fragments are long, the probability that two nicks occur at the same place on opposite strands to induce strand breakage is vanishingly small. Under the most severe conditions of DNase exposure described here, each fragment received an average of less than two nicks distributed over 155 base pairs (see Table I). Therefore, the population of full-length double-stranded fragments available for protein binding would not be reduced significantly by this exposure. Thus, we surmise that a less stringent definition of "one-hit" kinetics suffices for this purpose.

Another concern is that the number of backbone positions available to DNase remain constant throughout the DNase-catalyzed reaction and at all protein ligand concentrations. As described above, this is necessary for the applicability of the Poisson expression to describe the distribution of nicks among fragments of length B. If the protected site(s) were large compared to the length of the fragment, protected fragments would appear shorter and would receive a different average number of nicks. There are two elements of our approach that address this issue. (1) The total number of backbone positions available to DNase is maintained at approximately 3 μM. To achieve this, 2 μg/ml unlabeled nonspecific DNA is included in the reaction mixture. [We have verified by a filterbinding assay (see Table VII) that the presence of the unlabeled DNA does not

TABLE II
REPRESSOR BINDING TO FRAGMENTS OF
DIFFERENT LENGTHS

Fragment length[a] (bp)	Free energy of association[b] (ΔG, kcal/mol)	Standard deviation[c]
120	-13.3 ± 0.2	0.06
155	-12.7 ± 0.3	0.08
412	-12.9 ± 0.2	0.06

[a] The O_R1-containing DNA fragments are all derived from a 570 base pair precursor as shown in Fig. 2.
[b] Gibbs free energy of association (with 65% confidence interval) calculated from the best estimate of K_1, as determined by fitting each isotherm to Eq. (2) ($\Delta G = -RT \ln K_1$).
[c] Standard deviation of the fitted curve.

affect the specific binding of cI repressor at O_R1 under these conditions.] (2) The size of the protected region relative to the total length is used as a qualification for selecting an appropriate DNA fragment. We verified experimentally that the titration of cI repressor binding at O_R1 on fragments 155 base pairs or longer is not affected by the length of the fragment by conducting a footprint titration of a 412 base pair O_R1-containing fragment (Fig. 2). The resolved free energies listed in Table II are equivalent. However, the free energy resolved from a footprint titration conducted with a shorter (120 base pair) fragment was slightly higher (Table II). We will consider the implications of this observation in a separate study.

In the resolution of an isotherm, we make two fundamental comparisons to determine the degree of protection [Eq. (1)] at any ligand concentration. The correction for the total number of labeled DNA fragments that are loaded onto each lane is made by comparing the density of a band within the binding site to the density of a band outside the binding site (a standard band as defined in the section on "Quantitation of an Autoradiogram"). Comparison of the corrected site densities (at different ligand concentrations) determines the degree of protection. The correction for total DNA loaded is an interlane comparison whose accuracy is dependent on an intralane comparison; it assumes that the ratio of densities of any two bands (outside the protected region) is a constant independent of protein concentration. Some investigators have stated their preference that this constant the unity[7] so that interlane comparisons may be used to indicate degree of protection. We emphasize that both compari-

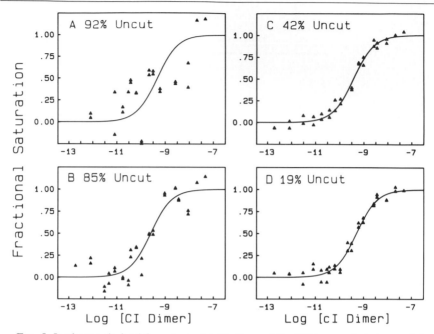

FIG. 5. Isotherms derived from footprint titrations of the 155 base pair O_R1-containing fragment conducted at four different DNase concentrations. The DNase exposure period was 1 min. The fraction of uncut DNA, calculated from the solid curve in Fig. 4, is indicated for each. Final concentrations of DNase were (A) 62.5 ng/ml; (B) 125 ng/ml; (C) 625 ng/ml; (D) 1.25 μg/ml.

sons must be made to determine correctly the degree of protection afforded by a bound protein.

As indicated above, if the ratio of densities of standard blocks is not constant for all lanes, substantial errors may be introduced into the resolved isotherm. We considered several experimental procedures that might effect such a deviation; they included DNase exposure that does not meet the stringent definition of one-hit kinetics given above. To test this possibility, a series of titrations of cI protein binding at O_R1 were conducted at different DNase concentrations (with the time of exposure kept constant (Fig. 5). Under conditions of extremely low nicking, where fragments nicked once dominate the small subset of nicked fragments (see Table I), the data were highly scattered. At higher extents of nicking, the precision of the data was greatly increased. However, it is significant that analysis of each of these isotherms yielded the same binding free energy (Table III). These results do not imply that an arbitrarily high extent of

TABLE III
RESOLVED FREE ENERGIES FROM O_R1 FOOTPRINT
TITRATIONS CONDUCTED WITH DIFFERENT
CONCENTRATIONS OF DNase[a]

[DNase]	Uncut[b] (%)	Free energy of association (ΔG, kcal/mol)	Standard deviation[c]
62.5 ng/ml	92	-12.5 ± 1.1	0.33
125 ng/ml	85	-12.7 ± 0.3	0.14
625 ng/ml	42	-12.7 ± 0.1	0.05
1.25 μg/ml	19	-12.5 ± 0.1	0.06

[a] Free energies of association for the resolved isotherms (solid curves) of Fig. 5, with 65% confidence
intervals.
[b] The percentage uncut at each DNase concentration
is taken from the curve fit to the points of Fig. 4.
[c] Standard deviation of the fitted curves.

nicking is desirable to improve precision; indeed, we know that is not
acceptable. Instead, the DNase exposure of the isotherms shown in Fig.
5B and C meet a less stringent definition of "one-hit" kinetics (as described above).

The increase in precision of footprint titrations in which a large percentage of the DNA has been nicked appears to be due to the greater
number of fragments that contribute to the exposure of the autoradiogram. In the formulation of an expression for site saturation as a function
of integrated optical densities [Eq. (1)], we implicitly assumed that all
bands exclusively represent labeled strands that have been shortened by
the action of DNase. However, we have observed that a small amount of
nicking is introduced during the preparation and purification of the DNA
fragments (e.g., see Lane 1, Fig. 1A). Fragments nicked in this way
contribute substantially to band density only when minimal cutting of the
DNA results in too few fragments. Thus, the autoradiogram of a footprint
titration conducted under these low DNase exposure conditions represents significant density that is independent of DNase.

As can be seen in Fig. 1, this problem is not significant with high
DNase exposure and fresh DNA. Thus, the optimum DNase exposure is a
compromise between (1) sufficiently high nicking by DNase exposure to
generate contrast between protected and standard bands and (2) sufficiently low DNase exposure to minimize the errors generated by imprecision in DNase exposure under conditions of multi-hit kinetics.

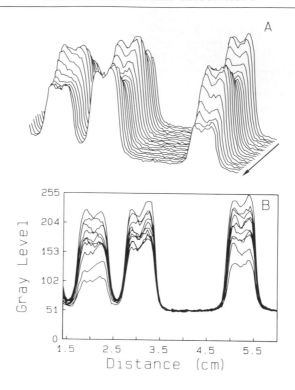

Quantitation of an Autoradiogram to Resolve Binding Isotherms

In this section we describe the rationale for the procedures by which we analyze an autoradiogram. These procedures include two-dimensional scanning of a film, the integration of optical density, the determination of local film backgrounds, and the relationship between the integrated optical density and the molar quantity of DNA represented by a region. We also compare the analyses of footprint titrations that rely on quantitation of individual bands to those that rely on blocks of contiguous bands.

Rationale for Two-Dimensional Scanning. Though gel electrophoresis provides a one-dimensional separation of DNA fragments, one-dimensional scanning of the resultant banding pattern is inadequate to quantitate the number of DNA fragments in any region of a gel lane. As Fig. 6 illustrates, the density within a band shows considerable variation in shape. In addition, the area of a band is highly correlated to its overall density (e.g., See Figs. 1, 3, 6C and D). Therefore, neither the maximum peak height nor a single dimension peak profile can provide an accurate representation of the density of a band. For this reason, complete two-dimensional analysis of the autoradiogram is mandatory. Each film is

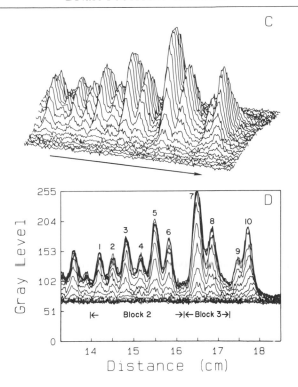

FIG. 6. Three-dimensional visualization of autoradiogram band separation. The digitized optical densities of regions of two autoradiograms show interlane and intralane separation of bands; the arrows indicate the direction of electrophoresis. Each region represents a portion of the 23 × 23 cm² scanned area of film. Distances are reported relative to the scanning origin in the upper left-hand corner of the film (as oriented in Fig. 1A). The optical density of each pixel is reported as a gray level value ranging from 0 to 255 (see text). (A) and (B) are two views of variable lane spacing. The region shown spans 8 mm in the direction of electrophoresis (beginning 2 cm from the top) and extends 4.4 cm across the gel. To represent compactly the optical density of this region, each of the 30 solid curves in (A) represents the average value of a 2 × 2 array of pixels; this is similar to the view that would be obtained by scanning with a 400 μm raster size. To allow quantitative comparison of the gray levels between lanes, 10 of these curves are plotted in (B) in a view collapsed along the axis parallel to the direction of electrophoresis. (C) and (D) are two views of part of the O_R1 footprint titration autoradiogram that is shown in Fig. 3. The lane represented here is that for which the boundaries of Blocks 2 and 3 are indicated in Fig. 3. This region of film spans 5 cm in the direction of electrophoresis (starting at 13.2 cm from the top of the scan); it extends 1.2 cm across the gel. In (C), each pair of pixels across the gel was averaged; however, there was no averaging of pixels along the direction of electrophoresis. Thus, the drawing illustrates the intralane resolution of bands possible with a 200 μm² pixel (see text for discussion). In (D), 15 of these curves are plotted in a view collapsed along the axis parallel to the direction of electrophoresis. The blocks of bands that were integrated for comparison with individual bands are indicated in (D) below the curves representing gray levels of interlane spacing regions. Block 1 is the sum of Blocks 2 and 3.

scanned as a matrix of individual picture elements (pixels). The integrated optical density of any region is the sum of the individual pixel densities contained therein. It is this integrated density that is correlated to moles of DNA fragments.

Integration of Optical Densities. Resolution requirements. Resolution of individual bands is a necessary criterion for accurate analysis. The smallest visible bands produced by our electrophoresis apparatus are of the order 1 mm (in the direction of electrophoresis) by 6 mm (the lane width). A 200 by 200 μm pixel size is found to give the largest pixel consistent with no loss of resolution of individual bands. Thus, an individual band contains at least 150 picture elements.

Criteria for distinguishing bands. We consider here the problem of defining correctly bands or blocks (i.e., groups of contiguous bands) to determine the integrated optical density of both standard and binding site regions. While there are few objective rules, some general guidelines are applicable.

Distinct bands or blocks that are clearly discernible across all lanes must be chosen. Accurate division of the overlapping density of neighboring bands of disparate density (e.g., bands 9 and 10 in Fig. 6D) is not possible. In principle, it is preferable to divide between widely separated bands, where the density between them approaches the background value (e.g., bands 6 and 7 in Fig. 6D). As can be seen in Fig. 6D, few such division points exist. However a good approximation is made when spatially distinct bands of similar density (e.g., bands 1 through 6) are divided at the point of the lowest optical density between them.

The inaccuracy inherent in defining boundaries makes a much smaller contribution to the contour definition of blocks (which are necessarily larger in area). Thus the integrated optical density of blocks is more accurate. A comparison of the precision of isotherms derived from analysis of blocks to the precision of isotherms derived from the analysis of single bands will be presented in a later section.

The choice of bands to be analyzed depends on the knowledge that the region of DNA protected from exposure to DNase by the presence of a protein ligand is, in general, larger than the actual binding site. With short spacer regions between adjacent specific sites in a multisite system (e.g., there are only three base pairs separating sites 2 and 3 of the wild-type lambda left operator[30,31]), there may be protection at the edge of one site due to bound ligand at an adjacent site. To avoid this potential error in the analysis we routinely integrate bands that represent only the central 10–12 base pairs of the 17 base pair lambda repressor binding sites.

Estimation of the Local Film Background Intensity. Accurate resolution of a binding isotherm is critically dependent on proper determination of the background film optical density. Variation in background density is commonly observed, both perpendicular to and parallel to the direction of electrophoresis. The magnitude of this variation may be as great as 0.04 optical density units. The lightest regions of an autoradiogram are found only in the protected site and at high saturation with ligand; they often have average pixel optical densities of the order 0.005 to 0.01 optical density units above the background level. Errors in the assumed background optical density will dominate the integrated optical density for such a region. Thus, the potential exists for a substantial systematic error to propagate into the data. It is for this reason that we find it necessary to properly determine the *local* background film density (as previously described) for all integrated regions of interest.

It is valid to define local background as the most probable pixel optical density (i.e., the peak of a histogram of individual pixel optical densities[27]), as long as only pixels representing background are considered. Significant contribution to this estimate from pixels with increased optical density due to ^{32}P exposure will skew the distribution, or, in extreme cases, will introduce additional peaks into the histogram. Therefore, it is critical that lanes be properly spaced such that real film background density, independent of ^{32}P exposure from adjacent lanes, is attained.

The necessary spacing between 6-mm-wide lanes was determined by electrophoresis of aliquots of a stock solution of the 155 base pair O_R1-containing DNA fragment (exposed to DNase under standard conditions). The gel was cast with a commercially available spacer comb (Bethesda Research Labs) that provided 3 mm wide spacing between the lanes. In addition, lane spacings of 12, 21, and 30 mm were obtained by loading samples into a subset of the available lanes.

The darkest region of the resulting autoradiogram is presented in perspective projection in Fig. 6A. This region contains individual pixel densities that approach the upper limit of resolvable optical density of the scanner. It represents the most dense region of any autoradiogram that can be analyzed. It is clear by inspection that the valley between the 3-mm spaced lanes on the left does not reach the background value defined by the broad valley between the 12-mm spaced lanes on the right. The collapsed view of these contours, shown in Fig. 6B, demonstrates this difference. Also, notice in this view that the background level is constant along the direction of electrophoresis in the valley with 12 mm spacing. In contrast, we observe a substantial slope of the valley floor along the direction of electrophoresis in the 3-mm valley. It is clear that the density in the 3-mm valley is influenced by the density of the bands in the lanes on either side.

Analysis of the region shown in Fig. 6 indicated that a spacing of 6 mm between 6-mm lanes is sufficient to ensure accurate determination of background values. A spacer comb was custom made to this specification; it gives 18 lanes in a 23-cm-wide scan width. Figure 3 shows a region of an O_R1 footprint titration in which the 6 mm spacing was used; Figure 6C shows a 12-mm-wide region of the autoradiogram shown in Fig. 3.

Calculation of Total DNA and Extent of Protection. Correlation between integrated optical density and amount of DNA. Our analysis of footprint titration experiments assumes a linear relationship between the integrated optical density determined for a band (or block), and the molar quantity of DNA represented by that band (or block). To assess the assumption of linearity, the following experiment was conducted. A solution of ^{32}P-labeled 155 base pair O_R1 DNA (see Fig. 2) was exposed to DNase in the absence of cI repressor. Serial dilution of this digested stock solution produced a logarithmic series of DNA concentrations; these were electrophoresed and autoradiographed using the procedures outlined in the section on "Experimental Design and Protocol." The integrated optical density (corrected for background) was determined for a series of bands and blocks whose density covers the full range resolvable by the P-1700 scanner. Any band or block necessarily contains a wide distribution of individual picture element optical densities, because the distribution of ^{32}P on a DNA sequencing gel is nonuniform. These densities were compared to the total ^{32}P applied to a given lane of the gel, as determined by liquid scintillation counting of each dilution of the digested stock DNA (see Fig. 7).

This experiment does not provide a simple measure of the response of the film to a known quantity of radioactive phosphorus. The film response has been measured in this way by others.[26,27] Rather, the experiment measures the relationship between the optical density of a region of an autoradiogram and the quantity of DNA represented by that region, under the conditions of a footprint titration experiment. Thus, the measured response was subject to every source of error at each step of a footprint titration experiment.

All of the DNA used in this experiment was derived from a single stock solution of nicked DNA; thus, each dilution contained the same fraction of DNA fragments of a given length. Since the ratio of DNA represented by any two arbitrarily chosen bands or blocks is, therefore, also constant, one need only evaluate this ratio in order to compare the experimental response of these different bands or blocks. This ratio is independent of the total DNA loaded per lane because each band or block represents a constant fraction of that total. In practice, the ordinate was scaled for one block relative to the other, applying the least-squares criterion to produce an unbroken smooth curve. The range of integrated opti-

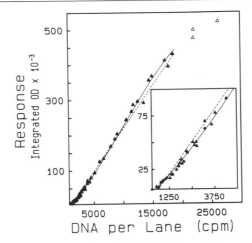

FIG. 7. Experimental response to ^{32}P-labeled DNA. The points are the integrated optical densities (corrected for background optical density) determined for standard blocks in lanes with known quantities of DNA. The three open symbols represent blocks in which individual pixel optical densities exceeded the maximum limit of resolution of the scanner. The solid line is the best cubic equation fit to the points; the dashed line is the best linear fit (see text). Inset shows data at low DNA (cpm) plotted on an expanded scale.

cal density described by the response curve is greatly increased by considering both light and dark blocks. The points in Fig. 7 are derived from integration of two blocks. The three open points represent blocks that contain individual picture elements with optical density above the upper limit for the film scanner. These were excluded from the analysis.

To evaluate the assumption of linearity, these data were fit to polynomial functions of varying order, scaled as described. A cubic equation, shown as the solid line in Fig. 7, gave the best description of the data. The dashed line through the points is the best linear fit, using the scale factors derived for the cubic fit. These curves suggest that the relative error involved in making the linear approximation is nowhere greater than a few percent (i.e., within the precision of a single point in a footprint titration experiment) except at very low values of the integrated optical density (see inset) where the error may approach 10%. Very low values of the integrated optical density occur only at high fractional saturation of the site (Fig. 1) where an error of 10% makes a negligible contribution when propagated into the fractional saturation [i.e., the error is 0.01 at $\bar{Y} = 0.9$, Eq. (1)]. Thus, the absolute error in fractional saturation is nowhere greater than about 0.02, which is within the precision of our current measurements.

The integrated densities of individual bands have also been analyzed

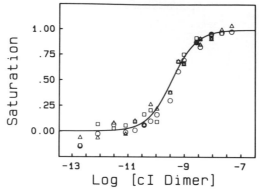

FIG. 8. O_R1 footprint titration isotherms derived from three sequential autoradiograms of a single gel. The identical block was analyzed for each. Symbols define exposure to ^{32}P disintegration relative to the darkest autoradiogram: circles, 0.35; triangles, 0.5; squares, 1.0. The free energy and standard deviation of the fit for each isotherm are listed in Table IV.

in the same manner. The results obtained (data not shown) are similar, though the departure from linearity is somewhat greater. There appears to be an advantage to the analysis of blocks, which offer a much broader distribution of individual pixel densities (e.g., see Figs. 1A, 3, and 6).

As a direct evaluation of the effect of the slight departure from a linear response, three sequential autoradiograms of a single O_R1 footprint titration gel were analyzed. One of these autoradiograms was the darkest that could be analyzed subject to the criterion that no individual pixel optical density exceed the maximum level resolvable by the scanner. The others were 50 and 35% of that exposure. It is clear that the three sets of points (Fig. 8) define the same isotherm. Both the free energy determined and the standard deviation of the fits (Table IV) were identical over nearly a 3-fold range of optical densities. Analysis of the residuals gave no evidence for systematic deviation of any one set of points from either of the others, or from the fitted curve. Therefore, we consider as acceptable any autoradiogram for which no pixel optical density exceeds the limit of resolvability of the scanner.

Analysis of individual bands and blocks. The integration and analysis of blocks composed of a contiguous group of bands (see Fig. 3) represent a great simplification in the analysis of autoradiograms. However, it was necessary to consider whether the extent of protection measured by a block was equivalent to that measured for an individual band.

This was demonstrated by analysis of a typical O_R1 footprint titration autoradiogram using 6 bands within the protein binding site that were well resolved in all lanes of the gel (Fig. 3). Each individual band was integrated and analyzed separately to yield the isotherms shown in Fig. 9A.

TABLE IV
FREE ENERGIES RESOLVED FROM SEQUENTIAL
AUTORADIOGRAMS OF A SINGLE O_R1
FOOTPRINT TITRATION[a]

Autoradiogram	Free energy of association (ΔG, kcal/mol)	Standard deviation
1	-12.7 ± 0.2	0.05
2	-12.7 ± 0.1	0.06
3	-12.5 ± 0.1	0.05

[a] Autoradiograms (with 3-fold different extent of exposure) of a single footprint titration of the binding of repressor dimers to the 155 base pair O_R1-containing DNA fragment. The data are shown in Fig. 8. Correspondence between the autoradiograms and the symbols of Fig. 8 is given by 1, circles; 2, triangles; 3, squares. Column 3 lists the standard deviations of the fitted curves.

These are equivalent isotherms; the resolved free energies, 65% confidence intervals, and standard deviations are strikingly similar (Table V). For comparison, a block encompassing the six bands was integrated and analyzed. The resulting isotherm is the same as that resolved from each of the individual bands; the resolved free energy is the same, and the standard deviation of the fitted curve is slightly better than for any single band (Fig. 9A and Table V). Simultaneous analysis of the data derived from individual analysis of the six bands yields a value of free energy whose standard deviation is comparable to that of the value determined by analysis of the isotherm resolved from block analysis (Table V). Thus, precision is not improved by analyzing each band individually. A plausible explanation is that some subjective judgments are required to define quadrilateral contours around bands (and blocks), as discussed above. An error in defining the contour around a band will result in a greater relative error in the integrated optical density than the same error in defining a block. It is also frequently possible to take advantage of the slight sequence sensitivity of DNase by choosing blocks bounded by bands that represent positions that are poorly cut [and are low in density (Fig. 6D)].

As an additional validation of the use of blocks to define isotherms, the self-consistency of isotherms resolved from different blocks was considered. Three blocks were analyzed: the block encompassing the 6 bands described above (Block 2), a block of 2 bands representing shorter frag-

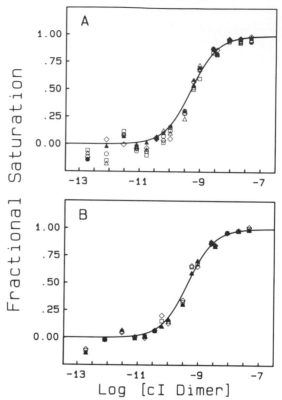

FIG. 9. Comparison of the analysis of an O_R1 footprint titration by individual bands and blocks of group of contiguous bands. The regions integrated and analyzed are shown in Fig. 3. (A) Isotherms derived from analysis of bands. For clarity, data are presented for only 4 of the 6 bands analyzed. Free energies resolved from analyisis of each band are listed in Table V. The curve shown is the fit to the band 2 data. The solid triangles show the data points for Block 2 for comparison. Other symbols are open triangles, Band 1; open circles, Band 2; open squares, Band 4; open diamonds, Band 5. (B) Isotherms derived from analysis of blocks. Symbols are open squares, Block 1; filled triangles, Block 2; open diamonds, Block 3. The solid curve is the best fit to Block 1 data. The resolved free energies are listed in Table V.

ments (Block 3), and the block that encompassed both of these regions (Block 1, see Figs. 3 and 6D). The three isotherms were equivalent to each other (Fig. 9A) and equivalent to those resolved from individual bands as judged by the resolved free energies (Table V). These results clearly show that it is possible to resolve an isotherm from analysis of bands or blocks chosen anywhere within a protected region.

TABLE V
FREE ENERGIES DERIVED FROM ANALYSIS OF
INDIVIDUAL BLOCKS AND BANDS OF A SINGLE
O_R1 FOOTPRINT TITRATION[a]

		Free energy of association (ΔG, kcal/mol)	Standard deviation
Band	1	-12.5 ± 0.2	0.09
Band	2	-12.5 ± 0.2	0.08
Band	3	-12.5 ± 0.2	0.07
Band	4	-12.4 ± 0.1	0.06
Band	5	-12.4 ± 0.1	0.06
Band	6	-12.5 ± 0.2	0.07
Bands	1–6[b]	-12.5 ± 0.1	0.07
Block	1	-12.5 ± 0.1	0.05
Block	2	-12.5 ± 0.1	0.06
Block	3	-12.5 ± 0.2	0.05

[a] Free energies (with 65% confidence intervals) for the resolved isotherms shown in Fig. 9. The bands and blocks analyzed are defined in Fig. 3. Column 3 lists the standard deviations of the fitted curves.
[b] Results when the data derived from all 6 bands is fit to Eq. (2).

Analysis of large blocks of bands within a protected region appears to be the optimal approach to the resolution of individual-site isotherms from a footprint titration because an entire titration can be analyzed by integrating one block for each site and one standard block. The time required for analysis is reduced by greater than an order of magnitude from the time required separately to analyze each band for each site.

In this section, all of our arguments have been developed for a DNA fragment that is protected from nicking when the protein ligand is bound. We are aware that in other systems, protein binding results in enhancement of nicking by DNase. In principle, enhancement data can be analyzed in a fashion analogous to protection data to resolve an isotherm. However, hypersensitive and protected positions must be analyzed separately; they cannot be integrated within a common block.

Resolution of Free Energies from Individual-Site Isotherms

In previous sections, we presented our development of the footprint titration technique to resolve isotherms of the binding of protein ligands to a single DNA site. However, the power of the footprint titration technique lies in its ability to resolve individual-site isotherms for a protein that binds to multiple, interacting sites. In this section we will demonstrate how to apply the analysis to a three-site system. We will also compare isotherms derived from footprint titration with those obtained from an independent technique, filterbinding titration.

Binding Isotherms

For a ligand binding to a single site on a macromolecule, the extent of saturation of the site is given by the familiar Langmuir isotherm

$$\overline{Y} = \frac{K_1[X]}{1 + K_1[X]} \tag{2}$$

where $[X]$ is the free ligand activity and K_1 is the equilibrium constant for the association of ligand. The active form of the cI repressor that binds to its operator sites is a dimer, that is in equilibrium with cI monomer in solution.[24,35] Therefore, in fitting Eq. (2) to the O_R1 isotherms, the free dimer concentration $[R_2]$ is equivalent to the free ligand activity. This was calculated from the total cI concentration using a known value for the association constant: $5 \times 10^7 \ M^{-1}$.[24] It should be emphasized that this value is only an estimate of the cI dimer association constant under our experimental conditions. Any error in this value will propagate into the free energies resolved from our studies.

The single-site case is simple because the system can exist in only two configurations. Thus, the distribution of macromolecules between states is uniquely determined by the extent of saturation of the site [Eq. (2)]. The half-saturation point of such a binding curve has a simple interpretation in terms of the association constant (i.e., $K_1 = 1/[X]$ at $\overline{Y} = 0.5$). These statements are not true for a cooperatively interacting, multisite system.

Consider, for example, a three-site system with cooperative interactions between sites. The extension of Eq. (2) to such a system is

$$\overline{Y}_t = \frac{(k_1[X] + 2k_2[X]^2 + 3k_3[X]^3)}{3(1 + k_1[X] + k_2[X]^2 + k_3[X]^3)} \tag{3}$$

where $\overline{Y}_t$ is the fraction of total sites with bound ligand when the ligand acceptor is in equilibrium with ligand at activity $[X]$. The equilibrium constants k_1, k_2, and k_3 are macroscopic binding constants which describe

[35] P. Chadwick, V. Pirrotta, R. Steinberg, N. Hopkins, and M. Ptashne, *Cold Spring Harbor Symp. Quant. Biol.* **35**, 283 (1970).

the binding of a total of 1, 2, or 3 ligands, respectively. Each is a combination of the microscopic constants that describe the intrinsic binding of ligand to a site in the absence of binding to any other site, and the interaction between ligands bound at interacting sites.

The isotherms described by Eq. (3) have been referred to as classical isotherms.[2] We will refer to techniques that only measure the fraction of total sites with bound ligand, as described by Eq. (3), as "classical" techniques. The microscopic constants that describe a cooperative, multisite system cannot, even in principle, be resolved from an isotherm of this type.[2]

An example of a multisite, cooperative system is the binding of the cI repressor dimer to the left operator region of the lambda phage. The left operator (or O_L) consists of three closely spaced, 17 base pair long, specific binding sites for repressor dimers.[30,31] Johnson[20] has made estimates of the ligand activities that half-saturate each of the sites of the wild type O_L, and of mutant operators in which single base pair substitutions have eliminated binding to one or more sites. Analysis of these data[20,36] by the method of Ackers *et al.*[3] (also see below) considered only pairwise cooperative interaction between adjacent sites. The analysis led to the development of a model in which a subset of 9 of the 13 theoretically possible energetic states is used to describe the system. These configurations are listed in Table VI.

The intrinsic free energies, ΔG_1, ΔG_2, and ΔG_3, of Table VI correspond to the microscopic association constants, K_1, K_2, and K_3, for the binding of repressor to any single site in the absence of binding to an adjacent site. The interaction or cooperative free energies, ΔG_{12} and ΔG_{23} (which correspond to the microscopic constants K_{12} and K_{23}) represent the additional energy required to load a site given that an adjacent site has already been loaded. The correspondence between the five macroscopic association constants, K_1, K_2, K_3, K_{12}, and K_{23}, and the three microscopic association constants of Eq. (3) is given by

$$k_1 = K_1 + K_2 + K_3 \tag{4a}$$
$$k_2 = K_1K_2K_{12} + K_1K_3 + K_2K_3K_{23} \tag{4b}$$
$$k_3 = K_1K_2K_3(K_{12} + K_{23}) \tag{4c}$$

Though the five free energies corresponding to the microscopic association constants of Eqs. (4) cannot be resolved from knowledge of the extent of saturation of the entire three-site system, they can be resolved from knowledge of the extent of saturation at each of the three sites.

[36] M. A. Shea, Ph.D. dissertation, The Johns Hopkins University, Baltimore, MD, 1983.

TABLE VI
OPERATOR CONFIGURATIONS AND ASSOCIATED FREE ENERGY STATES
OF BOUND REPRESSOR[a]

Species	Configurations of operator sites 1	2	3	Free energy contributions	Total free energy[b] (ΔG_s, kcal/mol)
1	0	0	0	(REF. STATE)	ΔG_{s1}
2	R_2	0	0	ΔG_1	ΔG_{s2}
3	0	R_2	0	ΔG_2	ΔG_{s3}
4	0	0	R_2	ΔG_3	ΔG_{s4}
5	R_2—R_2		0	$\Delta G_1 + \Delta G_2 + \Delta G_{12}$	ΔG_{s5}
6	R_2	0	R_2	$\Delta G_1 + \Delta G_3$	ΔG_{s6}
7	0	R_2—R_2		$\Delta G_2 + \Delta G_3 + \Delta G_{23}$	ΔG_{s7}
8	R_2—R_2		R_2	$\Delta G_1 + \Delta G_2 + \Delta G_3 + \Delta G_{12}$	ΔG_{s8}
9	R_2	R_2—R_2		$\Delta G_1 + \Delta G_2 + \Delta G_3 + \Delta G_{23}$	ΔG_{s9}

[a] This table represents the subset of all 13 theoretically possible configurations of the binding system that is pertinent to studies of lambda repressor proteins at O_L. (Species 1–8 are pertinent to O_R.) In the configuration list, O and R_2 represent free operator sites and operator sites with bound repressor dimers, respectively.

[b] ΔG_s, The total free energy of each configuration given by the sum of contributions from the five interaction terms: ΔG_1, ΔG_2, and ΔG_3, the free energies of binding at individual sites in the absence of bound ligands at any other sites, referred to as intrinsic binding energies, ΔG_{12} and ΔG_{23}, the free energies of cooperative interactions between adjacent DNA-bound repressor dimer molecules.

Consider the fractional probability of any specific configuration of the multisite O_L system. The probability may be expressed as

$$f_s = \frac{\exp(-\Delta G_s/RT) \times [R_2]^j}{\sum\limits_{s,j} \exp(-\Delta G_s/RT) \times [R_2]^j} \tag{5}$$

where ΔG_s is the sum of free energy contributions for species "s" (Table VI), R is the gas constant, T is the absolute temperature, $[R_2]$ is the concentration of unbound repressor dimers, and "j" is the number of dimers bound in a given operator configuration "s." In the denominator of Eq. (5), the summation over "s" is taken from 1 through 9 for wild-type operators and "j" takes on the value of 0, 1, 2, or 3 appropriate to the configuration of the species "s" (i.e., $j = 2$ for $s = 5$, 6, or 7). At a given value of $[R_2]$, the probability f_s represents the fraction of operators that have the configuration whose free energy ΔG_s appears in the numerator.

At any concentration of repressor protein, combinations of these terms describe the probability that corresponding combinations of sites one, two, or three are occupied. Therefore, these expressions permit evaluation of the free energies of interaction from individual-site binding isotherms. For the O_L wild-type operator, the fractional occupancy of sites is given by

$$\overline{Y}_{O_L1} = f_2 + f_5 + f_6 + f_8 + f_9 \qquad (6a)$$
$$\overline{Y}_{O_L2} = f_3 + f_5 + f_7 + f_8 + f_9 \qquad (6b)$$
$$\overline{Y}_{O_L3} = f_4 + f_6 + f_7 + f_8 + f_9 \qquad (6c)$$

where the subscripts refer to the configurations listed in Table VI.

In a cooperatively interacting system the ligand activity at half-saturation corresponds to no single binding constant. The probability of binding at an individual site in such a system depends upon the extent of occupancy of the neighboring sites. The mathematical formulation [Eqs. (6)] described above for cI repressor binding at wild-type O_L takes these interactions into account. Thus, it provides exact expressions for the fractional saturation at each of the binding sites regardless of the numerical values of the terms representing cooperativity.

The preceding statements are also true when mutant operators are analyzed by the general formulation given by Eq. (5). For mutant operators, equations analogous to Eqs. (6) are easily constructed by considering the subset of the configurations listed in Table VI that corresponds to the reduced valency of the mutant. Data for mutant and wild-type operators can also be analyzed simultaneously in a straightforward manner (see detailed discussion below). The set of equations to be solved include both Eqs. (6) and the analogous equations that describe binding to the individual sites of the mutants. The power to resolve microscopic association constants by footprint titrations is greatly expanded in this way.

Analysis of an O_L Footprint Titration

As demonstrated in Fig. 1, it is possible to resolve three complete individual-site isotherms from a single footprint titration of the binding of cI to wild type O_L. Each set of points comprising an individual-site isotherm relates the fractional saturation at one site to the varying concentration of unbound repressor dimer with which the sites are in equilibrium. The system of simultaneous equations that describe these isotherms is analyzed to give the best estimate to the five free energies (as described in the section on "Correction for the Variation of Total DNA per Lane") and the confidence limits. [In fitting individual-site binding data we assign equal weights to all points, based on the observation that the standard

TABLE VII
FREE ENERGIES FOR REPRESSOR BINDING TO
WILD-TYPE $O_L{}^a$

Parameter	Free energy of interaction (ΔG, kcal/mol)
ΔG_1	-14.8 ± 0.4
ΔG_2	-13.1 ± 0.3
ΔG_3	-13.0 ± 0.2
ΔG_{12}	-2.3
ΔG_{23}	-2.3
SD^b	0.055

[a] Interaction free energies (with 65% confidence intervals) resolved when the isotherms of Fig. 1B were analyzed according to Eqs. (6), where ΔG_{12} and ΔG_{23} were assumed to be -2.3 kcal/mol, as discussed in the text.
[b] Standard deviation of the fitted curves.

deviation appears constant across the entire isotherm (e.g., see Figs. 1, 5, 8, 9, and 10).] The 30 sets of points measured for the binding of cI repressor to O_L (shown in Fig. 1B) were analyzed by Eqs. (6) to estimate the five free energies specified by the model described in Table VI. The parameters resolved are listed in Table VII; these predict the solid curves shown in Fig. 1B.

It is possible in principle to resolve all five free energies from a single experiment. However, this was not possible for the isotherms of Fig. 1B. Extremely high correlation was observed between the parameters ΔG_{12}, ΔG_2, and ΔG_{23} when these data were fit for all five free energies. The quality of the fit is extremely sensitive to the sum of these free energies, but is relatively insensitive to their individual values over a wide range. This is partially a consequence of the model given in Table VI, in which binding at site 2 is dependent upon binding at both sites 1 and 3. Data from mutant operators in which the number of sites is reduced should be useful in resolving these parameters, as discussed above. Simulations[36] indicate that data for an $O_L1^- O_L3^-$ mutant taken together with data from the wild-type operator should be sufficient to resolve all five free energies.

The data of Fig. 1B could be analyzed to resolve the three intrinsic free energies (ΔG_1, ΔG_2, and ΔG_3), when values for the interaction free energies (ΔG_{12} and ΔG_{23}) were assumed. Data from the study of mutant O_L operators allow us to make an estimate of -2.3 kcal/mol for the value of these parameters. These values are derived from measurements of the ligand activity corresponding to half-saturation for the binding of repressor to mutant and wild-type operators at 37°.[20] The resolved free energies are listed in Table VII.

The pattern of the intrinsic free energies for the binding of repressor to sites 1, 2, and 3 of O_L that were resolved by our analysis of the data of Fig. 1B is the same as the pattern resolved from the analysis of the earlier studies of wild-type plus mutant half-maximal binding (Table X). The magnitude of the individual free energies is greater at 20° (Table VII) than at 37° (Table X) as expected from the difference in temperature.

The data of Fig. 1B dictate that the isotherms for sites 1 and 2 cross each other, despite the difference of 1.7 kcal/mol in intrinsic free energy for the binding of repressor to the two sites. This feature is observed for a wide range of assumed values of the cooperative free energies, ΔG_{12} and ΔG_{23}. Data simulations[36] have shown that this is a property of the nature of the cooperative interactions embodied in the system.

Comparison between Isotherms Derived from Footprint and Filterbinding Titrations

A standard technique for determination of protein–DNA binding constants is the filterbinding technique first described by Riggs et al.[37] This technique relies on the retention of protein–DNA complexes (but not unliganded DNA) by nitrocellulose filters, when an equilibrium mixture of protein plus DNA is drawn through the filter. The fraction of DNA retained by the filter as a function of free ligand concentration is described by Clore et al.[38]

$$\theta = (Z - 1)/Z \tag{7}$$

In Eq. (7), Z is the binding polynomial[39] given by

$$Z = \sum_{i=0}^{n} k_i[L]^i$$

where n is the number of binding sites on the macromolecule, k_i is the macroscopic association constant for binding of i ligands (define $k_0 = 1$), and $[L]$ is the free ligand activity.

This function is slightly different from Eq. (3) which was cited earlier as another example of a "classical" isotherm. The filterbinding technique measures the fraction of DNA with one or more bound ligands. Isotherms

[37] A. Riggs, H. Suzuki, and S. Bourgeois, J. Mol. Biol. 48, 67 (1970).
[38] G. M. Clore, A. M. Gronenborn, and R. W. Davies, J. Mol. Biol. 155, 447 (1982).
[39] J. Wyman, Jr., Adv. Protein Chem. 19, 224 (1964).

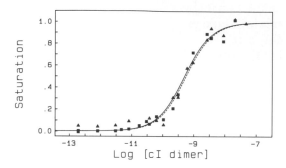

FIG. 10. Comparison of isotherms resolved from a footprint titration and from a filter-binding titration of repressor binding at O_R1. Solid triangles represent the footprint titration; solid squares represent the filterbinding titration. Solid (footprint) and dashed (filterbinding) curves are the best fit of Eq. (2) to each set of points. The resolved free energies are listed in Table VIII. Experimental conditions were 20°, pH 7.00, 0.20 M KCl (assay buffer). Filter-binding titrations are performed as follows. Protein is equilibrated with DNA as described for footprint titration experiments. To give replicate measurements, two aliquots of each equilibrated sample are drawn separately through 0.45-μm Millipore nitrocellulose filters. The filters are then washed twice with 0.4 ml of assay buffer at a flow rate of 50 ml/hr. The DNA retained is determined by scintillation counting of the samples (approximately 5000 cpm is applied to each filter). Under these conditions background retention in the absence of cI repressor is 1–2%; retention at saturating repressor levels is 10–13% for O_R1 and 80–90% for O_L.

of the type given by Eq. (3) describe the fraction of sites occupied by ligand. In terms of the binding polynomial, the general form of Eq. (3) can be written

$$\overline{Y} = (1/n) \left(\sum_{i=0}^{n} i k_i [L]^i \right) / Z \tag{8}$$

It is clear by inspection that for a single-site DNA fragment, $\overline{Y}$ and θ are equal; both are given by Eq. (2). Equation (2) also describes the isotherm derived from footprint titration analysis of binding to a single-site DNA fragment. Therefore, filterbinding titration of repressor binding to the O_R1 fragment provides an independent check on the validity of isotherms derived from individual-site, footprint titration analysis.

Figure 10 compares isotherms for the binding of cI repressor to O_R1 derived both from footprint titration and from filterbinding titration under identical experimental conditions. The two sets of points define nearly identical isotherms. The isotherms are shown as solid and dashed curves

TABLE VIII
COMPARISON BETWEEN FREE ENERGIES RESOLVED
FROM FOOTPRINT AND FILTERBINDING TITRATIONS
OF REPRESSOR BINDING TO O_R1[a]

Experiment	Free energy of association (ΔG, kcal/mol)	Standard deviation
Footprint titration	$-12.5 + 0.1$	0.06
Filterbinding titration[b] 2 μg/ml CT-DNA	$-12.4 + 0.2$	0.07
Filterbinding titration[b] without CT-DNA	$-12.5 + 0.3$	0.07

[a] Lines 1 and 2 list the resolved free energies (with 65% confidence intervals) for the isotherms of Fig. 10. Column 3 lists the standard deviation of the fitted curves.

[b] Lines 2 and 3 test the effect of the presence of calf thymus DNA (CT-DNA) on equilibrium between repressor protein and operator sites. Calf-thymus DNA is routinely added to footprint titration experiments for reasons discussed in the text (see section on "Range and Foundation of the Method").

in the figure. The free energies resolved are the same and are listed in Table VIII.

The filterbinding technique can be used in an analogous manner to provide an independent check on the validity of the individual-site isotherms determined for repressor binding to the three-site O_L system. (A complete discussion of the use of filterbinding titration for the study of specific, multisite, cooperative protein–DNA interactions will be presented elsewhere.) For a three-site system, Eq. (7) takes the form

$$\theta = \frac{k_1[L] + k_2[L]^2 + k_3[L]^3}{1 + k_1[L] + k_2[L]^2 + k_3[L]^3} \tag{9}$$

where the constants k_i are macroscopic association constants, identical to the constants of Eq. (3). Values for these macroscopic constants are easily calculated from the free energies that are resolved from an individual-site, footprint titration analysis of the binding of repressor to O_L. The relationships are given by Eqs. (4). Thus, the isotherm representing a

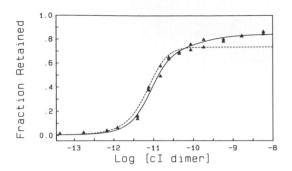

FIG. 11. Comparison between an isotherm resolved from a filterbinding titration of *c*I repressor binding at O_L and an isotherm predicted by analysis of footprint titration data. The set of replicate points represent a filterbinding titration of repressor binding to O_L (conditions and techniques are described in the legend to Fig. 10). The solid curve is the best fit of Eq. (9) to the points. Values of the free energies corresponding to the macroscopic association constants of Eq. (9) are listed in Table IX. The dashed curve is predicted by the microscopic free energies resolved from the footprint titration shown in Fig. 1B (see Table VII). The free energies corresponding to the calculated macroscopic association constants are listed in Table IX.

filterbinding titration of repressor binding to O_L is easily predicted from the results of individual-site analysis of footprint titration data.

Figure 11 makes a comparison between the measured and predicted isotherm for a filterbinding titration of repressor binding to wild-type O_L. The predicted curve is based on the footprint titration data of Fig. 1B. A complication in fitting the experimental filterbinding titration data of Fig. 11 is the observation that protein–DNA complexes are retained by the filters with less than 100% efficiency. To account for this, a retention efficiency parameter defined by Woodbury and von Hippel[40] has been propagated into the terms of the numerator of Eq. (9). Also, in the least-squares procedure used to fit Eq. (9) to the data of Fig. 11, weights proportional to $1/\theta_{obs}$ were assigned to the data. These were chosen based on the premise (confirmed by the data in Fig. 11) that the major contribution to the standard deviation in a filterbinding experiment is the constant relative variation in pipeting aliquots of binding mixtures onto the filters.

The best fit of Eq. (9) to the data of Fig. 11 is described by the solid curve in the figure. The predicted isotherm, calculated from the microscopic free energies resolved from the footprint titration data of Fig. 1B (values listed in Table VII) is the dashed curve in Fig. 11. The free ener-

[40] C. P. Woodbury and P. H. von Hippel, *Biochemistry* **22**, 4730 (1983).

TABLE IX
COMPARISON BETWEEN THE MACROSCOPIC FREE
ENERGIES DERIVED FROM A FOOTPRINT TITRATION
WITH VALUES RESOLVED FROM A FILTERBINDING
TITRATION OF O_L

Parameter	Footprint titration[a] (ΔG, kcal/mol)	Filterbinding[b] (ΔG, kcal/mol)
ΔG_1^M	-14.4	-14.6 ± 0.2
ΔG_2^M	-29.8	-29.5 ± 0.3
ΔG_3^M	-43.1	-41.2 ± 1.8
D^c	0.012	0.013

[a] Free energies corresponding to the macroscopic equilibrium constants of Eq. (9) are derived from the microscopic free energies of Table VII according to Eqs. (6).

[b] These predict the dashed curve in Fig. 11. Free energies (with 65% confidence limits) corresponding to the macroscopic equilibrium constants of Eq. (9) when fit (solid curve) to the data of Fig. 11.

[c] Weighted mean deviation of the theoretical and fitted curves is given by $D = (\Sigma_i w_i (\overline{Y}_i - \overline{Y})^2 / \Sigma_i w_i)^{1/2}$. [The weights assigned (w_i) were proportional to $1/\theta_{obs}$; see text.]

gies corresponding to the macroscopic association constants that describe each curve, as well as the weighted mean deviation of each curve, are listed in Table IX. It is clear from the weighted means deviation that the curves are equivalent descriptions of the filterbinding data. Further, each of the three free energies is the same for the two curves, to within the limits of experimental error.

Analysis of Mutant Operators to Determine Interaction Parameters

In the preceding section, parallel filterbinding and footprint titrations were shown to be consistent when judged on the basis of the resolved macroscopic and microscopic constants. For a multisite system, this comparison against results obtained by an independent method is limited by the different informational content of the classical and individual-site isotherms. Thus, we would like to show here that application of the footprint titration method to the study of a set of mutant and wild-type operators yields a self-consistent set of interaction constants.

Our initial analysis of footprint titration studies of the right operator of

TABLE X

FREE ENERGIES OF INTERACTION RESOLVED BY ANALYSIS OF ESTIMATES OF
HALF-SATURATION POINTS OF REPRESSOR BINDING TO MUTANT AND WILD-TYPE O_R[a]

Parameters[b]	O_R2^-	$O_R1^-O_R3^-$	$O_R1^-O_R2^-$	O_R3^-	O_R1^-	All[c]
ΔG_1	−11.67	—	—	(−11.7)	—	−11.69 ± 0.03
ΔG_2	—	−10.1	—	(−10.1)	(−10.1)	−10.19 ± 0.05
ΔG_3	−10.1	—	−10.1	—	(−10.1)	−10.09 ± 0.02
ΔG_{12}	—	—	—	−2.0	—	−1.99 ± 0.06
ΔG_{23}	—	—	—	—	−2.0	−1.94 ± 0.06

[a] Estimates of the repressor dimer concentration at half-saturation of the individual sites of wild-type and various mutant lambda right operators are from Johnson et al.[8] Columns 2–6 show the determination of intrinsic free energies for selected mutants (see text) (analysis of G. Ackers taken from Johnson[20] and Shea[36]). The cooperative free energies (ΔG_{12} and ΔG_{23}) are determined (see text) by analysis of O_R3^- and O_R1^- mutants, respectively; the values of ΔG_1, ΔG_2, and ΔG_3 were fixed at the values derived from analysis of the mutants in columns 2–4 that lack cooperative interaction. Values fixed for each analysis are enclosed in parentheses.

[b] Parameters are the free energies of binding of repressor to sites 1, 2, and 3 of O_R in the absence of binding to adjacent sites, and the additional (cooperative) free energy from binding at adjacent sites. These parameters are analogous to the free energy parameters described for O_L in Table VI.

[c] Interaction free energies (with 65% confidence intervals) resolved from the least-squares analysis of Ackers et al.[2] Data for the mutants listed in columns 2–6 and for wild-type O_R were fit simultaneously to the appropriate set of equations formulated on the basis of Eq. (5) (see text).

bacteriophage lambda[3] relied on knowledge of the concentrations of repressor dimers required to half-saturate sites of wild-type and mutant operators from experimental studies conducted at 37°.[8,20] These data were best described by a model in which pairwise interactions occur between adjacently bound cI dimers, except between dimers bound at sites O_R2 and O_R3 when sites O_R1 and O_R2 are occupied. This 8-species model for cI repressor interactions at O_R excludes configuration 9 (shown in Table VI) from the formulation of the expressions describing the binding isotherms. The reported values of intrinsic and cooperative Gibbs free energies were determined by simultaneous analysis of these data[8] and are listed in the last column of Table X.

It is also possible to evaluate mutant operators as macromolecules with a reduced number of binding sites and, therefore, fewer possible configurations. According to the binding isotherm expression for a single-site system [Eq. (2)], the equilibrium constant is given by the reciprocal of the activity of the free ligand at half-saturation. Thus, for some mutant operator regions, $[R_2]_{0.5}$ (the experimentally determined concentration at the half-saturation point) may be used to determine intrinsic free energies of binding.

This simple calculation for the O_R2^-, $O_R1^-O_R2^-$, and $O_R1^-O_R3^-$ mutants yielded the values for intrinsic binding constants and associated energies given in columns 2–4 of Table X. Note that these values are identical to the values determined from simultaneous analysis of all 11 data.

The procedure to determine the cooperative interaction energies from the mutant data was to fix the intrinsic binding energies at the values determined from the analysis of the "single-site" mutants and to analyze the binding data representing interacting sites solely for the cooperative energy of interaction (columns 4 and 5 Table X). The mutant operator O_R1^- is an example of a two-site system that retains cooperative interactions between repressor dimers bound at sites O_R2 and O_R3. An O_R1^- operator may assume four microscopic configurations (1, 3, 4, and 7 as shown in Table VI). Because the intrinsic binding energies ΔG_2 and ΔG_3 were determined from studies of an $O_R1^-O_R3^-$ and an $O_R1^-O_R2^-$ mutant, respectively, we could estimate directly the value of ΔG_{23} as follows. The fraction of operators with either O_R2 occupied or with O_R3 occupied was calculated using the appropriate 2-site binding expressions shown below:

$$\overline{Y}_{O_R2} = f_3 + f_7 \qquad (10a)$$
$$\overline{Y}_{O_R3} = f_4 + f_7 \qquad (10b)$$

where the fractional probability of ligand binding to a particular configuration ($\overline{Y}_n$) is related to the free energies ΔG_2, ΔG_3, and ΔG_{23} by Eq. (5). (Note that the sum over s is taken for values of $s = 1, 3, 4,$ and 7.) By setting $\overline{Y}_{O_R2}$ and $\overline{Y}_{O_R3}$ equal to 0.5, and introducing the corresponding value of $[R_2]_{0.5}$ required for half-saturation as well as the values of ΔG_2 and ΔG_3 (as listed in Table X), Eqs. (10a) and (10b) became expressions with a single unknown variable, ΔG_{23}. Available data also included those for an O_R3^- mutant; therefore, we could evaluate ΔG_{12} in as simple a manner as described above for ΔG_{23}. The values determined for ΔG_{12} and ΔG_{23} are shown in Table X. This independent analysis of the data for cI repressor binding at mutant operator regions highlights the self-consistency of the data and provides support for the footprinting method as one capable of measuring true equilibrium properties.

Concluding Remarks

In the preceding sections, the development of a set of experimental and analytical procedures, whereby individual-site binding information can be derived from quantitative footprint titration experiments, has been described in detail. Future articles will describe applications of this technique to relevant biological questions. In this section, the argument for the thermodynamic validity of the method will be presented, followed by a brief summary of its applicability to studies of gene regulation.

Thermodynamic Validity of the Footprint Titration Method

The DNase footprint titration method that has been presented is capable of quantitative characterization of the equilibrium properties of protein–DNA interaction. Several lines of evidence show that the method does yield thermodynamically valid individual-site isotherms. First, experimental studies demonstrate that the only experimental variable that affects the apparent degree of protection is the concentration of ligand. Second, the isotherms that are resolved are invariant both as a function of the time of equilibration of the protein and DNA prior to exposure to DNase, and as a function of the length of exposure to DNase. Therefore, the method probes an equilibrium rather than a kinetic phenomenon. Third, the free energies derived from the resolved individual-site isotherms are internally consistent, and are consistent with those resolved from an independent experimental technique. The arguments that support each of these conclusions are summarized below.

The results of the analysis of individual bands and of blocks (Fig. 3, Table V) show that every band or group of bands in a protected region responds in exactly the same way to the presence of the protein ligand. This indicates strongly that, for every lane of the gel, the relation of the density of any band (protected or not) to the total DNA in the equilibrium mixture is constant, except as modified by the single experimental variable: free ligand concentration. Ensuring this has required the development of rigorous experimental and analytical procedures. In particular, attention has been directed to establishing (1) experimental conditions that do not perturb the protein–DNA binding equilibrium, (2) electrophoretic procedures that allow resolution of boundaries between single bands on autoradiograms, (3) accurate quantitation of the spatial distribution of optical density, and (4) corrections that permit resolution of a binding isotherm. The critical technical considerations are summarized below.

We have shown that the equilibrium distribution of vacant and occupied binding sites is not perturbed under our conditions of DNase exposure. This conclusion is based on a series of footprint titration experiments conducted at different conditions of DNase exposure which yielded identical isotherms. Further, the potential errors that may be introduced by the characteristics of the DNase nicking statistics were considered. Some of these errors are avoided by including unlabeled, nonspecific

DNA, which maintains a constant number of substrate sites available to the DNase, regardless of the fractional saturation of the operator sites or of the concentration of the labelled operator DNA.

A substantial number of fragments must be generated by the DNase exposure to ensure sufficient contrast between protected and nonprotected bands. However, the calculations presented in Table I and the results presented in Fig. 5 suggest that it is not necessary to introduce a large average number of nicks per fragment to achieve this. Experiments are conducted under conditions such that the fraction of binding sites that are nicked during DNase exposure is sufficiently small that the binding equilibrium is not measurably perturbed by the introduction of nicks (Table I). With increased nicking, the chance of introducing error is increased; this is due in part to imprecision in exposure to DNase, which can cause variable size distribution of multiply nicked fragments.

We quantitate correctly the spatial distribution of optical density on an autoradiogram and make the corrections necessary to transform the degree of protection from exposure to DNase to fractional saturation. These procedures are independent of the system under study. Specific considerations include (1) the use of two-dimensional scanning at a spatial resolution sufficient to resolve single bands, (2) the use of sufficient lane spacing to ensure correct background determination, (3) consideration of the linearity of experimental response to the quantity of DNA, and (4) correction for the amount of DNA loaded into each lane.

The isotherms that have been resolved by the footprint titration technique do not change whether samples are incubated for 30 or 90 min prior to exposure to DNase. Therefore, there are no slow relaxation processes that affect the results. Also, we have shown that identical isotherms result when the time of exposure to DNase is varied (while the degree of exposure is held constant by proportional changes in the concentration of DNase). Thus, there are no kinetic processes induced by the addition of DNase, due to specific interactions of the DNase with bound or unbound cI repressor, or due to the small number of nicks introduced into the DNA by the DNase, as discussed above. This is a particularly important result, given the conclusion that the chemical relaxation of the binding equilibrium is a fast process on the time scale of the DNase exposure.

The careful control over all of these factors that is produced by our experimental protocol is highlighted by the striking reproducibility of the free energies derived from footprint titration experiments. The average of the free energies derived from all of the footprint titrations of the binding of cI repressor protein to O_R1, shown in this article is -12.7 kcal/mol, with a standard deviation of 0.2 kcal/mol. Reference to Tables II, III, IV, V, and VIII shows that this is similar to the confidence limits for any single experiment, conducted under any one of the variations in experimental procedure discussed.

While these observations are necessary to establish the thermodynamic validity of the free energies of interaction resolved from such binding isotherms, they are not sufficient to prove the validity. In addition, we must demonstrate that the interaction constants derived from footprint titrations are consistent with thermodynamic properties resolved by independent studies of protein–DNA interactions. As discussed above in detail, filterbinding titration, which measures the distribution of bound and free fragments in a different manner than footprint titration, yields isotherms that are entirely consistent with those resolved by footprint titration. The resolved free energies are also consistent for single sites within a multisite system when compared to the independent external measure of overall occupancy.

The comparison of the sequential and simultaneous analysis of wild-type and mutant O_R footprint titrations in Table X shows that the free energies resolved are also self-consistent. Taken together with the other arguments presented, these observations of both internal and external consistency of the results are sufficient to prove the thermodynamic validity of the quantitative footprint titration method.

Applications

In this article we have shown that individual-site binding isotherms are uniquely suited to permit the resolution of interaction parameters for systems exhibiting cooperative interactions between multiple sites. The analysis is completely general. Any number of specific sites can be analyzed regardless of the nature of the cooperative or anticooperative interactions between them. The development of the footprint titration method, which permits resolution of individual-site isotherms, permits quantitative characterization of systems that act as critical regulators of gene transcription.

The thermodynamic parameters that are resolved from footprint titration can be used in two important ways to further the understanding of gene regulation. First, the binding affinities of the various components of a gene regulatory system can be used to deduce the mechanism of the regulation. An example of this is the successful modeling of the switch from the lysogenic-to-lytic growth stage of the lambda phage.[1] Second, the range of precisely controlled experimental conditions over which the technique is applicable will allow us to measure other thermodynamic parameters (e.g., enthalpies and entropies) to study the roles of the various noncovalent forces of interaction involved in protein–DNA binding and site recognition. We intend to take advantage of the availability of mutants to correlate the functional and energetic properties of the macro-molecules with structural variations. In conjunction with information from other techniques, it may be possible in this way to understand the physical basis of site recognition and gene regulation.

Acknowledgments

We are very grateful to Ben Turner for his assistance and advice on virtually all aspects of this project. Much of the integration of densities of autoradiograms was performed by Kevin Bootes. We thank Marc Wold and Hillary Nelson for obtaining for us the plasmids used in this study and for many helpful suggestions. We thank Joan Sarkin for patiently word-processing the manuscript. This work was supported by NIH Grant GM 24486. The P-1700 film scanner is an NIH-Facility instrument available to us at The John Hopkins University Medical School.

Section II

Macromolecular Conformation: Spectroscopy

[10] Prediction of Protein Structure

By Patrick Argos and J. K. MohanaRao

Introduction

The goal of natural science is the comprehension and prediction of empirical phenomena through the use of models and a minimum set of axioms, though the approach to the end is probably asymptotic. Molecular biology does not escape this motive. More specifically, the prediction of a protein's secondary and tertiary architecture from knowledge of sequence alone is a present and future quest. Nature folds proteins repeatedly in fractions of a second given a physiologic milieu and properly coded amino acid sequence which itself justifies the entire genetic system. Attempts to fold proteins theoretically have been numerous and yet, not successful but, nonetheless, improving.

This introduction will relate a brief overview of the efforts at secondary and tertiary structural prediction. The "how to" of the chapter, which will comprise most of the material given, will concentrate on what to do with that newly determined sequence in terms of its theoretically predicted secondary structure and comparison with other sequences. The facility of cloning and nucleotide sequencing techniques has produced such a plethora of primary structures that prediction algorithms are becoming increasingly significant in deriving quick structural models that can be tested empirically. X-Ray crystallographic techniques[1] do yield atomic resolution architectures but they are tedious requiring considerable protein material, manpower, expense, and time which are not always readily available.

Methods to predict tertiary structure have not, as yet, been successful. After all, the techniques are concerned with the difficulties of a many-bodied problem where individual atoms from residue side chains interact with a fluid cytoplasmic environment or with other side group atoms. The methods center around energy minimization; a starting polypeptide chain model based on a simplified representation of the backbone (unit peptide preferences, virtual bonds, united-atom or residue approximations, spherical representation), or an empirically deduced approximation of the fold based on known structures, secondary structural predictions, and

---

[1] T. L. Blundell and L. N. Johnson, "Protein Crystallography." Academic Press, New York, 1976.

preferred residue interactions (cf. Refs. 2–4). The methods are not generally applicable and, when at least partially successful, involve specific structures or folding patterns. For example, Richmond and Richards[5] produced a starting model for myoglobin by locating residues central to helix–helix interactions from the amino acid sequence and predicting the secondary structure, which were followed by energy minimization and resulted in five or six plausible folds of which one was essentially correct. Cohen and Sternberg[6] added further constraints such as distances between helical termini or between specific histidine side chain atoms and the protoporphyrin iron. Cohen et al.[7,8] examined so-called nucleotide binding folds, a twice repeating $\beta\alpha\beta\alpha\beta$ pattern in α/β proteins which are composed of a sequential alternating mix of helices and strands.[9] One to several tertiary models are deduced from an amino acid sequence by a combinational approach which lists possible β-sheet structures by generating all β-strand topologies and alignments, followed by application of empirical constraints (derived from known α/β structures) on topology and location of nonpolar residues to mediate sheet/helix packing. If researchers have in their possession a sequence that is not a globin or at least an α/β protein, their present hope of predicting a tertiary structure is minimal.

Secondary structure prediction methods are easier to apply and generally about 60% correct.[10–12] However, the information derived is less; that is, only a topological string of possible secondary structures (helix, sheet, turn, and coil configurations[13,14]). The 60% figure shows predictive power as 25% would be expected by random predictions given that each of the four structural types generally comprises about one-fourth of all observed protein structure.[15] Excellent and up-to-date reviews of the secondary

[2] F. A. Momany, R. F. McGuire, A. W. Burgess, and H. A. Scheraga, J. Phys. Chem. **79**, 2361 (1975).
[3] M. Levitt, J. Mol. Biol. **104**, (1976).
[4] B. Robson and D. J. Osguthorpe, J. Mol. Biol. **132**, 19 (1979).
[5] T. J. Richmond and F. M. Richards, J. Mol. Biol. **119**, 537 (1978).
[6] F. E. Cohen and M. J. E. Sternberg, J. Mol. Biol. **137**, 9 (1980).
[7] F. E. Cohen, M. J. E. Sternberg, and W. R. Taylor, J. Mol. Biol. **156**, 821 (1982).
[8] F. E. Cohen, M. J. E. Sternberg, and W. R. Taylor, J. Mol. Biol. **148**, 253 (1981).
[9] M. Levitt and C. Chothia, Nature (London) **261**, 552 (1976).
[10] W. Kabsch and C. Sander, FEBS Lett. **155**, 179 (1983).
[11] J. A. Lenstra, Biochim. Biophys. Acta. **491**, 333 (1977).
[12] P. Argos, J. Schwarz, and J. Schwarz, Biochim. Biophys. Acta **439**, 261 (1976).
[13] J. S. Richardson, Adv. Protein Chem. **34**, 167 (1981).
[14] M. G. Rossmann and P. Argos, Annu. Rev. Biochem. **50**, 497 (1981).
[15] G. E. Schulz and R. H. Schirmer, in "Principles of Protein Structure," p. 108. Springer-Verlag, New York, 1979.

and tertiary methods have been given in two texts: one by Schulz and Schirmer[15] and the other by Ghelis and Yon.[16] Ptitsyn and Finkelstein also give a good discussion on the principles of protein folding.[17]

If a researcher has a fresh sequence, the following steps are recommended in its secondary structural analysis and comparison with other sequences. (1) Predict its secondary structure. Utilize an easily applied method (hand calculated or a simple computer program); for the more complex methods, write the author and ask for the program. (2) If it is a suspected or known lipid-bound protein, use special algorithms geared to delineate transmembrane helices and exposed turn segments connecting them. (3) Check if the pattern of secondary structures follows that of an already known tertiary structure (e.g., $\beta\alpha\beta$ in nucleotide binding folds) determined by X-ray diffraction methods (for a listing of known structures, see Rossmann and Argos[14]). (4) Try an amino acid sequence homology test between the new sequence and one already known that may have a similar fold or function. The author recommends the simpler Jukes and Cantor approach.[18] If specific functions are suspected for the new protein, such as DNA binding, try to find a homologous sequence stretch similar to a family of such stretches in proteins with various overall structures but a common functional binding mode (cf. Ref. 19). (5) If the homology criterion fails, try the residue physical characteristic comparison. It is now known that proteins lacking amino acid homology can nonetheless possess the same folds and could be divergently related in their evolution.[14] (6) When comparing the new protein to families containing several structurally homologous sequences, watch for certain clues in conserved structural and functional residues (e.g., the three Gly's following β_A in nucleotide folds or the Asp following β_B for hydrogen bonding to the NAD ribose hydroxyl[20]). (7) Use such aids as helical wheels to ascertain the closeness of certain residues (e.g., charged amino acids) on one helical side that may be important structurally and functionally (e.g., hydrophobic and hydrophilic sides indicating the membrane and protein-interior helical facedness in structures composed of bundled transmembrane helices[21] or the clustered basic residues in NH_2-terminal regions of plant viral

[16] C. Ghelis and J. Yon, "Protein Folding," p. 177. Academic Press, New York, 1982.

[17] O. B. Ptitsyn and A. V. Finkelstein, *Q. Rev. Biophys.* **13,** 339 (1980).

[18] T. H. Jukes and C. R. Cantor, *in* "Mammalian Protein Metabolism" (H. N. Munro, ed.), Vol. III, p. 21. Academic Press, New York, 1969.

[19] Y. Takeda, D. H. Ohlendorf, W. F. Anderson, and B. W. Matthews, *Science* **221,** 1020 (1983).

[20] J. Otto, P. Argos, and M. G. Rossmann, *Eur. J. Biochem.* **109,** 325 (1980).

[21] P. Argos, J. K. MohanaRao, and P. A. Hargrave, *Eur. J. Biochem.* **128,** 565 (1982).

coat proteins for possible nucleic acid interactions[22]). β-Strand sequences sometimes show alternating polar and nonpolar residues.[23,24]

Secondary Structure Prediction

Techniques for Soluble Proteins

The methods to delineate regions of ordered structures (α helix, β-strand, turn, and coil conformation[13,14,25]) in soluble protein amino acid sequences can be classified into the probabilistic, physicochemical, and information theory techniques.[15,16] The statistical approaches began with Ptitsyn and Finkelstein,[26] passed through further development by Dirkx,[27,28] Lewis et al.,[29] Crawford et al.,[30] Kabat and Wu,[31,32] and perhaps culminated with Chou and Fasman[33,34] and Nagano.[35,36] They all tend to examine frequencies with which residue singlets, doublets, and even triplets occur within known secondary structures and then perform some normalization to determine propensities or preferences for an amino acid to appear in a structural type. For example, Chou and Fasman[33] divide the compositional occurrence of a specific amino acid within a particular ordered structure by its composition within all the proteins examined such that conformational preferences above 1.00 show a structural bias and those below 1.00, a structural avoidance. These propensities can be used to predict structural sequence spans if a sequential cluster of generally four or more preferences greater than some threshold are found. Each of the researchers can also add their own particularized rules to the automated procedures.[33]

[22] P. Argos, Virology 110, 55 (1981).
[23] P. Argos and J. Palau, Int. J. Peptide Protein Res. 19, 380 (1982).
[24] J. Palau, P. Argos, and P. Puigdomenech, Int. J. Peptide Protein Res. 19, 394 (1982).
[25] G. E. Schulz and R. H. Schirmer, "Principles of Protein Structure," p. 66. Springer-Verlag, New York, 1979.
[26] O. B. Ptitsyn and A. V. Finkelstein, Biofizika (USSR) 15, 757 (1970).
[27] J. Dirkx, Arch. Int. Physiol. Biochim. 80, 185 (1972).
[28] F. Beghin and J. Dirkx, Arch. Int. Physiol. Biochim. 83, 167 (1975).
[29] P. N. Lewis, F. A. Momany, and H. A. Scheraga, Proc. Natl. Acad. Sci. U.S.A. 68, 2293 (1971).
[30] J. L. Crawford, W. N. Lipscomb, and C. G. Schellman, Proc. Natl. Acad. Sci. U.S.A. 70, 538 (1973).
[31] E. A. Kabat and T. T. Wu, Biopolymers 12, 751 (1973).
[32] T. T. Wu and E. A. Kabat, J. Mol. Biol. 75, 13 (1973).
[33] P. Y. Chou and G. D. Fasman, Biochemistry 13, 222 (1974).
[34] P. Y. Chou and G. D. Fasman, Adv. Enzymol. 47, 45 (1978).
[35] K. Nagano, J. Mol. Biol. 76, 241 (1977).
[36] K. Nagano, J. Mol. Biol. 75, 401 (1973).

The information approach of Robson and Pain,[37,38] Robson and Suzuki,[39] Garnier *et al.*,[40] and Maxfield and Scheraga[41] is perhaps the most mathematically sophisticated of the procedures. The benefit is the ability to examine residue interactions in determining the structural types as far as eight residues away in either terminal direction and to weight the validity of the interactive statistics according to the sufficiency of the information and data available.[40] As the data base of known structures increases, the reliability of this approach should also increase.

The physicochemical models utilize certain polar and nonpolar patterns of amino acids in secondary structures or rely on thermodynamic and statistical mechanical theory for determining conformational preferences. Lim,[42,43] Palau *et al.*,[24] Busetta and Hospital,[44] and Cohen *et al.*[45] examine the polar properties of residues in the relative positions 1–2–5, 1–4–5, 1–4, and 1–5 in possible α-helices where one side is likely polar to face the external hydrophilic environment and the other nonpolar to lie against the protein's globular structure. An alternating polar and nonpolar pattern can also be found in β-strands. In one mathematical form or another, rules and formulas are devised to search for spans with these characteristics to predict ordered structures. The statistical mechanical approach determines the probability of a residue to be in a given conformation based on experimentally and theoretically determined parameters from structural transitions of homo- and copolymers. The works of Tanaka and Scheraga,[46–49] Finkelstein and Ptitsyn,[50,51] and Burgess *et al.*[52] provide examples of this bent.

[37] B. Robson, *Biochem. J.* **141**, 853 (1974).
[38] B. Robson and R. H. Pain, *Biochem. J.* **141**, 883 (1974).
[39] B. Robson and E. Suzuki, *J. Mol. Biol.* **107**, 327 (1977).
[40] J. O. Garnier, D. J. Osguthorpe, and B. Robson, *J. Mol. Biol.* **120**, 97 (1978).
[41] F. R. Maxfield and H. A. Scheraga, *Biochemistry* **18**, 697 (1976).
[42] V. I. Lim, *J. Mol. Biol.* **88**, 857 (1974).
[43] V. I. Lim, *J. Mol. Biol.* **88**, 873 (1974).
[44] B. Busetta and M. Hospital, *Biochim. Biophys. Acta* **701**, 111 (1982).
[45] F. E. Cohen, R. M. Abarbanel, I. D. Kuntz, and R. J. Fletterick, *Biochemistry* **22**, 4894 (1983).
[46] S. Tanaka and H. A. Scheraga, *Proc. Natl. Acad. Sci. U.S.A.* **72**, 3802 (1975).
[47] S. Tanaka and H. A. Scheraga, *Macromolecules* **9**, 159 (1976).
[48] S. Tanaka and H. A. Scheraga, *Macromolecules* **9**, 812 (1976).
[49] S. Tanaka and H. A. Scheraga, *Macromolecules* **9**, 945 (1976).
[50] A. V. Finkelstein and O. B. Ptitsyn, *J. Mol. Biol.* **103**, 15 (1976).
[51] O. B. Ptitsyn and A. V. Finkelstein, *Biopolymers* **22**, 15 (1983).
[52] A. W. Burgess, P. K. Ponnuswamy, and H. A. Scheraga, *Isr. J. Chem.* **12**, 239 (1974).

Problems in the Prediction Techniques

There are problems associated with the techniques despite their 60% accuracy level.[10] (1) It is difficult to define exactly the residue span encompassing a certain ordered structure in known protein folds; the structures often make the transition from coil to order in a continuous fashion often requiring several residues. For example, it takes up to about four residues to start up a formal α-helix (cf. Ref. 23); there are also at least three turn types.[25] Crystallographers often use visually estimated hydrogen bonding patterns in their stick models to delineate the secondary structural regions. Others prefer automated algorithms applied to the crystallographic atomic coordinates and are based on dihedral angles (cf. Refs. 41,52) or C_α–C_α distances and α torsion angle patterns (e.g., Refs. 29,30,53). Levitt and Greer[54] have to date performed the most comprehensive automation for all the structural types; Kabsch and Sander[55] promise an even more consistent and extensive approach. Thus the inaccuracy of prediction methods is often found at the termini of sheets and helices which themselves are in dispute. Nonetheless, the methods can miss or overpredict secondary structural regions (e.g., Refs. 12,15,56,57). (2) There is a gamut of criteria to assess prediction accuracy ranging from the simple percentage correct to more sophisticated correlation coefficients. Kabsch and Sander[10] present the latest discussion on this problem but the bottom line generally leads to a rough 60% correctness. This value certainly outdoes the expected 25% by chance.[15] (3) Another controversy is associated with the size of the data base. It appears that the data sample used to determine singlet amino acid propensities is saturated[58] while there is an insufficient number of known structures to handle doublet and triplet frequencies though attempts have been made (e.g., Refs. 34,59). (4) The fourth enigma is the ambiguity of some prediction rules as well as that of the predictions themselves with overlapping helix and sheet determinations. Rules are often given to clear up the overlaps but they do not always work. Argos *et al.*[12] suggest examining several prediction methods with the consensus generally resolving the confusion.

[53] C. Ramakrishnan and K. V. Soman, *Int. J. Peptide Protein Res.* **20,** 218 (1982).
[54] M. Levitt and J. Greer, *J. Mol. Biol.* **114,** 181 (1977).
[55] W. Kabsch and C. Sander, *Biopolymers* **22,** 2577 (1983).
[56] G. E. Schultz, C. D. Barry, J. Friedman, P. Y. Chou, G. D. Fasman, A. V. Finkelstein, V. I. Lim, O. B. Ptitsyn, E. A. Kabat, T. T. Wu, M. Levitt, B. Robson, and K. Nagano, *Nature (London)* **250,** 140 (1974).
[57] B. W. Matthews, *Biochim. Biophys. Acta* **405,** 442 (1975).
[58] P. Argos, M. Hanei, and R. M. Garavito, *FEBS Lett.* **93,** 19 (1978).
[59] A. S. Kalaskar and V. Ramabrahman, *Int. J. Peptide Protein Res.* **22,** 83 (1983).

Which Method to Use

Which method to use is a matter of the applicant's ability to handle the computer and the degree of accuracy desired. Lenstra[60] recommends the methods of Nagano[35,36] and Argos *et al.*[12] while Kabsch and Sander[10] suggest Lim[42,43] and Garnier *et al.*[40] The better methods have about a 60% success rate. These techniques generally require a hefty computer program; the authors are often amenable to sending out their automated schemes if requested. Nonetheless, the Chou–Fasman technique[34] is the most popular and easiest to apply (can be done by hand) despite rule ambiguities and frustrating overlapping predictions. Kabsch and Sander[10] rate this method as 50% correct. If struggling with the computer is out of the question, then Chou–Fasman is desirable. The authors have found a simple approach that often overcomes overlapping predictions and application of ambiguous rules. The conformational preferences for sheet, helix, and turn configurations are listed under each amino acid of the sequence. Averages ("smoothing") can be taken for each successive cluster of three or five residues for each propensity type; this mean value is then assigned to the middle residue [e.g., the average for residues (i), $(i + 1)$, and $(i + 2)$ is given to $(i + 1)$]. This process can be repeated for the entire chain two to four times using the averages determined from the previous cycle for the next cycle; the terminal residues will simply maintain their original values. Peaks will form and the secondary structural spans chosen will correspond to the highest value of the three particular preferences. If the peak maximum is not greater than 1.0, the structure is unlikely for the considered span. Helices are likely to be six or more amino acids in length while strands and turns should be at least three residues. If a family of homologous sequences are known, more accurate predictions[12] are possible by averaging the smoothed values for all the sequences. An example is shown in Fig. 1 for three preprosomatostatins from man, catfish, and anglerfish. In this case, the predicted structure is largely helical and turn conformations. If the protein is likely to belong to one of the four structural classes (all-α, all-β, α/β, or $\alpha + \beta^{9,61}$) or contains all anti-parallel or all parallel β-strands, propensities derived from proteins known to possess the given structures may be used for more accuracy. The singlet amino acid preferences for the general data base are given by Chou and Fasman[33] and Levitt,[62] for the four protein classes by Palau *et al.*,[24] and for the two strand topologies by Lifson and Sander.[63]

[60] J. A. Lenstra, *Biochim. Biophys. Acta* **491**, 333 (1977).
[61] C. Chothia and J. Janin, *Nature (London)* **256**, 705 (1975).
[62] M. Levitt, *Biochemistry* **17**, 4277 (1978).
[63] S. Lifson and C. Sander, *Nature (London)* **282**, 109 (1979).

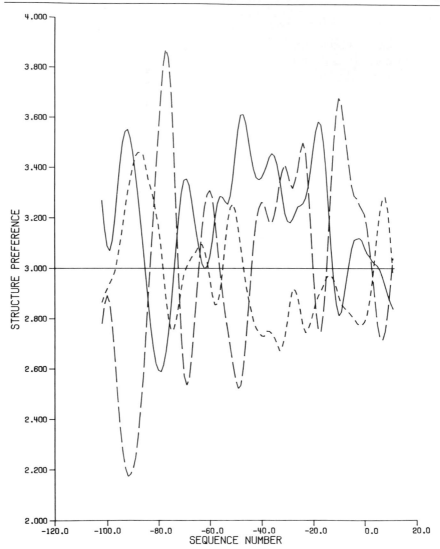

FIG. 1. Plots for the combined anglerfish, rat, and catfish preprosomatostatin smoothed curves for a particular secondary structure propensity *versus* the rat sequence number. The sequences have been aligned by Argos *et al.*[92] The parameters used include helix (——), turn (— —), and β-strand (– – –) potentials.

Prediction of Membrane-Bound Proteins

The only tertiary structure known to reasonably high resolution (around 6 Å) is that of the purple membrane protein of *Halobacterium halobium* (bacteriorhodopsin) determined from electron scattering techniques.[64,65] The structure is believed to consist of seven helices that traverse the membrane successively in a topological anti-parallel fashion and that are connected by exposed, largely turn segments on the two surfaces of the lipid bilayer. The helical spans are 20 to 25 residues in length and are largely composed of typically hydrophobic residues. Two principal methods have been devised to delineate the buried segments in sequences of other lipid-bound proteins. Kyte and Doolittle[66] define a residue hydropathy index which is plotted and smoothed (as previously discussed) against amino acid sequence number. High peaks covering a range of 20 or so residues suggest likely transmembrane helices with other spans exposed to the cellular or cytoplasmic millieu. Argos *et al.*[21] determine a weighted sum of five smoothed residue physical characteristic curves: bulk, polarity, hydration potential, turn propensity, and transfer free energy in burying a helix in a lipid environment. The curves are normalized to a value of 0.0. The weighting factors are determined to yield the best fit between predicted helices and those observed for bacteriorhodopsin. The procedures for delineating transmembrane helices in the weighted characteristic plot include (1) peak maxima greater than 1.0 and continuous positive values within the peak stretching for at least 13 residues with a subsequent expansion of the segment (if necessary) to the first encountered charged amino acid at either terminus, resulting in a region of at least 16 residues to allow about 25 Å to span the lipid portion of the bilayer or (2) peak maxima greater than 0.6 and continuous positive values stretching at least 18 residues, and (3) no more than three charged amino acids within any transmembrane span. The weighted curve for bacteriorhodopsin is shown in Fig. 2; the rules just described allowed the recognition of seven helices. Table I lists the predicted spans and those determined by Trewhella *et al.*[65] from the empirical data; their agreement is at the 85% level. The lipid-traversing segments determined for bovine rhodopsin are given in Fig. 3. The propensity values for amino acids to be within a buried helix are listed in Table II; the data base consisted of over 1100 residues. It should be noted that the transmembrane helices may contain residues not typical in helices of soluble proteins, as Pro and Gly.[67]

The clustered seven helices of bacteriorhodopsin can be enveloped in

[64] R. Henderson and P. N. T. Unwin, *Nature (London)* **271,** 15 (1975).
[65] J. Trewhella, S. Anderson, R. Fox, E. Gogal, S. Khan, and D. Engelman, *Biophys. J.* **42,** 233 (1983).
[66] J. Kyte and R. Doolittle, *J. Mol. Biol.* **157,** 105 (1982).
[67] J. K. MohanaRao, P. A. Hargrave, and P. Argos, *FEBS Lett.* **156,** 165 (1983).

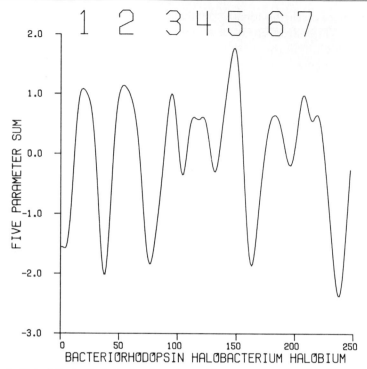

FIG. 2. Plot of the bacteriorhodopsin amino acid sequence number [H. G. Khorana, G. E. Gerber, W. C. Herlihy, C. P. Gray, R. J. Anderegg, K. Nihei, and K. Biemann, *Proc. Natl. Acad. Sci. U.S.A.* **76,** 5045 (1979)] *versus* a weighted five-parameter characteristic value for a given amino acid. The seven positive peaks suggest the transmembrane helical segments delineated from the curve by the rules discussed in the text. The troughs indicate exposed turn segments.

a distorted cylindrical surface with four helices lined up on one side and three fairly aligned helices on the other side such that the two helical layers come together (Fig. 4) forming packed side chains within the interior of the structure and the remaining side groups exposed to the lipid.[64,68] A "helical wheel" analysis appears to support this model. Wheels[69] are projections down the helical axis with successive C_α positions rotated to 100° along the projected peptide backbones. The direction of the projected side chains are indicated by spokes emanating from the wheel center. By noting residues within the helix that possess a transmembrane propensity value less than 1.0 (Table I), the helical side facing

[68] D. M. Engelman and G. Zaccai, *Proc. Natl. Acad. Sci. U.S.A.* **77,** 5894 (1980).
[69] M. Schiffer and A. B. Edmundson, *Biophys. J.* **7,** 121 (1967).

TABLE I
BACTERIORHODOPSIN HELICAL
SEQUENCE REGIONS[a]

Trewhella et al.	Argos et al.
9–29	10–29
42–62	44–69
79–99	86–101
107–127	108–128
136–156	135–156
176–196	176–193
204–224	203–224

[a] The bacteriorhodopsin helical sequence regions as determined by Trewhella et al.[65] from diffraction data and model analysis and as predicted by the residue physical characteristic algorithm of Argos et al.[21]

the protein core can be suggested. Figure 5 displays the wheel for a predicted helix in an *N. crassa* proteolipid subunit. Such facedness has been suggested for bacteriorhodopsin[68] and other membrane-bound proteins.[21] Other approaches include a consideration of all possible ways (nine) to halve the wheel and calculation of differences in the summed parameters (e.g., polarity or hydrophobicity) for residues on both sides. The sidedness can be selected by the register of the wheel halves producing the largest differences.[70]

With the proposed structural model at hand, various questions can be considered. Will the charges facing the protein core cancel each other through electrostatic interaction? Have all important functional side chains (e.g., retinal binding lysine in rhodopsins) been proposed within the protein core and within transmembrane segments? Does the molecule act as a dipole with more positively (or negatively) charged side chains on the exposed surfaces of one membrane side *versus* the other (e.g., bovine rhodopsin[71]?). Do the helices contain Thr and Ser which can hydrogen bond to the main chain and import stability? Are "stacked" interactions possible between Phe, Tyr, His, and Trp (e.g., Phe in relative helical positions 1 and 4 could allow $\pi-\pi$ electron overlap for structural stabil-

[70] P. A. Hargrave, J. H. McDowell, R. J. Feldman, P. H. Atkinson, J. K. MohanaRao, and P. Argos, *Vision Res.* **24**, 1487 (1984).
[71] P. A. Hargrave, J. H. McDowell, D. R. Curtis, J. Wang, E. Juszczak, S. L. Fong, J. K. MohanaRao, and P. Argos, *Biophys. Struct. Mech.* **9**, 235 (1983).

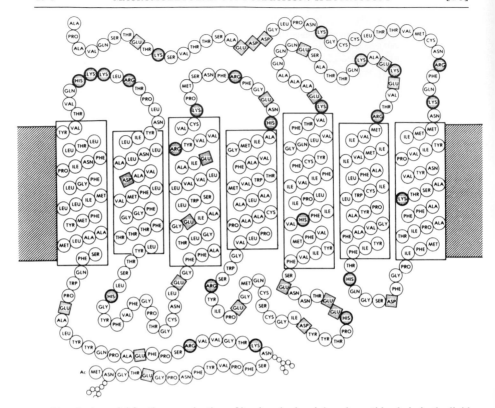

FIG. 3. A model for the organization of bovine rhodopsin's polypeptide chain in the lipid bilayer of the disc membrane.[71] The seven transmembrane helices were predicted by the algorithm of Argos *et al.*[21] The amino-terminal Met is exposed at the internal aqueous surface of the disc membrane while the carboxyl-terminal Ala is exposed at the external surface. Positively charged amino acids (Lys, Arg, His) are shown as shaded heavy circles and acidic residues (Glu, Asp) as shaded squares. Striped areas at each side of the drawing indicate the low polarity portion of the bilayer (fatty acid side chains and the lower half of the phospholipid head groups).

ity[70])? In a recent paper delineating a structural model for cytochrome *b*, Widger *et al.*[72] have found two transmembrane helices each containing two stacked histidine side chains separated by four helical turns which are proposed to provide octahedral ligands to two heme irons such that the protoporphyrin planes are aligned for electron transfer. What about basic and acidic residues at the helical termini that complement the main chain

[72] W. R. Widger, W. A. Cramer, R. G. Herrmann, and A. Trebst, *Proc. Natl. Acad. Sci. U.S.A.* **81,** 674 (1984).

TABLE II
THE MEMBRANE-BURIED PREFERENCES FOR THE
20 AMINO ACIDS CALCULATED FROM A
1125-RESIDUE DATA BASE[21a]

Amino acid	Normalized preference
Met	2.96
Leu	2.93
Phe	2.03
Ile	1.67
Ala	1.56
Cys	1.23
Val	1.14
Trp	1.08
Thr	0.91
Ser	0.81
Pro	0.76
Tyr	0.68
Gly	0.62
Gln	0.51
Arg	0.45
His	0.29
Asn	0.27
Glu	0.23
Lys	0.15
Asp	0.14

[a] The propensities are normalized to a value of 1.0 which indicates a neutral preference while those greater (or less) than 1.0 refer to strong (weak) propensities to be within a transmembrane helix.

dipole?[73] Are Glys and Pros contained within spatially consistent helical positions such that bends or kinks would allow room for an inserted prosthetic group?[74]

Amino Acid Sequence Homology

Knowledge of a protein's secondary structure is useful; but it is not a tertiary architecture with its attendant implications for active site mecha-

[73] O. B. Ptitsyn, *J. Mol. Biol.* **42,** 501 (1969).
[74] E. A. Dratz and P. A. Hargrave, *Trends Biochem. Sci.* **8,** 128 (1983).

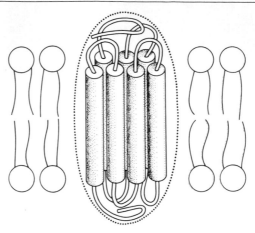

FIG. 4. Helical bundle model (4 + 3 arrangement) for bovine rhodopsin based on the bacteriorhodopsin model.[71] It must be noted that the helical connectivity and the choice of NH$_2$- and COOH-terminal helices for the model shown have not been experimentally demonstrated in bacteriorhodopsin.

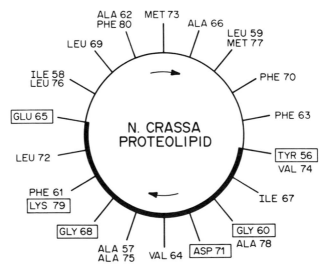

FIG. 5. Helical wheel for a predicted transmembrane segment[21] in the proteolipid subunit of the mitochondrial adenosinetriphosphatase from *Neurospora crassa* [W. Sebald, W. Machleidt, and E. Wachter, *Proc. Natl. Acad. Sci. U.S.A.* **77,** 785 (1980)]. Amino acids with membrane-buried propensities (Table I) less than 1.0 are boxed. The side of the helix suggested to face the protein interior is indicated by a thick line.

nisms, substrate binding pocket, and external recognition of interactive enzymes or receptors. Further insights into the tertiary architecture cannot be had from three-dimensional folding predictions as they are presently too specific and the methods are underdeveloped. However, if a given sequence can be related to other sequences for which the atomic resolution structure is known, some of the benefits of the structure–function relationship are possible. Even if the related sequences cannot be tied to already determined folds, conservation of certain residues in sequences aligned either through homology or similar secondary structural patterns can suggest mechanistic models for further experimentation, avoiding "stabs in the dark." Tests for homology can be one of the keys in searching for similar sequences.

The method most utilized by the authors for comparing primary structures is that of Jukes and Cantor[18] which is based on the minimum base change per codon (MBC/C) relationships between amino acids. Keim et al.[75] and Fitch and Smith[76] judge this criterion as one of the best and most sensitive, perhaps even more so than direct nucleotide sequence comparisons. Every amino acid span of length L residues from the first protein is aligned with all stretches of length L in the second protein. The total minimum base difference (MBD) for each of the possible oligopeptide alignments is determined by summing the MBC/C between paired amino acids in the matched L-residue spans. The length L is generally chosen as 10 to 20 residues, a number recommended by Jukes and Cantor to achieve statistical significance and yet make reasonable allowances for possible gaps or deletions. Significance is tested by calculating the ratio (P_{obs}/P_{calc}) for all possible MBD values resulting from a comparison of the two proteins. P_{obs} is the frequency with which a given MBD is observed in comparing all segments from the two proteins, while P_{calc} is the expected frequency calculated from the amino acid compositions of the proteins. As the frequence ratio becomes increasingly larger than 1, the significance of the homology would become greater.

If the P_{obs}/P_{calc} ratios are generally large for oligopeptide alignments with about the same stagger value (difference between the first sequence number of one oligopeptide and the corresponding aligned sequence position of the other), it is easy to align the sequences. Alignments are generally not significant if P_{obs}/P_{calc} is less than 2.0 and the method loses significance in matching complete sequences if the mean MBC/C is much greater than about 1.00 with a random relationship being around 1.45.[75] However, it happens often that only a few oligopeptides show a signifi-

[75] P. Keim, R. L. Heinrikson, and W. M. Fitch, *J. Mol. Biol.* **151,** 179 (1981).
[76] W. M. Fitch and T. F. Smith, *Proc. Natl. Acad. Sci. U.S.A.* **80,** 1382 (1983).

cant probability ratio at an expected stagger value. Alignment is unlikely if an NH-terminal portion of one molecule is aligned with a COOH-terminal region of the other, especially if the two sequences have roughly about the same number of residues. Nonetheless, a few homologous oligopeptides could indicate that the two proteins still possess a similar folding pattern but without recognizable homology for the entire sequences or that only certain spans important for function are structurally and sequentially homologous. The latter case is exemplified in the comparison of human hypoxanthine-guanine phosphoribosyltransferase and bacterial glutamine phosphoribosylpyrophosphate amidotransferase where only a 35-residue stretch, possibly important in the function of the two proteins, was found homologous.[77] Another example involves the DNA-binding region of several repressor proteins where a two-helix fold and sequence homology is functionally preserved.[78]

A sift is suggested through all the possible oligopeptide alignments with a mean MBC/C of 1.00 or less and with an expected stagger value and alignment of the sequences despite the noise from matches with inappropriate staggers or which would not allow a contiguous alignment of the residues with gaps of reasonable length. The comparison of bacterial arabinose and galactose binding proteins exemplifies this situation where secondary structure predictions were also utilized to effect the alignments with little homology.[79] Recent crystallographic evidence[80,81] has proved the sugar-binding protein match to be correct. Table III lists the homology statistics. It is important to watch for the conservation of Pro, Gly, and charged residues. An assessment of the degree of match can be performed through correlation coefficients where some residue physical characteristic (e.g., polarity or hydrophobicity) is attached to each of the aligned residues and correlations [see Eq. (1)] calculated between the two resulting parametric series.[82]

Sequence Comparisons Based on Residue Characteristics

There are numerous examples of proteins that have strong functional and structural correspondence despite a random relationship in the

[77] P. Argos, M. Hanei, J. M. Wilson, and W. N. Kelley, *J. Biol. Chem.* **258,** 6450 (1983).
[78] D. H. Ohlendorf, W. F. Anderson, M. Lewis, C. O. Pabo, and B. W. Matthews, *J. Mol. Biol.* **169,** 757 (1983).
[79] P. Argos, W. C. Mahoney, M. A. Hermodson, and M. Hanei, *J. Biol. Chem.* **256,** 4357 (1981).
[80] N. K. Vyas, M. N. Vyas, and F. A. Quiocho, *Proc. Natl. Acad. Sci. U.S.A.* **80,** 1792 (1983).
[81] S. L. Mowbray and G. A. Petsko, *J. Biol. Chem.* **258,** 7991 (1983).
[82] D. D. Jones, *J. Theor. Biol.* **50,** 167 (1975).

TABLE III

STATISTICS FOR ALIGNING THE RESIDUES OF
ARABINOSE (ABP) AND GALACTOSE (GBP)
BINDING PROTEINS FROM *E. coli*[79] USING THE
METHOD OF JUKES AND CANTOR[18a]

MBD	Total	Used	P_{obs}/P_{calc}
11	3	3	2.0
12	17	13	2.5
13	146	10	1.3
14	303	21	1.6

[a] MBD refers to the total minimum base difference which is the sum of the MBC/Cs in aligning 15-residue segments. P_{obs}/P_{calc} are the
probability ratios associated with the given
MBDs. "Total" refers to the number of times
the MBD was achieved in aligning all possible
15-residue spans in the two proteins, each
containing about 300 residues. "Used" indicates the number of 15-residue alignments that
were consistent with the overall sequence
match for the two binding proteins given by
Argos *et al.*[79] The "used" alignments provided homology for about one-fourth of the
residues (ABP amino acids 45 to 95 and 195 to
215); the remainder of the primary structures
was matched by secondary structure predictions and visual inspection which attempted to
preserve key residues as Pro and Gly. The
alignment has been shown to be essentially
correct.[80,81]

MBC/C for amino acids associated with spatially equivalenced peptide C_α
atoms.[83] Both domains of bovine liver rhodoanese show excellent structural homology, suggesting gene duplication, and yet the equivalenced
amino acids display a mean MBC/C of 1.27.[75,84] T4 phage and hen egg
white lysozymes bind six-membered oligosaccharides and possess good
structural correspondence while the mean MBC/C for structurally related
residues is near random.[85,86] Apparently the evolutionary memory of

[83] M. G. Rossmann and P. Argos, *Mol. Cell. Biochem.* **21,** 161 (1978).
[84] J. H. Ploegman, G. Drent, K. H. Kalk, and W. G. Hol, *J. Mol. Biol.* **123,** 557 (1978).
[85] M. G. Rossmann and P. Argos, *J. Mol. Biol.* **5,** 75 (1976).
[86] B. W. Matthews, M. G. Grutter, W. F. Anderson, and S. J. Remington, *Nature (London)*
290, 334 (1981).

structure is preserved over a longer time period than that for nucleotide sequence. It would be expected for two proteins with similar folds that their residue characteristics would be preserved sequentially in their primary structures. A method has been developed by Argos et al.[77] to recognize these similar characteristic trends.

The physical parameters selected to characterize the 20 amino acids include the experimental hydration potential of Wolfenden et al.,[87] the residue surrounding hydrophobicity statistically determined by Manavalan and Ponnuswamy,[88] the Chou–Fasman conformational preference parameters for α-helical, β-strand, and reverse turn configurations as calculated by Levitt,[62] and the residue polarity listed by Jones.[82] These characteristics represent the major forces thought to be required for proper protein folding (cf. Refs. 89,90). The parameters are listed and discussed by Argos et al.[77]

Plots of the amino acid sequence number *versus* a given parameter value for the particular residue in the sequence are calculated for the two proteins to be compared. The sequential, local trend in characteristic values is not easily discerned without plot smoothing accomplished by determining a least-square line for all successive five-point groups to calculate points for the smoothed graph. The best line is found through five points on the plot corresponding to residues (i) to ($i + 4$); the line is then used to calculate the new parametric value for residue ($i + 2$). The procedure is repeated for all possible five-residue clusters, constituting one cycle of smoothing. With newly calculated values, the entire process is repeated resulting in the second cycle. The plots for all six parameters are subjected to three cycles of smoothing. The least-square lines determined for the five residues at the NH_2 and COOH termini are used to determine the parametric values for the first and last two amino acids in the sequence, thereby mitigating end effects. Averaging the parameters over five residues as previously discussed is almost as effective.

Cross-correlation coefficients are calculated for a given residue characteristic between already smoothed plots resulting from the pair of amino acid sequences to be compared. The correlations (CCF) are given by[82]

$$\text{CCF} = \left[\sum_{j=1}^{n} (X_j - \bar{X})(Y_j - \bar{Y}) \right] \Big/ \left[\sum_{j=1}^{n} (X_j - \bar{X})^2 \sum_{j=1}^{n} (Y_j - \bar{Y})^2 \right]^{1/2} \quad (1)$$

[87] R. V. Wolfenden, P. M. Cullis, and C. C. F. Southgate, *Science* **206**, 575 (1979).
[88] P. Manavalan and P. K. Ponnuswamy, *Nature (London)* **275**, 673 (1978).
[89] T. E. Creighton, *Biophys. Mol. Biol.* **33**, 231 (1978).
[90] S. French and B. Robson, *J. Mol. Evol.* **19**, 171 (1983).

where X_j and Y_j are the parametric values taken from the respective characteristic curves for protein X and protein Y, n is the number of residues in the plot sequence segments to be correlated, and $\bar{X}$ and $\bar{Y}$ are the respective averages for the characteristic values used in the summations. The correlations are calculated for each residue parameter and for several successive lags (or phases) between the curves. For example, the smoothed hydrophobicities for the 120-residue protein Y can be correlated with all successive 120-residue hydrophobicity segments in the 299-residue protein X. A lag of zero would result in a cross-correlation between hydrophobicities where residue 1 of protein Y is matched with residue 1 of protein X, $2Y$ with $2X$, and so forth until $120Y$ with $120X$. A lag of $+10$ would pair X residue 1 with Y residue 11, $2X$ with $12Y$, and so forth to $120X$ with $130Y$. A lag of -5 would match $6X$ and $1Y$, $7X$ with $2Y$, ..., $120X$ with $115Y$. In this case, the first five residues of protein X would not be utilized in the summations. Cross-correlation values of 1.0, 0.0, and -1.0 would, respectively, refer to perfect, random, and oppositely phased relationships between the X and Y hydrophobicity curves.

In comparing two protein sequences for similar sequential residue characteristics, correlation-*versus*-lag curves are calculated over some successive lag range (e.g., -50 to $+100$) for each of the physical properties chosen appropriate for a particular case. In general, the six parameters listed here are adequate. The correlation-*versus*-lag plots are then summed for each parameter. A strong, dominating peak in the resulting curve could suggest a similar fold for the two proteins compared.

The summed characteristic correlation procedure has to date been applied to three protein types: phosphoribosyltransferases,[77] eye lens crystallins,[91] and somatostatin hormones.[92] Detailed results are given in each of the references. Figure 6 shows the six-parameter correlation-*versus*-lag curve in comparing the NH$_2$-terminal 120 residues of human hypoxanthine-guanine phosphoribosyltransferase with the 299-residue bacterial ATP phosphoribosyltransferase. The plot shows a good correlation peak for the NH$_2$-terminal residues of each protein which are likely to form a nucleotide binding domain as supported by secondary structure predictions and experiment.[77] Figure 7 displays the sum of the residue characteristics against sequence number for the two proteins. Control experiments were also run among proteins that are known to be structurally and functionally correlated as well as among those not related; it was found that a summed correlation maximum greater than 2.0 (average cor-

[91] P. Argos and R. J. Siezen, *Eur. J. Biochem.* **131**, 143 (1983).
[92] P. Argos, W. L. Taylor, C. D. Minth, and J. E. Dixon, *J. Biol. Chem.* **258**, 8788 (1983).

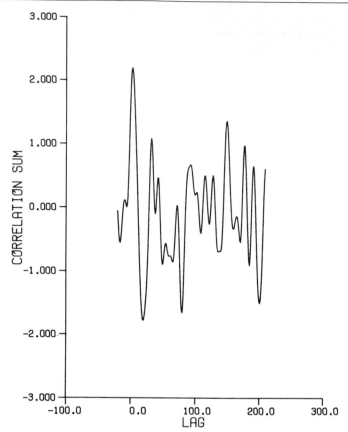

FIG. 6. Sum of the correlation coefficients *versus* the register (lag) of the sequence alignment for six amino acid physical characteristic curves in comparing the NH-terminal portion (residues 1 to 120) of human hypoxanthine-guanine phosphoribosyltransferase with the entire sequence of ATP phosphoribosyltransferase from *S. typhimurium*.[77]

relation for the six characteristics at 0.33) suggests a likelihood of similar folds while peaks less than 1.4 would indicate no such relationship.

Comparison of the eye lens α-, β-, and γ-crystallins by the correlation method affords a good example of recognizing internal folding repeats. X-Ray crystallography has shown that γ-crystallin[93] consists of four similarly folded domains (four anti-parallel β-strands and short connecting reverse turns in each motif), each about 45 residues in length. The pattern

---

[93] T. Blundell, P. Lindley, L. Miller, D. Moss, C. Slingsby, L. Tickle, B. Turnell, and C. Wistow, *Nature (London)* **289,** 771 (1981).

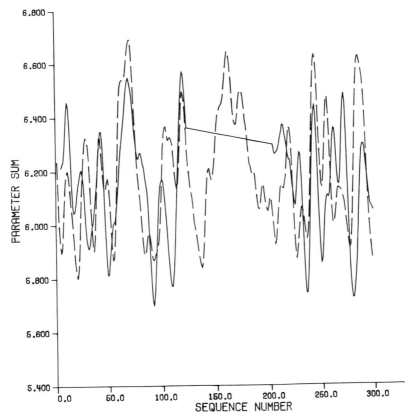

FIG. 7. The sum of the six, smoothed residue physical characteristic plots for human hypoxanthine-guanine phosphoribosyltransferase (solid curve) and bacterial ATP phosphoribosyltransferase (dashed curve) *versus* the amino acid sequence number of the ATP molecule. The correlation of the respective NH$_2$-terminal regions is clear; the straight line portion in the hypoxanthine-guanine plot indicates a possible gap in achieving the best alignment of the COOH-terminal regions of the two proteins.[77]

is suspected for β-crystallins[91,94,95]; α-crystallins (αA and αB) defy all sequence homology tests with β or γ despite their similar residue lengths and possible functions. Figure 8 displays the correlation-*versus*-lag curves for comparisons between the NH$_2$-terminal half of β-crystallin with the entire γ-chain and with the entire αA-crystallin. Peaks are found about

[94] G. Wistow, C. Slingsby, T. Blundell, H. Driessen, W. DeJong, and H. Bloemendal, *FEBS Lett.* **133,** 9 (1981).
[95] R. J. Siezen and P. Argos, *Biochim. Biophys. Acta* **748,** 56 (1983).

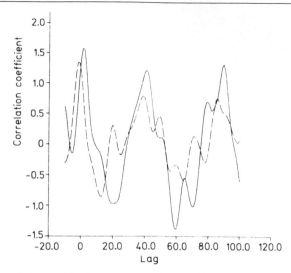

FIG. 8. Plots of the combined cross-correlation coefficients *versus* the lag for comparisons between the NH-terminal half of β-crystallin (residues 16 to 104) and the γ chain (solid line) or αA-crystallin (dashed line).[91]

every 45 residues suggesting a four-motif pattern for β- and α-crystallins. Minor peaks appear about every 10 amino acids, perhaps pointing to the repeating β-strand and reverse turn pattern. For the crystallin case only three physical parameters were utilized that were more appropriate to describe the known γ-crystallin structure (hydrophobicity, turn preference, and anti-parallel β-strand propensities[91]). Combined correlation-*versus*-lag plots can be added to clarify the results even further. Figure 9 shows plots of crystallin correlation curves resulting from the addition of the four lag profiles involving comparisons between the NH₂-terminal or COOH-terminal halves of γ-crystallin with the entire γ and β chains or with the complete αA and αB chains. Self-correlations are also possible. For example, if it is suspected that a protein has two repeating domains, the parametric profiles for the NH₂-terminal half can be run against the profile of the entire chain; a high correlation peak should appear at a lag value near the half-way sequence point. Another example might include searching for a suspected DNA-binding region within a sequence. Takeda *et al.*[19] have suggested several somewhat homologous DNA-binding stretches in repressor proteins. Smoothed physical characteristic curves could be calculated for each of the sequences; these curves could then be correlated against that of the suspected protein for all six parameters and the resulting lag profiles successively added for each homologous DNA-binding span. A large peak in the additive curve would suggest the most

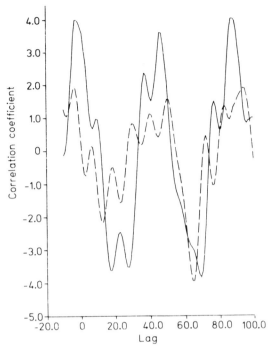

FIG. 9. Plots of the combined cross-correlation coefficients *versus* the lag resulting from the addition of the four lag profiles involving comparisons between the NH-terminal or COOH-terminal halves of γ-crystallin with the entire γ and β chains (solid line) or with the entire αA- and $\alpha\beta$-crystallins (dashed line).[91]

likely DNA-binding region assuming the suspected protein uses the same mode as the repressor proteins.

The physical characteristic approach has its cautions. The correlated segments should not be much smaller than 30 to 50 residues as the coefficients are unrealistically large when the parametric series contain few terms. Even though the hydrophobic cores of proteins are formed by similar folds, there often exist large (five or more residues) insertions and deletions at the protein surface.[14,83] In comparing the residue characteristics of such proteins, the relative insertions would result in out-of-phase curves and the related folds missed. The solution here would be to utilize shorter curve segments in analogy with the Jukes and Cantor[18] approach for amino acid sequences. However, this is prohibited by the artificially large correlations resulting from comparing profiles with only a few peaks and troughs.

[11] Calculation of Protein Conformation from Circular Dichroism

By JEN TSI YANG, CHUEN-SHANG C. WU, and HUGO M. MARTINEZ

Optical rotatory dispersion (ORD) and circular dichroism (CD) are two chiroptical phenomena which differentiate two enantiomers. They are caused by different interactions of left- and right-circularly polarized light with chiral molecules. Chirality is a geometric property of molecules; the corresponding substances are therefore optically active. The application of ORD and CD to the study of protein conformation was extensively reviewed in Volume XXVII of this series by Adler *et al.* in 1973.[1] Today CD has replaced ORD for the conformational analysis of proteins because CD bands that are characteristic of various secondary structures can be directly observed. In this chapter we will discuss some recent developments in the estimation of various conformations in a protein molecule from its CD spectrum in the ultraviolet region.

Historical Background

It is to Biot in 1812 that we owe the laws of rotatory polarization and rotatory dispersion. It was Cotton in 1896 who described the CD of solutions. Pasteur's separation of a racemic mixture of sodium ammonium tartrate in the 1840s was another milestone. However, the fertile era of ORD ended with the invention of the Bunsen burner in 1866, which made it almost too easy to work with the nearly monochromatic light of the sodium flame (for an account of the early history, see Lowry's classical monograph in 1935[2]). In the 1950s we witnessed a resurgence of interest in ORD, which stemmed from, first, conformational studies of steroids[3] and, second, the discovery of the helical ORD of synthetic polypeptides.[4-6] Progress was speeded up by the introduction of the first manual-type Rudolph spectropolarimeter in 1955.

In the early 1950s the idea that the Pauling–Corey α-helix must have a positive optical rotation at the sodium D line so that proteins become

[1] A. J. Adler, N. J. Greenfield, and G. D. Fasman, this series, Vol. 27, p. 675.

[2] T. M. Lowry, "Optical Rotatory Power." Longmans, Green, London, 1935; Dover, New York, 1964.

[3] C. Djerrassi, "Optical Rotatory Dispersion." McGraw-Hill, New York, 1960.

[4] P. Doty and J. T. Yang, *J. Am. Chem. Soc.* **78**, 498 (1956).

[5] W. Moffitt and J. T. Yang, *Proc. Natl. Acad. Sci. U.S.A.* **42**, 596 (1956).

[6] J. T. Yang and P. Doty, *J. Am. Chem. Soc.* **79**, 761 (1956).

more levorotatory upon denaturation (cf. Ref. 7) was received with some skepticism. A polypeptide chain was thought to have an equal probability of winding into a right- and left-handed helix, although earlier Huggins[8] had pointed out that L-polypeptides favor a right-handed helix because the β-carbon of each side chain sterically interferes with the carbonyl oxygen of the same residue in a left-handed helix. The dextrorotation of a helical polypeptide in the visible region was first observed with a crude polarimeter equipped with a Na lamp and a Hg lamp with green and blue filters for three wavelengths at 589, 546, and 436 nm.[9] Today modern instruments with a data processor seem to make it almost too easy to record data. Ironically, some American manufacturers in the early 1960s did not think it possible to design a precise and accurate circular dichrometer. At present the latest JASCO J-500 model spectropolarimeters virtually monopolize the market in this country.

Crick and Kendrew in 1957[10] stated that "this type of evidence [optical rotation of helices] is suggestive but falls short of being conclusive. It leads to a strong presumption that some sort of helical configuration is present, and of a single hand." But they went on to say that "It may be remarked that there is an encouraging parallelism between the X-ray and optical results." The structure of myoglobin was then not yet completely determined. Actually, Pasteur in 1860[11] foresaw the chiroptical phenomenon of a helix: "Imagine a winding stair, the steps of which shall be cubes, or any other object with a superposable image. Destroy the stair, and the dissymmetry will have disappeared. The dissymmetry of the stair was the result only of the mode of putting together its elementary steps." Fresnel in 1824 even predated Pasteur by anticipating the optical rotatory power of a helicoidal arrangement (see Ref. 2).

The late W. Moffitt developed an exciton theory of optical rotation for the α-helix[12] just for fun because he had frequently heard biochemists talk about this structure. The simplified form as proposed by Moffitt and Yang[5]

$$[m'] = a_0\lambda_0^2/(\lambda^2 - \lambda_0^2) + b_0\lambda_0^4/(\lambda^2 - \lambda_0^2)^2 \tag{1}$$

[7] C. Cohen, *Nature (London)* **175,** 129 (1955).

[8] M. L. Huggins, *J. Am. Chem. Soc.* **74,** 3963 (1952).

[9] J. T. Yang, *in* "Conformation of Biopolymers" (G. N. Ramachandran, ed.), p. 157. Academic Press, New York, 1967.

[10] F. H. C. Crick and J. C. Kendrew, *Adv. Protein Chem.* **12,** 133 (1957).

[11] L. Pasteur, *in* two lectures on "Researches on the Molecular Disymmetry of Natural Organic Products" presented to the Chemical Society of Paris on 20 January and 3 February, 1860. Translated from "Leçons de chimie professées en 1860," by W. S. W. Ruschenberger, *Am. J. Pharm.* **34** (Ser. 3, Vol. 10), 1, 97 (1862).

[12] W. Moffitt, *J. Chem. Phys.* **25,** 467 (1956).

was widely used for studying the conformation of polypeptides and proteins up to the 1970s. Here [*m'*] is the reduced mean residue rotation, which equals the mean residue rotation [*m*] multiplied by the Lorentz correction factor, $3/(n_\lambda + 2)$, *n* being the wavelength-dependent refractive index of the solvent. [Yang's minor contribution to the Moffitt equation was to replace frequency by wavelength and regroup all constants into two parameters, a_0 and b_0. Plotting $[m'](\lambda^2 - \lambda_0^2)$ against $1/(\lambda^2 - \lambda_0^2)$ yields a straight line, provided a third parameter λ_0 can be preset.[9] Moffitt was skeptical about the graphical determination of three parameters and had wanted to use the then novel computer at Harvard.] Simultaneously, Fitts and the late J. G. Kirkwood applied the polarizability theory of rotatory power[13] to explain the change of specific rotation in a helix-coil transition.[14,15] The dispute between Moffitt and Kirkwood was subsequently resolved in a joint publication.[16] Both theories require revisions and Eq. (1) is now regarded as empirical. Experimentally, with λ_0 preset at 212 nm a b_0 value of -630 deg cm² dmol⁻¹ is taken to represent a 100% right-handed helix; this ORD method still works reasonably well. Moffitt's prediction of a right-handed helix with a negative b_0 value happens to agree with the later X-ray results. This is a fortunate coincidence. Early experiments were done on poly(γ-benzyl-L-glutamate) in poor solvents; its helix turned out to be right-handed. Had poly(β-benzyl-L-aspartate) been used, its positive b_0 value of about +600 deg cm² dmol⁻¹ might have raised questions as to why the handedness of helices in polypeptides differs from that in proteins, which usually give a negative b_0 value. (For a comprehensive review of early ORD results of polypeptides and proteins, see Ref. 17.)

Because of the lack of circular dichrometers, CD remained to be overlooked until the middle 1960s. Holzwarth and Doty were the first to measure the CD spectrum of an *a*-helix.[18] It shows a typical double minimum at 222 and 208–210 nm and a maximum at 191–193 nm (Fig. 1),[19,20] which represent the $n-\pi^*$ transition,[21,22] and $\pi-\pi_\parallel^*$ and $\pi-\pi_\perp^*$ transitions,[12,22,23]

[13] J. G. Kirkwood, *J. Chem. Phys.* **5**, 479 (1937).
[14] D. D. Fitts and J. G. Kirkwood, *Proc. Natl. Acad. Sci. U.S.A.* **42**, 33 (1956).
[15] D. D. Fitts and J. G. Kirkwood, *J. Am. Chem. Soc.* **78**, 2650 (1946).
[16] W. Moffitt, D. D. Fitts, and J. G. Kirkwood, *Proc. Natl. Acad. Sci. U.S.A.* **43**, 723 (1957).
[17] P. Urnes and P. Doty, *Adv. Protein Chem.* **16**, 401 (1961).
[18] G. M. Holzwarth and P. Doty, *J. Am. Chem. Soc.* **87**, 218 (1965).
[19] N. Greenfield and G. D. Fasman, *Biochemistry* **8**, 4108 (1969).
[20] J. T. Yang and S. Kubota, *in* "Microdomains in Polymer Solutions" (P. L. Dubin, ed.), p. 311. Plenum, New York, 1985.
[21] J. A. Schellman and P. Oriel, *J. Chem. Phys.* **37**, 2114 (1962).
[22] I. Tinoco, Jr., R. W. Woody, and D. F. Bradley, *J. Chem. Phys.* **38**, 1317 (1963).
[23] R. W. Woody and I. Tinoco, Jr., *J. Chem. Phys.* **46**, 4927 (1967).

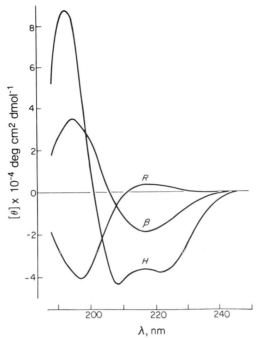

FIG. 1. CD spectra of the helix, β-form, and unordered form based on $(Lys)_n$ (M_r = 193,000) in water at 25°. Curves: R, unordered form at neutral pH; H, α-helix at pH 10.8; β, β-form at pH 11.1 after heating for 15 min at 52° and cooling back to 25°. Concentration of $(Lys)_n$: 0.07%. (From Yang and Kubota[20] with the permission of Plenum and copyrighted by Plenum.)

respectively. Current commercial instruments can record CD spectra down to about 184 nm. With a vacuum ultraviolet CD (VUCD) instrument the CD spectra can be further extended to 140 nm under ideal conditions; for aqueous solutions the lower limit is about 165 to 178 nm. At present there are only four such circular dichrometers in this country and no commercial VUCD instruments. The VUCD of the helix[24,25] shows a positive shoulder near 175 nm and one negative and one positive band at shorter wavelengths (Fig. 2).

In 1966 the CD spectrum of Pauling and Corey's β-pleated sheet (of β-form) was found for $(Lys)_n$ in sodium dodecyl sulfate solution,[26] $(Lys)_n$ at

[24] W. C. Johnson, Jr. and I. Tinoco, Jr., *J. Am. Chem. Soc.* **94**, 4389 (1972).
[25] M. A. Young and E. S. Pysh, *Macromolecules* **6**, 790 (1973).
[26] P. K. Sarkar and P. Doty, *Proc. Natl. Acad. Sci. U.S.A.* **55**, 981 (1966).

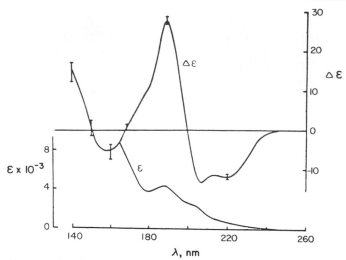

FIG. 2. Vacuum ultraviolet CD and absorption spectra of poly(γ-methyl-L-glutamate) in hexafluoroisopropanol. (Redrawn from Johnson and Tinoco[24] with the permission of the American Chemical Society, copyright 1972.)

alkaline pH (Fig. 1),[27] and silk fibroin in methanol–water.[28] It shows a negative band near 216–218 nm and a positive one between 195 and 200 nm, but the intensities of these CD bands are different in the three cases. For some homopolypeptides containing alkylated or acylated Ser, Thr, and Cys side chains, the extrema of the CD spectra of the β-form can be red-shifted; in particular, the n–π* transition was located above 225 nm, but the π–π* transition remained below 200 nm.[29,30] The similarity of the red-shifted n–π* transition to that predicted for β-turns (see below) suggests that these homopolypeptides may have a large fraction of residues in β-turns,[31] which also agree with the observation that most of these polypeptides form cross-β structures according to infrared dichroism criteria. Alternatively, the red-shifted n–π* transition could result from the coupling of a strong transition near 194 nm characteristic of dialkyl sulfides

[27] R. Townend, R. F. Kumosinski, S. N. Timasheff, G. D. Fasman, and B. Davidson, *Biochem. Biophys. Res. Commun.* **23,** 163 (1966).
[28] E. Iizuka and J. T. Yang, *Proc. Natl. Acad. Sci. U.S.A.* **55,** 1175 (1966).
[29] G. D. Fasman and J. Potter, *Biochem. Biophys. Res. Commun.* **27,** 209 (1967).
[30] L. Stevens, R. Townend, S. N. Timasheff, G. D. Fasman, and J. Potter, *Biochemistry* **7,** 3717 (1968).
[31] R. W. Woody, *in* "The Peptides" (V. Hrudy, ed.), Vol. 7, p. 15. Academic Press, New York, 1985.

with the peptide transitions.[32] The VUCD spectrum of the β-form has another negative band below 170 nm (not shown).[33-35] Parallel and antiparallel β-sheets have been predicted to have qualitatively similar CD spectra.[32,36-39]

The CD spectrum of the unordered form (Fig. 1) generally shows a strong negative band near 200 nm and a very weak band around 220 nm, which can be either a positive band or a negative shoulder. Theoretical attempts to explain this general feature have been reviewed by Woody.[40] Experimentally, the magnitude of the 200-nm band can vary with the polypeptides studied and the experimental conditions (see section on "Reference Spectra Based on Model Polypeptides").

Until the middle 1970s, the CD of β-turns (or β-bend, 3_{10}-bend, hairpin bend, β-loop, 1–4 bend, reverse turn, U-bend, etc.) had been neglected, although this conformation is now recognized to be abundant in protein molecules. Chou and Fasman[41] searched 29 proteins whose X-ray diffraction data were available and found that as many as one-third to one-fourth of the amino acid residues could be identified in β-turns. [For an extensive review on reverse turns in peptides and proteins, see Ref. 42. γ-Turns have a hydrogen bond between the C=O of the first residue and the NH of the third residue instead of the fourth residue as found in β-turns, but these turns occur infrequently in proteins and their CD spectrum is much less known than that of β-turns.] Earlier, Venkatachalam[43] classified 15 possible types of β-turns in globular proteins. By using more relaxed criteria, Lewis et al.[44,45] have grouped 10 types and termed three fundamental types as types I, II, and III, which account for about 80% of the occurrence in proteins (see Ref. 41). Types I', II', and III' are approxi-

[32] J. Applequist, Biopolymers 21, 779 (1982).
[33] J. S. Balcerski, E. S. Pysh, G. M. Bonora, and C. Toniolo, J. Am. Chem. Soc. 98, 3470 (1976).
[34] M. M. Kelly, E. S. Pysh, G. M. Bonora, and C. Toniolo, J. Am. Chem. Soc. 99, 3264 (1977).
[35] S. Brahms, J. Brahms, G. Spach, and A. Brack, Proc. Natl. Acad. Sci. U.S.A. 74, 3208 (1977).
[36] E. S. Pysh, Proc. Natl. Acad. Sci. U.S.A. 56, 825 (1966).
[37] R. W. Woody, Biopolymers 8, 669 (1969).
[38] E. S. Pysh, J. Chem. Phys. 52, 4723 (1970).
[39] V. Madison and J. A. Schellman, Biopolymers 11, 1041 (1972).
[40] R. W. Woody, J. Polymer Sci. Macromol. Rev. 12, 181 (1977).
[41] P. Y. Chou and G. D. Fasman, J. Mol. Biol. 115, 135 (1977).
[42] J. A. Smith and L. G. Pease, CRC Crit. Rev. Biochem. 8, 315 (1980).
[43] C. M. Venkatachalam, Biopolymers 6, 1425 (1968).
[44] P. N. Lewis, F. A. Momany, and H. A. Scheraga, Proc. Natl. Acad. Sci. U.S.A. 68, 2293 (1970).
[45] P. N. Lewis, F. A. Momany, and H. A. Scheraga, Biochim. Biophys. Acta 303, 221 (1973).

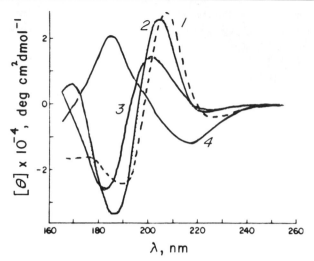

FIG. 3. Vacuum ultraviolet CD spectra of model β-turns. Curves: 1, $(\text{Ala}_2\text{-Gly}_2)_n$ in water; 2, N-isobutyl-L-proline-isopropylamide film cast from trifluoroethanol solution; 3, N-acetyl-L-Pro-Gly-L-Leu in trifluoroethanol at $-60°$; 4, cyclo(D-Ala-L-Ala-L-Ala-D-Ala-D-Ala-L-Ala) in D_2O. The intensities of $(\text{Ala}_2\text{-Gly}_2)_n$ have been multiplied by 0.5. (Redrawn from Brahms and Brahms.[47])

mate mirror images of types I, II, and III but they do not occur frequently. Woody's theoretical analysis[46] predicts that types I, II, and III should have simlar CD spectra, which generally resemble the spectrum of the β-form identified as a class A spectrum, but which have a weak negative band red-shifted to 220–230 nm, a positive band between 200 and 210 nm, and a strong negative band between 180 and 190 nm. This CD spectrum is referred to as a class B spectrum, the most common class associated with any β-turn conformation. Type II′ β-turns may have a helix-like CD spectrum, called a class C spectrum. Type I β-turns can have any L-amino acid as the second and third residues of the 4-residue reverse turn, i.e., the corner residues; type II has a glycine as the third residue; type III is a variant of type I in which the dihedral angles of the corner residues are roughly identical, thus forming one turn of 3_{10}-helix.

Experimentally, the CD spectra of β-turns generally support the theoretical predictions of Woody[46] as illustrated in Fig. 3.[47] [Brahms and Brahms have proposed that $(\text{Ala}_2\text{-Gly}_2)_n$ and N-isobutyl-L-Pro-D-Ala rep-

[46] R. W. Woody, in "Peptides, Polypeptides and Proteins" (E. R. Blout, F. A. Bovey, M. Goodman, and N. Lotan, eds.), p. 338. Wiley, New York, 1974.
[47] S. Brahms and J. Brahms, J. Mol. Biol. **138**, 149 (1980).

resented type II, N-acetyl-L-Pro-Gly-L-Leu type I (although a possible type II was not excluded), and cyclo(D-Ala-L-Ala$_2$-D-Ala$_2$-L-Ala) type IV' and partially type III.] As Woody[46] points out, many cyclic peptides whose type II structure has been established by NMR or X-ray diffraction studies show the class B spectrum,[48–51] as do several other systems for which β-turns have been proposed.[47,52–56] A number of peptides show the class C (or its image, C') spectrum.[48,49,57–59] However, Woody[46] indicates that the broad range of conformations in β-turns combined with the sensitivity of CD to the conformation imply that the CD spectra of some β-turns may be quite different from the consensus pattern and that one aspect of his theoretical calculations needs revision because the CD spectra of two well-authenticated type I β-turns, viz. cyclo(Gly-L-Pro-Ala)$_2$[49] and cyclo(L-Ala-L-Ala-ε-aminocaproyl),[50] resemble the spectrum of an α-helix (class C spectrum). According to the data compiled by Chou and Fasman,[41] the β-turns of 29 proteins are 45% type I, 22% type III, and 18% type II. Therefore, Woody[46] states that the crucial question is whether type I β-turns generally give a class C spectrum or whether because of their cyclic character the two authenticated systems show CD spectra deviated from class B spectra. The answer to this question is important in the CD calculations to be described in the section on "Methods of CD Analysis."

[48] C. A. Bush, S. K. Sackar, and K. D. Kopple, *Biochemistry* **17**, 4951 (1978).

[49] L. M. Gierasch, C. M. Deber, V. Madison, C.-H. Niu, and E. R. Blout, *Biochemistry* **20**, 4730 (1981).

[50] J. Bandekar, D. J. Evans, S. Krimm, S. J. Leach, S. Lee, J. R. McQuie, E. Minasian, G. Némethy, M. S. Pottle, H. A. Scheraga, E. R. Stimson, and R. W. Woody, *Int. J. Pept. Protein Res.* **19**, 187 (1982).

[51] R. Deslauriers, D. J. Evans, S. J. Leach, Y. C. Meinwald, E. Minasian, G. Némethy, I. D. Rae, H. A. Scheraga, R. L. Somorjai, E. R. Somorjai, E. R. Stimson, J. W. Van Nispen, and R. W. Woody, *Macromolecules* **14**, 985 (1981).

[52] D. W. Urry, M. M. Long, T. Ohnishi, and M. Jacobs, *Biochem. Biophys. Res. Commun.* **61**, 1427 (1974).

[53] S. Brahms, J. Brahms, G. Spach, and A. Brack, *Proc. Natl. Acad. Sci. U.S.A.* **74**, 3208 (1977).

[54] M. Kawai and G. D. Fasman, *J. Am. Chem. Soc.* **100**, 3630 (1978).

[55] S. K. Brahmachari, V. S. Ananthanarayanan, S. Brahms, J. Brahms, R. S. Rapaka, and R. S. Bhatnagar, *Biochem. Biophys. Res. Commun.* **86**, 605 (1979).

[56] S. K. Brahmachari, T. N. Bhat, V. Sudhakar, M. Vijayan, and V. S. Ananthanarayanan, *J. Am. Chem. Soc.* **103**, 1703 (1981).

[57] D. W. Urry, A. L. Ruiter, B. C. Starcher, and T. A. Hinners, *Antimicrob. Agents Chemother.* 87 (1968).

[58] S. Laiken, M. Printz, and L. C. Craig, *J. Biol. Chem.* **244**, 4454 (1969).

[59] V. S. Ananthanarayanan and N. Shyamasundar, *Biochem. Biophys. Res. Commun.* **102**, 295 (1981).

In the 1950s almost all biophysical studies of synthetic polypeptides were limited to two states, the helix and the coil. The theories of Zimm and Bragg and others (see Ref. 60) provide a solid foundation for the interpretation of experimental results. In retrospect it is obvious why the helix was overemphasized. Myoglobin and hemoglobin, the first two proteins studied by X-ray diffraction, are rich in helices and have no β-form. In fact, the then prevailing view was that the Pauling–Corey α-helix is the only regular secondary structure in globular proteins, and their β-pleated sheets were thought to exist in some fibrous proteins such as silk fibroin and perhaps in aggregates of polypeptides and proteins. Now we know that β-sheets and β-turns are abundant in many globular proteins. Therefore, the discovery of the CD spectra of these conformations in addition to the helix challenges us to improve and refine the analysis of the chiroptical properties of proteins.

Expression of Experimental Data[61]

Linearly polarized light can be resolved into two circularly polarized components. By definition CD is a measure of the differential absorbance between the left- and right-circularly polarized light, $A_L - A_R$. The emerging light from an optically active medium becomes elliptically polarized. Thus, an alternate measure of CD is the ellipticity, ψ. Because ψ is extremely small as compared with the absorbances, it simply becomes the ratio of the minor axis to the major axis of the resultant ellipse. Most commercial instruments, which directly measure $A_L - A_R$, are actually calibrated in units of ψ and many publications report CD in ellipticity rather than in differential absorbance. The two measures at wavelength λ are related by

$$\psi(\lambda) = 33(A_L - A_R)(\lambda) \tag{2}$$

where ψ is in degrees. (A more exact conversion factor is 32.982; however, 33.0 will suffice for three significant figures.)

In analogy to specific rotation, specific ellipticity, $[\psi]$, is defined as

$$[\psi](\lambda) = \psi(\lambda)/lc \tag{3}$$

Here l is the light path in dm and c the concentration in g ml^{-1}. The dimension of $[\psi]$ is deg cm^2 dg^{-1} (or 1.745×10^{-4} rad m^2 kg^{-1}). Biot in

[60] D. Polland and H. A. Scheraga, "Theory of Helix-Coil Transitions in Biopolymers." Academic Press, New York, 1970.

[61] J. T. Yang, *in* "A Laboratory Manual of Analytical Methods of Protein Chemistry Including Polypeptides (P. Alexander and H. P. Lundgren, eds.), Vol. 5, p. 25. Pergamon, New York, 1969.

1836 introduced the term specific rotation in which decimeters instead of centimeters were used "in order that the significant figures may not be uselessly preceded by two zeros." This awkward tradition is retained because the wealth of published data for more than 150 years forbids any new definition, which will merely cause confusion. The word "specific" is also not in accordance with IUPAC rules but is retained for historical reasons.

For biopolymers it is more convenient to express the data in terms of mean residues than in molar quantities. Thus, one measure of CD is simply the differential absorption coefficients, $\varepsilon_L - \varepsilon_R$, and its dimension is $cm^{-1}\ M^{-1}$ (or 0.1 $m^2\ mol^{-1}$) (it is understood that the molar concentration refers to mean residue moles per liter). The symbol $\Delta\varepsilon$ is often used when there is no risk of ambiguity; the same symbol has long been used for difference in absorption coefficients in absorbance spectroscopy. An alternate measure of CD is the mean residue ellipticity, $[\theta]$:

$$[\theta](\lambda) = M_0[\psi](\lambda)/100 \tag{4a}$$

or

$$[\theta](\lambda) = 100\psi(\lambda)/l'c' \tag{4b}$$

where M_0 is the mean residue weight, sometimes referred to as MRW, l' the light path in cm, and c' the concentration in mean residue moles per liter. The dimension of $[\theta]$ is deg cm^2 $dmol^{-1}$ (or 1.745×10^{-5} rad m^2 mol^{-1}). For most proteins M_0 is around 115, which can be used if the molar mass (or molecular weight) and the number of amino acid residues, or the amino acid composition of the protein is not known. In Eq. (4a) the factor 100 is again introduced by tradition; it merely reduces the magnitude by two orders of magnitude. The two measures of CD are related by

$$[\theta](\lambda) = 3300(\varepsilon_L - \varepsilon_R)(\lambda) \tag{5}$$

Sometimes molar CD instead of mean residue CD may also be used. In this case the concentration in $\varepsilon_L - \varepsilon_R$ is simply moles per liter. Likewise, in Eq. (4a) the molar mass (or molecular weight), M, replaces mean residue weight, M_0. Obviously, if the number of amino acid residues in a protein molecule is n, $M = nM_0$, that is, the molar ellipticity, $[\theta]$, is n times the mean residue ellipticity, $[\theta]$. The latter quantity is used to estimate the fractions of various conformations in a protein molecule regardless of its molar mass (or molecular weight).

Experimental Measurements

CD instruments have been greatly improved within the last decade. The Pockels' cell is now replaced by a photoelectric modulator, which

TABLE I

OPTICAL PROPERTIES OF d-10-CAMPHORSULFONIC ACID IN WATER^{a,b}

Quantity	Value	Quantity	Value
ε_{285}	34.5		
$(\varepsilon_L - \varepsilon_R)_{290.5}$	2.36	$[\theta]_{290.5}$	7,800
$(\varepsilon_L - \varepsilon_R)_{192.5}$	−4.72	$[\theta]_{192.5}$	−15,600
$(\varepsilon_L - \varepsilon_R)_{192.5}/(\varepsilon_L - \varepsilon_R)_{290.5}$	−2.00	$[\theta]_{192.5}/[\theta]_{290.5}$	−2.00

^{a} From Chen and Yang.[63]

^{b} Units: ε and $(\varepsilon_L - \varepsilon_R)$ are in cm^{-1} M^{-1} and $[\theta]$ is in deg cm^2 dmol^{-1}. All subscripts refer to wavelengths in nanometers.

eliminates much noise; multiple scannings are aided with a data processor. Nevertheless, an instrument must be properly calibrated; otherwise, any quantitative analysis of experimental data would be meaningless.

Assuming that the wavelength scale of commercial instruments is correctly adjusted by the manufacturer, the accuracy of the readings at various wavelengths must still be carefully checked. While a spectropolarimeter for ORD can easily be calibrated with a solution of sucrose (National Bureau of Standards sample), no standard for CD has been universally accepted. One frequently used compound is d-10-camphorsulfonic acid (this compound is dextrorotatory in the visible region and therefore has a prefix d by tradition). However, the compound is hygroscopic. Neglect of its water of hydration results in a smaller CD amplitude than the true value and has caused much confusion in the literature. Indeed, early values recommended by some manufacturers were incorrect. In addition, commercial products often contain colored impurities. In the late 1960s a computerized calibration of the circular dichrometer against a standardized spectropolarimeter was proposed.[62] This is made possible by using the Kronig–Kramers transform from CD to ORD. The ratio of molar ellipticity to molar rotation of d-10-camphorsulfonic acid at their extrema was found to be $[\theta]_{290.5}/[M]_{306} = 1.76$. The advantage of this method is that the ratio remains unaffected by the water of hydration and impurities, provided that the impurities are optically inactive.

Now that ORD is rarely used, a two-point calibration of the circular dichrometer by using purified, dried d-10-camphorsulfonic acid has been proposed (Table I).[63] One such procedure is to dissolve the compound in ethyl acetate at 75°; charcoal is added to remove colored impurities. The solution is filtered and the compound crystallized upon cooling. The im-

[62] J. Y. Cassim and J. T. Yang, *Biopolymers* **9**, 1475 (1969).
[63] G. C. Chen and J. T. Yang, *Anal. Lett.* **10**, 1195 (1977).

pure compound is recrystallized and washed several times with 1 : 1 ethyl acetate–ether. The crystals are vacuum dried over P_2O_5 in a desiccator at 50° and stored in the desiccator at room temperature. Alternately, d-10-camphorsulfonic acid can be converted to its monohydrate by storing the compound in a desiccator at 50% relative humidity for several days. To ensure accuracy the concentration of d-10-camphorsulfonic acid should be further determined spectroscopically (see Table I). The $[\theta]_{290.5}$ value (Table I) is several percent larger than that recommended by some manufacturers. During a 10-year period of testing, the ratio of $[\theta]_{192.5}/[\theta]_{290.5}$ of one instrument was found to vary between -1.96 and -2.04, i.e., -2.00 ± 0.04. Occasionally, the ratio could be as low as -1.90. This was usually traced to the aging of the xenon lamp, cleanness of optical cells (especially those of short pathlengths), or improper adjustment of the scale range of the instrument or combinations of these factors. Hennessey and Johnson[64] also reported $\Delta\varepsilon$ of 2.36 cm^{-1} M^{-1} at 290.5 nm, but their $\Delta\varepsilon$ at 192.5 nm was 4.9 instead of 4.72 cm^{-1} M^{-1} as listed in Table I. Thus, their ratio of $\Delta\varepsilon_{192.5}/\Delta\varepsilon_{290.5}$ was -2.08.

Users often take for granted that commercial instruments have been properly adjusted by the manufacturer. For routine probes of protein conformation operational errors in a particular instrument may be overlooked. However, if the CD spectrum of a protein is used to estimate the secondary structure, we must demand a high degree of accuracy for the experimental data collected. First, a blue or red shift of 1 or 2 nm in the wavelength scale could affect the CD analysis. This can easily be checked by observing the extrema of, say, d-10-camphorsulfonic acid (Table I) during calibration of the instrument. Since a CD spectrum is scanned from long to short wavelengths, care should also be taken to avoid backlash from the gears, that is, the scanning should begin several nanometers longer than the starting point. Hennessey and Johnson[64] tested a wavelength shift of up to 2 nm for the CD spectrum of lactate dehydrogenase and found that the total error in the estimated secondary structure could amount to 1% or more (see section on "Comparison of Methods for CD Analysis"). Second, the small difference in absorbance between left- and right-circularly polarized light is inherently noisy. This is particularly true when the scanning approaches the limits of the instrument. Thus, the signal-to-noise ratio decreases rapidly at wavelengths below 200 nm for commercial instruments. One way to reduce the noise is to increase the bandwidth so that more light will reach the photomultiplier tube. This operation is done at the expense of spectral purity, which in turn can distort the CD spectrum. As a rule of thumb the bandwidths should al-

[64] J. P. Hennessey, Jr. and W. C. Johnson, Jr., *Anal. Biochem.* **125**, 177 (1982).

ways be kept within 2 nm. Third, the noise of the CD spectrum can be reduced by raising the time constant. However, the rate of scanning must be slowed so that the instrument can respond to signal changes. An instrument equipped with a multiple scanning device and a data processor often can speed up the rate of scanning, but within limits. Hennessey and Johnson[64] recommend that the product of time constant and rate of scanning be kept below 0.33 nm. For instance, at the rate of scanning of 2 nm/min, the time constant should be set below 10 sec. On the other hand, a too high time constant may not be desirable because of possible instrumental drift. In our experience the product is usually kept below 0.1 nm if a data processor for repeated scannings is not used.

To record a spectrum, the circular dichrometer must first be warmed up for about 30 min to 1 hr so that the light source and the electric components of the instrument will reach a steady state of operation. To detect any possible shift the baseline should be measured before and after each sample spectrum. For many proteins the spectrum can be scanned from wavelengths longer than 240–250 nm where the CD of the sample and baseline coincides, provided of course the sample has no observable CD in the near ultraviolet region. Aromatic groups such as tyrosine and tryptophan of proteins often show small CD bands up to 300 nm, neglect of which will of course introduce errors in the alignment of the baseline.

Experimental errors by individual users are unavoidable, but they can be minimized by taking certain precautions. For routine measurements, fused quartz cells should be free of birefringence. Their path lengths must be accurately determined, especially for 1-mm or less cells. This can easily be done by measuring the absorbance of a solution of known absorption coefficient such as an aqueous solution of d-10-camphorsulfonic acid (Table I). Short path lengths can also be calculated by counting the interference fringes of the empty cell in a spectrometer. The cell must be thoroughly cleaned; a dirty inner wall can easily trap air bubbles and thereby distort the CD signal. If a constant temperature is critical, the cell must be thermostated. Whether a jacketed cell or a cell holder with a water jacket is used, the cell should be positioned firmly and reproducibly. If more than one cell is used, it is advisable to overlap portions of the spectrum obtained from two cells, which will indicate the precision of the data.

The choice of the protein concentration is a compromise between enough solute to improve the precision of the data and not too much absorbance at the wavelengths studied. Usually the absorbance of the solution in the cell is kept around one, that is, at least 10% of the light is transmitted. The absorbance should never be above two, for which less than 1% of the light is transmitted. If the supply of sample is limited, extremely dilute solutions may have to be used and loss of sample due to adsorption on the cell wall can introduce serious errors in the protein concentration. Water is the usual solvent for proteins. Buffers such as Tris and acetate will all absorb at low wavelengths and therefore increase the noise. Potassium fluoride or perchlorate instead of NaCl may be used for desired ionic strength because chloride ion absorbs strongly at low wavelengths (KF should not be used in acidic solution to avoid the production of HF).

Perhaps a major source of uncertainty is the preparation of proteins. The same protein studied by different laboratories can sometimes show considerable variations in its CD spectrum, probably because of the use of a poor preparation or of errors in the determination of protein concentration. But sometimes the positions of the extrema and even the entire profile can be different among published data. This poses a serious problem in the CD analysis to be described in the section on "Comparison of Methods for CD Analysis." The concentration of a protein solution must be known accurately, but this is not always easy to achieve. Thus, the method of determination of concentrations should be reported for the sake of comparison by other workers. If there is enough sample, the micro-Kjeldahl method can be used for determining the nitrogen content, provided that the amino acid composition of the protein is known and the solvent used does not contain nitrogen compounds. Colorimetric assays such as the Lowry method or ninhydrin test can be used with caution. These methods often vary from protein to protein even when they are standardized against the protein to be measured. Ideally, these assays should be calibrated against an amino acid analysis of the protein so that a correct absorption coefficient can be determined for routine determinations. Significant variations in the CD spectra reported by workers in various laboratories are disquieting and should be reinvestigated perhaps through collaborative efforts.

Methods of CD Analysis

The chiroptical properties of various conformations in a protein molecule are assumed to be additive. The contributions due to non-peptide chromophores below 250 nm are frequently neglected, presumably because aromatic groups and disulfide bonds account for only about 10% of the residues and apparently may not significantly perturb the CD spectrum due to various conformations. Thus, the experimental CD spectrum of a protein at each wavelength λ can be expressed as

$$X(\lambda) = \sum_{i=1}^{n} f_i X_i(\lambda) \tag{6}$$

where $X(\lambda)$ is the mean residue CD, that is, $\varepsilon_L - \varepsilon_R$ or $[\theta]$; f_i is the fraction of the ith conformation and $X_i(\lambda)$ is the corresponding reference CD. If the reference spectra can be evaluated, the f_i's can be solved from a series of simultaneous equations (one equation for each λ) by a least-squares method. One school uses synthetic polypeptides as model compounds to provide reference spectra. An alternative approach is to compute reference spectra from CD spectra of proteins of known secondary structure. Recently, a third approach was proposed: to directly analyze the CD spectrum of a protein as a linear combination of the CD spectra of proteins of known secondary structure, thus avoiding the problem of defining reference spectra of individual conformations.

Reference Spectra Based on Model Polypeptides

(Lys)$_n$ as a model compound is quite attractive because it can adopt three conformations merely by varying the pH and temperature of its aqueous solution (Fig. 1). The polypeptide behaves as a polyelectrolyte in neutral or acidic solution and is therefore a "random coil." Deprotonation of the polypeptide at pH above 10 induces a coil-to-helix transition. Mild heating of the helix around 50° for 10 or so min converts the helix to the β-form. In 1969 Greenfield and Fasman[19] first proposed estimation of various amounts of α-helix (H), β-form (β), and unordered form (R) in a protein molecule by utilizing three reference spectra of (Lys)$_n$ for Eq. (6):

$$X(\lambda) = f_H X_H(\lambda) + f_\beta X_\beta(\lambda) + f_R X_R(\lambda) \tag{7}$$

with $\Sigma f_i = 1$ (Table II).[65] Rosenkraz and Scholtan[66] used the same approach but substituted (Ser)$_n$ in 8 M LiCl for (Lys)$_n$ at neutral pH for the unordered form (Table II).

In 1980 Brahms and Brahms[47] included β-turns (t) in Eq. (6):

$$X(\lambda) = f_H X_H(\lambda) + f_\beta X_\beta(\lambda) + f_t X_t(\lambda) + f_R X_R(\lambda) \tag{8}$$

and have extended CD measurements into the vacuum ultraviolet region down to 165 nm. The reference spectrum for the helix was taken from the CD spectrum of myoglobin (79% helix, 6% β-turn, and no β-form) normal-

[65] J. T. Yang, G. C. Chen, and B. Jirgensons, in "Handbook of Biochemistry and Molecular Biology, Proteins" (G. D. Fasman, ed.), Vol. III, 3rd Ed., p. 3. CRC Press, Cleveland, Ohio, 1976.

[66] H. Rosenkraz and W. Scholtan, Hoppe-Seyler's Z. Physiol. Chem. 352, 896 (1971).

TABLE II

REFERENCE CD SPECTRA OF THREE CONFORMATIONS BASED ON SYNTHETIC POLYPEPTIDES:
MEAN RESIDUE ELLIPTICITIES IN deg cm² dmol⁻¹ [a]

	$(Lys)_n$ in water[b]						$(Lys)_n$ in SDS[c]	$(Ser)_n$ in 8 M LiCl[d]
	$[\theta]_H \times 10^{-3}$		$[\theta]_\beta \times 10^{-3}$		$[\theta]_R \times 10^{-3}$		$[\theta]_\beta \times 10^{-3}$	$[\theta]_R \times 10^{-3}$
λ(nm)	A	B	A	B	A	B		
250	0	0	0	0	0	0		
240	−3.30	−3.94	0.700	−1.16	−0.150	−0.125	−0.250	0
239	−3.80	−5.01		−1.43		−0.143	−0.333	−0.100
238	−4.30	−6.44	−1.40	−1.97	−0.140	−0.143	−0.500	−0.120
237	−6.20	−8.23		−2.33		−0.125	−0.670	−0.190
236	−8.00	−10.0		−2.86		−0.107	−0.750	−0.200
235	−10.0	−12.2		−3.58		−0.054	−0.834	−0.250
234	−11.4	−14.9	−3.60	−4.12	0	0.054	−1.17	−0.290
233	−14.0	−17.4		−4.83		0.143	−1.32	−0.300
232	−17.0	−20.0		−5.73		0.250	−1.67	−0.400
231	−19.2	−22.7		−6.80		0.446	−2.00	−0.450
230	−21.9	−25.2	−6.40	−7.70	0.800	0.643	−2.16	−0.500
229	−23.9	−27.9		−8.59		0.928	−2.50	−0.600
228	−26.0	−30.1		−9.67		1.29	−3.00	−0.700
227	−28.8	−31.9		−10.4		1.46	−3.50	−0.750
226	−30.8	−33.7		−11.8		1.79	−4.00	−0.800
225	−32.4	−35.4	−11.4	−12.8	2.70	2.18	−4.50	−0.900
224	−33.7	−36.3		−14.0		2.50	−5.00	−0.950
223	−35.0	−37.2		−14.9		2.82	−5.50	−1.00
222	−35.7	−37.6	−13.8	−15.8	3.90	3.00	−6.00	−1.10
221	−36.0	−37.2		−16.8		3.36	−6.50	−1.20
220	−35.3	−36.9	−15.7	−17.4	4.40	3.57	−6.83	−1.30
219	−35.5	−36.5		−17.9		3.71	−7.33	−1.40
218	−35.0	−36.2		−18.4		3.75	−7.66	−1.50
217	−33.1	−35.8	−18.4	−18.6	4.60	3.68	−7.83	−1.70
216	−32.5	−36.2		−18.3		3.50	−7.83	−1.80
215	−31.4	−36.2	−17.9	−17.6	4.10	3.14	−7.50	−1.90
214	−31.0	−36.5	−16.4	−17.0	3.50	2.50	−7.00	−2.10
213	−31.0	−37.2		−15.8		1.79	−6.16	−2.20
212	−31.0	−38.5		−14.1		0.928	−5.50	−2.30
211	−32.1	−40.1	−12.1	−12.5	0	−0.678	−4.83	−2.60
210	−32.4	−41.5	−10.8	−11.1	−1.40	−2.39	−4.00	−2.80
209	−32.3	−43.0		−8.06		−4.21	−3.33	−3.20
208	−32.6	−43.3	−4.70	−5.37	−3.40	−6.78	−2.50	−3.60
207	−32.5	−41.2		−2.15		−9.64	−0.834	−3.90
206	−29.5	−36.5		1.79		−12.9	0	−4.30
205	−25.0	−30.1	5.70	6.27	−14.5	−17.1	3.33	−4.80
204	−20.5	−23.3		8.95		−21.8	5.84	−5.30
203	−12.5	−16.1		13.2		−25.3	7.80	−5.90
202	0	−7.52	19.3	17.1	−25.6	−29.6	8.50	−6.40

(*continued*)

TABLE II (*continued*)

| λ(nm) | (Lys)$_n$ in water[b] | | | | | | (Lys)$_n$ in SDS[c] | (Ser)$_n$ in 8 M LiCl[d] |
| | $[\theta]_H \times 10^{-3}$ | | $[\theta]_\beta \times 10^{-3}$ | | $[\theta]_R \times 10^{-3}$ | | $[\theta]_\beta \times 10^{-3}$ | $[\theta]_R \times 10^{-3}$ |
	A	B	A	B	A	B		
201	6.00	0		21.4		−33.9	9.60	−6.70
200	14.3	13.2	24.3	24.6	−36.4	−36.1	10.4	−7.00
199	25.0	26.9		26.8		−38.9	11.4	−7.10
198	35.0	40.5		28.6		−40.3	12.0	−7.30
197	44.3	53.7	30.0	30.3	−41.9	−40.7	12.6	−7.40
196	55.0	66.2		32.8		−39.3	13.0	−7.50
195	64.3	80.1	31.9	33.9	−41.0	−38.2	13.4	−7.40
194		85.1		33.2		−36.4		
193		87.2		32.1		−33.9		
192.5	73.3		30.0		−37.5			
192		87.4		31.8		−32.0		
191	76.9	84.0	25.3	28.6	−34.7	−29.3		
190	74.8	80.6	22.4	25.0	−32.2	−26.1		

[a] From Yang *et al.*[65] by courtesy of CRC Press, and from Yang and Kubota[20] with the permission of Plenum (copyrighted by Plenum). Because circular dichrometers scan from long to short wavelengths, the λ values are listed in descending order.

[b] A, Helix, pH 11.1 at 22°; β-form, pH 11.1, heated at 52° for 15 min and cooled to 22°; unordered form, pH 5.7 at 22°. C = 0.01%. From Greenfield and Fasman.[19] Additional values for the helix in 0.1 M KF, pH 10.6 to 10.8 were taken from Holzwarth and Doty.[18] B, Repeated experiments[20] under the same conditions as those used by Greenfield and Fasman in A.

[c] From L. K. Li and S. Spector, *J. Am. Chem. Soc.* **91**, 220 (1969).

[d] From F. Quadrifoglio and D. W. Urry, *J. Am. Chem. Soc.* **90**, 2760 (1968).

ized to 100% helix. The model system for the β-form was (Lys-Leu)$_n$ in 0.1 M NaF at pH 7 and that for the unordered form was (Pro-Lys-Leu-Lys-Leu)$_n$ in salt-free solution. The β-turns were represented by L-Pro-D-Ala, (Ala$_2$-Gly$_2$)$_n$ and Pro-Gly-Leu (see Fig. 3). Brahms and Brahms tested 13 proteins comprising α-helix-rich, β-rich, α/β (alternating α and β segments), and α+β (mixtures of all-α and all-β segments) classes and reported a surprisingly good agreement between observed and calculated data, except for one protein, rubredoxin. They further indicated that the use of three sets of β-turns did not improve estimates of the secondary structure as compared with the use of (Ala$_2$-Gly$_2$)$_n$ alone. Thus, the CD spectrum of this polypeptide can be used as the reference spectrum for the β-turn, although intensities were arbitrarily reduced to one-half of their experimental data.

The major advantage of this approach is its simplicity. The reference spectra for the four conformations are directly measurable. The disadvantages are several-fold. First, the CD of the helix varies to some extent among different helical polypeptides.[1] In addition, uncharged polypeptides can easily aggregate; this will in turn distort the CD spectrum. The difference in numerical values for $(Lys)_n$ in water from two laboratories (Table II) may be partly due to the uncertainty in the degree of aggregation and partly due to experimental errors from different instruments, noting that instrumentation has been improved during the past decade. Second, the choice of the reference spectrum for the β-form is more problematic. $(Lys)_n$ in neutral SDS solution also adopts a β-form and yet its CD intensities are about one half of those obtained by mild heating at pH 11 (Table II). Which reference spectrum better represents the β-form is a matter of conjecture. Third, the so-called "random coil" is an extended polyion, which may not be an appropriate model for the unordered segments in proteins. The unordered form of $(Ser)_n$ in 8 M LiCl has a CD magnitude about one-fifth of that of $(Lys)_n$ in neutral solution (Table II). This uncertainty is again not yet resolved. In spite of these problems this method gives a reasonable estimate of the helicity mainly because the CD spectrum of the helix usually predominates over that of the β-form and the unordered form. However the use of helical $(Lys)_n$ does not take into consideration the chain-length dependence of the helical CD spectrum (see section below). Indeed, Straus et al.[67] resolved the CD spectra of myoglobin, hemoglobin, and lysozyme into several bands and found that the rotational strengths of the helical bands were always less than those of the helical polypeptides. Therefore, neglect of this chain-length factor will yield smaller fractions of helices than the true values.

Reference Spectra Based on Proteins

With proteins of known secondary structure, it is possible to determine the reference spectra from the CD spectra of these proteins by using Eq. (6). If only the helix, β-form, and unordered form are considered, solving the X parameters in Eq. (7) requires a minimum of three simultaneous equations with the f_is of three proteins deduced from X-ray diffraction studies. Saxena and Wetlaufer[68] used the CD spectra of myoglobin, lysozyme, and ribonuclease and solved the X_is at a series of wavelengths. The resultant reference spectra of the helix and β-form are quite similar to those obtained from synthetic polypeptides (Fig. 1). Independently, Chen

[67] J. H. Straus, A. S. Gordon, and D. F. H. Wallace, *Eur. J. Biochem.* **11**, 201 (1969).
[68] V. P. Saxena and D. B. Wetlaufer, *Proc. Natl. Acad. Sci. U.S.A.* **68**, 969 (1971).

and Yang[69,70] used five proteins (myoglobin, lysozyme, ribonuclease, papain, and lactate dehydrogenase) whose secondary structures were then known and solved five simultaneous equations [Eq. (7)] at each wavelength by a least-squares method with the constraint $\Sigma f_i = 1$. The reference spectra so obtained are again similar to those of model compounds. If only three proteins were used, the calculated reference spectra varied from one set to another, but they were heavily weighted by the inclusion of myoglobin, which has the highest helicity among the five proteins studied. [Inadvertently, Saxena and Wetlaufer[68] used reduced mean residue ellipticities of ribonuclease in their calculations. Thus, the numerical values of $[\theta]$ for this protein were about 25% lower than their true values because of the Lorentz correction factor (see footnote a in Ref. 70). However, their reference spectra were qualitatively correct.] This method of CD analysis has since been evaluated and modified by several laboratories.[71–75]

As X-ray diffraction results were updated, Chen et al.[76] reinvestigated the reference spectra of the helix, β-form, and unordered form based on five reference proteins according to Eq. (7). With the availability of new X-ray data on proteins, Chen et al.[76] also determined the reference spectra based on eight proteins (by adding insulin, cytochrome c, and nuclease). The reference spectrum of the helix so obtained was close to that reported previously,[70] but that of the β-form and unordered form changed significantly.

To account for the chain-length dependence of CD for the helix,[23,39] Chen et al.[76] replaced $X_H(\lambda)$ in Eq. (7) by $X_H^{\bar{n}}(\lambda)$ and introduced the relation

$$X_H^{\bar{n}}(\lambda) = X_H^{\infty}(\lambda)(1 - k/\bar{n}) \tag{9}$$

Here the superscripts $\bar{n}$ and ∞ refer to the average number of amino acid residues per helical segment in a protein molecule, mostly about 10 to 11, and the chain length of an infinite helix, respectively, and k is a wave-

[69] Y.-H. Chen and J. T. Yang, Biochem. Biophys. Res. Commun. 44, 1285 (1971).

[70] Y.-H. Chen, J. T. Yang, and H. M. Martiner, Biochemistry 11, 4120 (1972).

[71] R. Grosse, J. Malur, W. Meiske, J. G. Reich, and K. R. H. Repke, Acta Biol. Med. Germ. 29, 777 (1972).

[72] R. Grosse, J. Malur, W. Meiske, and K. R. H. Repke, Biochim. Biophys. Acta 359, 33 (1974).

[73] J. Markussen and A. Vølund, Int. J. Pept. Protein Res. 7, 47 (1975).

[74] J. B. Siegel, W. E. Steinmetz, and G. L. Long, Anal. Biochem. 104, 160 (1980).

[75] I. A. Bolotina, V. O. Chekhov, V. Lugauskas, A. V. Finkel'stein, and O. B. Ptitsyn, Mol. Biol. (USSR) 14, 891 (1980); English translation, p. 701, 1981.

[76] Y.-H. Chen, J. T. Yang, and K. H. Chau, Biochemistry 13, 3350 (1974).

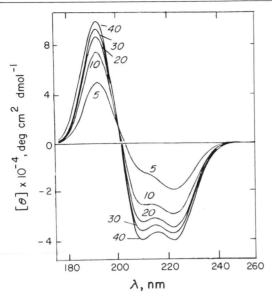

FIG. 4. Computed CD spectra for helices of different chain lengths. Numerals refer to the numbers of amino acid residues. (Redrawn from Chen et al.[76] with the permission of the American Chemical Society, copyright 1974.)

length-dependent constant. In practice an infinite helix applies to helical polypeptides such as $(Lys)_n$. On the basis of the CD spectrum of myoglobin[67,76] the reference spectrum of the helix can be calculated from its three Gaussian bands between 184 and 240 nm:

$$
\begin{aligned}
X_H^{\bar{n}}(\lambda) = &-3.73 \times 10^4(1 - 2.50/\bar{n})\exp[-(\lambda - 223.4)^2/10.8^2] \\
&-3.73 \times 10^4(1 - 3.50/\bar{n})\exp[-(\lambda - 206.6)^2/8.9^2] \\
&+10.1 \times 10^4(1 - 2.50/\bar{n})\exp[-(\lambda - 193.5)^2/8.4^2]
\end{aligned} \tag{10}
$$

(all λs are in nanometers). Thus, the CD magnitude of the helix decreases with decreasing chain length (Fig. 4).

In 1978 Chang et al.[77] enlarged the number of reference proteins to 15 (see footnote a in Table III) and added to Eq. (7) a term for the β-turn (t) (previously, the CD of β-turns was obscured in that of the other conformations):

$$
X(\lambda) = f_H X_H^{\bar{n}}(\lambda) + f_\beta X_\beta(\lambda) + f_t X_t(\lambda) + f_R X_R(\lambda) \tag{11}
$$

which is the same as Eq. (8) except that $X_H^{\bar{n}}$ substitutes for X_H. The net β-turn was used by subtracting from types I, II, and III their mirror images,

[77] C. T. Chang, C.-S. C. Wu, and J. T. Yang, *Anal. Biochem.* **91**, 12 (1978).

TABLE III
REFERENCE CD SPECTRA OF FOUR CONFORMATIONS BASED ON FIFTEEN PROTEINS: MEAN RESIDUE ELLIPTICITIES IN deg cm² dmol⁻¹ [a]

MEAN RESIDUE ELLIPTICITIES IN deg cm^2 $dmol^{-1}$ [a]

λ(nm)	$[\theta]_H^\infty \times 10^{-3}$ [b]	k_b	$[\theta]_\beta \times 10^{-3}$ [c]	$[\theta]_t \times 10^{-3}$ [c]	$[\theta]_R \times 10^{-3}$ [c]
240	−3.51	2.50	1.16	1.43	−2.04
239	−4.63	2.50	1.47	1.71	−2.46
238	−6.00	2.50	1.74	2.20	−2.81
237	−7.64	2.50	1.92	2.85	−3.20
236	−9.56	2.50	2.00	3.39	−3.45
235	−11.8	2.50	2.11	3.88	−3.72
234	−14.2	2.50	2.32	4.63	−4.33
233	−16.2	2.50	2.58	4.87	−5.30
232	−18.9	2.50	2.97	6.31	−6.62
231	−21.5	2.50	2.19	7.42	−7.33
230	−24.0	2.50	3.16	9.96	−9.67
229	−26.7	2.50	2.88	12.6	−11.0
228	−29.0	2.50	3.12	14.3	−12.8
227	−31.2	2.50	2.72	16.1	−13.7
226	−33.2	2.50	2.31	17.4	−14.3
225	−35.5	2.50	1.78	19.3	−14.7
224	−37.0	2.50	0.93	21.0	−14.6
223	−37.5	2.50	0.40	20.3	−14.4
222	−37.4	2.50	0.35	19.3	−13.8
221	−36.9	2.50	−1.04	18.0	−13.4
220	−36.3	2.60	−1.79	16.0	−12.7
219	−35.7	2.60	−2.49	14.5	−11.8
218	−35.3	2.70	−3.32	12.7	−10.8
217	−35.0	2.70	−3.97	11.6	−9.64
216	−34.8	2.80	−4.55	9.77	−8.58
215	−35.0	2.90	−4.93	8.55	−7.75
214	−35.3	3.00	−5.19	7.33	−7.17
213	−36.0	3.10	−5.17	6.58	−6.94
212	−36.8	3.10	−5.00	6.07	−6.64
211	−37.3	3.20	−4.44	4.91	−6.94
210	−38.0	3.30	−3.77	5.57	−8.02
209	−37.5	3.40	−3.45	6.00	−9.19
208	−36.0	3.50	−3.14	6.45	−9.87
207	−33.2	3.60	−2.40	8.63	−11.8
206	−28.8	3.70	−1.52	11.4	−14.0
205	−22.6	4.10	−0.62	13.9	−16.7
204	−14.5	4.80	0.55	16.3	−20.1
203	−4.53	9.40	1.35	17.9	−23.2
202	7.06	1.50	2.26	20.1	−25.1
201	20.0	1.20	4.93	18.0	−31.1
200	36.7	1.80	8.05	12.4	−36.0
199	47.6	2.10	9.39	11.3	−35.6
198	61.0	2.20	9.58	−1.17	−33.1
197	73.2	2.30	8.56	−13.3	−28.8
196	83.4	2.30	6.52	−27.7	−22.9

TABLE III (*continued*)

λ(nm)	$[\theta]_H^\infty \times 10^{-3\,b}$	k_b	$[\theta]_\beta \times 10^{-3\,c}$	$[\theta]_t \times 10^{-3\,c}$	$[\theta]_R \times 10^{-3\,c}$
195	91.0	2.40	4.23	−39.4	−16.5
194	95.6	2.40	0.52	−51.5	−9.64
193	97.0	2.40	−3.05	−49.9	0.28
192	95.3	2.40	−6.91	−70.3	6.41
191	90.7	2.40	−9.42	−75.7	13.4
190	83.8	2.40	−1.23	−78.4	20.1

[a] The proteins used were myoglobin, parvalbumin, insulin, lactate dehydrogenase, lysozyme, cytochrome c, carboxypeptidase A, thermolysin, subtilisin BPN′, papain, trypsin inhibitor, ribonuclease S, nuclease, ribonuclease A, and concanavalin A. Solvent: 0.01 M phosphate buffer (pH 7.0), except 0.01 M acetate buffer (pH 6.4) for thermolysin. See Chang *et al.*[77]

[b] $[\theta]_H^{\bar{n}} = [\theta]_H^\infty(1 - k/\bar{n})$. Based on the CD spectra of myoglobin [Eq. (10)]; $\bar{n} = 13.4$ for myoglobin.

[c] Solved by the least-squares method for Eq. (8) after subtracting $f_H[\theta]_H^{\bar{n}}$, s from corresponding experimental $[\theta]$s at each wavelength for the 15 proteins.

type I′, II′, and III′, respectively. With four reference spectra determined from reference proteins, the fractions of the helix (H), β-form (β), β-turn (t), and unordered form (R) can be estimated by a least-squares method. Chang *et al.*[77] introduced two constraints: $\Sigma f_i = 1$ and $1 \geq f_i \geq 0$. For experimental data between 190 and 240 nm, there are 51 data points at 1-nm intervals. Fifty-one simultaneous equations of Eq. (11) can be used to solve the f_is. Since the reference spectrum of the helix can be calculated from Eq. (10), Chang *et al.*[77] solved the reference spectra of the β-form, β-turn, and unordered form by first subtracting $f_H X_H^{\bar{n}}(\lambda)$ (assuming $\bar{n} = 10.4$) from the experimental $X(\lambda)$ for each of the 15 proteins used instead of directly determining the four reference spectra according to Eq. (11) (Table III). [Because the BMD07RT, UCLA computer program used previously[70] is not readily available for many computers, a program written in C-language is now listed in Appendix A.]

The CD spectrum of the β-form as computed by Chang *et al.*[77] resembles that shown in Fig. 1; it has a strong positive band near 200 nm and a negative one above 210 nm, which qualitatively agrees with Woody's theoretical treatment.[37] The computed CD spectrum for the β-turn has a positive band near 224 nm, another positive one around 200 nm, and an intense negative one below 190 nm. A positive rather than a negative band above 220 nm was unexpected, but the sign and positions of the other two bands agreed with Woody's treatment.[46] Chang *et al.*[77] also successively excluded two proteins with the highest content of unordered form from the set of reference proteins and determined the reference spectra based on 13, 11, 9, and 7 proteins. The spectra for the β-form, β-turn, and unordered form so obtained were qualitatively similar to those based on 15 proteins. These uncertainties in the reference spectra of the β-form, β-turn and unordered form will undoubtedly affect the CD analysis of proteins (see section on "Comparison of Methods for CD Analysis").

In an attempt to bypass the curve-fitting requirements, Baker and Isenberg[78] employed integrals over the CD data and calculated the helix, β-form, and unordered form of proteins from such integrals. But by matrix formulation this approach is equivalent to the least-squares method.[77] Similarly, Hammonds[79] has shown that a continuous least-squares fit is identical to a discrete least-squares fit.

The introduction of the constraint $f_i = 1$ reduces one variable by replacing f_R by $(1 - f_H - f_\beta - f_t)$. Removal of this constraint can provide the unity sum test. A CD analysis may be in doubt if it fails to pass this test. However, Chang et al.[77] could not detect a trend in their studies (to be discussed later). In some cases it made no difference whether this constraint was introduced or not. In other cases there were good agreements between CD estimates and X-ray results for the ordered structures, even though the sum of f_is could be much greater or less than unity. Conversely, poor results could be obtained even when the sum unity test was met.

The introduction of the constraint $1 \geq f_i \geq 0$ avoids a possible negative f_i value or one that is greater than unity, both of which are unacceptable. On the other hand, such unreasonable values can point up the uncertainty in the analysis, which would have otherwise escaped detection. Chang et al.[77] found that the introduction of this constraint often improved the estimation of various conformations. In all fairness the use of the two constraints or the lack of them is still an issue at present. Perhaps one compromise is to determine the secondary structures both with and without the two constraints. Any unusual disagreement between the two results will caution against a too literal interpretation of the results.

Recently Bolotina et al.[75,80,81] reinvestigated the method of Chen et al.[77] but counted the secondary structures from the X-ray diffraction data based on a single "rigid" criterion proposed by Finkel'stein et al.[82] The assigned f_i values are often smaller than those based on what they termed

[78] C. C. Baker and I. Isenberg, Biochemistry 15, 629 (1976).

[79] R. G. Hammonds, Jr., Eur. J. Biochem. 74, 421 (1977).

[80] I. A. Bolotina, V. O. Chekhov, V. Lugauskas, and O. B. Ptitsyn, Mol. Biol. (USSR) 14, 902 (1980); English translation, p. 709 (1981).

[81] I. A. Bolotina, V. O. Chekhov, V. Lugauskas, and O. B. Ptitsyn, Mol. Biol. (USSR) 15, 167 (1981); English translation, p. 130 (1981).

[82] A. V. Finkel'stein, O. B. Ptitsyn, and S. A. Kozitsyn, Biopolymers 16, 497 (1977).

the "mild" criterion proposed by Levitt and Greer.[83] In their first paper Bolotina et al.[75] used the same five proteins (myoglobin, lysozyme, ribonuclease A, papain, and lactate dehydrogenase) that had been studied by Chen et al.[69,70,76] The calculated reference spectra of the helix, β-form, and unordered form were found to agree well with the CD spectra of the corresponding conformations of (Lys)$_n$ (see Fig. 1). This observation contrasts with the finding of Straus et al.[67] that the rotational strengths and thereupon the intensities of the helical CD bands of proteins are always lower than those of helical synthetic polypeptides. It also disagrees with the chain-length dependence of the helical bands as found by Chen et al.[76] Nevertheless, the secondary structures of the five proteins as redetermined from their CD spectra were in good agreement with the X-ray data based on the "rigid" criterion.

In their second paper Bolotina et al.[80] also considered the contributions of the β-turns and determined the four reference spectra in Eq. (11) (Table IV). For the β-turns they used the data of Chou and Fasman on 29 proteins[41] but considered only the second and third residues of the four-residue β-turns to contribute to the CD reference spectrum. As in the method of Chang et al.,[77] only net β-turns were counted. The four reference spectra so obtained were used to analyze the secondary structures of five reference proteins and five additional proteins (subtilisin BPN', glyceraldehyde-3-phosphate dehydrogenase, insulin, concanavalin A, and cytochrome c). In their third paper Bolotina et al.[81] extended the number of reference proteins to six by adding subtilisin BPN' or seven by further adding glyceraldehyde-3-phosphate dehydrogenase. In addition, the β-form was separated into parallel and antiparallel β-forms. The reference spectra of the helix, β-turn, and unordered form remained the same regardless of whether five, six, or seven reference proteins were used. Whether the β-form in Eq. (11) was split into two terms, the changes were as small as 5% or less between 200 to 250 nm and well within the uncertainties of the method of analysis (O. B. Ptitsyn, personal communication).

The reference spectra, $X_i(\lambda)$ in Eq. (6) obtained from proteins depend on the assigned f_i values of the reference proteins. For a set of CD spectra, $X(\lambda)$, the use of smaller f_is would lead to large $X_i(\lambda)$s, which may partially account for the reference spectrum of the helix as found by Bolotina et al.[80] For instance, the mean residue ellipticities at 222 nm for helical (Lys)$_n$, an infinite helix ($[\theta]_H^\infty$), a helix with 10 residues ($[\theta]_H^{10}$), and the helix based on the "rigid" criterion were $-37,600$, $-37,400$, $-28,100$, and $-36,100$ deg cm^2 dmol^{-1}, respectively. O. B. Ptitsyn (personal com-

[83] M. Levitt and J. Greer, J. Mol. Biol. 114, 181 (1977).

TABLE IV
REFERENCE CD SPECTRA OF FOUR CONFORMATIONS BASED ON
FIVE PROTEINS: MEAN RESIDUE ELLIPTICITIES
IN deg cm^2 dmol^{-1} [a]

λ(nm)	$[\theta]_H \times 10^{-3}$	$[\theta]_\beta \times 10^{-3}$	$[\theta]_t \times 10^{-3}$	$[\theta]_R \times 10^{-3}$
250	0	0	0	0
248	−0.15	−0.02	0.92	−0.28
245	−0.85	0.30	0.64	−0.19
242	−2.14	0.71	2.42	−0.80
240	−3.83	1.97	1.24	−0.26
239	−4.62	2.70	0.99	−0.40
238	−6.03	2.52	1.95	−0.28
237	−7.29	3.46	1.86	−0.45
236	−8.87	3.62	1.93	−0.34
235	−10.7	5.12	−0.34	0.05
234	−12.8	5.57	−1.83	0.47
233	−15.4	7.16	−0.40	−0.07
232	−17.9	6.04	−0.32	0.22
231	−20.3	6.97	1.47	−0.69
230	−23.0	6.48	5.31	−1.55
229	−26.4	3.86	11.7	−2.33
228	−29.1	2.20	17.3	−3.27
227	−31.5	−1.17	22.2	−3.73
226	−33.4	−4.15	28.4	−4.81
225	−34.7	−6.71	30.9	−5.10
224	−35.4	−9.15	31.4	−5.13
223	−35.9	−11.4	31.8	−5.03
222	−36.1	−14.2	31.3	−4.49
221	−36.0	−15.8	30.6	−4.27
220	−35.6	−16.9	29.7	−4.21
219	−34.3	−17.0	27.5	−4.34
218	−33.4	−17.9	24.7	−3.87
217	−32.6	−18.7	23.0	−3.54
216	−32.2	−18.4	20.9	−3.17
215	−31.4	−18.4	20.1	−3.36
214	−30.6	−18.0	18.3	−3.50
213	−30.0	−16.7	16.0	−3.74
212	−29.9	−15.6	16.4	−4.38
211	−30.3	−13.6	14.6	−4.65
210	−30.4	−11.3	13.6	−5.41
209	−30.5	−9.61	11.9	−5.78
208	−29.7	−6.66	8.48	−6.47
207	−28.0	−3.99	5.62	−7.23
206	−25.2	0.69	0.60	−7.93
205	−20.6	4.47	−5.22	−8.53
204	−14.6	7.21	−9.40	−9.37
203	−8.03	11.5	−12.9	−10.7
202	−0.69	13.7	−10.8	−13.3
201	8.05	19.3	−6.0	−17.7
200	16.6	25.0	6.42	−23.7

[a] The proteins used were myoglobin, lactate dehydrogenase, lysozyme, papain, and ribonuclease A. From Bolotina et al.[75] with the permission of Plenum (copyrighted by Plenum).

munication) finds the chain-length dependence of the helix unnecessary in their empirical method. He further points out that the reference spectrum of the helix is practically the same for the "rigid" method of Bolotina *et al.*[80] and for the early results of Chen *et al.*[70] [The f_H values for papain (0.21) and lactate dehydrogenase (0.29) were lower than the later updated ones (0.28 and 0.45, respectively) used by Chen *et al.*[76] and Chang *et al.*[77]] That the counting of f_is in a protein molecule is a serious problem will be discussed in the section on "Comparison of Methods for CD Analysis."

O. B. Ptitsyn (personal communication) suggests that $(Lys)_n$ in aqueous solution at pH above 10 is not a good model for an ideal infinite helix. It may not be completely helical and the fluctuating helical regions are rather short and probably a little distorted (its low intrinsic viscosities were even smaller than the corresponding ones of the coiled form). Therefore, the reference spectra between helical $(Lys)_n$ and the helix based on the rigid method of Bolotina *et al.*[80] practically coincided. To what extent helical $(Lys)_n$ deviates from a perfect helix is not known. We tend to think that a helix consisting of long interrupted segments virtually approaches an infinite helix according to Eq. (9). On the other hand, the aggregation of an uncharged polypeptide can distort its CD spectrum.

The use of proteins for the determination of reference spectra is more realistic than the use of synthetic polypeptides. The disadvantages of this approach are several-fold. First, the choice of reference proteins is by necessity arbitrary. Initially, only proteins whose three-dimensional structures are available were used. Ideally, the set of reference proteins should cover a wide range of f_H, f_β, and f_t. This condition is difficult to meet. Today many proteins of known secondary structure can be used as reference proteins, but what constitutes a representative set of reference proteins remains obscure. Thus, different sets of reference proteins may give different intensities of reference spectra and even different profiles for the β-turn and unordered form.

Second, the determination of the secondary structures is not straightforward. X-Ray crystallographers often use different criteria for identifying the secondary structure. Users of X-ray diffraction data may also set their own criteria. For instance, Saxena and Wetlaufer[68] chose for f_H a mean value between the lower limit representing regular α-helix and the upper limit representing total helices, including 3_{10}- and distorted helices, a practice that is not followed by other workers. Chen *et al.*[70,76] and Chang *et al.*[77] used the values given by the X-ray crystallographers. No distinction was made between α-helix and 3_{10}- and distorted helices, nor between parallel and antiparallel β-forms. In the early days X-ray results were often subject to revision when data at higher resolutions were obtained. [In the mid-1960s ORD studies of cytochrome c suggested a helicity of 17–

27%, whereas X-ray diffraction at 4 A indicated little or no α-helix (see, for example, Ref. 84). Doubts were therefore cast on the ORD analysis, which after all is empirical. Now we know that cytochrome c has about 40% helix and no β-form, in full support of CD analyses of this protein.] Of the five proteins used by Chen et al.,[70] the f_H and f_β of lactate dehydrogenase were later updated from 0.29 to 0.45 and 0.20 to 0.24, respectively (even the number of amino acid residues was raised from 311 to 331) and those of papain from 0.21 to 0.28 and 0.05 to 0.14, respectively. Obviously, such revision will affect the reference spectra; while the CD of the helix did not alter much, that of the β-form and, in particular, the unordered form varied significantly. Bolotina et al.[75,80,81] preferred to use a "rigid" criterion for the secondary structures,[82] which differs from the criteria used by X-ray crystallographers or the relaxed criteria set by Levitt and Greer.[83] At present, there are no generally accepted criteria yet.

Third, the secondary structure of a protein molecule is usually far from ideal. At present only the reference spectrum of the helix is well understood. Even here the 3_{10}-helix and distorted helices are lumped together with the α-helix and the matter of chain-length dependence of the helix is still undergoing investigations. But these assumptions seem to be supported by experimental and theoretical calculations. Unlike the helix, the β-form is often twisted and nonplanar. The degree of twisting varies from one protein to another, but the sense of twist is always the same. The dependence of CD on the chain length and sheet width is still not settled. Even the separation of parallel and antiparallel β-forms can present problems. In the hypothetical case of a 3-strand sheet, two β-strands are parallel to each other and a third β-strand is antiparallel to the second strand. It is difficult to decide whether the middle strand can be considered as half parallel and half antiparallel. However, the β-form appears to have a single characteristic CD spectrum, albeit with considerable variations. The types of β-turns are more diverse and their CD spectra are not well characterized even for model compounds. The unordered form is more complicated than the ordered one; we can only assume a statistical average for the observed CD. All these problems will certainly complicate the CD analysis of proteins.

Linear Combination of CD Spectra of Proteins

Recently, Provencher and Glöckner[85] proposed direct analysis of the CD spectrum of a protein by linear combination of the CD spectra of

---

[84] S. Beychok, Annu. Rev. Biochem. 37, 437 (1968).
[85] S. W. Provencher and J. Glöckner, Biochemistry 20, 33 (1981).

reference proteins of known secondary structure. This approach may circumvent the dilemma between a stable but inadequate model and a realistic but unstable one (see section above). The instability of many parameters in Eq. (6) is controlled by a simple constrained regularization procedure. By introducing a coefficient γ_j to be determined and expressing $X(\lambda)$ in Eq. (6) as

$$X(\lambda) = \sum_{j=1}^{N_\gamma} \gamma_j R_j(\lambda) \tag{12}$$

the f_i values are simply

$$f_i = \sum_{j=1}^{N_\gamma} \gamma_j F_{ji} \tag{13}$$

Here $R_j(\lambda)$ is the CD of the jth reference protein at wavelength λ and N_γ is the number of reference proteins used. F_{ji} is the fraction of the ith conformation of the jth protein. The γ_j values are determined by minimizing a quantity ε:

$$\varepsilon = \sum_{\lambda=1}^{N_Y} [Y(\lambda - X(\lambda)]^2 + \alpha \sum_{j=1}^{N_\lambda} (\gamma_j - 1/N\gamma)^2 \tag{14}$$

again with the constraints: $\Sigma f_i = 1$ and $f_i \geq 0$. $X(\lambda)$ and $Y(\lambda)$ are the experimental and calculated values, respectively, and N_Y is the number of data points. For $\alpha = 0$, Eq. (14) is reduced to the least-squares solution. For $\alpha > 0$, the second term on the right side of Eq. (14) tends to stabilize the solution by keeping each γ_j small, i.e., near $1/N_\lambda$, unless the corresponding $R_j(\gamma)$ happens to have components that fit $X(\lambda)$ well and thereby significantly reduce the least-squares term, the first term on the right side of Eq. (14). A general FORTRAN regularization package has been developed and a user-oriented version is available upon request from the authors (see Appendix B). Provencher and Glöckner[85] used the same CD data of 18 proteins provided by Chang et al.,[77] except that the CD spectra of thermolysin and subtilisin BPN′ were excluded in their analysis in order to improve the correlation coefficients.

Independently, Hennessey and Johnson[86] used a similar approach but applied an eigenvector method of multicomponent analysis with an unconstrained least-squares method. Their analysis was based on matrix calculation of orthogonal basis CD spectra from the CD spectra of reference proteins. Fifteen proteins and one helical polypeptide, proton-

[86] J. P. Hennessey, Jr. and W. C. Johnson, Jr., *Biochemistry* **20**, 1085 (1981).

ated $(Glu)_n$, were studied (see footnote a of Table V) and their CD spectra were extended to 178 nm in the vacuum ultraviolet region. Hennessey and Johnson concluded that five most significant basis spectra were needed to reconstruct the original CD spectra of 15 proteins and one polypeptide (Fig. 5) and the remaining 11 basis spectra were probably noise. With only five basis CD spectra, no more than five independent types of secondary structure can be learned from a CD spectrum. However, Hennessey and Johnson actually considered eight types of secondary structures: the helix, the parallel and antiparallel β-forms, the four types of β-turns (types I, II, and III and type T, which combines the remaining β-turns), and the unordered form. They are able to do this because the same eigenvector method applied to the X-ray structural data gives only five important basis structure vectors. Thus only five independent secondary superstructures are necessary to describe the eight standard secondary structures, which are not independent to within a small error in the protein analyses. Since each basis CD spectrum corresponds to a mixture of secondary structures, the same coefficients for the reconstruction of the CD spectrum of a protein can be used to estimate the secondary structure of the protein. A FORTRAN program, "PROSTP," developed for this method of CD analysis of proteins is given in Appendix C.

The five basis spectra from 250 to 178 nm (Fig. 5) can only be used to analyze the conformation of a protein whose CD spectrum covers the same range of wavelengths. This limitation poses a practical problem because no commercial instruments for VUCD will be available in the foreseeable future unless there is such a demand. Consequently, the majority of users are compelled to limit their CD measurements to about 184 nm under ideal conditions; most reported CD spectra were cut off at 190 nm or even longer wavelengths. Hennessey and Johnson analyzed the effect of data truncation on the CD analysis and found that the 184-nm cut-off still made little difference in the results. As long as any major CD band is not eliminated, their method of analysis is fairly insensitive to the wavelength. However, truncation to 190 nm resulted in striking changes, which were reflected almost solely in the estimates of the β-form and unordered form as well as the totals (Σf_i). Further truncation to 200 nm accentuated some of these changes but actually corrected others. Table V cannot be used for the method of Hennessey and Johnson over wavelength ranges other than 250 to 178 nm. Instead, the basis CD spectra should be regenerated by using Program "BAVGEN" in Appendix C so that the vectors will be orthogonal.

Provencher and Glöckner[85] analyzed the CD spectrum of a protein from CD spectra of 16 proteins (Ref. 77), excluding one spectrum of the protein to be analyzed. This is equivalent to analysis with 15 basis spectra

TABLE V
FIVE BASIS CD SPECTRA ($\varepsilon_L - \varepsilon_R$) DERIVED
FROM FIFTEEN PROTEINS AND ONE
POLYPEPTIDE[a]

λ(nm)	1	2	3	4	5
250	0	0	0	0	0
248	0	0	0	0	0
246	−0.2	0.2	−0.1	−0.1	−0.1
244	−0.3	0.3	−0.2	−0.1	−0.2
242	−0.6	0.5	−0.3	−0.1	−0.3
240	−1.4	0.6	−0.4	−0.1	−0.4
238	−2.6	0.7	−0.5	−0.1	−0.5
236	−4.4	0.8	−0.6	0	−0.7
234	−6.8	0.9	−0.8	0.1	−0.7
232	−9.1	1.0	−1.1	0.3	−0.7
230	−11.7	1.2	−1.5	0.5	−0.8
228	−14.8	1.7	−1.8	0.9	−0.7
226	−17.2	1.7	−2.0	0.9	−0.8
224	−19.3	1.3	−2.1	0.8	0.8
222	−20.3	1.2	−2.3	0.7	−0.7
220	−20.1	1.4	−2.8	0.5	−0.6
218	−19.4	1.6	−3.1	0.5	−0.5
216	−18.9	1.4	−3.4	0.3	−0.3
214	−18.4	1.0	−3.6	0.3	−0.1
212	−18.3	0.6	−3.8	0.3	0.1
210	−18.5	0.2	−3.9	0.5	0.5
208	−18.3	−0.9	−4.0	0.5	1.0
206	−16.8	−2.6	−3.6	0.2	1.4
204	−11.7	−4.4	−3.3	−0.2	1.4
202	−3.8	−6.1	−3.0	−0.8	1.1
200	5.4	−8.5	−2.5	−1.1	0.2
198	17.4	−9.6	−2.5	−1.1	−0.9
196	28.0	−9.6	−2.1	−0.5	−1.1
194	36.3	−6.0	−2.5	0.9	−0.8
192	41.2	−1.2	−2.9	2.0	−0.2
190	40.0	2.3	−2.6	2.3	0.4
188	36.4	5.3	−2.5	1.9	0.5
186	30.8	6.6	−2.3	0.7	0.3
184	24.4	6.7	−2.1	−0.7	0.2
182	18.6	6.0	−2.3	−2.1	0.2
180	13.5	4.9	−2.9	−3.3	−0.1
178	10.4	4.6	−3.6	−3.6	−0.5

[a] The biopolymers used were α-chymotrypsin, cytochrome c, elastase, flavodoxin, glyceraldehyde-3-phosphate dehydrogenase, hemoglobin, lactate dehydrogenase, lysozyme, myoglobin, papain, prealbumin, ribonuclease, subtilisin BPN′, subtilisin Novo, triosephosphate isomerase, and helical poly(L-glutamic acid). All data are expressed as mean residue CD in $cm^{-1} M^{-1}$. From Hennessey and Johnson[86] with the permission of the American Chemical Society, copyright 1981.

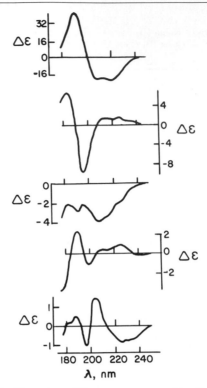

Fig. 5. Five most significant base CD spectra generated by the eigenvector method of multicomponent matrix analysis in descending order of significance. (Redrawn from Hennessey and Johnson[86] with the permission of American Chemical Society, copyright 1981.)

in the method of Hennessey and Johnson,[86] if the two groups had studied the same reference proteins. However, there are differences between the two methods. Provencher and Glöckner introduced the two constraints: $\Sigma f_i = 1$ and $f_i \geq 0$, but Hennessey and Johnson considered such constraints artificial. Instead, in their published work they normalized the calculated f_is to $\Sigma f_i = 1$. Hennessey and Johnson used only five most significant basis spectra (Fig. 5). The third spectrum has no nodes, the first one node, the second two nodes, the fourth essentially three nodes, and the fifth four nodes. The trend appears to be more nodes and lower intensities for less important basis spectra. The less significant spectra probably reflect noise, and are therefore eliminated from the CD analysis (W. C. Johnson, Jr., personal communication). With orthogonal basis spectra, the fit for the CD spectrum of a protein other than one of the original set of 16 proteins will be unique. Hennessey and Johnson did not take into consideration the chain-length dependence of the CD of the helix. How the exclusion of $(Glu)_n$, a helical polypeptide, will affect their basis spectra remains to be seen. However, W. C. Johnson, Jr. points out that their most significant basis spectra average any variability in the CD contributions due to other sources such as side chains, tertiary structure, chain length of the secondary structure, as well as errors in concentration measurements, CD measurement, etc. These variations are contained in the discarded basis spectra, which are merely "noise."[87]

Since the basis spectra are obtained from proteins of known secondary structure, the advantages and disadvantages of the Provencher–Glöckner and Hennessey–Johnson methods are the same as those described in the section on "Reference Spectra Based on Proteins." However, these two methods have one attractive feature: it circumvents the determination of reference spectra of the helix, β-form, β-turn, and unordered form. Of course these methods will only apply to proteins whose structural characteristics are well represented in the set of reference proteins used to construct the basis spectra. Hennessey and Johnson extended their CD measurements to the vacuum ultraviolet region just as Brahms and Brahms did in the section on "Reference Spectra Based on Model Polypeptides." Unfortunately, users of commercial instruments do not have access to wavelengths below 184 nm at present.

Comparison of Methods for CD Analysis

The very nature of any empirical method unavoidably invites adverse criticism, especially when various methods give different results. Before we rush to conclude that the methods of CD analysis are not suitable for determining protein conformation, it seems appropriate to compare the results based on some current methods and point out the uncertainties and possible improvements. Earlier, Rosenkranz[88] compared various methods that were available in 1974 and concluded that the reference spectra developed by Rosenkranz and Scholtan[66] and Chen et al.[70] gave good agreement between experimental and calculated CD spectra for several proteins tested. He further indicated that these two methods were highly suitable for determining the conformation of proteins having CD bands of the helix or β-form or both. He also found that the new reference spectra obtained by Chen et al.[76] were less satisfactory than those used previously.[70] Suffice it to say, a good fit between the experimental and calculated curves does not guarantee a correct solution of the secondary structure of a protein, but a poor fit often points up the imperfection in the

[87] W. C. Johnson, Jr., in "Methods of Biochemical Analysis" (D. Glick, ed.), p. 61. Wiley (Interscience), New York, 1985.
[88] H. Rosenkranz, Z. Klin. Chem. Klin. Biochem. 12, 415 (1974).

method of analysis. Thus, we should also compare CD estimates with X-ray results, if possible.

All current methods for CD analysis are developed to determine not only the helix and β-form but also the β-turn and unordered forms. The correlation coefficients between CD and X-ray diffraction analysis from five laboratories can be calculated, although the proteins used were not all identical (Table VI): (I) Chang et al.[77] and (II) Bolotina et al.[80] determined reference spectra of proteins for the four conformations, (III) Provencher and Glöckner[85] and (IV) Hennessey and Johnson[86] used a linear combination of basis spectra of proteins, and (V) Brahms and Brahms[47] computed the CD spectrum of a protein from the spectra of model compounds. A correlation coefficient, r, near 1 indicates a successful prediction, whereas an r value near zero predicts no better than a random assignment and an r value close to -1 indicates a total disagreement between experimental and computed results. On this basis the estimated f_H values correlated highly between CD and X-ray results among all five laboratories. The CD estimates for the β-form show significant correlation, especially for the Hennessey–Johnson and Brahms–Brahms results. But in all five methods the correlation coefficient for β-turns was poor when unbiased results were compared.

The results from proteins used to calculate correlation coefficients by Chang et al.[77] and Bolotina et al.[80] had a bias effect, viz. the reference proteins included the protein to be analyzed. Provencher and Glöckner[85] evaluated the data of Chang et al.[77] and showed that the unbiased correlation coefficient for the β-form dropped sharply and that for the β-turn became negative (Table VI). Hennessey and Johnson determined the correlation coefficients both with and without the CD spectrum of the protein to be analyzed in the construction of their basis spectra.[86] The r values for the β-forms and β-turns were poorer with the unbiased data than with the biased ones. Bolotina et al.[80] found that the CD spectrum of concanavalin A was poorly described by their reference spectra and they attributed this anomaly to large CD contributions from aromatic chromophores in accordance with the work of Hermann et al.[89] Correction for these nonpeptide contributions led to good correlation coefficients, but when such corrections should be applied to an unknown protein and how they can be applied become problematic. The method of Brahms and Brahms is also unbiased because they used model peptides and myoglobin for the reference spectra, and their arbitrary reduction of the CD magnitude of $(Ala_2-Gly_2)_n$ for the β-turn by one-half is subject to mention.

[89] M. S. Hermann, C. E. Richardson, L. M. Setzler, W. D. Behnke, and R. E. Thompson, Biopolymers 17, 2107 (1978).

TABLE VI

CORRELATION COEFFICIENTS BETWEEN CD ESTIMATE AND X-RAY RESULTS
OF THE f_i VALUES

Method	Helix	β-Form	β-Turn	Unordered form
Chang et al. (1978)[a]				
1	0.87	0.61	0.15	—
2	0.92	0.83	0.23	0.37
3	0.85	0.25	−0.31	0.46
Bolotina et al. (1980)[b]				
1	0.93	0.93	0.84	—
2	0.95	0.57	0.39	0.39
Provencher and Glöckner (1981)[c]	0.96	0.94	0.31	0.49
Hennessey and Johnson (1981)[d]				
1	0.98	0.83; 0.71	0.78	0.88
2	0.95	0.66; 0.51	0.25	0.72
Brahms and Brahms (1980)[e]	0.92	0.93	0.33	0.65

[a] (1) From Chang et al.[77] based on 18 proteins (see footnote a of Table III plus adenylate kinase, α-chymotrypsin, and elastase). The Pearson product–moment correlation coefficient, r, is defined as

$$r = [\Sigma X_i Y_i - \Sigma X_i \Sigma Y_i / n] / \{[\Sigma X_i^2 - (\Sigma X_i)^2 / n] \times [\Sigma Y_i^2 - (\Sigma Y_i)^2 / n]\}^{1/2}$$

Here X_i and Y_i are the experimental and calculated values, respectively, and n is the number of samples studied. (2) From Provencher and Glöckner.[85] The data on 16 of the 18 proteins used by Chang et al. in (1), excluding subtilisin BPN' and thermolysin, were recalculated. (3) Same as in (2) but with 15 proteins. The protein to be analyzed was excluded from the reference proteins.

[b] (1) From Table 3 of Bolotina et al.[81] based on 13 proteins (see the text). The CD spectrum of concanavalin A was corrected. Glyceraldehyde-3-phosphate dehydrogenase was omitted in the calculation of the r value for the β-turn. (2) From Woody.[31] The data for 10 proteins from Table 3 of Bolotina et al.[80] including the uncorrected data for concanavalin A (see the text) were recalculated. Again glyceraldehyde-3-phosphate dehydrogenase was excluded for the β-turn calculation.

[c] From Table II of Provencher and Glöckner in footnote a(2) based on the CD spectra of Chang et al. in footnote a(1). The protein to be analyzed was not used in the construction of base spectra.

[d] From Table VIII of Hennessey and Johnson.[86] The two numbers under the β-form refer to antiparallel and parallel β-forms. The r value for the β-turn was based on all β-turns. (1) Analyzed from basis CD spectra; (2) analyzed from basis CD spectra constructed without the spectrum of the protein to be analyzed. The individual r values for the β-turns (types I, II, and III and other β-turns combined) were (1) 0.53, 0.73, 0.36, and 0.71, respectively, and (2) −0.07, 0.51, −0.44, and 0.38, respectively.

[e] From Woody[31] in footnote b(2) based on data from 17 proteins in Table 2 of Brahms and Brahms,[47] excluding rubredoxin and two other proteins. Brahms and Brahms suggested that the secondary structure of rubredoxin was different in crystals and in solution. The two other proteins were not analyzed by Levitt and Greer whose method was used by Woody for determining the secondary structures.

A further test of various methods can be made by comparing the CD estimates of the secondary structure of several proteins, most of which had been studied by the five laboratories (Table VII). Several features emerge. First, there are considerable differences in the way secondary structures are totalled from X-ray diffraction results by different laboratories. Following Bolotina et al.,[75] we illustrate in Fig. 6 the secondary structures of five proteins from X-ray analysis. X-Ray crystallographers inspect their models and often determine the local conformation of a residue in relation to neighboring residues and the pattern of hydrogen bonds involving these residues. According to Levitt and Greer,[83] the most successful criteria should be based on peptide hydrogen bonds, inter-C^α distances, and inter-C^α torsion angles. Small errors in the dihedral angles could lead to quite a different assignment. Accordingly, these authors have used precise rules and developed a computer program to automatically analyze the atomic coordinates of many globular proteins. A list of α-helices, β-sheets, and β-turns of almost all proteins of known secondary and tertiary structure before 1977 was compiled. The assignment of Levitt and Greer does not always correspond to that by X-ray crystallographers. Bolotina et al.[80] used rigid criteria proposed by Finkel'stein et al.[82] for the helix and β-form and counted only the second and third residues of the 4-residue β-turns based on the Chou–Fasman work with 29 proteins. Chang et al.,[77] Bolotina et al.,[80] and Provencher and Glöckner[85] considered only the net β-turns by canceling types I', II', and III' from their mirror images, types I, II, and III. Hennessey and Johnson[86] lumped together all β-turns other than types I, II, and III in one class. Clearly, these different procedures will yield different f_i values. As Fig. 6 illustrates, myoglobin has an f_H of 0.88 according to Levitt and Greer, but most workers count it slightly less than 0.80. The f_H and f_β values for lysozyme are 0.45 and 0.19, respectively, by the method of Levitt and Greer,[83] but only 0.30 and 0.09, respectively, according to Bolotina et al.[80]; the f_t value for lysozyme is taken as 0.32 by Hennessey and Johnson[86] and 0.19 by Bolotina et al.[80] Likewise, the f_β value varies from 0.23 found by Bolotina et al.[80] to 0.40 by Chang et al.[77] and to 0.46 by Levitt and Greer.[83] The f_i values used by five laboratories (Table VII) are not exactly identical in most cases. These differences will in turn affect the reference or base spectra of proteins of known structure, not to mention that different sets of reference proteins may also alter these reference or base spectra.

Another complication involving the counting of secondary structure often seems to have been overlooked. Because the CD bands of various conformations arise from peptide chromophores, peptide bonds rather than amino acid residues should be used to determine the fractions of helix, β-form, and β-turn. As far as we are aware, Hennessey and John-

TABLE VII
COMPARISON OF THE SECONDARY STRUCTURE OF TEN PROTEINS BETWEEN CD
ESTIMATES AND X-RAY RESULTS

Protein	Method[a]		f_H	f_β	f_t	f_R	Σf_i[b]
Myoglobin	I	X-ray	0.79	0	0.05	0.16	
		CD (1)	0.80	0	0.02	0.18	
		CD (2)	0.83	0.05	0.07	0.33	1.18
	II	X-ray	0.76	0	0.06	0.18	
		CD	0.79	0.01	0.07	0.13	
	III	X-ray	0.79	0	0.05	0.16	
		CD	0.86	0	0	0.14	
	IV	X-ray	0.78	0	0.12	0.10	
		CD (1)	0.75	−0.01	0.08	0.07	0.89
		CD (2)	0.75	−0.03	0.06	0.07	0.87
Lactate dehydrogenase	I	X-ray	0.45	0.24	0.06	0.25	
		CD (1)	0.45	0.18	0.13	0.24	
		CD (2)	0.47	0	0.13	0.22	0.81
	II	X-ray	0.39	0.14	0.11	0.36	
		CD	0.32	0.11	0.09	0.48	
	III	X-ray	0.45	0.24	0.06	0.25	
		CD	0.40	0.22	0.13	0.26	
	IV	X-ray	0.41	0.17	0.11	0.31	
		CD (1)	0.40	0.19	0.13	0.29	1.01
		CD (2)	0.42	0.04	0.11	0.24	0.81
	V	X-ray	0.42	0.26	0.19	0.13	
		CD	0.41	0.22	0.17	0.20	
Lysozyme	I	X-ray	0.41	0.16	0.23	0.20	
		CD (1)	0.32	0.29	0.08	0.31	
		CD (2)	0.32	0.33	0.08	0.33	1.06
	II	X-ray	0.30	0.07	0.19	0.44	
		CD	0.30	0.07	0.19	0.44	
	III	X-ray	0.41	0.16	0.23	0.20	
		CD	0.45	0.21	0.26	0.08	
	IV	X-ray	0.36	0.09	0.32	0.23	
		CD (1)	0.34	0.32	0.26	0.25	1.17
		CD (2)	0.36	0.20	0.28	0.18	1.02
	V	X-ray	0.45	0.19	0.14	0.22	
		CD	0.43	0.12	0.08	0.37	
Cytochrome c	I	X-ray	0.39	0	0.24	0.37	
		CD (1)	0.44	0	0.28	0.28	
		CD (2)	0.48	−0.16	0.33	0.31	0.96
	II	X-ray	0.43	0	0.14	0.43	
		CD	0.33	0.09	0.08	0.50	
	III	X-ray	0.39	0	0.24	0.37	
		CD	0.33	0.09	0.17	0.41	
	IV	X-ray	0.38	0	0.17	0.45	
		CD (1)	0.40	0.04	0.14	0.36	0.93
		CD (2)	0.32	−0.01	0	0.16	0.46

(*continued*)

TABLE VII (continued)

Protein	Method[a]		f_H	f_β	f_t	f_R	$\Sigma f_i{}^b$
	V	X-ray	0.43–0.48	0.09	0.23	0.25–0.20	
		CD	0.46	0	0.22	0.31	
Subtilisin BPN′	I	X-ray	0.31	0.10	0.22	0.37	
		CD (1)	0.15	0.58	0.04	0.23	
		CD (2)	0.19	0.15	0.06	0.18	0.58
	II	X-ray	0.27	0.16	0.13	0.44	
		CD	0.26	0.16	0.12	0.46	
	III	X-ray	0.31	0.10	0.22	0.37	
		CD	0.15	0.48	0.18	0.18	
	IV	X-ray	0.30	0.09	0.21	0.40	
		CD (1)	0.24	0.22	0.14	0.29	0.89
		CD (2)	0.26	0.19	0.14	0.40	1.00
Papain	I	X-ray	0.28	0.14	0.17	0.41	
		CD (1)	0.29	0	0.15	0.56	
		CD (2)	0.30	0.06	0.18	0.60	1.14
	II	X-ray	0.24	0.11	0.10	0.55	
		CD	0.25	0.11	0.10	0.54	
	III	X-ray	0.28	0.14	0.17	0.41	
		CD	0.27	0.05	0.31	0.36	
	IV	X-ray	0.28	0.09	0.14	0.49	
		CD (1)	0.27	0.14	0.15	0.50	1.06
		CD (2)	0.25	0.19	0.17	0.45	1.06
Ribonuclease A	I	X-ray	0.23	0.40	0.13	0.24	
		CD (1)	0.21	0.39	0.10	0.30	
		CD (2)	0.17	0.79	0.09	0.36	1.41
	II	X-ray	0.17	0.23	0.15	0.45	
		CD	0.21	0.25	0.16	0.38	
	III	X-ray	0.23	0.40	0.13	0.24	
		CD	0.26	0.44	0.11	0.19	
	IV	X-ray	0.24	0.33	0.14	0.29	
		CD (1)	0.25	0.18	0.20	0.22	0.86
		CD (2)	0.22	0.38	0.15	0.26	1.02
α-Chymotrypsin	I	X-ray	0.09	0.34	0.34	0.23	
		CD (1)	0.05	0.53	0.02	0.40	
		CD (2)	0.07	0.35	0.04	0.39	0.85
	III	X-ray	0.09	0.34	0.34	0.23	
		CD	0.09	0.29	0.22	0.40	
	IV	X-ray	0.10	0.34	0.20	0.36	
		CD (1)	0.13	0.25	0.19	0.32	0.89
		CD (2)	0.12	0.09	0.15	0.20	0.56
Elastase	I	X-ray	0.07	0.52	0.26	0.15	
		CD (1)	0	0.46	0.07	0.47	
		CD (2)	−0.08	0.72	−0.01	0.44	1.07
	III	X-ray	0.07	0.52	0.26	0.15	
		CD	0.04	0.49	0.14	0.32	
	IV	X-ray	0.10	0.37	0.22	0.31	

TABLE VII (continued)

Protein	Method[a]	f_H	f_β	f_t	f_R	Σf_i[b]
		CD (1)	0.09	0.30	0.22	0.36	0.97
		CD (2)	0.09	0.35	0.23	0.38	1.05
Concanavalin A	I	X-ray	0.02	0.51	0.09	0.38	
		CD (1)	0.25	0.46	0.20	0.09	
		CD (2)	0.33	−0.56	0.25	−0.02	0
	II	X-ray	0.02	0.46	0.16	0.34	
		CD	0	0.12	0.02	0.86	
	III	X-ray	0.02	0.51	0.09	0.38	
		CD	0.08	0.41	0.15	0.36	
	V	X-ray	0.06	0.59	0.20	0.25	
		CD	0.03	0.49	0.19	0.29	

[a] From Table I of Chang et al.[77] (1) With the constraints: $\Sigma f_i = 1$ and $1 \geq f_i \geq 0$; (2) without the constraints. II, From Table 3 of Bolotina et al.[80] If the β-form was split into parallel and antiparallel β-forms, the results were slightly different (see Table 1 of Bolotina et al.[81] III, From Table I of Provencher and Glöckner,[85] based on the same CD spectra and X-ray results used by Chang et al. in footnote aI. IV, Based on Table IX of Hennessey and Johnson.[86] The results listed here were provided by W. C. Johnson, Jr. (private communication) who pointed out that the f_i values by X-ray diffraction were slightly different from the published work. Also unlike the published results, the CD analysis was not adjusted for intensity so that $\Sigma f_i = 1.00$. The f_β value represents the sum of f_β(parallel) and f_β(antiparallel). The f_t value represents the sum of f_t(I), f_t(II), f_t(III), and f_t(T). However, the results listed here have the bias effect, i.e., the protein to be analyzed was included in the set of reference proteins. V, From Table 2 of Brahms and Brahms.[47] The f_i values listed here were based on Levitt and Greer.[83] Method III was unbiased; the protein to be analyzed was removed from the set of reference proteins. Method V also had no bias effect since the reference spectra were based on model peptides and myoglobin.

[b] For X-ray results, Σf_i is always unity and is therefore not listed. For CD results, the absence of a number in this column indicates that the constraint $\Sigma f_i = 1$ was used.

son did use peptide bonds as basic units.[86] For a protein having i helical segments, the number of bonds is the total number of amino acid residues in these segments minus i residues. The same is true for counting β-strands. The number of bonds is the total number of residues minus i residues for i strands. Chen et al.[76] also recognized this discrepancy in the counting of β-turns. Each β-turn involves four residues but only three peptide bonds are considered in theoretical calculations; thus, for n β-turns, which have $4n$ residues, the number of peptide bonds is $3n$, which is a 25% reduction. The situation is even more confusing if a β-turn is linked to a helix or a β-strand. How to identify and distribute these residues among the three conformations remains problematic. For an unknown protein we do not know the number of helical segments, β-strands,

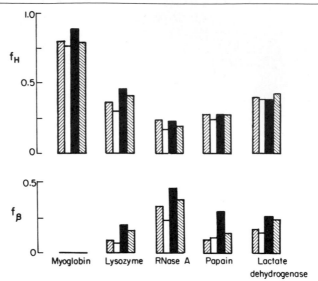

Fig. 6. The fractions of helix and β-form of five proteins based on X-ray diffraction results. Open bar, according to "rigid" criteria proposed by Finkel'stein et al.[82] and used in Bolotina et al.[75,80] Solid bar, according to the criteria of Levitt and Greer.[83] Stippled bars, based on the data provided by X-ray crystallographers and used in Hennessey and Johnson[86] (left) and Chen et al.[76] (right).

and β-turns. Thus, the secondary structures are usually counted by the number of amino acid residues.

Second, CD estimates for the helix are usually excellent regardless of the method used. This is perhaps due to the high intensities of the helical CD bands, which usually predominate over the CD contributions of other conformations if the protein contains a moderate amount of helix. In addition, the helical segments found in proteins are mostly close to the ideal α-helix, and do not have many 3_{10}-helices or distorted helices. Chen et al.[70] found that the reference spectra varied from one set of reference proteins to another, but such variations were minimized by the inclusion of myoglobin, which has the highest helicity among the reference proteins studied. Chang et al.[77] found that their method did not apply to concanavalin A, which has almost no helixes; however, CD estimates for the low helicity in α-chymotrypsin and elastase were good. Bolotina et al.[80] reported a good estimate for the helicity, but not for the β-forms and β-turns of concanavalin A after correction for the CD contributions of aromatic chromophores. On the other hand, Provencher and Glöckner[85] obtained a good fit between the observed and calculated CD spectrum of concanavalin A without resorting to such corrections, but their estimates appeared to be low in relation to the X-ray results. Results for the β-forms and β-turns of the proteins listed in Table VII were often unsatisfactory by the method of Chang *et al.*[77] However, no single method can guarantee reliable results in every case. If $\Delta f_i \geq \pm 0.10$ between the CD estimate and X-ray result for any conformation was considered to be unsatisfactory, the method of Bolotina *et al.*[80] gave poor results for cytochrome c, thermolysin (not shown), and concanavalin A (based on an uncorrected CD spectrum). The method of Provencher and Glöckner[85] was unsatisfactory for β-turn estimates in papain, α-chymotrypsin, and elastase (Table VII) and also in insulin, parvalbumin, and trypsin inhibitor (not shown). Estimates for the helix and β-forms of subtilisin BPN′ were also poor, probably due to inaccurate CD data for this protein used in their analysis. The method of Hennessey and Johnson[86] did not do too well for β-estimates in α-chymotrypsin, lysozyme, subtilisin BPN′, and ribonuclease (Table VII) and in flavodoxin and prealbumin (not shown). Indeed, there are uncertainties in each method. However attractive a particular method may be, it must be subjected to extensive tests on many proteins of known secondary and tertiary structure. Whether and how the use of a different set of reference proteins or additional proteins to the current sets will alter the CD estimates remain to be investigated.

Third, Chen *et al.*[69,70] first introduced two constraints: $\Sigma f_i = 1$ and $1 \geq f_i \geq 0$. This will reduce one parameter, f_R, to be determined and also will avoid any negative numbers in the computed f_i values. Such constraints are of course artificial, but Chang *et al.*[77] found that their CD estimates improved with the constraints. Subtilisin BPN′ was a notable exception, but the calculated Σf_i without the constraints was too far from unity. The results with concanavalin A without the constraints were even worse (Σf_i became zero). However, the sum unity test did not seem to guarantee the correctness of the CD estimates. For instance, Σf_i was close to one for lysozyme, but the β-estimate was too high and the β_t-estimate too low without the two constraints. On the other hand, the α-estimate for myoglobin was excellent, even though it failed the sum unity test. Provencher and Glöckner[85] used the two constraints, whereas Hennessey and Johnson[86] did not. It is not known whether the removal of the two constraints would affect the Provencher–Glöckner analysis.

Fourth, Hennessey and Johnson[86] advocated the use of vacuum ultraviolet CD and suggested that even the basis spectra measured to 178 nm might not be enough for all independent pieces of information.[87] But CD estimates by the Hennessey–Johnson method with a 200-nm cut-off were rather close to those with a 178-nm cut-off in several cases (Table VII). Notable exceptions were the β-estimates for lactate dehydrogenase, lysozyme, and α-chymotrypsin and the β_t-estimate for cytochrome c. On the other hand, the β-estimate for ribonuclease A was better with the 200-nm cut-off. Truncation at 190 nm led to some CD estimates that were not as good as with truncation at 200 nm (not shown). This could be merely a fortuitous coincidence, but it deserves further investigation. In the spectral analysis by Provencher and Glöckner,[85] the data below 190 or 210 nm were given lower statistical weights or completely discarded and yet their results were quite insensitive to this adjustment.

The CD analysis of protein conformation has made considerable progress during the past decade, but the methods described here are still not perfect and need extensive tests on proteins of known secondary structure. Some methods appear to be good in most cases, but there is no guarantee that CD estimates of the secondary structure of an unknown protein will always be correct. Because β-forms have a much broader range of conformations than helices, the CD spectra of β-forms are expected to be more variable than those of helices. The geometry of β-turns in proteins is even more variable than that of either helices or β-forms. What constitutes representative CD spectra for β-turns is still an open question. Perhaps there is a whole range of CD curves for β-turns, which depend on the types of β-turns and also the amino acid residues in them. One can only hope that the reference proteins used for the determination of reference or basis spectra happen to adequately represent an unknown protein to be analyzed. If possible, calculations should be complemented by other evidence, such as the sequence-predictive method. If a CD spectrum of a protein shows a well-defined double minimum at 222 and 208–210 nm or a single minimum between 210 and 220 nm the CD estimate of the amount of helix or the β-forms may be accepted with some confidence. On the other hand, if the negative CD bands between 200 and 240 nm are broad, or skewed, or have shoulders instead of distinct minima and the positive band below 200 nm is unusually weak, the CD estimates may be very uncertain. At present the estimation of β-turns by CD analysis is most uncertain. For proteins containing a mixture of β-forms and β-turns and no helix, such as cobra neurotoxin[90] and α-bungarotoxin,[91] the CD contributions due to nonpeptide chromophores seem no longer to be negligible. Therefore, these "atypical" CD spectra are difficult to analyze by current CD calculations. Occasionally, one finds CD estimates in the literature good to the first decimal. Such practice goes beyond the limits of CD analysis.

The methodology of CD analysis also deserves further investigation.

[90] Y.-H. Chen, T.-B. Lo, and J. T. Yang, *Biochemistry* **16,** 1826 (1977).
[91] Y.-H. Chen, J.-C. Tai, W.-J. Huang, M.-Z. Lai, M.-C. Hung, M.-D. Lai, and J. T. Yang, *Biochemistry* **21,** 2592 (1982).

Among the uncertainties totaling of secondary structure from X-ray data remains problematic. The rigid criteria used by Bolotina et al.[75,80] versus the relaxed criteria proposed by Levitt and Greer[83] or some other criteria to be developed must be further tested; eventually, a general consensus may be reached for CD analysis. The use of peptide bonds rather than amino acid residues as the basic units in the counting is logical, but the number of helical segments, β-strands, and β-turns of an unknown protein cannot be predetermined and this problem must be overcome. The use of VUCD is attractive, but for practical purposes whether commercial circular dichrometers with a cut-off around 184 nm will be equally adequate for the CD analysis of proteins must yet be decided. The assumption that the CD contributions due to nonpeptide chromophores can be neglected may be permissible for most proteins because the mole fractions of aromatic groups and cystine residues are usually small. Yet there are bound to be some exceptions and failure to detect them will affect CD estimates of various conformations in a protein (for an early discussion on this subject, see Ref. 92). Despite these reservations, we feel confident that current methods for the CD analysis of proteins still can be refined and improved. It is immaterial whether one method is better than another. What is important is that one or more methods can provide a reasonable estimate of the secondary structure of proteins with confidence.

Appendix A lists a computer program for the Chang–Wu–Yang method; appendix B provides information concerning the Provencher–Glöckner program; appendix C is the Hennessey–Johnson program. Tables VIII and IX list experimental CD data for 25 proteins from two laboratories. The purpose is to illustrate differences for the same proteins studied by different laboratories and also to provide for future comparisons. Requests from many researchers for these data have convinced us that they are being used for testing new methods of CD analysis. The numerical values for the proteins studied by Brahms and Brahms[47] and Bolotina et al.[75,80,81] are not available.

Concluding Remarks

Since X-ray diffraction provides the three-dimensional structure of crystalline proteins, one may justifiably question whether the CD analysis of proteins is useful or perhaps futile. The chiroptical method does not challenge but can complement the precise determination of the secondary structure of proteins by crystallography. First, not all proteins are easily crystallized. More important, crystallographic analysis may be insuffi-

---

[92] G. D. Fasman, *PAABS Rev.* **2,** 587 (1974).

TABLE VIII

NUMERICAL VALUES OF CD: MEAN RESIDUE ELLIPTICITIES OF EIGHTEEN PROTEINS: $[\theta] \times 10^{-3}$, deg cm^2 dmol^{-1} ^{a}

λ (nm)	Myoglobin	Parvalbumin	Adenylate kinase	Insulin	Lactate dehydrogenase	Lysozyme	Cytochrome c	Carboxy-peptidase A	Thermolysin
240	-2.29	-1.77	-1.45	-0.93	-1.79	-1.40	-0.71	-1.45	-0.85
239	-2.86	-2.24	-1.77	-1.28	-2.22	-1.76	-1.03	-1.83	-1.06
238	-3.84	-2.74	-2.21	-1.58	-2.73	-2.21	-1.33	-2.28	-1.30
237	-4.78	-3.36	-2.78	-1.94	-3.37	-2.71	-1.73	-2.76	-1.58
236	-5.89	-4.07	-3.43	-2.46	-4.01	-3.29	-2.25	-3.30	-1.98
235	-7.14	-4.89	-4.22	-3.20	-4.71	-4.06	-2.92	-3.91	-2.41
234	-8.80	-5.75	-5.07	-3.92	-5.54	-4.91	-3.68	-4.49	-2.99
233	-10.4	-6.76	-5.97	-4.52	-6.37	-5.82	-4.62	-5.05	-3.60
232	-12.3	-7.76	-6.84	-5.29	-7.27	-6.54	-5.78	-5.50	-4.36
231	-14.1	-8.92	-7.79	-6.13	-8.21	-7.17	-6.85	-5.86	-5.13
230	-15.7	-10.0	-8.78	-6.97	-9.17	-7.66	-8.08	-6.14	-5.92
229	-17.9	-11.0	-9.73	-7.94	-10.1	-8.11	-9.06	-6.39	-6.69
228	-19.8	-12.0	-10.5	-8.68	-11.0	-8.43	-10.0	-6.64	-7.48
227	-21.4	-12.8	-11.3	-9.53	-11.7	-8.75	-10.6	-6.79	-8.14
226	-22.5	-13.5	-12.0	-10.3	-12.3	-9.02	-11.5	-6.97	-8.70
225	-23.8	-14.0	-12.5	-10.9	-12.8	-9.29	-12.0	-7.12	-9.26
224	-24.3	-14.4	-13.0	-11.1	-13.1	-9.51	-12.1	-7.30	-9.78
223	-24.9	-14.5	-13.3	-11.3	-13.2	-9.65	-12.2	-7.43	-10.0
222	-24.9	-14.5	-13.5	-11.3	-13.2	-9.83	-12.2	-7.63	-10.2
221	-24.7	-14.4	-13.4	-11.3	-13.1	-9.96	-12.2	-7.86	-10.4
220	-24.4	-14.2	-13.4	-11.2	-12.9	-10.1	-11.9	-8.11	-10.5
219	-24.0	-14.0	-13.4	-11.1	-12.7	-10.1	-11.6	-8.42	-10.6
218	-23.5	-13.7	-13.3	-11.1	-12.5	-10.2	-11.1	-8.72	-10.8
217	-22.8	-13.4	-13.3	-11.2	-12.2	-10.2	-10.7	-8.95	-10.9
216	-22.3	-13.2	-13.2	-11.4	-12.1	-10.4	-10.3	-9.23	-11.0
215	-21.7	-12.9	-13.2	-11.5	-11.9	-10.7	-10.0	-9.43	-11.0
214	-21.7	-12.9	-13.3	-11.8	-11.9	-11.0	-9.82	-9.63	-11.2
213	-21.9	-13.0	-13.4	-12.2	-11.9	-11.4	-9.72	-9.76	-11.4
212	-22.1	-13.3	-13.6	-12.8	-12.0	-11.9	-9.73	-9.89	-11.6
211	-22.6	-13.7	-13.8	-13.1	-12.1	-12.4	-9.83	-9.94	-11.8
210	-22.8	-14.2	-14.0	-13.3	-12.3	-12.9	-9.88	-9.51	-12.1
209	-22.9	-14.4	-13.9	-13.9	-12.0	-13.2	-9.62	-8.77	-12.0
208	-21.8	-14.1	-13.2	-13.6	-11.3	-13.4	-9.07	-7.71	-11.4
207	-20.0	-13.3	-11.9	-12.4	-10.4	-13.3	-8.23	-5.98	-10.4

λ (nm)	Subtilisin BPN'	Papain	Trypsin inhibitor	Ribonuclease S	Nuclease	Ribonuclease A	α-Chymotrypsin	Elastase	Concanavalin A
240	-1.32	-1.05	-0.88	0.173	-1.16	0.18	-0.67	-0.42	-0.79
239	-1.58	-1.67	-0.99	0.096	-1.41	0.13	-0.81	-0.50	-0.91
238	-1.83	-1.97	-1.15	0	-1.78	0.05	-0.92	-0.63	-1.08
237	-2.18	-2.43	-1.39	-0.21	-2.14	-0.10	-1.09	-0.75	-1.30
236	-2.52	-2.87	-1.63	-0.48	-2.61	-0.38	-1.40	-0.80	-1.56
235	-2.95	-3.46	-1.95	-0.79	-3.09	-0.67	-1.84	-0.84	-1.87
234	-3.42	-4.16	-2.30	-1.15	-3.70	-1.05	-2.40	-0.84	-2.21
233	-3.97	-5.04	-2.70	-1.62	-4.37	-1.54	-3.17	-0.84	-2.66
232	-4.57	-6.05	-3.14	-2.17	-5.14	-2.03	-3.99	-0.75	-3.17
231	-5.13	-7.14	-3.67	-3.85	-6.00	-2.59	-4.57	-0.84	-3.88
230	-5.64	-8.15	-4.20	-3.36	-6.77	-3.29	-4.78	-0.92	-4.59
229	-6.32	-9.29	-4.77	-4.06	-7.54	-4.00	-4.60	-1.00	-5.24
228	-6.84	-10.1	-5.35	-4.79	-8.29	-4.75	-4.39	-1.17	-5.78
227	-7.43	-10.8	-5.95	-5.56	-8.91	-5.49	-4.26	-1.34	-6.17
226	-7.91	-11.3	-6.52	-6.27	-9.45	-6.24	-4.14	-1.55	-6.48
225	-8.33	-11.6	-7.07	-7.04	-9.96	-6.97	-4.11	-1.76	-6.68
224	-8.72	-11.8	-7.56	-7.67	-10.3	-7.64	-4.11	-2.01	-6.71
223	-8.93	-11.8	-8.03	-8.32	-10.6	-8.31	-4.16	-2.26	-6.60
206	-17.6	-11.8	-10.2	-10.8	-8.99	-12.6	-7.00	-3.65	-8.78
205	-14.0	-9.60	-8.55	-8.26	-7.11	-11.6	-5.39	-0.53	-7.05
204	-10.0	-6.66	-6.10	-5.57	-4.87	-10.2	-3.02	2.37	-4.89
203	-3.86	-2.86	-3.47	-1.47	-2.35	-8.30	0	5.26	-2.24
202	4.17	0.46	3.10	2.20	-0.16	-5.23	2.88	7.85	0
201	8.66	5.43	6.52	6.61	3.78	-1.00	5.35	10.6	3.44
200	16.2	10.1	10.4	11.8	7.07	2.04	7.54	12.6	7.92
199	25.1	14.3	13.9	17.6	11.0	4.91	9.60	13.9	12.0
198	32.9	17.1	17.6	22.0	14.7	7.36	11.2	14.8	15.9
197	40.1	18.8	20.9	25.0	18.2	9.82	12.6	14.7	19.9
196	47.0	19.6	23.4	25.7	20.9	12.7	13.0	13.9	22.6
195	51.1	19.6	25.6	25.8	22.5	15.1	12.9	12.8	24.1
194	53.1	19.1	26.9	23.5	23.8	16.4	12.3	11.5	24.8
193	54.9	17.5	27.9	20.6	24.1	17.4	11.8	9.88	24.9
192	55.1	14.8	28.3	16.9	24.0	17.6	11.0	8.64	24.0
191	54.4	10.2	28.0	15.3	23.4	16.8	10.3	7.22	23.4
190	52.9	6.01	27.5	12.2	22.2	16.0	9.46	5.78	21.1

(continued)

TABLE VIII (continued)

λ (nm)	Subtilisin BPN'	Papain	Trypsin inhibitor	Ribonuclease S	Nuclease	Ribonuclease A	α-Chymotrypsin	Elastase	Concanavalin A
222	-9.14	-11.7	-8.40	-8.90	-10.7	-8.85	-4.26	-2.51	-6.37
221	-9.23	-11.6	-8.78	-9.30	-10.8	-9.34	-4.36	-2.80	-6.17
220	-9.23	-11.4	-9.07	-9.73	-10.8	-9.72	-4.49	-3.14	-5.89
219	-9.19	-11.2	-9.31	-10.1	-10.8	-10.0	-4.66	-3.47	-5.55
218	-9.10	-11.1	-9.53	-10.3	-10.7	-10.3	-4.85	-3.81	-5.18
217	-8.97	-11.0	-9.71	-10.5	-10.5	-10.4	-5.06	-4.09	-4.76
216	-8.80	-11.0	-9.86	-10.6	-10.3	-10.5	-5.30	-4.36	-4.30
215	-8.72	-11.1	-10.1	-10.7	-10.2	-10.7	-5.54	-4.83	-3.85
214	-8.63	-11.3	-10.3	-10.8	-10.2	-10.8	-5.83	-5.31	-3.34
213	-8.63	-11.6	-10.7	-11.0	-10.3	-11.0	-6.20	-5.75	-2.72
212	-8.68	-11.9	-11.2	-11.0	-10.5	-11.0	-6.52	-6.20	-2.04
211	-8.72	-12.3	-11.6	-11.1	-10.7	-11.1	-6.89	-6.70	-1.25
210	-8.72	-12.6	-12.5	-11.0	-11.1	-11.1	-7.27	-7.21	-0.45
209	-8.63	-12.6	-13.7	-10.8	-11.3	-10.9	-7.67	-7.65	0.26
208	-8.12	-12.3	-14.7	-10.4	-11.1	-10.5	-8.08	-8.08	1.10
207	-7.95	-11.5	-16.1	-9.69	-10.7	-9.88	-8.49	-8.08	2.20
206	-7.52	-10.5	-17.1	-8.73	-10.1	-8.98	-8.96	-8.71	2.97
205	-6.80	-9.27	-17.8	-7.50	-9.24	-7.90	-9.32	-8.89	3.87
204	-6.03	-8.78	-18.5	-6.00	-8.20	-6.52	-9.67	-9.53	4.90
203	-4.96	-8.39	-18.7	-4.27	-6.88	-5.38	-9.82	-10.5	6.00
202	-3.80	-8.19	-18.4	-2.36	-5.10	-3.31	-9.73	-11.3	7.22
201	-2.39	-8.49	-17.5	-0.88	-3.38	-1.39	-9.26	-12.0	8.39
200	-0.56	-8.88	-16.5	2.01	-1.07	0.47	-8.83	-12.4	9.68
199	1.61	-8.39	-15.3	3.81	1.65	2.16	-7.25	-12.6	11.0
198	4.15	-6.99	-12.4	5.04	4.12	3.50	-6.20	-12.1	11.7
197	7.20	-4.94	-9.73	5.78	6.09	4.72	-4.98	-11.4	11.9
196	9.75	-2.30	-7.15	6.35	7.98	5.66	-3.67	-10.5	11.9
195	12.3	0.90	-4.49	6.57	9.54	6.30	-2.53	-9.34	11.5
194	13.6	4.29	-2.66	6.57	10.6	6.47	-1.22	-8.01	10.5
193	14.9	7.59	-0.42	6.26	11.3	6.41	0.87	-6.67	8.64
192	15.7	10.5	2.41	5.87	11.9	6.18	2.27	-4.77	6.45
191	15.7	12.9	4.08	5.30	12.1	5.77	3.41	-3.43	4.64
190	15.2	14.8	4.99	4.60	12.4	5.13	4.11	-2.10	2.77

[a] See Chang et al.[77]

TABLE IX

Numerical Values of CD: $\varepsilon_L - \varepsilon_R$ of Fifteen Proteins and One Helical Polypeptide: $\Delta\varepsilon$ in cm^{-1} M^{-1} [a]

λ (nm)	α-Chymotrypsin	Cytochrome c	Elastase	Hemoglobin	Lactate dehydrogenase	Lysozyme	Myoglobin	Papain
260	-0.04	0	0	0	0	0	0	0
258	-0.04	0	0	0	0	0	0	0
256	-0.04	0	0	0	0	0	0	0
254	-0.04	0	0	0	0	0	0	0
252	-0.04	0	0	0	0	-0.03	0	0
250	-0.04	0	0	0	0	-0.09	0	0
248	-0.04	0	0	0	-0.06	-0.13	0	0
246	-0.04	0	-0.04	0	-0.19	-0.16	-0.15	0
244	-0.06	0	-0.07	-0.15	-0.34	-0.20	-0.20	0
242	-0.11	-0.08	-0.09	-0.35	-0.52	-0.32	-0.35	-0.14
240	-0.18	-0.20	-0.15	-0.65	-0.79	-0.50	-0.55	-0.29
238	-0.26	-0.38	-0.19	-1.15	-1.17	-0.71	-1.00	-0.46
236	-0.40	-0.64	-0.20	-1.80	-1.61	-1.14	-1.70	-0.69
234	-0.58	-1.07	-0.22	-2.70	-2.13	-1.59	-2.75	-0.96
232	-0.91	-1.55	-0.24	-3.60	-2.72	-2.09	-3.75	-1.31
230	-1.22	-2.16	-0.28	-4.60	-3.36	-2.43	-4.80	-1.72
228	-1.27	-2.83	-0.34	-5.85	-3.85	-2.69	-6.55	-2.13
226	-1.21	-3.28	-0.47	-6.50	-3.14	-2.88	-7.55	-2.47
224	-1.16	-3.59	-0.60	-7.00	-4.22	-3.03	-8.05	-2.67
222	-1.14	-3.78	-0.76	-7.20	-4.15	-3.25	-8.30	-2.73
220	-1.19	-3.71	-0.93	-7.15	-4.05	-3.45	-8.10	-2.71
218	-1.25	-3.58	-1.12	-7.00	-3.85	-3.50	-7.80	-2.69
216	-1.35	-3.38	-1.32	-6.75	-3.65	-3.56	-7.60	-2.66
214	-1.51	-3.19	-1.54	-6.55	-3.62	-3.66	-7.35	-2.69
212	-1.67	-2.98	-1.76	-6.55	-3.55	-3.89	-7.25	-2.84
210	-1.89	-2.91	-2.00	-6.60	-3.27	-4.18	-7.55	-2.98
208	-2.13	-2.82	-2.20	-6.15	-3.27	-4.57	-7.40	-3.06
206	-2.38	-2.30	-2.37	-5.05	-2.49	-4.40	-6.40	-2.74
204	-2.61	-1.40	-2.49	-2.30	-1.21	-3.70	-4.00	-2.28
202	-2.66	-0.16	-2.64	1.50	0.24	-2.25	-0.60	-1.97

(continued)

TABLE IX (*continued*)

λ (nm)	α-Chymotrypsin	Cytochrome c	Elastase	Hemoglobin	Lactate dehydrogenase	Lysozyme	Myoglobin	Papain
200	-2.51	1.19	-2.95	5.90	2.70	-0.10	3.60	-1.91
198	-2.10	2.56	-3.20	11.00	5.22	2.10	9.00	-1.71
196	-1.51	3.68	-2.82	13.75	7.61	4.00	13.75	-1.07
194	-0.75	3.72	-2.19	14.10	8.69	5.05	16.45	0.35
192	0	3.29	-1.40	13.05	8.99	5.18	17.40	2.07
190	0.72	2.81	-0.65	10.70	8.21	4.90	16.45	3.50
188	1.11	2.46	0.19	8.15	6.35	4.20	14.65	4.03
186	1.18	2.32	0.55	6.00	4.59	3.40	12.40	3.92
184	1.02	2.30	0.55	4.80	2.84	2.30	9.60	3.45
182	0.60	2.31	0.19	3.95	1.48	1.22	7.15	2.76
180	0.11	2.00	-0.50	3.45	0.52	0.40	5.90	2.24
178	-0.43	1.42	-0.99	2.85	-0.25	-0.20	4.65	1.48

λ (nm)	Ribonuclease	Subtilisin BPN'	Flavodoxin	GPD[b]	Prealbumin	Subtilisin Novo	TPI[c]	(Glu)$_n$
260	0	0	0	0	0	0	0	0
258	-0.03	0	0	0	0	0	0	0
256	0.04	0	0	0	0	0	0	0
254	0.04	0	0	0	0	0	0	0
252	0.05	0	0	0	0	0	0	0
250	0.05	0	0	0	0	0	0	0
248	0.06	-0.02	-0.10	-0.10	0	0	0	0
246	0.06	-0.04	-0.20	-0.20	0	-0.10	-0.10	0
244	0.07	-0.13	-0.30	-0.30	0	-0.20	-0.20	0
242	0.09	-0.20	-0.40	-0.40	0	-0.30	-0.30	0
240	0.07	-0.30	-0.50	-0.50	0	-0.50	-0.60	-0.50
238	0.05	-0.47	-0.60	-0.70	-0.10	-0.80	-0.90	-1.30
236	-0.02	-0.67	-0.80	-1.00	-0.10	-1.20	-1.30	-2.50
234	-0.25	-0.92	-1.10	-1.30	-0.20	-1.60	-1.90	-4.10
232	-0.50	-1.21	-1.30	-1.60	-0.30	-1.90	-2.50	-5.50
230	-0.82	-1.52	-1.60	-2.00	-0.50	-2.20	-3.20	-7.20

228	−1.22	−1.86	−2.00	−2.40	−0.50	−2.60	−3.90	−8.80
226	−1.63	−2.19	−2.40	−2.80	−1.00	−2.90	−4.40	−10.60
224	−2.04	−2.43	−2.70	−3.00	−1.20	−3.20	−4.80	−12.40
222	−2.39	−2.60	−2.90	−3.10	−1.40	−3.30	−5.10	−13.30
220	−2.68	−2.70	−3.00	−3.20	−1.60	−3.40	−5.20	−13.00
218	−2.87	−2.72	−2.90	−3.10	−1.70	−3.30	−5.20	−12.30
216	−3.02	−2.70	−2.80	−3.00	−1.80	−3.20	−5.00	−12.00
214	−3.09	−2.61	−2.80	−2.90	−1.70	−3.00	−4.70	−11.80
212	−3.14	−2.51	−2.60	−2.80	−1.60	−2.90	−4.50	−11.90
210	−3.21	−2.39	−2.40	−2.80	−1.50	−2.50	−4.40	−12.10
208	−3.18	−2.21	−2.10	−2.50	−1.30	−1.80	−4.00	−12.50
206	−2.81	−1.90	−1.60	−2.20	−1.00	−1.00	−3.00	−12.70
204	−2.20	−1.41	−0.60	−1.60	−0.40	0.20	−1.20	−10.20
202	−1.34	−0.78	0.90	−0.80	0.30	1.70	1.10	−5.50
200	−0.22	0.07	2.10	0.60	1.30	3.10	4.00	−0.50
198	0.80	1.61	−3.20	1.90	2.30	4.60	6.60	7.50
196	1.72	3.31	4.00	3.30	2.90	5.80	9.40	14.70
194	2.00	4.54	4.20	4.10	2.90	6.30	10.50	23.10
192	2.00	5.07	3.80	4.40	2.50	6.00	10.30	29.70
190	1.79	4.85	3.00	4.10	1.90	5.10	9.10	30.60
188	1.49	3.80	2.20	3.40	1.00	3.80	7.60	29.70
186	1.39	2.84	1.40	2.30	−0.10	2.50	6.10	26.20
184	1.30	1.79	0.60	1.50	−1.10	1.60	4.30	21.60
182	1.00	0.79	−0.10	0.80	−1.90	1.00	3.00	17.10
180	0.59	0.03	−1.00	0.20	−2.70	0.40	1.80	12.50
178	0.30	−0.55	−1.20	−0.30	−3.20	−0.40	1.20	10.30

[a] W. C. Johnson (private communication).
[b] GPD, Glyceraldehyde-3-phosphate dehydrogenase.
[c] TPI, Triosephosphate isomerase.

cient in extrapolating static information to the dynamic properties of proteins in solution. It is here that physical techniques such as CD will realize their full potential.

The very simplicity of the CD method and the short time required for the measurements are extremely attractive. To study conformation and conformational changes of proteins in solution, we must first know that the CD analysis of native proteins is fairly reliable, albeit empirical. This is the raison d'etre for numerous attempts to improve the estimates of secondary structure of proteins from their CD spectra. To further ensure reliability, it is often advantageous to compare the CD analysis with empirical predictions of secondary structure of proteins from sequences, if such are available. The very nature of any empirical method does caution us against a too literal interpretation of experimental analysis. At present CD is a powerful tool for studying protein conformation. It will remain so and continue to be refined or improved unless a better method to determine the secondary structure of proteins in solution can be found.

Appendix A. Program for the Chang–Wu–Yang Method[77]

The program in C language consists of two parts. (1) It generates the reference spectra of the helix, β-form, β-turn, and unordered form or, if the reference spectrum of the helix is predetermined from the CD spectrum of myoglobin [Eq. (10)], the reference spectra of the other three conformations, from the CD spectra of a set of reference proteins according to Eq. (11) in the text. (2) It uses the four reference spectra to analyze the CD spectrum of a protein for secondary structure.

Appendix B. Program for the Provencher–Glöckner Method[85]

S. W. Provencher (private communication) has developed a constrained regularization method for inverting data represented by linear algebraic or integral equations. A portable FORTRAN IV package called CONTIN was written for converting noisy linear operator equations and seeking an optimal solution. The program for CD analysis of proteins is one of the USSR subprograms in the general purpose package, which has nearly 6000 lines of code. Application packages are documented in S. W. Provencher, CONTIN Users Manual, EMBL Technical Report DA05, European Molecular Biology Laboratory (1982). The program was also described in detail in two recent publications: S. W. Provencher, *Computer Phys. Commun.* **27**, 213–227, 229–242 (1982). A listing or tape is available upon request from the author.

```
/*          Copyright by the Regents of the University of California, Nov. 16,
 *          1983.  All rights reserved.

 *          Prepared in the Biomathematics Computation Laboratory, Dept. of
 *          Biochemistry and Biophysics, Univ. of Calif. at San Francisco,
 *          San Francisco, Calif. 94131, by Hugo M. Martinez.

 *          The program is written in the C language.


 *          CIRCULAR DICHROISM ANALYSIS OF PROTEIN CONFORMATION PARAMETERS

 *          Given the CD spectrum X() of a protein, estimates of the helical H,
 *          beta B, turns T and random R fractional amounts are obtained.
 *          The estimation procedure is the method of least squares and is
 *          based on the assumption that the CD value X(l) of the protein
 *          at the wavelength l can be expressed as

 *              X(l) = H*h(l)  + B*b(l)  + T*t(l)  + R*r(l)

 *          in which h() is the CD spectrum of a protein for which H = 1,
 *          b() is the CD spectrum of a protein for which B =1, etc.
 *          Minimizing the sum of the squares of the deviations between
 *          the experimental X(l) and predicted values H*h(l) + B*b(l)
 *          + T*t(l) + R*r(l) is subject to the constraint that the fraction
 *          unknowns H, B, T and R are non-negative and must sum to the
 *          value 1.  The required minimization is obtained by a stepwise
 *          exhaustive search strategy.  Taking three of the unknowns as
 *          independent, the minimizing solution lies in a 3-dimensional
 *          cube in which each coordinate is restricted to values between
 *          0 and 1.  Using a step size of .05 for each coordinate, the best
 *          point of the unit cube is first found.  This estimate is then
 *          refined by adopting a step size of .005 and resricting the
 *          exhaustive search to a coordinate variation of less than .05
 *          about the first estimate.  The assured accuracy of the solution
 *          is therefore to within plus or minus .05 for each fraction and
 *          very likely accurate to within plus or minus .005.

 *          The CD spectrum of a sample protein is assumed to be in a
 *          user-designated file.  It is to be in a 2-column format, with
 *          the first column containing the wave numbers in decreasing order
 *          (normally from 240 to 190 nm) and the second column containing
 *          the corresponding CD values.  The number of wave numbers is
 *          given by the parameter NLAM.  Its value is defined in the source
 *          code and is normally set to 51 corresponding to the normal
 *          range of 240 to 190 for the wave numbers.  If a different range
 *          is to be used, the value of NLAM must be redefined and the source
 *          code recompiled.

 *          The reference "pure" spectra h(), b(), t() and r() constitute
 *          a data base contained in a user-designated file.  The program
 *          gives the option to construct such a file by providing it with
 *          the name of a directory in which there is a file for each
 *          reference protein.  Each reference protein file contains the
 *          H, B, T and R values for the corresponding protein, followed
 *          by its NLAM CD spectrum values X(l) in decreasing order of l.
 *          It is separately prepared by the user as a text file with a text editor.

 *          Construction of the reference pure spectra data base, given the
 *          reference protein directory, is by the method of least squares
 *          based on the above linear relation.  Since there are no constraints
 *          on the values of the pure reference spectra, the normal equations
 *          of the least squares problem are solved directly using the
 *          Gauss-Seidel method of solving linear equations.

 *          There are two construction procedures of the reference spectra.
 *          In construction procedure A all four reference spectra are computed.
 *          In construction procedure B only the b(), t() and r() spectra
 *          are computed since h() is regarded as corresponding to the spectrum
 *          of myoglobin modified by a wavelength and protein dependent factor
 *          (1 - k/n).  Thus, the above relation becomes

 *              X(l) = H*hmyo(l) * (1-k/n)  + B*b(l)  + T*t(l)  + R*r(l)
```

```
*          in which hmyo() is the myoglobin spectrum, k(l) is the wavelength
*          dependent parameter and n is the protein dependent parameter.
*          In order to carry out construction procedure B, there is a data
*          file called myoinf.tbl in which successive lines contain the
*          l, hmyo(l) and k(l) values in decreasing order of l from 240 to
*          190.  Additionally, the first line of each reference protein file
*          contains a fifth entry corresponding to its n value.  For an
*          interpretation of the k and n parameters, refer to the article by
*          Chang, C.T., Wu, C.-S.C. and Yang, J.T. (1978) Anal. Biochem. 91,
*          13-31.
*          Note: If a range different from 240 to 190 for the wave numbers
*          is to be used, the file myoing.tbl must be edited accordingly.
*/


#include <stdio.h>

float spectra[52][5];      /* the reference pure spectra */

float w[5][5], p[5];       /* working arrays used by estimate() function */
int nref;                  /* number of referecnce proteins */
float f[20][5];            /* Reference conformation fractions.  H = f[j][1],
                              B = f[j][2], T = f[j][3] and R = f[j][4] for the
                              jth reference protein. */
float cd[20][60];          /* CD spectra of the reference proteins.   cd[j][l]
                              is X(l) for the jth reference protein. */
float a[5][5];             /* Augmented matrix used in solving the normal
                              equations of the "pure" spectra problem. */

#define NLAM 51            /* number of wavelengths (wave numbers) */

int cp;                    /* construction procedure flag; cp is 0 for procedure
                              A and 1 for procedure B */

float hmyo[52]; /* this is the cd spectrum for myoglobin, also
                              known as theta inf */
float kmyo[52];            /* this is the wavelength dependent parameter k in
                              the factor (1 - k(l)/n)   */
main()
{
        char option();


        title();
        if (option() == 'e')
                estimate();
        else
                construct();
}

estimate()
{


        char refname[50],sampname[50];
        int lam, k, i, j, np;
        float eave, emin, ecurrent;
        float error();
        float f[5], l[5], u[5];
        float step, step1;
        char c, ans[5];
        float dum;
        FILE *fopen(), *fp;


        printf("\n* It is assumed that the reference spectra exist in an");
        printf("\n* ascii file of NLAM rows corresponding to the NLAM wavelengths");
        printf("\n* in decreasing order.  Each row has four");
        printf("\n* columns corresponding, respectively, to the helical, beta,");
        printf("\n* turn and random reference CD values.\n");
        printf("\n* The reference spectra file is normally prepared with");
        printf("\n* the 'construction' option and given the name 'refcdA.out'");
        printf("\n* or 'refcdB.out' depending upon whether construction");
```

```
printf("\n* procedure A or B was used.");
printf("\n");

printf("\nEnter name of reference spectra file: ");
scanf("%s",refname);

if ((fp = fopen(refname,"r")) == NULL) {
        printf("\n\7The reference spectra file %s does not exist!",
                refname);
        exit(0);
}

for (k = 1; k <= NLAM; k++) {
        fscanf(fp,"%f%f%f%f",&spectra[k][1],&spectra[k][2],
                        &spectra[k][3],&spectra[k][4]);
}
fclose(fp);

printf("\n* In the following request for the sample protein file");
printf("\n* it is assumed that it is an ascii file with the format:");
printf("\n* There are 51 lines of two entries each.  The");
printf("\n* 1st is the wave number and the 2nd is the CD value.");
printf("\n* Wave numbers are in decreasing order.");
printf("\n");
printf("\nEnter name of data file for the sample protein: ");
scanf("%s",sampname);
if ((fp = fopen(sampname,"r")) == NULL) {
        printf("\n\7The sample file %s does not exist!",sampname);
        exit(0);
}
for (k = 1; k <= NLAM; k++)
        fscanf(fp,"%f%f",&dum,&spectra[k][0]);
fclose(fp);

fprintf(stderr,"\nComputation in progress.\n");
fp = fopen("confpar.out", "w");

fprintf(fp,"\n* Conformation parameter estimates of the protein");
fprintf(fp,"\n* %s using the ref. spectra file %s.",sampname,
        refname);
fprintf(fp,"\n\n");
/* compute coefficients of error function */

for (i = 0; i <= 4; i++)
        for (j = 0; j <= 4; j++)
                w[i][j] = 0;

for (i = 0; i <= 4; i++)
        for (j = 0; j <= 4; j++)
                for (k = 1; k <= NLAM; k++)
                        w[i][j] += spectra[k][i]*spectra[k][j];

/* 1st order scan */


eave = 0; emin = 1.e20;
p[0] = -1;
f[1] = f[2] = f[3] = 0;  f[4] = 1;
np = 0;
step = 0.05;

for (p[1] = 0; p[1] <= 1; p[1] += step)
        for (p[2] = 0; p[2] <= 1-p[1]; p[2] += step)
                for (p[3] = 0; p[3] <= 1-p[1]-p[2]; p[3] += step)
                {
                        p[4] = 1-p[1]-p[2]-p[3];
                        np++;
                        ecurrent = error();
                        eave += ecurrent;
                        if (ecurrent < emin)
                        {
                                for (i = 1;i <= 4; i++) f[i] = p[i];
                                emin = ecurrent;
                        }
                }
```

```
          /* output results of 1st order scan */

          eave = eave/np;
          fprintf(fp,"\n\n First Order Scan Results with step = %f", step);
          fprintf(fp,"\n\n\t average rms error = %f, minimum rms error = %f",
                  eave, emin);
          fprintf(fp,"\n\n\t Fh = %6.4f, Fb = %6.4f, Ft = %6.4f, Fr = %6.4f",
                  f[1],f[2],f[3],f[4]);

          /* 2nd order scan */

          step1 = step/10;

          for (k = 1; k <= 3; k++)
          {
                  if (f[k] - step < 0) l[k] = 0;
                  else l[k] = f[k] - step;

                  if (f[k] + step > 1) u[k] = 1;
                  else u[k] = f[k] + step;
          }

          for (p[1] = l[1]; p[1] <= u[1]; p[1] += step1)
          for (p[2] = l[2]; p[2] <= 1-p[1] && p[2] <= u[2]; p[2] += step1)
          for (p[3] = l[3]; p[3] <= 1-p[1]-p[2] && p[3] <= u[3]; p[3] += step1)
          {
                  p[4] = 1-p[1]-p[2]-p[3];
                  ecurrent = error();

                  if (ecurrent < emin)
                  {
                          for (k = 1; k <= 4; k++) f[k] = p[k];
                          emin = ecurrent;
                  }
          }

          /* print results of 2nd order scan */

          fprintf(fp,"\n\n Second Order Scan Results with step = %f", step1);
          fprintf(fp,"\n\n\t minimum rms error = %f", emin);
          fprintf(fp,"\n\n\t Fh = %6.4f, Fb = %6.4f, Ft = %6.4f, Fr = %6.4f",
                  f[1],f[2],f[3],f[4]);
          fprintf(stderr,"Conformation parameter estimates are in the ");
          fprintf(stderr,"'confpar.out' file.\n");

}

float error()
{
          float err = 0.;
          double sqrt();

          int i, j;

          for (i = 0; i <= 4; i++)
                  for (j = 0; j <= 4; j++)
                          err += p[i]*p[j]*w[i][j];
                          err = sqrt(err/(NLAM-3));
          return(err);

}
title()
{
          printf("\n\n* Program for estimating conformation paramaters (fractions)");
          printf("\n* of a protein from its CD spectrum, or for constructing a");
          printf("\n* reference 'pure' spectra data base.\n");
}
char option()
{
          char ans[5];
```

```
        while (1) {
                printf("\nConformation estimate (e) or reference spectra ");
                printf("construction (c) ? ");
                scanf("%s",ans);
                if (*ans != 'e' && *ans != 'c') {
                        printf("\7");    /* sound bell */
                        continue;
                }
                else break;
        }
        return(*ans);
}

construct()     /* construct the reference spectra */
{
        int l,k,j;
        char pans[5];
        FILE *fpout,*fopen();
        float nave,h;

        printf("\n* There are two construction procedures: (A) for the");
        printf("\n* construction of all 4 reference spectra, or (B) for");
        printf("\n* the construction of just the beta, turn and randdom,");
        printf("\n* with the helical being computed from the myoglobin ");
        printf("\n* spectrum contained in the thetainf.tbl file.\n");
        printf("\nProcedure A or B? ");
        scanf("%s",pans);
        if (*pans == 'A' || *pans == 'a')
                cp = 0;
        else
                cp = 1;
        getdata();      /* components and spectra of the reference proteins */

        fprintf(stderr,"\nComputation in progress.\n");
        if (cp) {       /* if construction procedure B */
                fpout = fopen("refcdB.out","w");
                nave = 0;
                for (k = 1; k <= nref; k++)
                        nave += f[k][5];
                nave /= nref;
        }
        else
                fpout = fopen("refcdA.out","w");

        for (l = 1; l <= NLAM; l++) {
                mkeqns(l);       /* make the normal equations */
                solve(4-cp);     /* solve the normal equations */

                if (cp) {                /* if procedure B */
                        h = hmyo[l]*(1-kmyo[l]/nave);
                        fprintf(fpout,"%8.2f ",h);
                }
                for (k = 1; k <= 4-cp; k++)
                        fprintf(fpout,"%8.2f ",a[k-1][4-cp]);
                fprintf(fpout,"\n");
        }
        if (cp)
                printf("\nThe ref. spectra are in the file 'refcdB.out'.");
        else
                printf("\nThe ref. spectra are in the file 'refcdA.out'.");

}
getdata()       /* get the reference data: fractional components of the
                   reference proteins and their cd spectra */
{
        int j,k;
        char prodir[50];             /* reference protein directory */
        char command[80];
        char pname[25];
        char rpname[100];
        float dum;
        FILE *fp,*fpx,*fopen();
```

```
        printf("\n* It is assumed that the reference protein data are all");
        printf("\n* in a single directory containing a file for each protein.");
        printf("\n* Each reference protein file is an ascii file of NLAM+1 lines.");
        printf("\n* Line one are the helical, beta, turn and random confor-");
        printf("\n* mation fraction values followed by its n value.");
        printf("\n* Subsequent lines contain a wave");
        printf("\n* number and the corresponding CD value.  The wave numbers");
        printf("\n* are in decreasing order.\n");

        printf("\nEnter name of the reference protein directory: ");
        scanf("%s",prodir);
        strcpy(command,"ls ");
        strcat(command,prodir);
        strcat(command," > tprodir");
        system(command);

        fp = fopen("tprodir","r");
        j = 1;
        while (fscanf(fp,"%s",pname) != EOF) {
                strcpy(rpname,prodir);
                strcat(rpname,"/");
                strcat(rpname,pname);
                fpx = fopen(rpname,"r");
                if (fpx == NULL)
                        fprintf(stderr,"cannot open %s",rpname);
                for (k = 1; k <= 5; k++)
                        fscanf(fpx,"%f",&f[j][k]);
                k = 1;
                while (fscanf(fpx,"%f%f",&dum,&cd[j][k]) != EOF)   {
                        k++;
                }
                fclose(fpx);
                j++;
        }
        nref = j-1;

        if (cp) {                        /* if construction procedure B */
                k = 1;
                fpx = fopen("thetainf.tbl","r");
                for (k = 1; k <= NLAM; k++) {
                fscanf(fpx,"%f %f %f ",&dum,&hmyo[k],&kmyo[k]);
                }
                fclose(fpx);
        }
}
mkeqns(l)       /* make the normal equations in the form of deriving
                   an augmented matrix to pass on to the solve() routine;
                   the value of l is the wavelength number */
{
        int i,j,k;
        float cdfact;

        for (i = 1; i <= 4-cp; i++) {
                for (j = 1; j <= 4-cp; j++) {
                        a[i-1][j-1] = 0;
                        for (k = 1; k <= nref; k++)
                                a[i-1][j-1] += f[k][i+cp]*f[k][j+cp];
                }
                a[i-1][j-1] = 0;
                for (k = 1; k <= nref; k++) {
                        if (cp) /* if procedure B */
                                cdfact = cd[k][l]-f[k][1]*hmyo[l]*
                                        (1 - kmyo[l]/f[k][5]);
                        else
                                cdfact = cd[k][l];
                        a[i-1][j-1] += f[k][i+cp]*cdfact;
                }
        }

}
```

```
solve(neq)                      /* For solving neq linear equations having
                                augmented matrix a. Solution is left in a[][neq] .
                                The Gauss-Jordan elimination method is used. */
{
        int  i, j, k, l;
        float z;

                /* implement the Gauss-Jordan algorithm */

        for (k = 0; k < neq; k++)
        {
                for (l = k; l <  neq && a[l][k] == 0; l++);
                if (l == neq)
                        return(-1);      /* to indicate eqns are singular */

                if (l != k)      /* interchange rows k and l */
                {
                        for (j = 0; j <= neq; j++)
                        {
                                z = a[k][j];
                                a[k][j] = a[l][j];
                                a[l][j] = z;
                        }
                }

                z = a[k][k];     /* normalize row k */
                if (!z) {
                        fprintf(stderr,"\nz is zero!");
                        exit(0);
                }
                for (j = 0; j <= neq; j++) a[k][j] = a[k][j]/z;

                        /* insert 0's above and below k-th row in k-th col. */

                for (i = 0; i < neq; i++)
                {
                        z = a[i][k];
                        if (i != k)
                                for (j = 0; j <= neq; j++)
                                        a[i][j] -= z*a[k][j];
                }
        }
        return(0);       /* to indicate that solution OK */
}
```

Appendix C. The Hennessey–Johnson Program[86]

The program BAVGEN generates the basis CD spectra and their corresponding secondary structures from the CD spectra of the reference proteins and their secondary structures determined from X-ray analysis. The program PROSTP uses the basis CD spectra and their corresponding secondary structures to analyze the CD spectrum of a protein for secondary structure.

```
C... ------------------------------------------------------------------
C--------------------------------------------------------------------
C      PROGRAM NAME: BAVGEN (BASIS VECTORS GENERATION)
C
C   COMPUTATION OF SINGULAR VALUE DECOMPOSITION OF A MATRIX AND
C   GENERATION OF BASIS VECTORS
C
C   THIS PROGRAM CALLS THE SUBROUTINES SVD AND BASSP. SVD HAS BEEN TAKEN
C   FROM "COMPUTER METHODS FOR MATHEMATICAL COMPUTATION" BY G.E. FORSYTHE,
C   M.A. MALCOLM AND C.B. MOLER (1977) CHAPTER 9, PRENTICE-HALL INC. N.J.
C   THIS SUBROUTINE COMPUTES THE SINGULAR VALUE DECOMPOSITION OF A
```

```
C     RECTANGULAR MATRIX.
C
C     INPUT DATA VARIABLES
C     --------------------
C
C
C     CD:      CD DATA MATRIX.
C     INT:     WAVELENGTH INTERVAL BETWEEN TWO DATA POINTS.
C     IWF:     WAVELENGTH OF THE LAST DATA POINT.
C     IWS:     WAVELENGTH OF THE FIRST DATA POINT.
C     MATXU:   MATXU=0(OR 1) WOULD ASSIGN MATU 'FALSE'(OR 'TRUE').
C              THIS SHOULD BE SET TO 'TRUE' IF OUTPUT MATRIX 'U' IS DESIRED.
C     MATXV:   MATXV=0(OR 1) WOULD ASSIGN MATV 'FALSE'(OR 'TRUE').
C              THIS MUST SET TO 'TRUE' TO GET THE OUTPUT MATRIX 'V' AND TO
C              GENERATE THE BASIS VECTORS.
C     NBASV:   NUMBER OF BASIS VECTORS TO BE GENERATED FROM 'CD' AND 'V'
C              MATRICES. PROGRAM GENERATES THE MOST IMPORTANT BASIS VECTORS
C              FROM FIRST TO A MAXIMUM OF NBASV='NSAM'.
C              NBASV=0 WOULD NOT GENERATE ANY BASIS VECTORS AND
C              TERMINATE THE PROGRAM AFTER COMPUTING THE 'U' AND 'V' MATRIX.
C     NPTS:    NUMBER OF CD DATA POINTS.
C     NSAM:    NUMBER OF PROTEINS.
C     NSTR:    NUMBER OF DIFFERENT SECONDARY STRUCTURES TO BE CONSIDERED
C              (EG. HX,BA,BP,TN,OT). NSTR=0 WOULD NOT COMPUTE 'BVSS' MATRIX
C              AND TERMINATE THE PROGRAM AFTER COMPUTING 'U', 'V' AND
C              'BVCD' MATRICES.
C     PRNAME:  NAME OF PROTEINS INCLUDED IN THE 'CD' MATRIX. THE MAXIMUM
C              LENGTH OF THE TITLE IS 40 CHARACTERS.
C     SS:      PROTEIN SECONDARY STRUCTURE MATRIX CORRESPONDING TO CD DATA
C              MATRIX FROM X-RAY DATA.
C     SSTR:    TYPE OF SECONDARY STRUCTURE.
C              (HX:HELIX;BA:ANTIPARALLEL BETA SHEET;BP:PARALLEL BETA SHEET;
C              TN:TURNS;OT:OTHER STRUCTURES).
C
C              OUTPUT DATA VARIABLES
C              ---------------------
C
C     BVCD:    BASIS CD VECTORS.
C     BVSS:    SECONDARY STRUCTURE MATRIX CORRESPONDING TO THE BASIS CD
C              VECTORS DEFINED BY 'BVCD'.
C     SIGMA:   SINGULAR VALUE MATRIX. THE SIZE OF THE SINGULAR VALUE INDICATES
C              THE IMPORTANCE OF THE CORRRESPONDING CD BASIS VECTOR.
C              ONLY CD BASIS VECTORS WITH SIGNIFICANT SINGULAR VALUES SHOULD
C              BE USED IN AN ANALYSIS. SEE HENNESSEY & JOHNSON (1981)
C              BIOCHEMISTRY, VOL.20, 1085-1094.
C     U:       U MATRIX AND WILL BE PRINTED WHEN MATXU=1.
C     V:       EIGEN VECTOR MATRIX V AND WILL BE PRINTED WHEN MATXV=1.
C
C---------------------------------------------------------------------------
C---------------------------------------------------------------------------
C
      REAL CD(50,25),U(50,25),V(50,25),SIGMA(25),WORK(25)
      INTEGER I,IERR,J,NSAM,NPTS
      DIMENSION BVCD(25,50),BVSS(25,50),SSTR(10),TOT(25),SS(50,25)
      DIMENSION PRNAME(40)
      LOGICAL MATU,MATV
C
C        INITIAL INPUT DATA FOR SINGULAR VALUE DECOMPOSITION
C
      WRITE(6,2000)
      WRITE(6,1000)
      READ(5,1001)NSAM,IWS,IWF,INT,NPTS,NBASV,NSTR,MATXU,MATXV
      IM=1
      MATU=IM.EQ.MATXU
      MATV=IM.EQ.MATXV
      WRITE(6,1002)NSAM,IWS,IWF,INT,NPTS
      WRITE(6,1005)
      DO 10 I=1,NSAM
      READ(5,1006)PRNAME
      READ(5,1010)(CD(J,I),J=1,NPTS)
      WRITE(6,1008)PRNAME
      WRITE(6,1020)(CD(J,I),J=1,NPTS)
10    CONTINUE
```

```
C
C        SUBROUTINE SVD IS CALLED FOR SINGULAR VALUE DECOMPOSITION
C
C        THE VARIABLE NM MUST BE SET ATLEAST AS LARGE AS THE MAXIMUM OF
C        ROW DIMENSION OF 'CD'.SEE SUBROUTINE SVD FOR FURTHER COMMENTS.
C
         NM=50
C
         CALL SVD(NM,NPTS,NSAM,CD,SIGMA,MATU,U,MATV,V,IERR,WORK)
         IF(IERR .NE. 0)WRITE(6,1025) IERR
         WRITE(6,1030)
         WRITE(6,1035)(SIGMA(I),I=1,NSAM)
         IF(.NOT. MATU) GO TO 17
         WRITE(6,1040)
         DO 15 I=1,NSAM
15       WRITE(6,1020)(U(J,I),J=1,NPTS)
17       IF(.NOT. MATV) GO TO 22
         WRITE(6,1045)
         DO 20I=1,NSAM
20       WRITE(6,1020)(V(I,J),J=1,NSAM)
C
C        GENERATION OF BASIS CD VECTORS FROM CD DATA AND V MATRIX
C
22       IF(NBASV.EQ.0) GO TO 40
         WRITE(6,2000)
         CALL BASSP (CD,V,NSAM,NBASV,NPTS,BVCD)
         WRITE(6,1050)
         WRITE(6,1055)NSAM,NBASV
         DO 25 I=1,NBASV
         WRITE(6,1060)I,SIGMA(I)
25       WRITE(6,1020)(BVCD(I,J),J=1,NPTS)
C
C        GENERATION OF SECONDARY STRUCTURES CORRESPONDING TO BASIS CD VECTORS
C        USING PROTEIN STRUCTURE DATA AND V MATRIX
C
         IF (NSTR.EQ.0) GO TO 40
         WRITE(6,2000)
         WRITE(6,1065)
         WRITE(6,1070)
         READ(5,1075)(SSTR(I),I=1,NSTR)
         WRITE(6,1080)(SSTR(I),I=1,NSTR)
         DO 30 I=1,NSAM
         READ(5,1006)PRNAME
         READ(5,1010)(SS(J,I),J=1,NSTR)
30       WRITE(6,1090)PRNAME,(SS(J,I),J=1,NSTR)
         CALL BASSP(SS,V,NSAM,NBASV,NSTR,BVSS)
         DO 35 I=1,NBASV
         WRITE(6,1085)I,SIGMA(I)
35       WRITE(6,1020)(BVSS(I,J),J=1,NSTR)
40       CONTINUE
C
C        END OF PROGRAM
C
1000  FORMAT(/,2X,'COMPUTATION OF SINGULAR VALUE DECOMPOSITION FOR A
     $ RECTANGULAR MATRIX',//)
1001  FORMAT(10I4)
1002  FORMAT(10X,'NUMBER OF CD SPECTRA =',I5,/,
     $10X,'WAVELENGTH RANGE IN NM=',I5,' TO',I5,' AT',I5,' NM INTERVAL.'
     $,/,10X,'TOTAL NUMBER OF CD DATA POINTS =',I5,///)
1005  FORMAT(10X,' PROTEIN CD MATRIX: CD ',/)
1006  FORMAT(1X,40A1)
1008  FORMAT(5X,40A1)
1010  FORMAT(F10.5)
1020  FORMAT(10X,15F6.2)
1025  FORMAT(' TROUBLE, IERR=',I4)
1030  FORMAT(////,10X,' LIST OF SINGULAR VALUES',//)
1035  FORMAT(5X,10F10.2)
1040  FORMAT(////,10X,'U MATRIX',/)
1045  FORMAT(////,10X,'V MATRIX',/)
1050  FORMAT(10X,'GENERATION OF BASIS CD VECTORS BVCD
     $ USING MATRICES CD AND V',////)
```

```
1035    FORMAT(10X,'NUMBER OF CD SPECTRA=',I5,/,
       $10X,'NUMBER OF BASIS VECTORS TO BE GENERATED=',I5,///)
1060    FORMAT(//,10X,'THIS IS BASIS CD VECTOR NUMBER',I3,
       $13X,'SIGMA=',F10.3,/)
1065    FORMAT(2X,'GENERATION OF SECONDARY STRUCTURE MATRIX BVSS CORRES
       $PONDING TO BASIS CD VECTORS USING MATRICES SS AND V'////)
1070    FORMAT(10X,'PROTEIN SECONDARY STRUCTURE MATRIX FROM X-RAY DATA'/)
1075    FORMAT(A4)
1080    FORMAT(/50X,5(2X,A4))
1085    FORMAT(//,10X,'THIS IS THE STRUCTURE FOR BASIS CD VECTOR
       $NUMBER',I3,13X,'SIGMA=',F10.3,/)
1090    FORMAT(10X,40A1,10F6.2)
2000    FORMAT(1H1)
        END
        SUBROUTINE BASSP(S,V,N,NBASV,NPTS,BSPEC)
C
        DIMENSION S(50,25),V(50,25),BSPEC(25,50),TOT(25)
C
C       THIS PROGRAM IS DESIGNED TO COMPUTE THE BASIS VECTORS FOR
C       A SAMPLE MATRIX USING DATA S AND V.
C
        DO 20 I=1,N
        TOT(I)=0.0
        DO 20 J=1,NPTS
20      BSPEC(I,J)=0.0
C
C       COMPUTE BASIS VECTORS.
C
        DO 700 I=1,NBASV
        DO 70 J=1,NPTS
        DO 70 K=1,N
        BSPEC(I,J)=BSPEC(I,J)+V(K,I)*S(J,K)
70      CONTINUE
        DO 700 K=1,NPTS
700     TOT(I)=TOT(I)+BSPEC(I,K)
        RETURN
        END
C ----------------------------------------------------------------------
C ----------------------------------------------------------------------
C PROGRAM NAME: PPOSTP (PROTEIN STRUCTURE PREDICTION)
C
C THIS PROGRAM COMPUTES THE SECONDARY STRUCTURE FOR A NUMBER OF PROTEINS
C FROM THE MATRIX OF THEIR CD DATA 'SAMV'. THE MATRIX OF BASIS CD VECTORS
C 'BVCD' AND THEIR CORRESPONDING STRUCTURE MATRIX 'BVSS' USED FOR THIS
C CALCULATION  ARE GENERATED BY A SEPARATE PROGRAM CALLED 'BAVGEN'.
C WHENEVER A NEW 'BVCD' MATRIX IS GENERATED, THE CORRESPONDING STRUCTURE
C MATRIX 'BVSS' MUST ALSO BE COMPUTED. SEE HENNESSEY & JOHNSON (1981)
C BIOCHEMISTRY, VOL.20, 1085-1094 FOR MORE DETAILS ABOUT THE METHOD.
C
C THE NUMBER OF DATA POINTS, WAVELENGTH RANGE AND WAVELENGTH INTERVAL
C BETWEEN TWO DATA POINTS SHOULD BE SAME FOR 'SAMV' AND 'BVCD'.
C
C THIS PROGRAM CALLS THE SUBROUTINES LSTSQ,LINV1F AND MATPRO. LINV1F
C IS THE STANDARD LIBRARY SUBOROUTINE TO INVERT A SQUARE MATRIX
C DEVELOPED BY IMSL, INC.
C
C THE MAXIMUM NUMBER OF PROTEINS AND DATA POINTS HAVE BEEN DIMENSIONED TO
C 20 AND 50 RESPECTIVELY. THIS CAN BE EXTENDED BY CHANGING THE DIMENSIONS.
C
C          INPUT DATA VARIABLES
C          --------------------
C
C BVCD:    BASIS CD SPECTRA MATRIX.
C BVSS:    SECONDARY STRUCTURE MATRIX CORRESPONDING TO THE SPECTRA
C          DEFINED BY 'BVCD'.
C INT:     WAVELENGTH INTERVAL BETWEEN TWO DATA POINTS.
C IWF:     WAVELENGTH OF THE LAST DATA POINT.
C IWS:     WAVELENGTH OF THE FIRST DATA POINT.
C NBASV:   NUMBER OF BASIS CD VECTORS. ONLY CD BASIS VECTORS WITH
C          SIGNIFICANT SINGULAR VALUES SHOULD BE USED IN AN ANALYSIS.
C          THESE VALUES ARE COMPUTED IN THE PROGRAM 'BAVGEN'. SEE
C          HENNESSEY & JOHNSON (1981), BIOCHEMISTRY, VOL.20, 1085-1094.
```

```
C  NSAMV: NUMBER OF SAMPLE PROTEINS FOR WHICH SECONDARY STRUCUTRES
C         ARE TO BE DETERMINED.
C  NPTS:  NUMBER OF CD DATA POINTS.
C  NSTR:  NUMBER OF DIFFERENT SECONDARY STRUCTURES(EG. HX,BA,BP,TN,OT).
C  PRNAME:NAME OF THE PROTEINS INCLUDED IN 'SAMV' MATRIX. MAXIMUM LENGTH
C         FOR EACH TITLE IS 40 CHARACTERS.
C  SAMV:  CD DATA MATRIX OF CD SPECTRA FOR 'NSAMV'SAMPLE PROTEINS.
C  SSTR:  TYPE OF SECONDARY STRUCTURE.
C         (HX:HELIX;BA:ANTIPARALLEL B-SHEET;BP:PARALLEL B-SHEET
C         TN:TURNS;OT:OTHERS).TOT(TOTAL) SHOULD ALSO BE INCLUDED IN THE END.
C
C         OUTPUT DATA VARIABLES
C         ---------------------
C
C  PROST: SECONDARY STRUCTURE MATRIX FOR THE 'NSAMV' SAMPLE PROTEINS
C         COMPUTED FROM 'BVSS' AND'X'. THIS IS CARRIED OUT BY THE
C         SUBROUTINE 'MATPRO'.
C  TOTAL: SUM OF THE PREDICTED SECONDARY STRUCTURE OF A GIVEN PROTEIN.
C         THIS NEED NOT BE EQUAL TO ONE AS THE METHOD IS UNCONSTRAINED.
C  X:     THE COEFFICIENTS DETERMINED USING THE MATRICES 'BVCD' AND 'SAMV'.
C         THIS IS CARRIED OUT USING 'LSTSQ' AND 'LINV1F' SUBROUTINES.
C
C------------------------------------------------------------------------------
C------------------------------------------------------------------------------
C
C
      DIMENSION BVCD(20,50),SAMV(20,50),X(20,20),BVSS(20,20),
     *PROST(20,20),TOTAL(20),SSTR(10),PRNAME(20,40)
C
      WRITE(6,2000)
C
      WRITE(6,1000)
C
C  READS THE INITIAL DATA
C
      READ(5,1005)NSAMV,NPTS,IWS,IWF,INT,NBASV,NSTR
C
C  READS THE BASIS CD VECTORS
C
      WRITE(6,1010)NSAMV,IWS,IWF,INT,NPTS,NBASV,NSTR
      WRITE(6,1012)
      DO 10 I=1,NBASV
      READ(5,1020)(BVCD(I,J),J=1,NPTS)
10    WRITE(6,1030)(BVCD(I,J),J=1,NPTS)
C
C  READS THE CD DATA OF PROTEINS TO BE PREDICTED
C
      WRITE(6,1015)NSAMV
      DO 20 I=1,NSAMV
      READ(5,1016)(PRNAME(I,K),K=1,40)
      READ(5,1020)(SAMV(I,J),J=1,NPTS)
      WRITE(6,1018)(PRNAME(I,K),K=1,40)
20    WRITE(6,1030)(SAMV(I,J),J=1,NPTS)
C
      CALL LSTSQ(BVCD,SAMV,NPTS,NBASV,NSAMV,X)
C
C  PRINT OUT RESULTS
C
      WRITE(6,1040)
      DO 50 I=1,NSAMV
50    WRITE(6,1030)(X(I,J),J=1,NBASV)
C
C  READS THE SECONDARY STRUCTURE CORRESPONDING TO BASIS CD VECTORS
C
      WRITE(6,1045)NSTR
      READ(5,1065)(SSTR(I),I=1,NSTR+1)
      WRITE(6,1058)(SSTR(I),I=1,NSTR)
      WRITE(6,1070)
      DO 60 I=1,NBASV
      READ(5,1020)(BVSS(I,J),J=1,NSTR)
60    WRITE(6,1030)(BVSS(I,J),J=1,NSTR)
```

```
C
      CALL MATPRO(NSAMV,NBASV,NSTR,X,BVSS,PROST,TOTAL)
C
C   PRINT OUT RESULTS
C
      WRITE(6,1055)
      WRITE(6,1060)(SSTR(I),I=1,NSTR+1)
      DO 70 I=1,NSAMV
70    WRITE(6,1080)(PRNAME(I,K),K=1,40),(PROST(I,J),J=1,NSTR),
     $TOTAL(I)
C
C   END OF PROGRAM
C
1000  FORMAT(//15X,'COMPUTATION OF SECONDARY STRUCTURE OF PROTEINS FROM
     $ CD DATA',///)
1005  FORMAT(10I4)
1010  FORMAT(10X,'NUMBER OF PROTEINS FOR PREDICTION =',I5,/,
     $10X,'WAVELENGTH RANGE IN NM =',I5,' TO',I5,' AT',I5,' NM
     $ INTERVAL',/,10X,'NUMBER OF DATA POINTS=',I5,/,
     $10X,'NUMBER OF BASIS CD VECTORS=',I5,/,
     $10X,'NUMBER OF SECONDARY STRUCTURES=',I5,//////)
1012  FORMAT(10X,'BASIS CD VECTORS',//)
1015  FORMAT(////,10X,'CD DATA OF',I5,'   PROTEINS FOR PREDICTION',//)
1016  FORMAT(1X,40A1)
1018  FORMAT(10X,40A1)
1020  FORMAT(F10.5)
1030  FORMAT(10X,15F6.2)
1040  FORMAT(////,10X,'MATRIX X',/)
1045  FORMAT(////,10X,'PROTEIN STRUCTURE VECTORS CORRESPONDING
     $ TO BASIS CD VECTORS'/,10X,'NUMBER OF SECONDARY STRUCTURE=',I5,/)
1055  FORMAT(//////10X,'SECONDARY STRUCTURE PREDICTION'//)
1058  FORMAT(//10X,10(2X,A4))
1060  FORMAT(//50X,10(2X,A4),/)
1065  FORMAT(A4)
1070  FORMAT(/)
1080  FORMAT(10X,40A1,10F6.2)
2000  FORMAT(1H1)
      END
C
C
C
C
      SUBROUTINE LSTSQ(BASV,SAMV,NPTS,NBASV,NSAMV,BCOEF)
      DIMENSION BASV(20,50),SAMV(20,50),BCOEF(20,20),BST(20,20),
     $BBT(20,20),BBTI(20,20),WK(20)
C
C   INITIALIZE REGISTERS
C
      DO 3I =1,20
      DO 2J =1,20
      BBT(I,J)=0.0
      BBTI(I,J)=0.0
      BCOEF(I,J)=0.0
2     BST(I,J)=0.0
3     WK(I)=0.0
C
C     SET UP 'PRO' MATRIX
C
      DO 30 J=1,NBASV
      DO 20 K=1,NBASV
      DO 10 L=1,NPTS
10    BBT(J,K)=BBT(J,K)+BASV(J,L)*BASV(K,L)
20    CONTINUE
30    CONTINUE
      IDGT=3
C
C   INVERT 'PRO' MATRIX
C
      CALL LINV1F (BBT,NBASV,20,BBTI,IDGT,WK,IER)
C
      DO 500 I=1,NSAMV
```

```
        DO 100 J=1,NBASV
        DO 90 K=1,NPTS
90      BST(I,J)=BST(I,J)+SAMV(I,K)*BASV(J,K)
100     CONTINUE
        DO 200 J=1,NBASV
        DO 200 K=1,NBASV
        BCOEF(I,J)=BCOEF(I,J)+BST(I,K)*BBTI(K,J)
200     CONTINUE
500     CONTINUE
        RETURN
        END
C
C
C
C
C       PROGRAM MATPRO.FOR
        SUBROUTINE MATPRO(IX,IY,IZ,A,B,C,SUM)
        DIMENSION A(20,20),B(20,20),C(20,20),SUM(20)
C
C       THIS PROGRAM CALCULATES THE PRODUCT OF TWO INPUT MATRICES.
C
C       INITIALIZATION OF REGISTERS.
C
        DO 3 I=1,20
        DO 2 J=1,20
2       C(I,J)=0.0
3       SUM(I)=0.0
C
C       CALCULATE PRODUCT OF MATRICES AND SUMS OF THE PRODUCT ROWS.
C
        DO 50 I=1,IX
        DO 49 J=1,IZ
        DO 48 K=1,IY
48      C(I,J)=C(I,J) + A(I,K)*B(K,J)
49      CONTINUE
50      CONTINUE
        DO 53 I=1,IX
        DO 52 J=1,IZ
52      SUM(I)=SUM(I) + C(I,J)
53      CONTINUE
C

        RETURN
        END
```

Addendum

This review was completed in December 1983. L. A. Compton and W. C. Johnson, Jr. [*Biophys. J.* **49**, 494a (1986)] have now reported a simplified method of computing the least-squares solution to the CD spectra of proteins by using a simple matrix multiplication. It is based on the generalized (Moore–Penrose) inverse matrix theorem, which does not depend on standard matrix diagonalization or inversion subroutines as described in Appendix C.

Acknowledgments

We thank Professor W. C. Johnson, Jr. for providing us with his computer program and unpublished data quoted in this work and Professor R. W. Woody for sending us a preprint of his review and the English translations of three USSR publications. We are indebted to both of them for their valuable comments and discussion. Thanks are also due Mrs. Y. M. L. Yang for her assistance in the preparation of this chapter. This work was supported by U.S. Public Health Service Grant GM-10880-24 and National Science Foundation grant PCM 83-14716.

[12] Magnetic Circular Dichroism

By BARTON HOLMQUIST

Introduction

Magnetic circular dichroism (MCD) has in recent years added significantly to our knowledge of the electronic structure of a wide variety of molecules, from simple gases to complex macromolecules. Early uses of MCD were centered primarily on inorganic materials, but instrumental progress and the development of theoretical considerations have stimulated expansion of MCD to many areas of biological chemistry. The first biological material examined by modern techniques was the heme protein cytochrome c[1] where V. Shashoua first demonstrated that magnetic optical activity could add to the armamentarium of methods for the spectroscopic characterization of proteins, in this case the sensitivity of the method to the oxidized and reduced states of the protein. As a result extensive studies of the porphyrins and related pyrrole heterocycles and porphyrin-containing proteins and enzymes have resulted. Hemeproteins have continued to attract the greatest attention with several recent reviews devoted exclusively to hemoprotein[2-6] and porphyrin[3,4] analysis. Similarly, non-heme iron proteins were early surveyed by MCD,[7] but it is only in the last few years that the value of MCD for this large group of proteins has been recognized, particularly for determination of the iron–sulfur cluster type.[8-10] Recent reviews of this facet of MCD spectroscopy[10,10a] as well as reviews on MCD in general,[11] the application of MCD

[1] V. E. Shashoua, this series, Vol. 27, p. 96.
[2] B. Holmquist, *in* "The Porphyrins" (D. Dolphin, ed.), Vol. III, p. 249. Academic Press, New York, 1978.
[3] M. Hatano and T. Nozawa, *Adv. Biophys.* **II**, 95 (1978).
[4] J. C. Sutherland, *in* "The Porphyrins" (D. Dolphin, ed.), Vol. III, p. 225. Academic Press, New York, 1978.
[5] L. E. Vickery, this series, Vol. 54, p. 284.
[6] B. Holmquist, *Adv. Inorg. Biochem.* **2**, 75 (1980).
[7] D. D. Ulmer, B. Holmquist, and B. L. Vallee, *Biochem. Biophys. Res. Commun.* **51**, 1054 (1973).
[8] P. J. Stephens, C. E. McKenna, B. E. Smith, H. T. Nguyen, M.-C. McKenna, A. J. Thomson, F. Devlin, and J. B. Jones, *Proc. Natl. Acad. Sci. U.S.A.* **76**, 2585 (1979).
[9] P. J. Stephens, C. E. McKenna, M.-C. McKenna, H. T. Nguyen, and F. Devlin, *Biochemistry* **20**, 2857 (1981).
[10] M. K. Johnson, A. E. Robinson, and A. J. Thomson, *in* "Iron-Sulfur Proteins" (T. G. Spiro, ed.), p. 368. Wiley, New York, 1982.
[10a] H. Beinert and A. J. Thomson, *Arch. Biochem. Biophys.* **222**, 333 (1983).
[11] J. C. Sutherland and B. Holmquist, *Annu. Rev. Biophys. Bioeng.* **9**, 293 (1980).

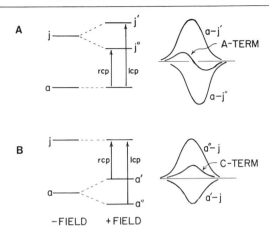

Fig. 1. Origin of MCD A and C terms. In A, Zeeman splitting of the excited state j by the magnetic field into j' and j'' results in the two absorption curves for left and right circularly polarized light $a-j$ and $a-j'$. The resultant MCD curve is the biphasic A term. In B, the magnetic field splits the degenerate ground state into a' and a'' resulting in the corresponding absorption bands. The net difference in this case is an MCD C term of symmetrical line shape.

to cobalt-substituted proteins,[12] metalloproteins in general,[12a] and on various theoretical aspects[3,13] have also appeared. Considered here will be instrumental and application developments of this technique largely appearing since the last report in this series.[14]

The Phenomenon

MCD arises when an external applied magnetic field perturbs the ground or excited states of a molecule resulting in differential absorption of polarized light. Three types of MCD effects—A, B, and C terms—can be differentiated based on the nature of the transition involved. In the case of a transition which has a degenerate excited state (Fig. 1) the absorption band ε representing the transition from the ground state a to the excited state j centered at λ_{max} is observed in the absence of a magnetic field. It absorbs left and right circularly polarized light equally with no net differential absorption. In a magnetic field the excited state j will undergo Zeeman splitting into equally higher and lower energy levels j'

[12] B. L. Vallee and B. Holmquist, *Adv. Inorg. Biochem.* **2,** 27 (1980).
[12a] D. M. Dooley and J. H. Dawson, *Coord. Chem. Rev.* **60,** 1 (1984).
[13] P. J. Stephens, *Adv. Chem. Phys.* **35,** 197 (1976).
[14] B. Holmquist and B. L. Vallee, this series, Vol. 49, p. 149.

and j''. These magnetically distinct transitions differentially absorb left and right circularly polarized light with equal intensity but of opposite sign as indicated. The difference between these two bands manifests as a symmetrical, biphasic signal crossing the zero axis at λ_{max}; the resulting band is called an A term.

In the case of a transition with a degenerate ground state the Zeeman splitting results in a different effect, the C term (Fig. 1). The magnetic field splits the ground state a' into a' and a'', in a manner similar to that just discussed. These split transitions absorb circularly polarized light differentially with the result being bands of opposite sign. In this instance the electronic population of a' and a'' is temperature dependent being a function of the Boltzmann distribution. The resulting transitions a' to j and a'' to j exhibit different intensities and the net magnetic circular dichroic band is Gaussian with an intensity inversely dependent on temperature.

A third type of MCD effect is commonly observed, especially in complex molecules with many transitions and low symmetry such as those observed in most biological chromophores. It is called a B term and arises from magnetically induced mixing of states that are not necessarily degenerate. Its shape is similar to that of the C term but it does not exhibit the inverse temperature dependence on the C term.

MCD spectra can be analyzed by various techniques including Gaussian curve fitting or moment analysis,[12,13] and the appropriate terms can then be assigned to the various transitions. The methods employed to extract the various parameters which serve to describe MCD spectra in general have been detailed by Stephens[13] and specifically for cobalt complexes and proteins by Kaden et al.[15] Briefly, they include fitting of spectra to Gaussian curves of damped oscillator functions for which A and B + C/kT terms can be identified. Temperature dependence or independence of spectral intensities separate Gaussian shaped C terms from the B terms. Where complete spectra are unavailable, moment analysis can be employed.[15] This technique has been utilized in conjunction with curve fitting procedures to analyze, e.g., Co(II) MCD spectra of metalloenzymes.

Measurement and Instrumentation

Magnetic circular dichroism, a measure of the Faraday effect, is the determination of the differential absorption of left and right circularly polarized light, as is circular dichroism, except that in MCD a magnetic field (H) either parallel (the standard convention) or antiparallel to the

[15] T. A. Kaden, B. Holmquist, and B. L. Vallee, *Inorg. Chem.* **13**, 2585 (1974).

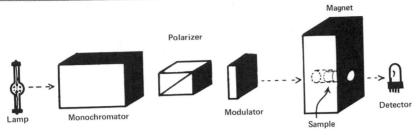

FIG. 2. Block diagram of the optical components of a typical spectropolarimeter for measurement of MCD.

direction of propagation of the beam of light is applied to the sample, thus inducing magnetic optical activity. A block diagram of a typical MCD instrument is shown in Fig. 2. The instrument is a CD instrument to which a suitable magnet (electro-, permanent, or superconducting) is mounted such that only the sample resides within the core of the magnet where highest flux density occurs. Commercial instruments are available from Jasco (Easton, MD) and Roussel-Jovan (France) which have been designed to incorporate various types of magnets. The Cary Model 61 (Varian Instruments, Palo Alto, CA) is no longer manufactured. Several home-built machines have been described.[16–18]

The intensity of an MCD Cotton effect is proportional to the field strength (with the exception of saturation phenomena at very low temperature, see below) and hence use of high fields such as those available from superconductors (up to 15 T) have been stressed. They require liquid helium and liquid nitrogen and demand 2–4 hr of preparation before measurements can be made. However, cost, convenience, and ease of operation associated with permanent or electromagnets (up to 1.5 T) dictates their frequent use.

Several home-built instruments and cryostats have been described all of which basically employ the scheme of Fig. 2.[16–18] The CD-MCD instrument detailed by Hatano et al.[16] uses a new optical element, the acoustic optical filter, rather than a conventional monochrometer to produce monochromatic light while simultaneously linearly polarizing the transmitted light, eliminating the need for a conventional polarizer. The device consists of a TeO_2 crystal coupled to an audio transducer. Introduction of an acoustic wave through the crystal causes anisotropic Bragg diffraction.

[16] M. Hatano, T. Nozawa, T. Murakami, T. Yamamoto, M. Shigehisu, S. Kimura, T. Takakuwa, N. Sakayangi, T. Yano, and A. Watanebe, *Rev. Sci. Instrum.* **52**, 1311 (1981).

[17] J. C. Sutherland, *Anal. Biochem.* **113**, 108 (1981).

[18] E. R. Zygowicz, B. R. Hollebone, and H. R. Perkins, *Clin. Chem.* **26**, 1413 (1980).

The acoustic optical filter offers a means for rapid scanning MCD since the scanning speed is limited by the transit time of the acoustic wave across the optical beam and is thus quite rapid. Hence a stopped-flow mixing system has been included in their design (Fig. 3). A scanning rate of 5 nm/msec is possible with the current instrument although the fastest reaction illustrated, the reaction of cytochrome c with ascorbate, was at a rate of 50 nm/sec. The stopped-flow components consist of a four-jet mixer mounted to a 70-μl cuvette which is mounted in a 1.5-T electromagnet. The high resolution, 0.1 nm at 400 nm at a scanning rate of 5 nm/ msec, and sensitivity of this instrument appears comparable, if not better, than commercial slow-scanning dichrometers.

Details of the procedures involved in routine MCD spectroscopy have been described[14] and only pertinent points are reviewed here. In general, the same considerations regarding solvents, pathlengths, concentrations, and other parameters involved in CD[19] also pertain to MCD. Solutions should be filtered or centrifuged to remove insolubles. An absorbance of 2 should not be exceeded throughout the spectral region investigated to avoid spurious signals when low levels of transmitted light are measured. Optimal signal-to-noise ratios are obtained at an absorbance of 0.86 through a measured band. Since MCD is theoretically a property of all matter, due caution as concerns solvent selection is required.

In an MCD experiment the signal measured is the sum of the natural CD of the sample and that induced by the magnetic field such that

$$\Delta A = A_L - A_R = \Delta A_{CD} + H\Delta A_{MCD}$$

where A_L and A_R are the respective absorbances of left and right circularly polarized light and H is the magnetic field strength. Consequently the MCD measurement involves measurement of the CD spectrum in the absence of the magnetic field ΔA_{CD} followed by that in its presence $\Delta A_{CD} + H\Delta A_{MCD}$. The true MCD is then calculated as the difference in the apparent CD with and without the field. If solvent base lines are flat or identical for both measurements, point by point subtraction is possible along the wavelength axis. Correction for CD is directly made through the use of computer data handling with modern instruments. For noncomputer interfaced instruments such as the Cary 61 a calculator-aided data analysis system has been described.[14] The availability of microcomputers and plug-in analog to digital boards makes it now easy and inexpensive to place the Cary 61 "on line." A voltage proportional to the differential absorption is available (test point 10 on the synchronous demodulator board J16, 5 V for full scale deflection) which with suitable attenuation

[19] A. T. Adler, N. J. Greenfield, and G. K. Fasman, this series, Vol. 27, p. 675.

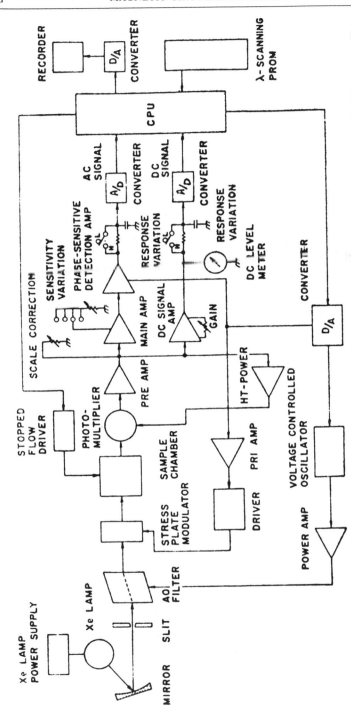

FIG. 3. Optical and signal processing diagram of the rapid scanning-stopped-flow CD/MCD spectropolarimeter designed by Hatano *et al.*[16] An acoustic optical tunable filter (AO) provides both rapid wavelength scanning and linear polarization.

can be directly input to a computer. An Apple IIe and the 12-bit Adalab interface board (Microware Inc. Box 771, State College, PA) make up the hardware for our system. Hormann and Peake[20] have described a similar system for automating the Varian E9 EPR spectrometer and supplied a program for data storage, spectral subtraction, smoothing, etc. which is directly applicable for use with the Cary 61.

The units of MCD are expressed variously as some function related to the differential absorption of left and right circularly polarized light ΔA, $\Delta\varepsilon$, or θ (degrees ellipticity) normalized to unit field strength. Ellipticity in degrees is related to ΔA by $\theta = 33\Delta A$. It has been recommended[11] that MCD be reported in units of differential absorption per unit of applied field, $\Delta\varepsilon_M/H$, in units of M^{-1} cm^{-1} T^{-1} where T is the field strength in Tesla (1 T = 10,000 G). The relationship between θ_{obsd}, the degrees ellipticity displayed by some instruments, to observed differential absorbance is

$$\Delta\varepsilon_M = \Delta\varepsilon/H = \frac{\Delta A}{bcH} = \frac{\theta_{obsd}}{33bcH} \qquad (M^{-1}\ \text{cm}^{-1}\ \text{T}^{-1})$$

where $\Delta\varepsilon$, θ_{obsd}, and ΔA are values induced by the magnetic field, c is the molar concentration, b is the path length in cm, and H is the field strength in Tesla. MCD data are frequently reported in degrees ellipticity per unit field. When such units are employed θ_{obsd} is converted to molar magnetic ellipticity $[\theta]_m$ by

$$[\theta]_m = \frac{100\theta_{obsd}}{bcH} \qquad (\text{deg cm}^2\ \text{dmol}^{-1}\ \text{T}^{-1})$$

Calibration of MCD requires two standards, one for the CD and a second for the MCD. Sublimed 10-camphorsulfonic acid is commonly employed for CD using the value $\Delta\varepsilon = 2.37\ M^{-1}$ cm^{-1} ($\theta = 7821$ deg cm^2 dmol^{-1}) at 290 nm. Several materials have been used to standardize MCD by actual measurement of samples of known $\Delta\varepsilon_M$. Cobalt sulfate in water, 0.05 M, exhibits a $\Delta\varepsilon_M = -0.0185\ M^{-1}$ cm^{-1} T^{-1} at 510 nm. This standard does not exhibit natural CD and has long-term stability.[12]

Analytical Applications

Two frequently quoted advantages of MCD over absorption spectroscopy which have led to its analytical application are the ability to resolve transitions hidden under absorption envelopes and, with certain chromo-

[20] G. Hormann and B. Peake, *J. Magn. Res.* **53**, 121 (1983).

phores such as those containing the tetrapyrrole nucleus (heme proteins, porphyrins etc.), an increased detection sensitivity. The analysis of tryptophan in proteins[21] is a good example of the former in that the indole system is the only group normally present in proteins that exhibits a positive MCD band. The band is at a relatively isolated wavelength, 293 nm, and its intensity can be used as a direct monitor of tryptophan content of proteins merely by measuring induced ellipticity of the protein at that wavelength.

The extreme amplitudes of the Soret A and Q bands of the porphyrins and heme proteins have drawn attention to the potential measurement of porphyrin-containing compounds in clinical samples. Myoglobin is elevated in serum in acute myocardial infarction and can be measured directly by MCD at a sensitivity of 0.1 μg/ml, a value sufficiently low for diagnostic purposes (typical infarct values are 0.3–50 μg/ml). By reduction of the hemochrome with dithionite and complexation with pyridine an 80-fold increase in the MCD amplitude at 555 nm accrues, providing the requisite sensitivity.[22]

Of course other heme components of biological fluids can also be measured, i.e., those not necessarily protein bound, and use has been made of this capacity to monitor occupational exposure to lead based on the measurement of protoporphyrin in serum,[18] measurement of the total porphyrin content in urine,[23] and to measure both uroporphyrin I and coporphyrin II in urine, this latter measurement useful for distinguishing between different porphyrias.[23,23a,23b] Figure 4 shows the absorption and MCD spectra of a typical serum sample comparing the information obtained by absorption and MCD. The Soret absorption near 400 nm is barely visible as a small shoulder superimposed on the background absorbance while the characteristic A term of the Soret band is clearly resolved and quantifiable, a clear illustration of the specificity frequently achieved in MCD.

Currently such analytical measurements are less or equal in sensitivity to presently applied methods, e.g., immunochemical, fluorescence, etc., but they are more precise. In view of the type of instrumentation employed in these initial investigations costing in excess of $50,000, such

[21] B. Holmquist and B. L. Vallee, *Biochemistry* **12**, 4409 (1973).
[22] G. Barth, R. Records, R. E. Linder, E. Bunnenberg, and C. Djerassi, *Anal. Biochem.* **80**, 20 (1977).
[23] S. M. Kalman, G. Barth, R. E. Linder, E. Bunnenberg, and C. Djerassi, *Anal. Biochem.* **52**, 83 (1973).
[23a] R. S. Day, *S. Afr. Med. J.* **65**, 713 (1984).
[23b] K. M. Ivanetich, C. Movsowitz, and M. R. Moore, *Clin. Chem.* **30**, 391 (1984).

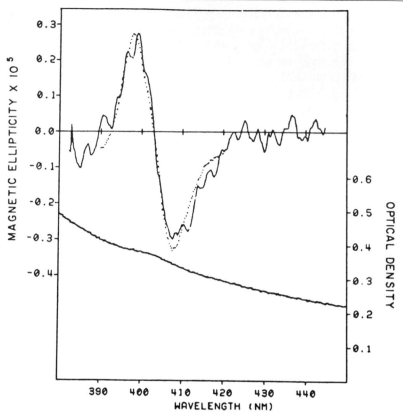

FIG. 4. MCD and absorption (lower curve) of urine illustrating the sensitivity of MCD over absorption for detecting small amounts of uroporphyrin (13 μg/liter) and coproporphyrin (17 μg/liter). The dotted line is a fit to the MCD spectrum. From Barth et al.[22]

analyses are unlikely to be routinely applied. However, as pointed out by Barth et al.,[22] instruments designed at reduced cost, specifically for such measurements, could be developed by incorporating a single-wavelength source with optimized resolution, small bore high field electromagnets, modern electronics, and computerization for data acquisition and manipulation.

Transition Metals

The study of the MCD properties of cobalt and to a minor extent nickel when substituted into metalloproteins has continued to reveal features of the coordination, geometry, and roles of metals at specific sites in proteins and enzymes. These metals, especially cobalt, substitute readily for zinc providing a spectral probe in the visible region arising from its d → d transitions. The spectra reflect the environment of the metal and its response to ligand exchange or addition, to conformational changes induced by a multiple of factors such as inhibitor binding at or near the metal, and even to events involving metal participation in enzymatic catalysis. The established ability of MCD spectra of model compounds to characterize the metal coordination, systematically investigated by several groups,[15,24–27] has provided a firm basis for the interpretation and assignment of the coordination of the metal binding site. Based on such comparisons and additional spectral evidence, octahedral cobalt sites have been observed in concanavalin A[28] where Co(II) substitutes for Mn, in yeast enolase,[29] a Co(II) for Zn substitution, in the structural sites of alkaline phosphatase[30,31] where Co(II) occupies the two Mg sites, in a regulatory site in *Aeromonas* aminopeptidase[31a] and in both metal sites in *Bacillus aureus* phospholipase.[31b] In each, a very weak negative MCD band centered near 510 nm is observed, much like that exhibited for cobalt hexahydrate.

Tetrahedral cobalt sites, with their very characteristic MCD spectrum consisting of a large negative transition near 600 nm, a shoulder to lower wavelength, and one or two small positive bands near 550 nm have been assigned for cobalt when substituted for zinc at the active sites of the metalloenzymes thermolysin,[32,33] superoxide dismutase,[34,34a,34b] bovine carboxypeptidase A,[26,33,35] *S. griseus* carboxypeptidase,[36] and in certain forms of carbonic anhydrase,[25,26] alkaline phosphatase,[30,31] alcohol dehy-

[24] T. A. Kaden, B. Holmquist, and B. L. Vallee, *Biochem. Biophys. Res. Commun.* **46,** 1654 (1972).
[25] J. E. Coleman and R. V. Coleman, *J. Biol. Chem.* **247,** 4718 (1972).
[26] B. Holmquist, T. A. Kaden, and B. L. Vallee, *Biochemistry* **14,** 1454 (1975).
[27] H. Katô and K. Akimoto, *J. Am. Chem. Soc.* **96,** 1351 (1974).
[28] C. E. Richardson and W. D. Behnke, *J. Mol. Biol.* **102,** 441 (1976).
[29] K. M. Collins and J. M. Brewer, *J. Inorg. Biochem.* **17,** 15 (1982).
[30] R. A. Anderson, F. S. Kennedy, and B. L. Vallee, *Biochemistry* **15,** 3710 (1976).
[31] R. A. Anderson and B. L. Vallee, *Biochemistry* **16,** 4388 (1977).
[31a] J. M. Prescott, F. W. Wagner, B. Holmquist, and B. L. Vallee, *Biochemistry* **24,** 5356 (1985).
[31b] R. Bicknell, G. R. Hanson, B. Holmquist, and C. Little, *Biochemistry* (in press).
[32] B. Holmquist, *Biochemistry* **16,** 4591 (1977).
[33] B. Holmquist and B. L. Vallee, *Proc. Natl. Acad. Sci. U.S.A.* **76,** 6216 (1979).
[34] G. Rotilio, L. Calabrese, and J. E. Coleman, *J. Biol. Chem.* **248,** 3855 (1973).
[34a] J. C. Dunbar, B. Holmquist, and J. T. Johansen, *Biochemistry* **23,** 4324 (1984).
[34b] J. C. Dunbar, B. Holmquist, and J. T. Johansen, *Biochemistry* **23,** 4330 (1984).
[35] K. F. Geoghegan, B. Holmquist, C. A. Spilburg, and B. L. Vallee, *Biochemistry* **22,** 1847 (1983).
[36] K. Breddam, T. Bazzone, B. Holmquist, and B. L. Vallee, *Biochemistry* **18,** 1563 (1979).

drogenase,[37] D-lactate dehydrogenase,[38] and the catalytic sites of *Aeromonas* aminopeptidase.[31a] Several of these assignments have been supported by X-ray diffraction analysis of the native zinc proteins, e.g., thermolysin, carboxypeptidase A, and horse liver alcohol dehydrogenase. The fact that the ligands and geometry of the cobalt in the latter two enzymes are identical to those in the native zinc enzyme[39,40] would seem to allay any doubt that using cobalt spectra to define the metal sites in metalloenzymes is equivocated by significant distortion induced by the substitution and that mechanistic and functional deductions based on the cobalt derivatives are valid.

The use of cobalt as a probe of metal sites in proteins has been extended to the examination of discrete states during catalysis by recent studies with cobalt carboxypeptidase A at low temperature ($-40°$). Kinetic studies have revealed remarkable and characteristic transient changes in the cobalt atom d → d transitions by absorption, EPR, and MCD spectroscopy.[41] Stabilization by low temperature has enabled various spectral techniques to be used to characterize these cryotrapped intermediates occurring in the reaction sequence. The status of the metal atom in one such stage of catalysis, a stabilized intermediate likely representing a species whose transformation to product is rate limiting, is shown in Fig. 5. The MCD spectrum reflects a spectral response analogous to the absorption changes with the intermediate exhibiting two prominent negative extrema rather than the single negative band associated with the free enzyme. Warming restores the MCD to that of the enzyme plus products. Though involvement of the metal in catalysis is evident, the exact nature of the distortion of the metal remains to be clarified. The preliminary MCD data suggest distortion toward 5-coordinate geometry.

Cobalt also substitutes for copper in various copper proteins and for the metals found in metallothionein. Thus far all these substitutions have shown four-coordinate type MCD spectra. Solomon *et al.*[42] investigated the MCD of cobalt-substituted stellacyanin, azurin, and plastocyanin,

[37] B. L. Vallee, D. L. Drum, F. S. Kennedy, *in* "Alcohol and Aldehyde Metabolizing Systems" (R. G. Thurman *et al.*, eds.), Vol. 1, p. 55. Academic Press, New York, 1974.

[38] F. F. Morpeth and V. Massey, *Biochemistry* **21**, 1381 (1982).

[39] G. Schneider, H. Eklund, E. C. Zeppezauer, and M. Zeppezauer, *Proc. Natl. Acad. Sci. U.S.A.* **80**, 5289 (1983).

[40] K. D. Hardman and W. N. Lipscomb, *J. Am. Chem. Soc.* **106**, 463 (1984).

[41] K. F. Geoghegan, A. Galdes, R. A. Martinelli, B. Holmquist, D. S. Auld, and B. L. Vallee, *Biochemistry* **22**, 2255 (1983).

[42] E. I. Solomon, J. W. Hare, D. M. Dooley, J. H. Dawson, P. J. Stephens, and H. B. Gray, *J. Am. Chem. Soc.* **102**, 168 (1980).

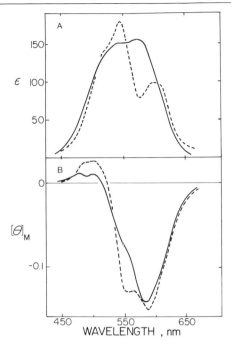

FIG. 5. Absorption (A) and MCD (B) spectra of cobalt carboxypeptidase A (———) and a transient intermediate formed in its reaction with the ester dansyl-Ala-Ala-phenyllactate (– – –). The spectra are at −40° in the cryosolvent 40% ethylene glycol–20% methanol–40% water and are measured within 2 min of mixing. From Geoghegan et al.[41]

each of which gives nearly identical tetrahedral-like spectra. Two forms of hemocyanin, those from squid and horseshoe crab,[43,44] and plasma amine oxidase[45] also exhibit tetrahedral spectra though the cobalt derivatives in these cases are not thoroughly characterized. Cobalt metallothionein[46–48] exhibits a rather complex spectrum due to the presence of multiple cobalt atoms, seven per molecule, and of extensive charge transfer bands (see below). Yet the characteristic tetrahedral cobalt MCD is apparent above 500 nm (Fig. 6).

[43] S. Suzuki, J. Kino, M. Kimura, W. Mori, and A. Nakahara, Inorg. Chim. Acta 66, 41 (1982).
[44] S. Suzuki, J. Kino, and A. Nakahara, Bull. Chem. Soc. Jpn. 55, 212 (1982).
[45] S. Suzuki, W. Mori, J. Kino, Y. Nakao, and A. Nakahara, J. Biochem. (Tokyo) 88, 1207 (1980).
[46] M. Vasák, J. H. R. Kagi, B. Holmquist, and B. L. Vallee, Biochemistry 20, 6659 (1981).
[47] A. Y. C. Law and M. J. Stillman, Biochem. Biophys. Res. Commun. 102, 397 (1981).
[48] G. A. Carson, P. A. W. Dean, and M. J. Stillman, Inorg. Chim. Acta 56, 59 (1981).

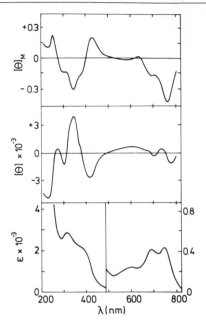

FIG. 6. Electronic spectral properties of cobalt-substituted metallothionein. In the MCD (top panel) the tetrahedral cobalt atom contributions are apparent above 500 nm while the S → Co(II) charge transfer bands are those observed below 500 nm. From Vasák et al.[46]

The MCD spectrum of nickel(II) metallothionein[46] includes a distinct band at 710 nm assigned to the $^3T_1(F) \rightarrow {}^3T_1(P)$ electronic transition. No distinct transitions are observable by absorption because of the superimposition of the red edge of low-energy charge transfer bands which extend into the visible. This band is diagnostic of tetrahedral coordination. This first example of nickel MCD in a biological system provides impetus for studies with other nickel species.

Nickel is a natural component of numerous hydrogenases. Its electronic transitions are obscured by Fe–S centers invariably present limiting studies of the Ni atom to EPR and XAS methods. The MCD spectrum of *Methanobacterium thermoautotrophicum* hydrogenase[48a] reveals d–d transitions assigned to Ni(III) in the regions 530–670 and 300–460 nm. Further studies with characterized Ni(III) complexes will be required to allow assignments relating to coordination number and geometry.

[48a] M. K. Johnson, I. C. Zambrano, M. H. Czechowski, H. D. Peck, Jr., D. V. Der Vartanian, and J. Le Gall, *Biochem. Biophys. Res. Commun.* **128**, 220 (1985).

Sulfur as Ligands to Metals

In the last decade it has become evident that in numerous metalloproteins sulfur serves as a ligand to the metal. Documentation of this has been by X-ray structure analysis in for example alcohol dehydrogenase,[49] and plastocyanin,[50] but strong evidence had been obtained for sulfur ligands in these and other proteins based on spectral data when cobalt (and in some cases nickel) is substituted for the native metal, either zinc or copper. These data are based on the observation of S → Co(II) ligand to metal charge transfer (LMCT) transitions in the MCD of such proteins. These transitions occur in the wavelength region 230–400 nm and for distorted tetrahedral geometry four transitions are possible, $S_{\pi,\alpha} \to d_x^2$, $d_{x^2-y^2}$, of which in practice two are frequently observed. While often obscured in absorption by the aromatic amino acid residues of proteins their intensity in MCD has proven of value for their identification and characterization. They were first observed in alcohol dehydrogenase,[37] in the very complex MCD spectrum of the cobalt enzyme where it is now known that two types of cobalt sites with different sulfur complements and their attendant LMCT and d → d transitions exist. LMCT bands have now been observed in the cobalt derivatives of metallothionein,[46,47] rubredoxin,[51] and β-lactamase II.[52] Analysis of the charge transfer bands in Co(II) (Fig. 6) and Ni(II) metallothionein was greatly aided by the very clear MCD structure of transitions, which in this protein are not obscured by aromatic absorption since it contains no aromatic residues. Of unproven function, the protein contains 20 residues of cysteine out of a total of 61 and binds 7 atoms of Zn or Cd that can be substituted with Co(II) or Ni(II). All 20 cysteine residues participate in metal binding via mercaptide linkages possibly in polynuclear metal clusters with some mercaptides serving as bridging ligands.

The MCD spectra of Cd, Zn metallothionein[46-48] and derivatives containing Bi, Cu, and Hg[53,54] also exhibit LMCT in the 200 to 400 nm region. The MCD of the Cd, Zn derivative[47] is dominated by a derivative-shaped

[49] H. Eklund, B. Nordström, E. Zeppezauer, G. Söderland, I. Ohlsson, T. Borwly, Osoderberg, O. Papier, C.-I. Brändén, and A. Akerson, *J. Mol. Biol.* **102**, 27 (1976).

[50] P. M. Colman, H. C. Freeman, J. M. Guss, M. Murata, V. A. Norris, J. A. M. Ramshaw, and M. P. Venkatappa, *Nature (London)* **272**, 319 (1978).

[51] S. W. May and J-Y. Kuo, *Biochemistry* **17**, 3333 (1978).

[52] G. S. Baldwin, A. Galdes, H. A. O. Hill, S. G. Waley, and E. P. Abraham, *J. Inorg. Biochem.* **13**, 189 (1980).

[53] J. A. Szymanska and M. J. Stillman, *Biochem. Biophys. Res. Commun.* **108**, 919 (1982).

[54] J. A. Szymanska, A. J. Zelazowski, and M. J. Stillman, *Biochem. Biophys. Res. Commun.* **115**, 167 (1983).

signal and it compares favorably with model Cd–thiolate complexes. Mercury–sulfide charge transfer transitions dominate the high energy regions of the Cu, Hg form of the protein which led to the suggestion that the Hg site is not tetrahedral.

Templeton and Tsai[55] have identified charge transfer from thiolate to a nonmetal species, namely imidazolium ion, at the active site of lipoamide dehydrogenase. The very weak A term centered near 550 nm in the MCD of the reduced enzyme was eliminated upon photooxidation to destroy histidine. The flavin chromophore, also a cofactor for this enzyme, is not responsible for the observed effects since the flavin ring system has only C_s symmetry and the A term can only arise from spin degeneracy of unpaired electrons from the charge transfer complex.

The ability of MCD to establish sulfur ligation in proteins has been used to confirm that the large class of inhibitors of zinc metalloproteases containing mercaptans function through a sulfur to metal ligation mode. Captopril, D-3-mercapto-2-methylpropanoyl-L-proline, best illustrates this class of compound. It is an orally active antihypertensive drug that operates by inhibiting the zinc protease angiotensin converting enzyme (K_i = 20 nM). The analogous inhibitors of carboxypeptidase A and thermolysin, N-mercaptoacetyl-L-phenylalanine and N-mercaptoacetyl-L-phenylalanyl-L-alanine, respectively, form tight complexes with the cobalt derivatives of these enzymes.[33] The carboxypeptidase inhibitor, K_i = 2×10^{-7} M, markedly perturbs the d $\rightarrow$ d transitions to produce a resultant spectrum very characteristic of tetrahedral geometry. A second effect is apparent in the vicinity of 340 nm, i.e., a large band with ε = 1000 M^{-1} cm^{-1} that is clearly due to LMCT between the cobalt atom at the active site and the bound inhibitor (Fig. 7). While the band is seen in the absorption as well, the MCD provides much better resolution. Indeed, in the analogous experiment with the related protease thermolysin the LMCT band appears only as a shoulder in the absorption but it is clearly resolved in the MCD; the lower wavelength of the transition, 325 nm, compared to its position in carboxypeptidase, 340 nm, precludes unambiguous identification by absorption due to extensive overlap with the protein absorption. This illustrates further the capacity of MCD to uncover transitions which are obscured by other spectral methods.

Neocarzinostatin

MCD was found to complement other spectral methods (absorption and fluorescence) in characterizing the active chromophoric conjugate in

55 D. M. Templeton and C. S. Tsai, *Biochem. Biophys. Res. Commun.* **90**, 1085 (1979).

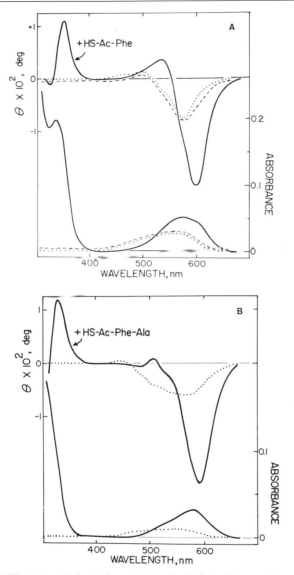

FIG. 7. (A) MCD (top) and absorption spectra of cobalt carboxypeptidase A (1.9×10^{-4} M) in the absence ($\cdots$) and presence (——) of the metal liganding inhibitor mercaptoacetyl-L-phenylalanine. Displacement of the mercapton inhibitor by benzylsuccinate (---) collapses the LMCT. (B) Similar experiment with cobalt thermolysin (3.2×10^{-4} M) ($\cdots$) using mercaptoacetyl-L-phenylalanyl-L-alanine. MCD are measured at 4 T. From Holmquist and Vallee.[33]

neocarzinostatin.[56–58] Earlier investigations had concluded that this antitumor antibiotic, which causes damage to DNA by strand scission and single strand breaks, was a pure protein consisting only of amino acids (totally sequenced) of molecular weight 10,000. The MCD of the native drug clearly indicated the presence of a nonprotein component whose visible absorption had previously been ascribed to tryptophan. By extraction in methanol, the chromophore was separated from the protein, shown to be the active component of the drug, and partial structure analysis has indicated the presence of 2-hydroxy-5-methoxy-7-methyl naphthoate which accounts for most of the spectral properties of the native drug as well as the isolated chromophore.

Iron–Sulfur Proteins

In the previous entry in this series[14] the utility of MCD in determining the spin state of the iron in hemeproteins illustrated a clear and unique application where MCD becomes a method of choice for an analysis particularly difficult by alternative means. Extensive use of this analysis has been made and hemeprotein studies by MCD analysis continues with considerable success. Several recent extensive reviews of the results of MCD studies of hemeproteins and models have appeared[2–6] and this subject will not be considered further.

From extensive investigations of various iron–sulfur proteins containing 1 to 4 iron atom clusters it has become evident that MCD is a valuable tool to ascertain the structure type of such centers in proteins of unknown composition.[8–10,59,60] The MCD of 1-, 2-, and 4-iron center systems exemplified here by rubredoxin,[7] *Spirulina maxima* ferredoxin,[60] and *Chromatium* high potential iron protein,[59] respectively, each has characteristic features in their oxidized and reduced states which enable the distinction of cluster type (Figs. 8 and 9). The reduced and oxidized iron clusters in, e.g., rubredoxin and several models for this class[61,62] have MCD consider-

[56] M. A. Napier, B. Holmquist, D. J. Strydom, and I. H. Golberg, *Biochemistry* **20**, 5602 (1981).
[57] M. A. Napier, B. Holmquist, D. J. Strydom, and I. H. Goldberg, *Biochem. Biophys. Res. Commun.* **89**, 635 (1979).
[58] M. A. Napier and I. H. Goldberg, *Mol. Pharmacol.* **23**, 500 (1983).
[59] P. J. Stephens, A. J. Thomson, T. A. Keiderling, J. Rawlings, K. K. Rao, and D. O. Hall, *Proc. Natl. Acad. Sci. U.S.A.* **75**, 5273 (1978).
[60] P. J. Stephens, A. J. Thomson, J. B. R. Dunn, T. A. Keiderling, J. Rawlings, K. K. Rao, and D. O. Hall, *Biochemistry* **17**, 4770 (1978).
[61] T. Muraoka, T. Nozawa, and M. Hatano, *Chem. Lett.* 1373 (1976).
[62] T. Muraoka, T. Nozawa, and M. Hatano, *Bioinorg. Chem.* **8**, 45 (1978).

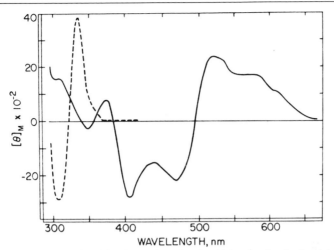

FIG. 8. MCD spectra of oxidized (——) and reduced (---) rubredoxin. Adapted from Ulmer *et al.*[1]

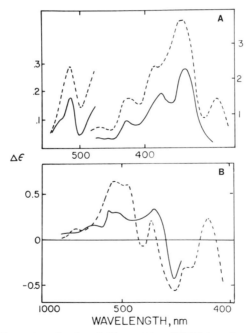

FIG. 9. (A) MCD spectra of reduced *Chromatium* HIPIP (——) and oxidized *Bacillus stearothermophilus* ferredoxin (---). Adapted from Stephens *et al.*[58] (B) MCD spectrum of oxidized (---) and reduced (——) *Spirulina maxima* ferredoxin. Adapted from Stephens *et al.*[59]

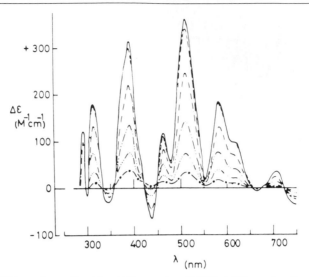

FIG. 10. MCD spectrum of oxidized *Chromatium* HIPIP at 5 T as a function of temperature illustrating the large intensity enhancement at low temperature: 50 K (−·−), 20 K (−····−), 10 K (−·····−), 4.22 K (−··−), 2.07 K (−−−), 1.48 K (——). From Johnson *et al.*[10]

ably more intense than the other cluster types and more importantly in the reduced state no features appear above 400 nm. Two iron center clusters exhibit weak MCD but both redox states are easily detectable. Contributions both negative and positive are characteristic. Above about 400 nm several positive bands are seen but below this wavelength only negative contributions are observed. This class is transparent in the near IR. Iron–sulfur proteins with 4-iron centers invariably exhibit only positive MCD above 300 nm and this applies to all oxidation states. Such spectral comparisons enabled Ulmer *et al.*[7] to suggest that the iron in xanthine oxidase is of the 2-iron class and more recently Stephens *et al.*[60] have shown that the nitrogenase iron proteins of *Azotobacter vinelandii* and *Klebsiella pneumoniae* are unquestionably of the 4-iron type.

A further dimension of MCD illustrated at the hands of the iron–sulfur proteins is its use as an optical probe of the electronic ground state of a metalloprotein when measurements are made near liquid helium temperatures. This is based on the fact that paramagnetic chromophores invariably exhibit temperature-dependent spectra, frequently increasing the intensity as much as 70-fold (Fig. 10). At such low temperatures and at high magnetic fields (near 5 T) saturation and full magnetization of a paramagnet occurs, the MCD becoming field strength independent. Such measurements allow determination of ground state *g* values and hence supply

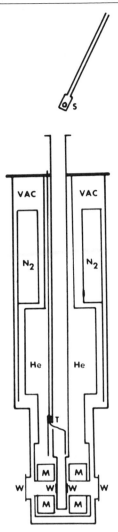

FIG. 11. Cryostat for measuring MCD at temperatures as low as 1.5 K. It is an Oxford Instruments split-coil superconducting magnet type SM 4. (S) Sample holder; (W) optical windows; (M) superconducting magnet coils; (T) transfer line for He. From Johnson *et al.*[10]

information similar to EPR, Mössbauer, and magnetic susceptibility. The cryostat used for such measurements (Fig. 11) meets the need for measurements at subhelium temperatures by providing for pumping of the helium reservoir during measurement.

Low temperature MCD studies are being extensively employed to further characterize the iron–sulfur proteins and in combination with EPR and other methods have confirmed the presence of 3Fe centers in *Azotobacter vinelandii* ferredoxin I,[63] *E. coli* nitrate reductase,[64] succinate dehydrogenase,[65] and *E. coli* fumarate reductase.[66] The identification, with the aid of low temperature MCD, that the latter two enzymes have 3 different FeS clusters, one each of 2Fe, 3Fe, and 4Fe, provides the first rational picture of the clusters in these enzymes clarifying much of past data.

[63] T. V. Morgan, P. J. Stephens, F. Devlin, B. K. Burgess, and C. D. Stout, *FEBS Lett.* **183,** 206 (1985).
[64] M. K. Johnson, D. E. Bennett, J. E. Morningstar, M. W. Adams, and L.E. Mortenson, *J. Biol. Chem.* **260,** 5456 (1984).
[65] T. B. Singer and M. K. Johnson, *FEBS Lett.* **190,** 189 (1985).
[66] J. E. Morningstar, M. K. Johnson, G. Cecchini, B. A. Ackrell, and E. B. Kearney, *J. Biol. Chem.* **260,** 13631 (1985).

[13] Resolution-Enhanced Fourier Transform Infrared Spectroscopy of Enzymes

By Heino Susi and D. Michael Byler

General Background

Infrared spectroscopy constitutes one of the oldest methods for studying the secondary structure of polypeptides and proteins. As early as 1950, before any detailed X-ray results were available, let alone circular dichroism or optical rotatory dispersion data, Elliott and Ambrose showed that the "amide I" band is observed around 1650–1660 cm^{-1} for the α-helical conformation and around 1630–1640 cm^{-1} for β-strands.[1] Since then basic theoretical work on the subject has been carried out primarily by Miyazawa and co-workers[2,3] and by Krimm and his colleagues.[4–5b] A number of reviews and summaries have been concerned

[1] A. Elliott and E. J. Ambrose, *Nature (London)* **4206,** 921 (1950).
[2] T. Miyazawa, T. Shimanouchi, and S. Mizushima, *J. Chem. Phys.* **24,** 408 (1956).
[3] T. Miyazawa, *J. Chem. Phys.* **32,** 1647 (1960).
[4] S. Krimm, *J. Mol. Biol.* **4,** 528 (1962).
[5a] S. Krimm and Y. Abe, *Proc. Natl. Acad. Sci. U.S.A.* **69,** 2788 (1972).
[5b] S. Krimm and J. Bandekar, *Biopolymers* **19,** 1 (1980).

with practical applications, including seven published in this series.[6-10] The latest one, in 1982, concerns measurement of peptide hydrogen exchange in rhodopsin.

Polypeptides and proteins exhibit a total of nine characteristic absorption bands in the infrared region. These are usually termed the amide A, B, and amide I–VII bands.[2,8b] The amide I ($\sim$1630–1690 cm^{-1}) band has been found to be the most useful for protein structure studies by infrared spectroscopy.[3-6,8b,10-11c] For deuterated proteins the designations amide I', II', etc., are employed.[8]

For proteins—as distinct from many synthetic polypeptides—each characteristic absorption band is generally a composite, consisting of overlapping components representing α-helical segments, β-sheet sections, turns, and unordered regions.[12a-15] These subbands usually cannot be resolved by conventional spectroscopic techniques because their inherent widths are greater than the instrumental resolution. Infrared spectroscopy, until recently,[14d-g] has therefore been essentially a qualitative

[6] D. L. Wood, this series, Vol. 4 [3], p. 104.

[7] W. P. Jencks, this series, Vol. 6 [125], p. 914.

[8a] H. Susi, this series, Vol. 26 [17], p. 381.

[8b] H. Susi, this series, Vol. 26 [22], p. 455.

[8c] S. N. Timasheff, H. Susi, and J. A. Rupley, this series, Vol. 27 [23], p. 548.

[9] D. F. H. Wallach and A. R. Oseroff, this series, Vol. 33 [22], p. 247.

[10] H. B. Osborne and E. Nabedryk-Viala, this series, Vol. 88 [81], p. 676.

[11a] N. Miwa, J. Antibiot. **35**, 1553 (1982).

[11b] R. J. Jakobsen, L. L. Brown, T. B. Hutson, D. J. Fink, and A. Veis, Science **220**, 1288 (1983).

[11c] R. Mendelsohn, G. Anderle, M. Jaworsky, H. H. Mantsch, and R. A. Dluhy, Biochim. Biophys. Acta **775**, 215 (1984).

[12a] R. M. Gendreau, Proc. SPIE Int. Soc. Opt. Eng. **553**, 4 (1985).

[12b] K. B. Smith, C. A. Penkowski, and R. J. Jakobsen, Proc. SPIE Int. Soc. Opt. Eng. **553**, 178 (1985).

[13a] V. E. Koteliansky, M. A. Glukhova, M. V. Bejanian, V. N. Smirnov, V. V. Filimonov, O. M. Zalite, and S. Yu. Venyaminov, Eur. J. Biochem. **119**, 619 (1981).

[13b] S. Yu. Venyaminov, M. L. Metsis, M. A. Chernousov, and V. E. Koteliansky, Eur. J. Biochem. **135**, 485 (1983).

[14a] H. Susi and D. M. Byler, Biochem. Biophys. Res. Commun. **115**, 391 (1983).

[14b] J. M. Purcell and H. Susi, J. Biochem. Biophys. Methods **9**, 193 (1984).

[14c] W. J. Yang, P. R. Griffiths, D. M. Byler, and H. Susi, Appl. Spectrosc. **39**, 382 (1985).

[14d] H. Susi, D. M. Byler, and J. M. Purcell, J. Biochem. Biophys. Methods **11**, 235 (1985).

[14e] D. M. Byler and H. Susi, Proc. SPIE Int. Soc. Opt. Eng. **553**, 289 (1985).

[14f] D. M. Byler and H. Susi, Biopolymers **25**, 469 (1986).

[14g] D. M. Byler, J. N. Brouillette, and H. Susi, Spectroscopy **1**, 29 (1986).

[15] Yu. N. Chirgadze, O. V. Federov, and N. P. Trushina, Biopolymers **14**, 679 (1975).

tool for conformational studies of proteins, although computerized techniques did make some semiquantitative estimates possible.[13,15,16]

The use of Fourier transform infrared spectroscopy (FTIR) has led to major improvements in this regard. In principle, FTIR provides several advantages over conventional dispersive techniques[17a-19]: higher (1) resolution, (2) sensitivity, (3) signal-to-noise ratio (S/N), and (4) frequency accuracy. Any one of the first three advantages can be emphasized at the expense of the other two. For protein structure studies, high sensitivity makes it possible to acquire usable infrared spectra of aqueous solutions[20a-e]; such spectra are always notoriously difficult to obtain. The improved S/N ratio facilitates resolution enhancement of observed protein spectra through the application of (1) second derivative[14a-b,21] and (2) Fourier self-deconvolution[12a,14b-g] techniques. The latter is of particular importance because it offers the possibility of obtaining quantitative information on the conformation of proteins from infrared spectra.[14d-g] Figure 1 shows (1) the original, (2) the deconvolved, and (3) the second derivative spectra of bovine α-chymotrypsin in D_2O solution from 1250 to 1800 cm^{-1}.[14g] The strong spectroscopically unresolvable amide I' band, centering around 1640 cm^{-1} in the original spectrum, is resolved into seven components by both deconvolution and second derivative techniques. This newly discovered fine structure reflects different conformational entities. The sections which follow describe the basic principles, techniques, and applications of these methods in more detail.

Basic Theory

Fourier transform infrared spectrometers are fundamentally different in construction and operation than conventional dispersive instru-

[16] M. Rüegg, V. Metzger, and H. Susi, *Biopolymers* **14**, 1465 (1975).

[17a] P. R. Griffiths, in "Analytical Applications of FT-IR to Molecular and Biological Systems" (J. R. Durig, ed.), p. 11. Reidel, Dordrecht, 1980.

[17b] P. R. Griffiths, *Science* **222**, 297 (1983).

[18] J. E. Bertie, in "Vibrational Spectra and Structure" (J. R. Durig, ed.), Vol. 14, p. 221. Elsevier, New York, 1985.

[19] J. L. Koenig, *Adv. Polym. Sci.* **54**, 87 (1984).

[20a] L. D'Esposito and J. L. Koenig, in "FTIR Spectroscopy" (J. F. Ferraro and L. J. Basile, ed.), Vol. 1, p. 61. Academic Press, New York, 1978.

[20b] J. L. Koenig and D. L. Tabb, in "Analytical Applications of FT-IR to Molecular and Biological Systems" (J. R. Durig, ed.), p. 241. Reidel, Dordrecht, 1980.

[20c] R. M. Gendreau, S. Winters, R. I. Leininger, D. Fink, C. R. Hassler, and R. J. Jakobsen, *Appl. Spectrosc.* **35**, 353 (1981).

[20d] S. Winters, R. R. Gendreau, R. I. Leininger, and R. J. Jakobsen, *Appl. Spectrosc.* **36**, 404 (1982).

[20e] M. Therrein, M. Lafleur, and M. Pézolet, *Proc. SPIE Int. Soc. Opt. Eng.* **553**, 173 (1985).

[21] D. C. Lee, D. A. Elliott, S. A. Baldwin, and D. Chapman, *Biochem. Soc. Trans.* **13**, 684 (1985).

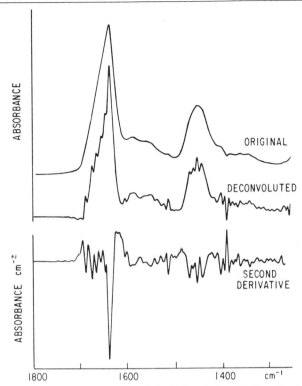

FIG. 1. FTIR spectrum of α-chymotrypsin in D_2O solution. Pathlength, 0.075 mm. Constants used for deconvolved spectrum: $\sigma = 6.5$ cm^{-1}, $K = 2.4$. Second derivative of original spectrum, obtained by Eq. (1) (Byler et al.[14g]).

ments.[17a–19] In the latter, a grating or prism disperses a collimated beam of infrared light onto a slit which effectively blocks all but a narrow range of frequencies from reaching the detector. By continuously changing the angle of the grating with respect to the incident light beam, a complete spectrum can be scanned, one spectral resolution element at a time. The FTIR instrument, by contrast, is nondispersive and makes use of an interferometer to encode data from the whole spectral range simultaneously. In general, the interferometer is some variation of the original design by Michelson.[22] (A few commercial instruments now employ a refractively scanned interferometer,[17b] which differs markedly from the Michelson interferometer.)

Figure 2 depicts the principal features of a typical Michelson interferometer. Set at right angles to one another is a pair of plane mirrors, F and

[22] A. A. Michelson, Philos. Mag. **31,** 256 (1891); **34,** 28 (1892).

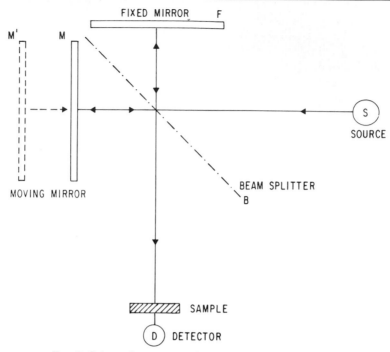

FIXED MIRROR F

M'   M

MOVING MIRROR

SOURCE S

BEAM SPLITTER B

SAMPLE

D DETECTOR

FIG. 2. Schematic representation of a Michelson interferometer.

M; mounted between is a semireflecting film, or beamsplitter B, with its plane at a 45° angle to the mirror faces. As a collimated beam of light from the source S impinges on the beamsplitter B, half is reflected to mirror F and half is transmitted to mirror M. After reflection at the mirrors, the two beams reconverge at the beamsplitter B, where again each is 50% reflected and 50% transmitted. For simplicity, first consider a monochromatic light source of wavelength λ. Twice the difference in the distances from the beamsplitter B to each of the two mirrors is designated as x, the optical retardation or optical path difference. If x is zero or an integral multiple of λ, the two beams will recombine at B in-phase. Due to constructive interference, the signal at the detector will be of maximum intensity. On the other hand, if x has any other value, the two beams of light will be partially out-of-phase, resulting in destructive interference and decreased detector signal. When $x = (n + 1/2)\lambda$ ($n = 0, 1, 2, \ldots$), the signal is zero. If mirror F is fixed in position while mirror M moves at a constant velocity v through some distance r, the signal observed at the detector will be a cosine wave whose frequency is represented by $f = v/\lambda = vk$. (Here k is the wavenumber frequency of the incident radiation.)

The amplitude or intensity of this signal as a function of x is $I(x)$ and is called an interferogram. $I(x)$ is proportional to $\cos(2\pi kx)$.

In the case of polychromatic light, all frequencies will be in-phase only at zero path difference ($x = 0$). At all other values of x, varying degrees of destructive interference will occur. Now the interferogram resulting from one sweep of the moving mirror M will be proportional to the sum of the cosine waves, $\Sigma_i A_i \cos(2\pi k_i x)$. A_i is the maximum amplitude of the cosine for each incident frequency, k_i. (For a frequency continuum, i becomes infinite and the sum is replaced by an integral.) This interferogram $I(x)$ now has maximum amplitude at $x = 0$. If certain frequencies of the incident radiation are absorbed by a sample, the interferogram changes because the amplitudes of the cosine waves of the absorbed frequencies decrease. Even when such changes are readily apparent upon visual inspection of the interferogram they are difficult to interpret. The interferometric data from the "time domain" are therefore Fourier transformed into the "frequency domain" to give an uncorrected, "single-beam" absorption spectrum of the sample, $E(k)$. Dividing $E(k)$ by the spectrum of the incident beam with no sample present, $E(k)_0$, one obtains the spectrum of the sample in percentage transmittance. This is commonly called a "ratioed" spectrum.

FTIR spectrometry offers several theoretical advantages over dispersive infrared measurements[17a–19]:

1. Because the throughput of the incident light is not slit limited, FTIR spectrometers are inherently more sensitive than dispersive instruments (*Jacquinot's advantage*). For a given source more energy will reach the detector, resulting in a greater signal-to-noise (S/N) ratio. (This advantage is somewhat offset because dispersive instruments can use slow-response thermocouple detectors which have higher detectivities than the most sensitive, low-temperature FTIR detectors for the mid-infrared region.)

2. FTIR spectrometers simultaneously encode all spectral frequencies to give a complete spectrum in a matter of seconds (*Fellgett's* or multiplex *advantage*). Because the S/N ratio is proportional to the square root of the number of scans, signal averaging the data from a large number of scans can significantly increase the S/N ratio within a relatively short time. In addition, rapidly occurring changes may be monitored rather easily.

3. Modern, high-speed FTIR spectrometers contain a laser reference interferometer to facilitate digitization of data and for frequency calibration. Because the laser frequency is known to at least seven significant figures, its interferogram frequency is used to calibrate the frequencies of the digitized infrared data points to better than 0.01 cm^{-1} (*Conne's advan-*

tage). The peak positions of observed infrared bands of condensed-phase samples generally cannot be measured with such accuracy, particularly if the bands overlap or are broad. Use of Fourier self-deconvolution and of second derivatives, however, can increase substantially the accuracy of the measured peak positions.

4. Finally, because such good S/N ratios and high wavenumber precision are possible, FTIR spectra can be manipulated easily and efficiently by a computer. This facilitates such mathematical data treatments as precise measurement of peak maxima,[23] interactive spectral subtraction,[20e] calculation of second derivatives,[14a–b,21] Fourier self-deconvolution,[12a,14b–g] and curve fitting.[14d–g]

One additional distinction between FTIR and dispersive spectroscopy is worth noting. For the former, resolution remains constant across the spectral range, but the S/N ratio does not. In dispersive spectrometry, just the reverse is generally true: constant S/N can be achieved, but only at the expense of varying resolution.

Several recent reviews give more detailed treatment of FTIR theory and instrumentation, as well as additional background references.[17a–19]

Second Derivative Spectra

Because FTIR spectra are digitally encoded with one data point n every $\Delta k/(2^m) = \Delta W$ frequency units, where Δk is the nominal instrumental resolution selected (in cm^{-1}) and m is the number of times the interferogram is zero-filled prior to Fourier transformation, the second derivative of the spectrum may be calculated by a modification of a straightforward analytical method described by Butler and Hopkins.[24a] In particular, at data point n the value of the second derivative A_n'' in absorbance units/wavenumber^2 is

$$A_n'' = (A_{n+1} - 2A_n + A_{n-1})/(\Delta W)^2 \qquad (1)$$

where A_n is the absorbance at data point n of the original spectrum. (Note that this function gives the second derivative spectrum without any smoothing.) Smoothing may then be done as needed. Prior to plotting and frequency measurement, we occasionally employ for this purpose a seven- or a nine-point Savitsky–Golay function.[24b]

The intrinsic shape of a single infrared absorption line of an isolated

[23] D. G. Cameron, J. K. Kauppinen, D. J. Moffatt, and H. H. Mantsch, *Appl. Spectrosc.* **36**, 245 (1982).

[24a] W. L. Butler and D. W. Hopkins, *Photochem. Photobiol.* **12**, 439, 452 (1970).

[24b] A. Savitsky and M. J. E. Golay, *Anal. Chem.* **36**, 1627 (1964).

molecule may be approximated by a Lorentzian function,[25a-28] i.e.,

$$A = \frac{\sigma}{\pi(\sigma^2 + k'^2)} \tag{2a}$$

$$= \frac{1}{\sigma\pi(1 + Bk'^2)} \tag{2b}$$

where A is the absorbance, 2σ is the width at half height, k' $(= k - k_0)$ is the frequency referred to the band center at k_0, and $B = 1/\sigma^2$. Substituting Eq. (2b) into Eq. (1) gives the second derivative of this function:[25a-26b]

$$A'' = \frac{-2B(1 - 3Bk'^2)}{\sigma\pi(1 + Bk'^2)^3} \tag{3}$$

The peak frequency for the second derivative is identical with the frequency of the original band center, k_0. The half width of the second derivative, σ^{II}, is related to the half width of the original line[26a-27] by

$$\sigma^{II} = \sigma/2.7 \tag{4}$$

and the peak intensity of the second derivative, A_0'', to that of the original intensity A_0 by

$$A_0'' = -2A_0/\sigma_2 \tag{5}$$

[Eq. (5) is derived by dividing Eq. (3) by Eq. (2) and noting that $k' = 0$ at the band center.] The peak height of the second derivative is thus proportional to the original peak height and inversely proportional to the square of the original half width. Thus weak, but sharp lines, such as arise from water vapor, noise, or interference fringes are greatly accentuated relative to the much broader lines of the condensed phase sample. For real spectra with overlapping bands which deviate from Lorentzian shape, the relationships are more complex. Nonetheless, the above formulas do provide a reasonable approximation for interpreting second derivative spectra.

The original method of Butler and Hopkins for analytically determining second derivative spectra[23,25a-26b] was to take the first derivative of the

[25a] W. F. Maddams and W. L. Mead, *Spectrochim. Acta* **38A**, 437 (1982).

[25b] W. F. Maddams and M. J. Southon, *Spectrochim. Acta* **38A**, 459 (1982).

[26a] J. K. Kauppinen, D. J. Moffatt, H. H. Mantsch, and D. G. Cameron, *Appl. Spectrosc.* **35**, 271 (1981).

[26b] J. K. Kauppinen, D. J. Moffatt, H. H. Mantsch, and D. G. Cameron, *Anal. Chem.* **53**, 145 (1981).

[27] J. K. Kauppinen, *Spectrom. Tech.* **3**, 199 (1983).

[28] H. H. Mantsch, H. L. Casal, and R. N. Jones, *Adv. Spectrosc.* **13**, in press (1986).

original spectrum twice, using the function

$$A'_n = [A_{n+1} - A_{n-1}]/\Delta W \tag{6}$$

This method, however, results in a second derivative function which differs somewhat from Eq. (1):

$$A''_n = (A_{n+2} - 2A_n + A_{n-2})/(\Delta W)^2 \tag{7}$$

In this case, the second derivative spectrum is smoothed compared to that calculated by Eq. (1). Smoothing occurs because now the calculation of A''_n spans five data points including A_n whereas in Eq. (1) only three data points are involved.

Savitsky and Golay have described another approach which combines smoothing and derivatization.[24b] Here the spectral curve is fitted to a polynomial by a least-squares calculation. With the polynomial coefficients in hand, derivatization is straightforward. One additional method for obtaining derivative spectra involves the use of Fourier transformation,[26a–27,29a–b] again with some simultaneous smoothing.

Fourier Self-Deconvolution

The Fourier transformation of a Lorentzian band, $E(k)$ (Fig. 3A), of half-width at half-height, σ (cm^{-1}), results in a time-domain function, or interferogram, $I(x)$, which is an exponentially damped cosine wave (Fig. 3A).[26a–29b] The envelope of this function is a decay curve whose exponent is directly proportional to σ. If $I(x)$ is multiplied by the product of an apodization function $D(x)$ and an exponentially increasing weighting function, the exponent of the decay curve and thus the rate of decay of $I(x)$ will be decreased. Therefore, when the reverse Fourier transform is calculated, the width of the resulting spectral band, $E_L(k)$, will be less than that of the original undeconvolved band, $E(k)$ (Figs. 3B–D).

If the positive exponent of the weighting function has the same absolute value as the negative exponent in the decay function, the resulting interferogram $I_L(x)$ will be a cosine wave truncated at $x = L$. The value of L for $I_L(x)$ is determined by the point L where the apodization function $D(x)$ goes to zero. Instrument resolution, Δk, at which the original spectrum was measured places an approximate upper limit on L of $1/\Delta k$.[27,29a–b] The reverse Fourier transform of $I_L(x)$ gives a sinc function (sinc $x = $ [sin x]/x) (Fig. 3B). The characteristic side-lobes found on either side of the central maximum of this function clearly show that resolution enhancement of bands beyond $L = 1/\Delta k$ is not possible, even for perfectly

[29a] W.-J. Yang and P. R. Griffiths, *Proc. SPIE Int. Soc. Opt. Eng.* **289**, 263 (1981).
[29b] W.-J. Yang and P. R. Griffiths, *Compt. Enhanced Spectrosc.* **1**, 157 (1983).

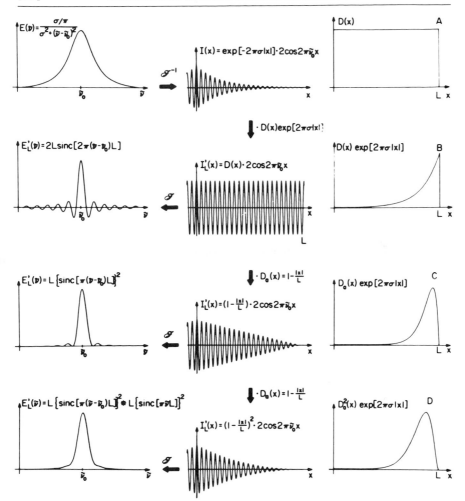

FIG. 3. Illustration of the various steps of the Fourier self-deconvolution procedure starting with a Lorentzian line at $\bar{\nu}_0$ and using different apodization functions, $D_g(x)$: in (B), $D_g(x)$ equals $D(x)$, the "boxcar function" [see Eq. (8) of text]; in (C), $D_g(x)$ equals $D_a(x)$, the "triangular function"; in (D), $D_g(x) = D_a^2(x)$. Except for (A), the middle column shows the interferograms, $I_L'(x)$, resulting from the application of the functions, $D_g(x) \exp(2\pi\bar{\nu}_0|x|)$ (right-hand column), to the interferogram $I(x)$. These functions are all scaled to the same height. In the left-hand column, except for (A), the lineshapes $E_L'(\bar{\nu})$, resulting from the self-deconvolution are shown. Note: throughout the text, k is used for wavenumber instead of $\bar{\nu}$ (Kauppinen *et al.*[26a]).

noise-free data. Indeed, to avoid the appearance of side-lobes in the deconvolved spectrum, alternative apodization functions to the frequently used boxcar function,

$$D(x) = \begin{cases} 1, \text{ for } x \le L \\ 0, \text{ for } x > L \end{cases} \tag{8}$$

are commonly used. These functions reduce the size of the side lobes, but at the cost of decreased resolution enhancement. Yang and Griffiths suggest as a rule-of-thumb that, at best, the full width at half-height of bands after deconvolution cannot be reduced to less than about $1.5(\Delta k)$.[29a-b]

The necessity of truncating the interferogram after Fourier transformation at $x = L$ tends to introduce new, nonrandom noise with a periodicity of about $1/L$.[26a-29b] This periodic noise will appear across the whole spectrum, in contrast to the side-lobes mentioned above which are observed only along either side of an over-deconvolved band. With care in the choice of deconvolution parameters, such noise will have minimal intensity. Nonetheless, because it is generally present to some degree in all deconvolved spectra, one must firmly resist the temptation to attempt to obtain increased resolution enhancement of deconvolved data by taking the second derivative of the deconvolved spectrum. Such a procedure will intensify the otherwise only faintly perceptible periodic noise with a concomitant loss in S/N. This, in turn, will make objective discrimination between real, but weak, spectral components and the periodic artifacts difficult, if not impossible.

FTIR Spectroscopy of Aqueous Solutions

Strong absorption by water throughout much of the mid-infrared spectral region has always made the acquisition of usable infrared data from aqueous solutions difficult. Studies of aqueous protein solutions are further complicated because the bending vibration of water absorbs strongly near 1640 cm^{-1}, right in the midst of the region (1630–1690 cm^{-1}) where the conformation-sensitive amide I protein vibrations occur. With non-computerized dispersive instruments, extremely careful differential work was necessary to obtain any useful results.[8b] The high frequency accuracy and reproducibility, and high S/N ratios, obtainable with FTIR instrumentation now make such measurements somewhat less difficult experimentally.[12a,20a-21] Instead of using differential techniques with very carefully matched cells, digitized FTIR spectra of the solution and of the solvent are obtained separately, and the latter subtracted from the former. Neither subtraction nor differential procedures are as straightforward as they first appear. Whenever a solute is present, changes occur in the frequency, width, and height of the solvent band, even for weakly interact-

TABLE I

CHARACTERISTIC AMIDE I AND AMIDE II FREQUENCIES
$(cm^{-1})^a$ FOR PROTEINS IN H_2O SOLUTION AS MEASURED BY
FTIR SPECTROSCOPY

Protein	Predominant conformation	Amide I	Amide II
Hemoglobin	α-Helix[b]	1656	1547
Ribonuclease A	β-Strands[b]	$\begin{cases} 1656^c \\ 1646^c \end{cases}$	1548
β-Lactoglobulin A	β-Strands[c]	1632	1551
α-Casein	Unordered[d]	1655	1551

[a] Koenig and Tabb.[20b]
[b] Levitt and Greer.[31]
[c] Sawyer et al.[33]
[d] H. Susi, S. N. Timasheff, and L. Stevens, *J. Biol. Chem.* **242**, 5460 (1967) and S. N. Timasheff, H. Susi, and L. Stevens, *J. Biol. Chem.* **242**, 5467 (1967).
[e] Frequencies too high for a typical β-protein.

ing, low-polarity solvent systems. For water and other polar, hydrogen-bonding solvents, such effects become much more pronounced. When strong, broad bands of the solute and the solvent overlap, as in the case of the protein amide I band and the bending mode of water, the residual features of the solvent spectrum which remain after subtraction often distort the solute bands so that accurate measurement of their frequency and intensity is no longer possible. Because absorption by water is significantly lower in the amide II region (1530–1560 cm^{-1}), only minimal difficulty is encountered with solvent subtraction for these protein bands. Side-chain bands, associated with CH_2, CH_3, COO^-, NH_2, and SH groups can also be easily observed in H_2O solution; the latter,[30] though very weak, falls within the relatively clear spectral region around 2560 cm^{-1}.

Koenig and Tabb[20b] have tabulated a number of observed infrared frequencies for selected proteins in H_2O solution. Table I lists some of these values. Although the amide II frequencies can be observed with considerable accuracy (because of minimal interference by water), all appear within 4 cm^{-1} of one another and thus provide no reliable conformational correlations. Frequency shifts of the amide II band accompanying dissolution, as reported by these authors,[20b] are such that these bands become much less useful as a conformational indicator in H_2O solution than they are in the solid state.[3,4] The amide I frequencies, on the other hand, are uncertain because of the difficulties inherent in compensating

[30] J. O. Alben and G. H. Bare, *J. Biol. Chem.* **255**, 3892 (1980).

for the strong water band about 1640 cm^{-1}. They agree well with earlier work in the case of β-lactoglobulin and α-casein,[8b] but the hemoglobin frequency (1656 cm^{-1}) is 4 cm^{-1} higher than the old value for myoglobin,[8b] although both proteins have an almost identical secondary structure for their helical segments.[31] The frequencies given for ribonuclease seem altogether too high for a protein with a high β-structure content[31] (see below).

Both the older differential technique and the computerized FTIR subtraction procedure are prone to serious errors for aqueous solutions in the region around 1640 cm^{-1}. Thus, we conclude that even with the improved accuracy and sensitivity of FTIR spectroscopy, conformational studies of proteins are best carried out in D$_2$O solution,[14] where no strong solvent bands appear close to the amide I′ frequency region.

Application of Second Derivative FTIR Spectra

The intrinsic shape of an ideal infrared absorption line is approximated by a Lorentzian function, as described in the theoretical section. In the second derivative spectrum of such a line the peak frequency is identical with the original peak frequency and the half width is reduced by a factor of 2.7.[26a,27] The peak height is proportional to the original peak height (with opposite sign), and inversely proportional to the square of the original half-width [see Eq. (5)]. There are side-lobes on either side of the principal peak.[25,26] The net result is a very much sharper line accompanied by side-lobes. For real spectra the relationships are more complex, but the above simplified relationships provide a good starting point for interpreting the second derivative spectra of proteins.[14a-b,21] The main value of second derivative spectra lies in the ease with which the peak frequencies of unresolved components can be identified. For side-chain bands a considerable sharpening is also observed, which makes identification and assignments much easier. Quantitative information, however, is difficult to obtain from second derivative spectroscopy because of the complex patterns created by overlapping peaks and side-lobes.

The bottom curve in Fig. 1 depicts the second derivative FTIR spectrum of α-chymotrypsin.[14g] The original unresolved strong amide I band around 1640 cm^{-1} now exhibits six sharp peaks. The increased resolution of the side-chain bands below 1600 cm^{-1} is also clearly evident. Figure 4 shows the second derivative spectra of two other proteins[14a] which further illustrate the power of the method. The original FTIR spectra of these proteins are as nondescript as the top curve in Fig. 1.

In the second derivative spectrum of bovine hemoglobin (Fig. 4A)[14a]

[31] M. Levitt and J. Greer, *J. Mol. Biol.* **114**, 181 (1977).

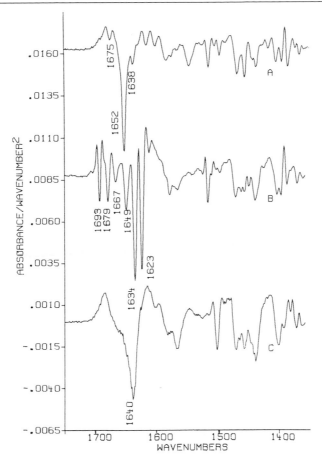

FIG. 4. Smoothed second derivative FTIR spectra in D_2O: (A) hemoglobin, pD 7; (B) native β-lactoglobulin A, pD 7; (C) denatured β-lactoglobulin A, pD 13; concentration 5% w/v (Susi and Byler[14a]).

there is a single strong peak at 1652 cm^{-1} in the amide I region, obviously representing the α-helix (Table II). Hemoglobin is about 80% α-helical and contains no β-structure.[31] The weak peaks at ~1638 and 1675 cm^{-1} can therefore be assigned to short extended chains connecting the helical cylinders.[14f]

Figure 4B gives the second derivative spectrum of native bovine β-lactoglobulin A. Circular dichroism[32] and infrared studies[14f] suggest an α-helix content of 10–15% and a β structure content of about 50%. A recently published note on the crystal structure of this protein confirms

---

[32] S. N. Timasheff, R. Townend, and L. Mescanti, *J. Biol. Chem.* **241,** 1863 (1966).

TABLE II
CHARACTERISTIC FREQUENCIES (cm^{-1}) AND
ASSIGNMENTS OF AMIDE I COMPONENTS[a] FOR 19
GLOBULAR PROTEINS

Mean frequency	Assignment
1624 ± 4[b]	Extended chain
1631 ± 3	Extended chain
1637 ± 3	Extended chain
1645 ± 4	Unordered
1653 ± 4	Helix
1663 ± 4	Turns
1671 ± 3	Turns
1675 ± 5	Extended chain
1683 ± 2	Turns
1689 ± 2	Turns
1694 ± 2	Turns

[a] Byler and Susi[14f]; Byler et al.[14g]
[b] Maximum range of the observed frequency for each amide I band component.

these results.[33] The strong 1634 cm^{-1} peak is evidently associated with the β substructure, and the bands between 1667 and 1693 cm^{-1} with the second β structure band and with turns (Table II), as in bovine ribonuclease A.[14a] It is interesting to note that the 1623 cm^{-1} peak is absent in ribonuclease A but present in concanavalin A[14b] which also has a very high β content.[31] Perhaps this peak represents a variation of the β structure present in Jack Bean concanavalin A and β-lactoglobulin A, but not in ribonuclease A. Further study is evidently required. Figure 4C gives the second derivative spectrum of denatured β-lactoglobulin A, which is assumed to be in a "random form."[14a-b] All sharp peaks associated with helix, extended chain, and turns have disappeared.

Other spectral features are not as easily interpreted but are not related to the secondary structure. The weak peak around 1600 cm^{-1} is too low for an amide I' component[1,2]; it is probably caused by aromatic side chain groups.[14a] Other side chain groups of histidine, tryptophan, and phenylalanine, as well as the asymmetric stretching mode of side chain COO$^-$ groups also absorb in the 1550 to 1610 cm^{-1} region.[14a] The very stable, sharp 1515 cm^{-1} band is associated with tyrosine residues.[16] From ~1430 to 1480 cm^{-1} we have overlapping bands caused by (1) CH$_2$ and CH$_3$

---

[33] L. Sawyer, M. Z. Papiz, A. C. T. North, and E. E. Eliopoulos, *Biochem. Soc. Trans.* **13**, 265 (1985).

bending modes of side chains, (2) the amide II' mode, essentially ND bending, and (3) bending modes of traces of HOD.[14a,34] Between 1360–1380 cm^{-1} the symmetric CH_3 bending vibrations of side chains are expected.[34] More detailed assignments must wait a thorough study of more proteins. It is evident, nevertheless, that second derivative spectra furnish new information about the side chains as well as the secondary structure of proteins.

Application of Fourier Self-Deconvolution and Curve Fitting

Figure 5 shows the self deconvolved FTIR spectra of ribonuclease A and α-chymotrypsin in the amide I' region, as obtained in D_2O solution, along with the original spectra. The deconvolved spectra are resolved into Gaussian components by means of a computer program which uses Gauss–Newton iteration. Deconvolution, as carried out by the described method,[14f,26a–28] requires two constants as computer input: the estimated half-width at half-height of the unresolved bands, σ, and the resolution "efficiency" factor, K, which reflects the narrowing of the unresolved bands, i.e., the "effective increase in resolution."[27] The choice of these constants is of utmost importance for obtaining meaningful results.[14b–g,26a–29b] The constant σ should, ideally, be as close to the true half-width of the unresolved components as possible. When unresolved components with different half-widths must be deconvolved simultaneously, a compromise becomes necessary. This is frequently the case for protein spectra. A choice of σ = 6.5 has been found to be satisfactory,[14f] although lower values might be necessary for quantitative work. A maximum value for K can be estimated by the approximate relationship: $K_{max} \approx \log(S/N)$. A K value of 2.4 would thus require an S/N ratio of at least 260. In practice, K values higher than 3 are rarely used.[29a] It must be strongly emphasized that an improper selection of σ and/or K can lead to serious errors and artifacts. Deconvolution increases with increasing σ, but too large a value of σ leads to "overdeconvolution," resulting in sidelobes and distorted spectra. Too high a K value, in turn, leads to excessive nonrandom noise with a period of about $1/L$.[26a,27] This, in turn, can distort the shape of weak bands and also introduce spurious peaks. Self-deconvolution, like second derivative spectroscopy, enhances noise, interference fringes, impurity bands, and water vapor bands, as well as the true bands of the sample. A high S/N ratio, high nominal instrument resolution, and well purged instruments are therefore essential for good results.

---

[34] L. J. Bellamy, "The Infra-red Spectra of Complex Molecules," Vol. 1, 3rd Ed., pp. 6, 8, 21–27, 198–200. Chapman & Hall, London, 1975.

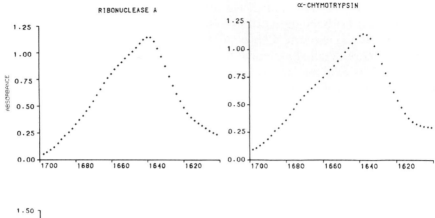

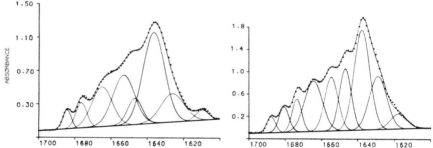

Fig. 5. Amide I' bands of ribonuclease A and α-chymotrypsin. Upper curve: digitized original FTIR spectrum (crosses), 5% w/v in D₂O, pathlength 0.075 mm. Lower curve: deconvolved spectrum (crosses) and individual Gaussian components (solid lines) (deconvolution constants: $\sigma = 6.5$ cm⁻¹, $K = 2.4$). Component bands were calculated by least-squares fit using a Gauss–Newton iteration. Solid overall curve is obtained by summation of the components. Root mean square (RMS) error: 0.01 absorbance units (Byler *et al.*[14g]).

The interpretation of deconvolved spectra[14f] (Fig. 5) is quite analogous to the interpretation of the second derivative spectra discussed in the previous section. Ribonuclease has an approximate α-helix content of 22% and a β-structure content of about 46%.[31] The 1636 cm⁻¹ band can be associated with β segments and the much weaker band near 1660 cm⁻¹ with helical segments (Table II).[8b,14f] The weak peaks between 1655 and 1700 cm⁻¹ evidently correspond to the second β structure band and to turns. Comparison with the spectrum of bovine ribonuclease S,[14f] which has a very high β structure content and relatively few turns,[31] suggests that the 1676 cm⁻¹ band (Fig. 5) can be associated with the β structure while the remaining weak bands are due to turns.

In the spectrum of α-chymotrypsin the bands close to 1627, 1637, and 1674 cm⁻¹ can be associated with β-strands, the band near 1653 cm⁻¹ with

TABLE III
PERCENTAGE HELIX AND EXTENDED CHAIN BY FTIR AND
X RAY FOR SIX TYPICAL PROTEINS

Protein	Helix (%)		Extended chain (%)	
	FTIR[a]	X ray[b]	FTIR[a]	X ray[b]
Carboxypeptidase	40	39	33	30
α-Chymotrypsin	12	10	50	49
Concanavalin A	4	2	60	60
Lysozyme	41	45	21	19
Papain	27	29	32	29
Ribonuclease A	21	22	50	46

[a] Byler and Susi.[14f]
[b] Levitt and Greer.[31]

the α-helix and the bands close to 1687, 1681, and 1665 cm^{-1} with turns (Table II).[14f–g]

Figure 5 demonstrates that deconvolved amide I' spectra can be resolved into components with easily measurable areas. Herein lie the seeds for quantitative conformational analysis by FTIR spectroscopy.

Based on recent systematic studies of the frequency distribution of 138 observed amide I band components for 19 globular proteins with widely varying proportions of α-helix and β structures,[14f–g] a recurring pattern of bands at a limited number of distinct frequencies becomes apparent. As shown in Table II by the mean frequency value and narrow absolute range of each, these 11 characteristic amide I components show relatively little overlap with one another. The table also presents proposed assignments for each characteristic frequency to a particular type of secondary structure.[14f–g]

Table III gives the percentage of helix content and extended chain content for six typical proteins as calculated from FTIR data.[14f–g] Corresponding values for the secondary structure of these proteins taken from Levitt and Greer's comprehensive examination of protein X-ray data[31] are given for comparison. Their study represents one of the most thorough and consistent interpretations of protein crystallographic data in terms of conformation of which we are aware, and covers a total of 62 proteins. It is not possible to give similar comparisons for bends and turns, because these protein substructures do not have the same type of periodic, repeating structures found in helices and sheets. In addition, bends and turns usually consist of only a small number (four or less) of adjacent residues. For these reasons, whether a specific amino acid residue in a given protein belongs to a turn rather than the end of a helix or β-strand continues to elude precise definition.

As seen in Table III, the agreement between the FTIR results and values derived from X-ray crystallography is quite good. In Table III FTIR results from solution data[14f-g] are compared with values from solid state X-ray studies.[31] At present, this is unavoidable because no uniform, comprehensive set of secondary structure data for proteins in aqueous solution has appeared in the literature.

One cannot emphasize too strongly that values for "percentage helix" or "percentage extended structure" are always somewhat subjective even when based on accurate bond lengths and angles. Ambiguity arises from the uncertainty in the choice of the exact point along the peptide chain where one segment of secondary structure begins and another ends. This choice depends not only on the manner in which an ideal helix or sheet has been defined by the various investigators[14f-g,31] but also on just how regular any of these regions of protein substructure really are. With these uncertainties in mind, the agreement between the reported X-ray values[31] and the FTIR values[14f-g] is particularly encouraging.

Despite such caveats and limitations, analysis of protein secondary structure in aqueous solution by judicious curve fitting of deconvolved FTIR spectra offers great promise for future studies of protein secondary structure including conformational changes which result when the biomolecule is subjected to changes in its environment.

Solvent Denaturation Studies

FTIR spectra, resolution enhanced by second derivative techniques and Fourier self-deconvolution, permit quite detailed conformational studies related to solvent denaturation.[14b,35] Figure 6 shows the deconvolved spectrum, the original FTIR spectrum, and the second derivative spectrum of native bovine chymotrypsinogen A in D_2O solution, and the corresponding spectra as obtained in 60% (v/v) O-deuterated methanol (MeOD) in D_2O. The spectrum of the native protein (Fig. 7A), which contains about 45% β structure and 11% α-helix[31] is somewhat similar to the ribonuclease spectrum shown in Fig. 5. β structure bands are observed at 1636 (strong) and 1674 (weak) cm^{-1}. The 1647 cm^{-1} band is probably associated with unordered sections and the remaining bands with turns (Table II).[14a-b,f-g] The denatured protein exhibits bands at 1617 cm^{-1} (strong) and 1686 cm^{-1} (weak). Quite similar results are obtained in isopropanol-d solution.[14b] The bands at 1617 and 1686 cm^{-1} correspond to

[35] F. M. Wasacz, *Proc. SPIE Int. Soc. Opt. Eng.* **553,** 183 (1985).

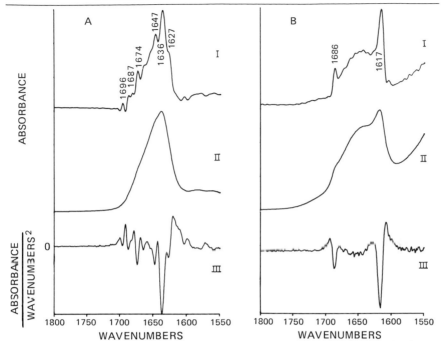

FIG. 6. FTIR spectra of chymotrypsinogen A in various solvents: (I) deconvolved spectrum (deconvolution constants: $\sigma = 6.5$ cm^{-1}, $K = 2.4$); (II) original spectrum; (III) second derivative spectrum. (A) native protein in D$_2$O, 4.7% w/v, pD 7; (B) in 60% v/v MeOD solution in D$_2$O, 0.47% w/v protein at pD 3 (Purcell and Susi[14b]).

no conformation previously observed in the infrared spectra of globular proteins. They are, nevertheless, close to usually observed β structure bands and probably indicate the presence of extended strands of some kind. The ill-defined absorption between the two sharp bands could be associated with unordered hydrated segments and/or irregular helices.[14b]

The usefulness of FTIR spectroscopy as a tool for denaturation studies is further illustrated in Fig. 7, which shows the spectra of the native form and three different denatured forms of the protein β-lactoglobulin A. Figure 7A, the spectrum of the native material, is typical for a protein with a high β structure content; the assignment was discussed above in conjunction with the second derivative spectrum shown in Fig. 4B. Figure 7B shows the spectrum of alkaline-denatured β-lactoglobulin A at pD 13. A single strong, broad band is observed around 1640 cm^{-1}, suggesting hydrated unordered chains.[14a,f] Figure 7C shows the spectrum obtained in acidic MeOD solution. Here one observes a strong band at 1647 cm^{-1} and two weaker ones at 1618 and 1687 cm^{-1}. The latter suggest an extended chain conformation similar to the one observed for chymotrypsinogen A

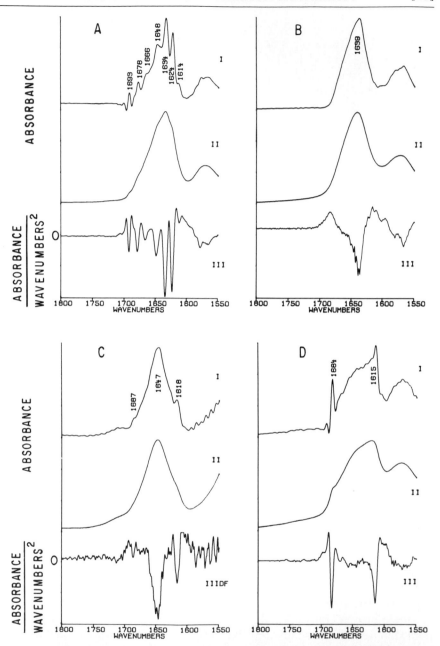

FIG. 7. FTIR spectra of β-lactoglobulin A in various solvents. (I), (II), (III) as in Fig. 6. (A) Native protein in D_2O, 5% w/v, pD 7; (B) in D_2O, 5% w/v, denatured at pD 13; (C) 60% v/v MeOD in D_2O, 0.4% w/v protein, pD 4 (inordinate noise is apparent above 1600 cm^{-1}); (D) 40% v/v isopropanol-d in D_2O, 0.35% w/v protein, pD 7 (Purcell and Susi[14b]).

under similar conditions (Fig. 1B). It is interesting to note that acidic MeOD has a different effect on the two proteins and that this difference is clearly reflected in the spectra obtained. (The sharp peaks in Fig. 7C observed below 1600 cm^{-1} are noise due to the weak signal-to-noise ratio of this rather dilute protein solution.) In isopropanol-d solution, as shown in Fig. 7D, β-lactoglobulin A behaves like chymotrypsinogen A. Figure 7 thus shows how FTIR spectra can clearly distinguish between the native form and three denatured forms of the same protein. We know of no other spectroscopic technique that will accomplish this in quite as much detail.

Acknowledgments

The authors are grateful to their colleague James M. Purcell for his collaboration on this project, particularly with respect to the studies on protein denaturation. They also thank Janine N. Brouillette for her technical assistance in preparing samples and obtaining their FTIR spectra.

[14] Protein Secondary Structure Analysis Using Raman Amide I and Amide III Spectra

By ROBERT W. WILLIAMS

Introduction

This chapter describes methods for the estimation of protein secondary structure content—in terms of percentage helix, β-strand, and reverse turn—from a least-squares analysis of Raman amide I and amide III spectra. A statistical analysis of these estimates for proteins with known structures is included to establish the degree of confidence that may be placed on results for other proteins.

The amide I analysis here is a refinement of earlier work,[1-3] while the amide III analysis is new. Most of these procedures have been automated and Fortran programs for their implementation may be obtained from the author. These programs call well-documented subroutines written by Lawson and Hanson.[4]

[1] R. W. Williams, A. K. Dunker, and W. L. Peticolas, *Biophys. J.* **32,** 232 (1980).
[2] R. W. Williams and A. K. Dunker, *J. Mol. Biol.* **152,** 783 (1981).
[3] R. W. Williams, *J. Mol. Biol.* **166,** 581 (1983).
[4] C. L. Lawson and R. J. Hanson, "Solving Least Squares Problems." Prentice-Hall, New York, 1974.

The Least-Squares Problem

Both the amide I and amide III spectra are analyzed in the same way, as a linear combination of the spectra of other proteins whose structures are known:

$$Ax \cong b \tag{1}$$

A is a p by n matrix containing the normalized spectra of n reference proteins at p wavenumbers. See Tables I, VI, and IX, which show matrix A for the amide I and amide III analysis. Column matrix b contains the normalized spectrum of the protein being analyzed.

The solution, column matrix x, is used to calculate fractions of secondary structure as a weighted sum of the structures of proteins represented in matrix A:

$$f = Fx \tag{2}$$

where F is an m by n matrix containing m classes of secondary structure of the n reference proteins. See Tables II, VII, and X, which list matrix F. Column matrix f then contains m elements which are the fractions of secondary structure for the protein being analyzed.

Reference Proteins

The source and state of the proteins used here are described elsewhere[3] with the exception of triosephosphate isomerase (TPI) and adenylate kinase (ADK). The TPI was from Sigma Lot 72F-9645 and was further purified on a Sephacryl S-200 column followed by vacuum dialysis concentration. The buffer was 0.05 M phosphate, pH 7.5, 0.14 M NaCl (PBS). The ADK, from Sigma Lot 42F-9680, was dialyzed and vacuum dialysis concentrated against PBS. Both samples showed a single band on SDS gels. An earlier data set[3] included spectra of TPI and ADK crystals in 3.2 M $(NH_4)_2SO_4$. The subtraction of the very intense $(NH_4)_2SO_4$ spectrum from the amide I region was suspected to give inaccurate results. Also, these proteins contain the highest fractions of parallel β-sheet in the reference set, and the evidence that the amide I analysis could distinguish between parallel and antiparallel β-sheet depended upon these samples. The new spectra of TPI and ADK in PBS give better overall results than previous samples. However, when relatively small variations in the baseline adjustment of these spectra were made, the results for parallel β-sheet were quite variable. The value of the correlation coefficient, as in Table III, for parallel β-sheet was 0.4 in a typical experiment of this kind, which is not sufficiently significant. Therefore, parallel β-sheet is not

TABLE I

NORMALIZED AMIDE I INTENSITIES[a] FOR 15 PROTEINS AND POLY-L-LYSINE:
THE ELEMENTS OF MATRIX A IN EQ. (1)

	Protein[b]							
cm^{-1}	RNA	LDH	INS	DHE	PTI	MEL	CNA	CPA
1615	80	151	91	105	97	124	61	107
1620	96	263	113	210	122	259	98	128
1625	165	326	177	308	200	332	140	217
1630	269	427	307	447	308	473	217	346
1635	367	518	441	607	418	629	311	474
1640	468	616	590	744	502	843	403	588
1645	606	767	775	913	600	1064	506	740
1650	734	912	983	1194	706	1221	620	918
1655	859	1012	1114	1237	846	1166	763	1001
1660	983	1023	1048	1003	982	974	926	1023
1665	1085	974	893	797	1043	793	1132	1024
1670	1077	840	780	649	1046	646	1254	956
1675	973	679	736	527	982	507	1161	811
1680	805	564	707	425	819	416	889	646
1685	613	466	585	350	593	329	630	485
1690	432	324	407	273	391	238	435	343
1695	280	236	237	203	253	140	291	201
1700	151	138	122	145	144	90	175	117
1705	81	94	62	100	99	80	95	70
1710	46	84	38	78	68	59	52	41
S[c]	3	10	2	4	8	8	6	4

	Protein[b]							
cm^{-1}	ADH	LZM	CBP	CRA	THL	PLL	fd	AVN
1615	75	84	118	92	104	256	105	49
1620	126	128	211	172	120	378	182	131
1625	234	199	344	268	207	555	282	190
1630	370	322	482	422	347	783	452	253
1635	507	489	590	584	491	1075	672	341
1640	627	646	730	689	639	1303	936	427
1645	743	802	885	800	798	1425	1237	519
1650	860	988	1070	896	960	1182	1382	605
1655	948	1146	1179	947	1052	887	1200	726
1660	1008	1145	1091	982	1030	680	912	912
1665	1046	1027	890	1006	964	546	705	1116
1670	959	851	751	833	877	444	567	1117
1675	775	665	618	652	730	352	462	1019
1680	603	530	512	591	597	284	352	865
1685	460	423	353	476	450	195	270	642
1690	343	319	208	347	333	117	200	457
1695	236	211	130	232	229	83	146	322
1700	146	125	90	141	152	50	103	220
1705	82	69	53	77	91	22	72	155
1710	53	42	22	56	52	19	50	115
S[c]	4	3	8	6	4	8	4	7

[a] Normalized intensity = (baseline adjusted intensity/sum of intensities from 1625 to 1710 cm^{-1} in 5 cm^{-1} intervals) times 10^4.
[b] See Table II for abbreviations of proteins.
[c] S, One standard deviation in experimental noise.

included in the present analysis. The remaining reference spectra were from earlier work[3] and these spectra were treated as described below.

Instrumentation

Two different Raman instruments were used in this work. One is described elsewhere.[3] The instrument at USUHS includes a Coherent Innova 90 argon ion laser operating at 514.5 nm. For the solutions of ADK and TPI, 200 mW of light power was focused to a beam about 0.1 mm diameter at the sample. The spectrometer consists of a 300-mm focal length $f/0.6$ aspheric lens for light collection optics, a Spex 1403 double monochromater with 1800 groove/mm holographic gratings, and an RCA 31034 PMT cooled to $-30°$. Spectral band pass was set to about 6.0 cm^{-1}. Repetitive scanning was controlled by a Spex Datamate coupled to a Cromenco Z80 microcomputer. Data points were taken at every 1 cm^{-1} at 0.5 cm^{-1}/sec and about 50 scans were collected and averaged.

Amide I Analysis

Baseline Adjustment

Measurement of the amide I spectrum may be restricted to the region between 1500 and 1800 cm^{-1} in order to save time and increase the signal-to-noise ratio.

The spectrum of buffer is subtracted from the spectrum of the protein to satisfy two criteria: the spectrum between 1730 and 1800 cm^{-1} must be linear or free of curvature, and linear extrapolation of this region must intersect close to the baseline at 1500 to 1520 cm^{-1}.

When the protein sample is very fluorescent the spectrum of the fluorescence band may introduce considerable curvature to the baseline in the amide I region. Under these circumstances it becomes impossible to satisfy the two criteria cited above. When this happens the spectrum between 800 or 1500 and 2500 cm^{-1} may be collected at a relatively low signal-to-noise ratio, the buffer spectrum may be subtracted to a first approximation, and the curvature of the fluorescence band between 1520 and 1730 may be estimated from an interpolation of the curve between 1730 and 2500 to 1500 or 800 cm^{-1}. A cubic spline program is used to do this. A line is drawn by hand to fit the fluorescence band above 1730 cm^{-1}. Points along this line and at or below 1500 cm^{-1} are digitized and the curve connecting these points is computed using a cubic spline algorithm. Ultimately some subjective judgment must be used to evaluate the results of this calculation, since a reasonable cubic spline interpolation depends

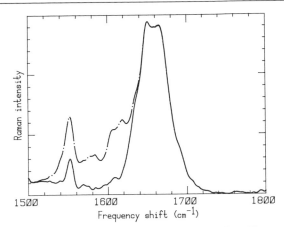

FIG. 1. Raman spectrum of triosephosphate isomerase solution. The spectrum of buffer has been subtracted to make the 1730 to 1800 cm^{-1} region linear. This figure shows the amide I band before (dash–dot line) and after (solid line) the fitting and subtraction of side chain bands.

on the spacing and location of the digitized points, and on the curvature of the line being interpolated. The 1500 to 1800 cm^{-1} region of this curve is then subtracted from the amide I spectrum so that the buffer subtraction criteria can be satisfied. Variations in the somewhat subjective solutions obtained for this fluorescence problem have been found to be within 1 standard deviation for the structure measurements. All of the spectra listed in Table I were processed without resorting to this fluorescence correction with the exception of PTI which required only minor adjustment.

After the two water subtraction criteria are satisfied a linear baseline with a variable slope must be subtracted so that the 1730 to 1800 cm^{-1} region has a zero slope and is close to zero. The region 1500 to 1520 cm^{-1} may be somewhat greater than zero depending on the intensity of the 1555 cm^{-1} Trp band and of the 1440 to 1460 cm^{-1} CH bend. The criterion that the 1730 to 1800 cm^{-1} region be linear and close to zero takes priority in this case. Again this involves a subjective judgment which has not led to errors greater than the standard deviation in secondary structure estimates. Figure 1 shows a typical result of baseline adjustment using the rules given above.

The amide I spectrum is smoothed using the algorithm by Savitsky and Golay.[5] The data here were 5-point smoothed, but experience has shown that no resolution is lost with a 7- or 9-point smooth.

[5] A. Savitsky and M. Golay, *Anal. Chem.* **36**, 1627 (1967).

Next, side chain bands due to Tyr, Phe, and Trp at 1618 cm^{-1}, 1606 cm^{-1}, and below are subtracted from the spectrum to facilitate accurate normalization of the amide I band. This is done using an automated curve fitting program which employs an approach similar to that described by Roberts *et al.*[6] This program may be obtained from the author. What follows here is a brief description of essential features of this program. The forcing factor is set to 1.1, so that the main diagonal elements of the coefficient matrix are multiplied by 1.1. A sum of approximate Gaussian–Lorenzian product functions is used to fit the band shapes. The function has the form $f = A\{\exp[-B(x - x_0)]\}^2/\{1.0 + [RB(x - x_0)]^2\}$. Elements of the symmetrical coefficient matrix are calculated from the partial derivatives of this function with respect to A, B, and x_0. x is the wavenumber, A is the intensity, and B is a function of the band width. R is a factor which controls the relative contribution of Gaussian and Lorenzian band shapes and is set to 2.5. Initial values of x_0, A, and B for five bands, respectively, are set to 1555, $0.9I_{1555}$, 0.044; 1585, $0.7I_{1585}$, 0.044; 1606, $0.7I_{1606}$, 0.044; 1618, I_{1618}, 0.044; 1650, $1.1I_{1650}$, 0.021 where I_{1585} means the digitized intensity at 1585 cm^{-1}.

The fitting and subtracting procedure is iterated twice. The calculated bands are multiplied by a factor of 0.7 each iteration so that the entire band is not subtracted in only one iteration. The band at 1650 cm^{-1} is not subtracted. Excursions of the calculated parameters x_0, A, and B are limited on each iteration as follows: Intensity change is limited to not greater than 0.3 times A. The wavenumber change is limited to not greater than 4 cm^{-1}. The width factor, B, is limited to not more than 20% change. For bands at 1555 and 1585 cm^{-1}, B is limited to greater than 0.28. For bands at 1606 and 1618 cm^{-1}, B is limited to greater than 0.39. For the band at 1650 cm^{-1}, B is limited to values greater than 0.020. Bands other than at 1650 cm^{-1} are limited to have x_0 not greater than 1619 cm^{-1}. Figure 1 shows a typical result of this procedure.

After the side chain bands are subtracted the intensity values from 1615 to 1710 cm^{-1} at 5 cm^{-1} intervals are entered into matrix b, b and matrix A are normalized as described in Table I. Matrix A from Table I is used to obtain x from Eq. (1), and matrix F from Table II is used to obtain the secondary structure estimates from Eq. (2).

Interpretation of Results

Table III shows a statistical analysis of the estimation of five defined classes of structure plus the undefined class for the reference proteins

[6] S. M. Roberts, D. H. Wilkinson, and L. R. Walker, *Anal. Chem.* **42**, 886 (1970).

TABLE II

COMPARISON OF SECONDARY STRUCTURE CONTENT (%) AS ESTIMATED FROM RAMAN AMIDE I AND X-RAY DATA; THE x AND A ROWS ARE THE COLUMNS OF MATRIX F IN EQ. (2) EXCEPT FOR TPI AND ADK

Protein[a]	Method[b]	Structure types[c]						
		Ht	S	T	U	Ho	Hd	Krank[d]
1. RNA	R1	21	49	20	11	11	10	5
	R2	20	48	20	12	10	10	4
	x	23	46	21	10	10	12	
2. LDH	R1	47	22	17	12	29	18	6
	R2	46	22	17	13	28	18	7
	x	42	24	20	14	21	21	
3. INS	R1	52	23	16	10	35	17	2
	R2	57	20	14	11	36	22	6
	x	55	21	12	13	31	24	
4. DHE	R1	76	5	10	8	53	23	5
	R2	77	2	11	10	49	28	7
	x	80	0	11	10	58	21	
5. PTI	R1	23	48	20	10	13	10	5
	R2	25	46	19	10	13	12	7
	x	26	45	16	14	12	14	
6. MEL	R1	78	4	10	6	60	18	7
	R2	77	4	10	7	60	18	5
	x	85	0	8	8	62	23	
7. CNA	R1	11	59	23	11	5	6	2
	R2	3	65	22	7	4	−1	9
	x	2	65	22	10	0	2	
8. CPA	R1	41	31	19	10	24	17	6
	R2	39	32	19	10	22	16	7
	x	40	30	21	9	26	14	
9. ADH	R1	37	32	19	13	23	14	5
	R2	33	33	22	12	21	12	9
	x	29	37	19	12	18	12	
10. LZM	R1	48	25	18	11	30	18	6
	R2	46	19	21	14	21	25	7
	x	46	19	22	12	24	22	
11. CBP	R1	64	11	16	10	44	20	5
	R2	67	8	14	11	40	27	9
	x	71	8	15	7	41	29	
12. CRA	R1	44	28	18	12	28	15	5
	R2	49	25	15	10	34	15	1
	x	46	22	22	11	28	17	
13. THL	R1	47	25	18	10	31	16	7
	R2	49	30	16	11	31	18	3
	x	40	24	19	12	26	14	

(continued)

TABLE II (*continued*)

Protein[a]	Method[b]	Structure types[c]						
		Ht	S	T	U	Ho	Hd	Krank[d]
14. PLL	R1	85	0	6	4	72	13	1
	R2	92	−1	0	9	90	2	5
	A	100	0	0	0	100	0	
15. fd	R1	92	0	6	6	72	20	2
	R2	94	−2	6	2	83	10	7
	A	90	0	6	4	76	14	
16. AVN	R1	10	56	20	11	4	6	3
	R2	4	58	21	16	1	4	7
	A	10	57	22	11	4	6	
17. TPI	R1	44	28	16	10	32	12	2
	R2	53	23	15	10	37	16	2
	x	52	24	12	12	36	16	
18. ADK	R1	61	17	14	9	44	17	6
	R2	62	16	13	8	45	18	5
	x	62	19	15	5	40	22	

[a] Abbreviations for proteins are as follows: RNA, ribonuclease A; LDH, lactate dehydrogenase; INS, insulin; DHE, deoxyhemerythrin; PTI, pancreatic trypsin inhibitor; MEL, melittin; CNA, concanavalin A; CPA, carboxypeptidase A; ADH, alcohol dehydrogenase; LZM, lysozyme; CBP, calcium binding protein; CRA, crambin; THL, thermolysin; PLL, poly-L-lysine; fd, fd phage; AVN, avidin; TPI, triosephosphate isomerase; ADK, adenylate kinase.

[b] R1, Used NNLS to analyze Raman spectra. R2, used SVA to analyze Raman spectra. x, estimates derived from α-carbon coordinates (Levitt and Greer[7]) except MEL, CBP, and CRA where estimates were obtained from inspection of α-carbon coordinates (R. Williams, unpublished, 1983). A, these values are used in place of X-ray values in matrix F and are based on the combined evidence from low resolution X-ray studies, CD, and Raman studies. The primary sources are: for PLL, N. Greenfield and G. D. Fasman, *Biochemisry* **8**, 4108 (1969); for fd, D. A. Marvin and E. J. Wachtel, *Nature* (*London*) **253**, 19 (1975); for AVN, R. B. Honzatko and R. W. Williams, *Biochemistry* **24**, 6201 (1982).

[c] See Table III.

[d] For R1 Krank is the number of different protein spectra used by NNLS in the fit. For R2 Krank is the number of orthogonal components used by SVA in the fit.

listed in Table II. This analysis was done by excluding each protein being analyzed from the reference set. Detailed assignments of most of these classes and the criteria for choosing them are described by Levitt and Greer.[7] The distinction between ordered and disordered helix is a rationalization of substantial differences in the spectra of helical proteins. A

[7] M. Levitt and J. Greer, *J. Mol. Biol.* **114**, 181 (1977).

TABLE III

STANDARD DEVIATIONS AND CORRELATION COEFFICIENTS FROM A COMPARISON OF
AMIDE I AND X-RAY-DERIVED ESTIMATES OF SECONDARY STRUCTURES
FOR 15 PROTEINS[a]

| | | Types of secondary structure[c] | | | | | |
Statistic	Method[b]	Ht	S	T	U	Ho	Hd
Standard	R1	5	4	2	2	4	4
deviation[d] (%)	R2	4	3	2	3	5	4
Correlation	R1	0.98	0.97	0.84	0.61	0.98	0.81
coefficient[e]	R2	0.98	0.99	0.87	0.31	0.96	0.88

[a] All proteins in Table II are included in the calculation of statistics except PLL, fd, and AVN. TPI and ADK were excluded from the reference set.

[b] R1, NNLS was used to solve the least-squares problem. R2, Singular value analysis was used to find a solution.

[c] Abbreviations: Ht, total helix; S, β-strand; T, turn; U, undefined; Ho, ordered α-helix; Hd, disordered helix.

[d] Standard deviation, $S = \{[\Sigma D_i^2 - (1/N)(\Sigma D_i)^2]/(n - 1)\}^{1/2}$, where $D_i = $ (Raman %$_i$ structure $-$ X-ray %$_i$ structure).

[e] Correlation coefficient, $r = [\Sigma(x_i - \bar{x})(y_i - \bar{y})]/nS_xS_y$, where $x_i = $ Raman % structure, $y_i = $ X-ray % structure, and S_x and S_y are standard deviations.

convention was chosen[3] to define disordered helix as a combination of 3_{10}, α_{II}, and non-hydrogen-bonded helical residues, including the 2 residues on each end of helical segments. Ordered helix includes all other helical residues, which are assumed to be α-helix. This distinction has repeatedly been supported by the statistical analysis through several refinements of this method and of the reference set.

The objective criteria for assignments of structure made by Levitt and Greer consistently detect more β-strand in proteins than do crystallographers reporting their own results. For proteins with significant amounts of β-sheet this excess above the X-ray structures has an average of 7% with a standard deviation of 6%. This should be considered when interpreting estimates of β-strand content. The criteria[7] assign β-strand to residues with appropriate α-carbon coordinates and which may not be hydrogen bonded into true β-sheets. This may be fortuitous for Raman measurements, since extended or unfolded polypeptide strands with β-sheetlike dihedral angles probably have considerable amide I intensity at the β-sheet frequency, 1670 cm^{-1}.

The range of the errors, or differences, between Raman and X-ray structure is reasonably small. The worst estimate is for the β-strand content of concanavalin A, which is 9 percentage points different from the X-ray value. A few other differences are at 8%, but none is greater.

TABLE IV

SOME PARAMETERS CALCULATED FROM A SINGULAR
VALUE ANALYSIS OF THE LEAST-SQUARES PROBLEM OF
FITTING THE REFERENCE PROTEINS TO LYSOZYME
(METHOD R2)

| Krank[a] | G_{coef}[b] | | Solutions from Eq. (2) | |
			Helix	β-Strand
		X-ray =	46	19
1	−2730		51	27
2	22		52	26
3	−219		55	18
4	107		52	17
5	−35		51	18
6	−15		50	18
7	67		46	19
8	−6		46	19
9	−57		38	24
10	−22		33	28
11	25		34	28
12	−25		29	30
13	13		25	32
14	−10		30	29
15	9		27	33

[a] Krank is the pseudorank—or number of eigenvectors used in the fit.

[b] G_{coef}: the transformed spectrum g where $g = U^T b$, scaled up by 10^4. The magnitude of the eighth element of g is less than twice the average standard deviation in the noise ($2S = 10.4$ here) so Krank is taken to be 7 and the corresponding solutions for helix and β-strand are 46 and 19%, respectively.

It may be observed that the side chains of asparagine and glutamine have amide groups which contribute to the amide I intensity, but which are not directly involved in the backbone secondary structure. When the analysis was corrected for the fraction of amide groups due to side chains there was no improvement in the results. The complications introduced by this correction were not justified. The analysis of both amide I and amide III bands described here avoids this side chain correction.

Some interpretation of the singular value analysis is required for the solution of Eq. (1). A straightforward criterion is used to select a solution from the set of candidates supplied by the subprogram SVA.[4] This crite-

TABLE V

SPECTRUM OF WATER FOR FREQUENCY
CALIBRATION OF SPECTRA COLLECTED ON
OTHER INSTRUMENTS[a]

20	600	1467	304
49	676	1409	262
53	742	1324	221
84	836	1249	183
102	917	1162	143
139	1004	1103	125
170	1116	1002	108
227	1210	903	81
229	1292	807	63
279	1383	729	62
328	1454	641	40
370	1490	545	35
430	1514	486	23
456	1504	418	26
524	1493	362	17
			14

[a] This average spectrum was collected at each 1
cm^{-1} with a 6 cm^{-1} bandpass at 15° and 13-
point smoothed.[5] The intensity values here are
listed by columns from 1500 to 1800 cm^{-1} in 5
cm^{-1} intervals.

rion is illustrated somewhat ideally in Table IV. The rule is as follows: if a value of g, or G_{coef}, corresponding to Krank $= k$ is less than twice the standard deviation in the experimental noise, then the solution corresponding to Krank $= k - 1$ is chosen.

Frequency Standard

A 1 cm^{-1} shift from the calibration setting at which the reference spectra were collected will produce a 3% shift in the relative content of β-strand and helix. A check of the spectrometer's calibration is required periodically. This check may be made by comparing the spectrum of water at 15°, between 1500 and 1800 cm^{-1} with a standard water spectrum used for calibration during collection of reference spectra. This spectrum is listed in Table V. At this calibration, the 546.074 and 576.960 nm mercury vapor lines are observed at 1124.4 and 2104.7 Δcm^{-1} on the spectrometer, with 19435.2 cm^{-1} set to 0.0 Δcm^{-1}. The reference spectra are shifted 1.7 Δcm^{-1} higher than they would be if the spectrometer were

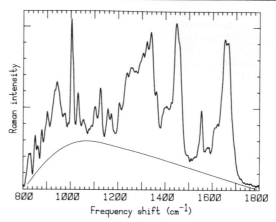

FIG. 2. Raman spectrum of triosephosphate isomerase after subtraction of buffer, showing the baseline which is to be subtracted from the spectrum to obtain the adjusted amide III region.

perfectly adjusted. When shifts of 2 cm^{-1} or less are detected a mechanical correction of the shift is generally not attempted. Instead, spectra are shifted by the computer to conform with the standard.

Amide III Analysis

Baseline Adjustment

The spectrum of buffer is subtracted from the amide III region to satisfy the criteria described for the amide I analysis. However, if the protein concentration is sufficiently high (2%) solvent correction may not be necessary since the spectrum of water in this region is relatively linear.

The baseline is then corrected for the broad fluorescence curve as follows: a smooth curve is hand drawn through minima in the plotted spectrum near 610 or 800 cm^{-1}, depending on how much of the spectrum is collected, 1065 and 1145 cm^{-1}, 1520 cm^{-1}, and through the 1730 to 1800 cm^{-1} region. Points along this curve at about every 100 cm^{-1} are then digitized and a cubic spline function is interpolated through these points and subsequently subtracted from the spectrum. This procedure is subjective. However, the resulting correction is often not great, and possible variations in the results are also relatively small. These variations have led to changes in secondary structure estimates not greater than one standard deviation.

Figure 2 shows the spectrum of a protein after buffer subtraction and before subtraction of the baseline. The baseline is drawn in to illustrate a typical curve to be subtracted from the spectrum. The bend in the spectrum, and in the baseline correction near 1000 cm^{-1} is due to a difference between the spectrum of water in a buffer solution and the spectrum of water with a high concentration of protein present.

When samples are not particularly fluorescent, good results can be obtained by collecting the spectrum from 1100 to 1500 cm^{-1} and by fitting a straight line through the minima.

The amide III spectrum is then 9-point smoothed using the algorithm by Savitsky and Golay.[5]

Interpretation of Results

The amide III region contains bands which are not due to amide III modes. When the entire amide III region is analyzed for secondary structural information the results are disappointing. However, when the regions containing non-amide III bands are eliminated from the analysis, the correlations between X-ray and Raman estimates of secondary structure content improve substantially.

For this reason column matrix b is constructed from discontinuous parts of the amide III region. After baseline subtraction and smoothing, the digitized intensities from 1225 to 1245 cm^{-1}, from 1275 to 1310 cm^{-1}, in 5 cm^{-1} intervals, and again at 1235 and 1280 cm^{-1} are entered into b. The values at 1235 and 1280 cm^{-1} are repeated since this was found to improve the results. Matrix A is taken from Table VI, A and b are normalized, x is obtained from Eq. (1), matrix F is taken from Table VII, and the secondary structure f is given by Eq. (2).

Table VIII shows a statistical analysis of the estimation of three classes of structure plus the undefined class for the reference proteins listed in Table VII. This analysis, as for the amide I band, was done by excluding each protein being analyzed from the reference set. Only two classes of structure are successfully measured according to the correlation coefficients given in Table VIII, helix and β-strand. Although the standard deviation for turn is reasonably small, the correlation coefficient is not significantly different from zero. This result is consistent with the fact that intensity above 1310 cm^{-1} is not included here, and with theoretical calculations for the amide III frequencies of many turns which put much of the intensity from these bands above 1310 cm^{-1}.[8]

The structure assignments listed in Table VII are different from those listed in Table II since proline has no amide III mode and these values were adjusted to account for this fact.

[8] S. Krimm and J. Bandekar, *Biopolymers* **19**, 1 (1980).

TABLE VI

NORMALIZED AMIDE III INTENSITIES[a] FOR 12 PROTEINS AND
POLY-L-LYSINE: THE ELEMENTS OF MATRIX A IN EQ. (1)

| | Protein[b] | | | | | | |
cm⁻¹	INS	DHE	PTI	CNA	CPA	LZM	CBP
1225	41	26	34	70	43	35	27
1230	53	34	55	88	59	53	31
1235	62	43	79	102	69	63	36
1240	71	46	89	102	77	63	46
1245	75	48	88	98	76	68	49
1275[c]	82	80	74	55	68	69	73
1280	80	82	65	57	77	64	74
1285	74	81	69	56	76	66	81
1290	62	75	73	48	58	68	89
1295	57	76	61	39	58	74	92
1300	59	94	52	37	63	81	91
1305	67	93	53	41	65	82	94
1310	73	97	64	47	64	86	106
1235[d]	62	43	79	102	69	63	36
1280	80	82	65	57	77	64	74

| | Protein[b] | | | | | |
cm⁻¹	CRA	THL	RNS	PLL	fd	AVN
1225	53	38	45	24	22	76
1230	54	55	65	22	30	91
1235	57	62	87	23	38	99
1240	61	67	94	29	44	99
1245	68	72	84	32	51	86
1275[c]	79	71	68	57	76	56
1280	69	70	63	67	75	52
1285	68	69	59	74	66	43
1290	65	65	56	86	75	46
1295	74	67	52	103	90	47
1300	72	77	59	132	105	53
1305	78	73	53	118	98	47
1310	76	82	65	142	117	54
1235[d]	57	62	87	23	38	99
1280	69	70	63	67	75	52

[a] Normalized intensity = (baseline adjusted/intensity/sum of intensities at wave numbers listed above) times 10^3.

[b] Abbreviations are as in Table II except RNS, ribonuclease S.

[c] The spectrum from 1250 to 1270 cm⁻¹ is left out since its inclusion causes poor results. This is probably due to a non-amide III band at 1260 cm⁻¹.

[d] Intensities at 1235 and 1280 cm⁻¹ are weighted doubly by including them twice.

TABLE VII
COMPARISON OF SECONDARY STRUCTURE CONTENT (%)[a] AS
ESTIMATED FROM RAMAN AMIDE III AND X-RAY DATA; THE x
AND AI ROWS ARE THE COLUMNS OF MATRIX F IN EQ. (2)

Protein[b]	Method[c]	Structure types[d]				
		H	S	T	U	Krank[e]
1. INS	R1	46	28	17	10	4
	x	49	26	13	12	
2. DHE	R1	74	10	12	5	3
	x	82	0	8	9	
3. PTI	R1	32	46	14	8	3
	x	26	49	11	13	
4. CNA	R1	9	59	21	11	2
	x	3	67	22	8	
5. CPA	R1	43	32	14	10	3
	x	40	31	21	9	
6. LZM	R1	48	29	13	10	6
	x	46	19	22	13	
7. CBP	R1	77	7	6	10	3
	x	74	9	16	1	
8. CRA	R1	52	25	15	8	4
	x	50	22	20	8	
9. THL	R1	47	24	18	11	6
	x	40	30	18	11	
10. RNS	R1	23	50	16	11	6
	x	24	54	14	8	
11. PLL	R1	95	0	6	4	1
	AI	90	0	0	10	
12. fd	R1	78	6	6	10	5
	AI	90	0	6	4	
13. AVN	R1	8	63	8	2	
	AI	8	59	21	10	

[a] Prolines are not counted here since they have no amide III vibration. X-Ray values are adjusted accordingly.

[b] Abbreviations are as in Table II except RNS for ribonuclease S.

[c] R1, Nonnegative least squares, NNLS; x, X-ray (Levitt and Greer[7]); AI, these values are based on the Raman amide I derived estimates of structure, and are consistent with other estimates. See Table II.

[d] See Table VIII.

[e] Number of other protein amide III spectra selected by NNLS for best fit.

TABLE VIII
STANDARD DEVIATIONS AND CORRELATION
COEFFICIENTS FROM A COMPARISON OF RAMAN
AMIDE III AND X-RAY-DERIVED ESTIMATES OF
SECONDARY STRUCTURE FOR 10 PROTEINS[a]

	Type of secondary structure[c]			
Statistic[b]	H	S	T	U
Standard deviation (%)	5	6	5	4
Correlation coefficient	0.98	0.97	0.28	−0.08

[a] Proteins in Table VI are included except PLL, fd, and AVN. NNLS was used to solve the least-squares problem.
[b] Statistics are defined in Table I.
[c] H, Helix; S, β-strand; T, turn; U, undefined.

Amide III Analysis Using Amide Proton Exchange

Introduction

A preliminary account of structure estimation using amide proton exchange has appeared elsewhere.[9] When a protein is dissolved in D_2O, amide protons which are exposed to solvent exchange with deuterium and the amide III vibrations shift from between 1200 and 1330 cm^{-1} to a lower frequency between 900 and 1030 cm^{-1}. When a protein has all of its amide protons exchanged in this way, the region from 1200 to 1330 cm^{-1} contains only those non-amide III bands which interfere with the amide III analysis described above. When this spectrum which contains the interfering bands is subtracted from the spectrum of protein in H_2O, only the structurally sensitive amide III bands remain in the 1200 to 1330 cm^{-1} region. In principle, then, this approach offers an advantage over the analysis described above for the amide III spectrum.

In practice there are serious problems with this method. Proteins in the reference set and proteins being analyzed for total secondary structure content must undergo nearly complete exchange. Many proteins are resistant to complete exchange within a reasonable period of time under native conditions, and when the rate of exchange is increased with heat or chemically induced unfolding the protein may not return to a fully native

[9] R. W. Williams, T. Cutrera, A. K. Dunker, and W. L. Peticolas, *FEBS Lett.* **115,** 306 (1980).

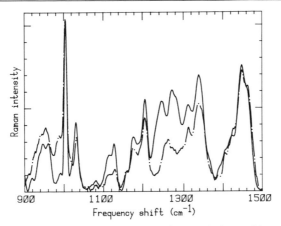

FIG. 3. Raman spectra of insulin: in H_2O (solid line), and after amide proton exchange in D_2O (dash–dot line). These spectra have had solvent and baseline subtracted.

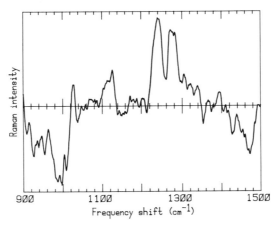

FIG. 4. Difference of the two insulin spectra shown in Fig. 3, showing the minimization of the band at 1205 cm^{-1}, and the near zero difference proximal to 1060, 1145, 1380, and 1500 cm^{-1}.

structure. Another problem is to prove that a protein has been completely exchanged.

If these problems can be overcome even partly this approach to the amide III analysis offers the following advantages. The secondary structure of nonexchanged amide groups can be measured separately from total secondary structure. With rapid mixing and quenching it may be possible to probe secondary structure formation in protein folding pathways on a millisecond time scale.[10] Also, if the hydrophobic region of a

---

[10] See P. S. Kim, this series, Vol. 131 [8].

TABLE IX

NORMALIZED AMIDE III INTENSITIES[a] FROM H_2O MINUS D_2O
SPECTRA FOR 6 PROTEINS AND HELICAL POLY-L-LYSINE: THESE
ARE THE ELEMENTS OF MATRIX A IN EQ. (1)

cm⁻¹	Protein[b]						
	CNA	INS	RNS	PTI	fd	fds	PLL
1225	77	43	54	46	25	92	11
1230	103	66	80	77	34	108	15
1235	115	80	102	116	40	109	18
1240	111	88	102	122	41	98	20
1245	102	83	79	103	46	86	24
1250	86	56	61	81	46	74	28
1255	70	30	58	61	40	63	29
1260	61	27	67	56	34	56	32
1265	54	55	70	62	22	49	36
1270	46	76	69	61	30	44	48
1275	41	74	58	47	54	38	65
1280	46	71	54	33	64	40	71
1285	40	66	44	34	54	38	69
1290	26	41	26	48	50	34	66
1295	10	24	7	43	59	21	70
1300	2	23	3	30	69	22	75
1305	0	32	11	26	75	16	83
1310	2	25	20	23	75	9	84
1315	4	18	19	22	71	4	80
1320	2	22	15	13	70	0	74

[a] See Table VII.
[b] See Table X.

membrane significantly slows the rate of deuterium exchange of proteins
in transmembrane domains relative to surface domains, then the secondary structures of these separate domains can be measured.[9]

Subtraction Criteria

The spectrum of D_2O (buffer) is subtracted from the spectrum of a
protein in D_2O in two steps to minimize the contribution of the D_2O band
around 1200 cm⁻¹. The first subtraction should make the baseline of the
resulting difference spectrum appear much like the baseline of the pro-
tein-in-H_2O spectrum. The second step is described below.

Baselines of the spectra of a protein in H_2O and in D_2O are then
adjusted as described in the amide III section. Figure 3 illustrates the
appearance of these spectra at this step.

The protein-in-D_2O spectrum is subtracted from the protein-in-H_2O
spectrum to satisfy two criteria: one is to minimize the tyrosine and

TABLE X

COMPARISON OF SECONDARY STRUCTURE CONTENT
(%)[a] AS ESTIMATED FROM RAMAN AMIDE III (H_2O
MINUS D_2O SPECTRA) AND X-RAY DATA; THE METHOD
X ROWS ARE THE COLUMNS OF MATRIX F IN EQ. (2)

Protein[b]	Method[c]	Structure types[d]			
		H	S	T	U
CNA	R1	18	57	20	11
	x	3	67	22	8
INS	R1	42	42	11	6
	x	49	26	13	12
RNS	R1	23	49	17	10
	x	24	54	14	8
PTI	R1	27	48	17	8
	x	26	49	11	13
fd	R1	83	11	4	2
	x	90	0	6	4
fds	R1	11	60	20	8
	AI	15	53	21	11
PLL	R1	93	0	6	4
	AI	100	0	0	0

[a] Prolines are not counted here since they have no amide III vibration. X-Ray values are adjusted accordingly.
[b] Abbreviations are as in Table II except RNS for ribonuclease S and fds for sonicated fd phage.
[c] R1, NNLS; x, X-ray diffraction (Levitt and Greer[7]); AI, these values are based on the Raman amide I derived estimates of structure.
[d] See Table VIII.

phenylalanine band at 1205 cm^{-1}. The other is to minimize the difference due to the phenylalanine band at 1003 cm^{-1}. The later criterion is complicated by the amide III band from β-sheet which appears in the difference spectrum from about 980 to 1020 cm^{-1}. When this later criterion gives ambiguous results the former receives priority.

The resulting difference spectrum is then corrected to satisfy the following observations. The intensity from about 1180 to 1220 cm^{-1} is close to zero. This will not be the case when too much or too little of the D_2O spectrum has been subtracted and if required the second step of D_2O subtraction occurs at this time to satisfy this rule. Other regions of the difference spectrum are close to zero. They are around 1050 to 1100 cm^{-1}, 1360 to 1400 cm^{-1}, and immediately above 1500 cm^{-1}. These four regions are often not all close to zero at this step, since the fluorescence and

TABLE XI

STANDARD DEVIATIONS AND CORRELATION
COEFFICIENTS FROM A COMPARISON OF AMIDE III
(H_2O MINUS D_2O SPECTRA) DERIVED ESTIMATES OF
SECONDARY STRUCTURES WITH ESTIMATES OF
STRUCTURE DERIVED FROM X-RAY OR OTHER
METHODS FOR 5 PROTEINS[a]

Statistic[b]	Types of secondary structure[c]			
	H	S	T	U
Standard deviation (%)	9	10	3	4
Correlation coefficient	0.89	0.80	0.72	−0.82

[a] Proteins in Table X are included, except fds and PLL. NNLS was used to solve the least-squares problem.
[b] Statistics are defined in Table I.
[c] H, Helix; S, β-strand; T, turn; U, undefined.

solvent backgrounds for the H_2O and D_2O samples differ. If they are not, the final step is to fit a cubic spline curve through these regions and subtract this curve from the difference spectrum so that this criterion is satisfied. The difference spectrum for insulin shown in Fig. 4 is typical of the final results.

The digitized intensities from 1225 to 1320 cm^{-1} are entered into column matrix b; matrix A is taken from Table IX. A and b are normalized, x is obtained from Eq. (1), matrix F is taken from Table X, and the secondary structure is given by Eq. (2).

Interpretation of Results

Table XI shows a statistical analysis of the estimation of three classes of structure plus the undefined class for reference proteins listed in Table X. Again each protein being analyzed was excluded from the reference set. The standard deviation (error) for helix and β-sheet is high. This could be due to the low number of proteins in the reference set and to the absence of certain secondary structure types in the reference set during the analysis of two proteins, the fd coat and concanavalin A. When concanavalin A is analyzed, there are no high-β-sheet proteins in the reference set and the fit obtained is poor. Likewise, poly-L-lysine is a dubious model for helical proteins in this application, and when fd phage is analyzed there are no other proteins in the reference set with a very high helix content.

When this entire set of reference proteins is used to analyze secondary structure for another protein, the actual errors are likely to be smaller than those shown in Table XI, and as β-sheet and helical proteins are added to this set of data the statistics are more likely to reflect this lower error.

It is noteworthy that the correlation coefficient for turn is relatively high in Table XI. This may be due to the inclusion of data above 1310 cm^{-1}.[8]

Acknowledgments

I am pleased to have this opportunity to thank Dr. Warner Peticolas without whose help and ideas this work would not have been started. It is also a pleasure for me to thank Dr. Keith Dunker for his ideas and support in the beginning of this work. I thank also Richard Honzatko, Janet Smith, Wayne Hendrickson, Marinn Szebenyi, Alex Wlodawer, Brian Matthews, Martha Teeter, Terry Cutrera, Richard Priest, and Joel Schnur. This work was supported by a National Research Council–Naval Research Laboratory Research Associateship, NSF Grant PCM-8302893, ONR Grant WR30342, USUHS Grant R07139, and USUHS Grant C07147 awarded to Robert Williams.

[15] Ultraviolet Resonance Raman Spectroscopy of Biopolymers

By Bruce Hudson and Leland Mayne

Introduction

Raman spectroscopy of biopolymers is conveniently divided into two types, resonant and nonresonant, on the basis of proximity of the excitation radiation frequency to electronic transitions of components of the sample. Nonresonant, or classical, Raman spectroscopy has proven very useful in the determination of protein, nucleic acid, and membrane conformational states. Resonance Raman spectroscopy has provided considerable information on heme proteins and other metalloproteins and visual pigments. These substances have chromophores absorbing in the visible region readily accessible to conventional lasers commonly used for Raman spectroscopy. They are also nonfluorescent so that the relatively weak Raman signals can be observed without contamination from this broad fluorescence emission.

Resonance of the excitation frequency with an electronic transition results in several important changes in the Raman spectrum of a biopolymer. Under resonance conditions, the intensity is often greatly enhanced permitting studies of relatively dilute solutions. The enhancement of Raman scattering is restricted to vibrational motions of the component with an electronic transition and, thus, the resonance Raman spectrum is often greatly simplified relative to that observed off-resonance. Resonance Raman spectra are often dominated by the properties of the chromophore in the particular excited electronic state resulting in the resonance. This means that the resonance Raman spectrum may be quite different from that observed off-resonance. The observation of combinations and overtones, for example, is common under resonance conditions but not off-resonance. Furthermore, the dominance of the resonance Raman spectrum by the proximal electronic state means that the spectrum will depend on which electronic state is nearest resonance. Thus, one expects distinct spectra depending on excitation conditions. This will be illustrated below. These features of resonance Raman spectra make it a very useful technique for the study of electronic excited states as well as for the establishment of conformational features of chromophoric biopolymer components.

Ultraviolet resonance Raman scattering in the 250 to 300 nm region has been used to investigate the electronic excitations of the nucleic acid bases.[1-7] Use of radiation in this region to investigate proteins results in intense fluorescence from the aromatic residues. The level of light scattered by the Raman process is generally very weak by the standards used in fluorescence spectroscopy. This means that the broad, featureless fluorescence emission can completely obscure the vibrational features of a Raman spectrum. Subtraction of the background due to the fluorescence is generally not sufficient to reveal the Raman peaks because the noise contributed by statistical contributions to the fluorescence cause the background-subtracted spectra to have more noise than signal by orders of magnitude.

We have recently applied methods for generation of ultraviolet radia-

---

[1] Y. Nishimura, A. Y. Hirakawa, and M. Tsuboi, *Adv. Infrared Raman Spectrosc.* **5**, 217 (1979).

[2] D. C. Blazej and W. L. Peticolas, *Proc. Natl. Acad. Sci. U.S.A.* **74**, 2639 (1977).

[3] W. L. Peticolas and D. C. Blazej, *Chem. Phys. Lett.* **63**, 604 (1979).

[4] Y. Nishimura, H. Haruyama, K. Nomura, A. Hirakawa, and M. Tsuboi, *Bull Chem. Soc. Jpn.* **52**, 1340 (1979).

[5] Y. Nishimura, A. Y. Hirakawa, and M. Tsuboi, *Chem. Lett.* 907 (1977).

[6] M. Tsuboi, S. Takahashi, and I. Harada, *in* "Physico-chemical Properties of Nucleic Acids" (J. Duchesne, ed.), Vol. 2, p. 91. Academic Press, New York, 1973.

[7] L. Chinsky, A. Laigle, W. L. Peticolas, and P.-Y. Turpin, *J. Chem. Phys.* **76**, 1 (1982).

tion in the 200–250 nm region to Raman spectroscopy.[8–21] This has been used in studies of small molecules[8–15] as well as biopolymers.[16–19] This work has been reviewed.[20,21] The use of radiation in this spectral region has several significant advantages. The first is that it provides resonance enhancement, and thus sensitivity and selectivity, for a new set of "chromophores" that are common in biopolymers. The problem of fluorescence from protein components is eliminated because this emission is at longer wavelengths than those used to collect the Raman spectrum. Thus, a 4000 cm⁻¹ interval to longer wavelengths of, say, 215 nm ends at 235 nm, well short of the onset of fluorescence. Another advantage of this procedure is that it often results in new Raman peaks and therefore increases the total information available. This is due, in some cases, to resonance with electronic states higher than that responsible for the lowest transition. In others, it is due to the fact that the off-resonance spectrum is dominated by contributions from many excited states at very high energies with properties different from those of states resonant in this UV excitation region.

This article describes the technology needed to perform these UV resonance Raman experiments. Emphasis is placed on the laser methods used to generate the excitation radiation. This is followed by some discussion of the primitive state of our present sample handling methods. The signal collection methods for the rest of the Raman experiment are, for the most part, conventional and will only be outlined. This method will then be illustrated with several examples including ultraviolet resonance Raman spectra of the nucleic acid bases[16,17] and several protein compo-

[8] L. D. Ziegler and B. Hudson, *J. Chem. Phys.* **74**, 982 (1981).

[9] L. D. Ziegler and B. Hudson, *J. Chem. Phys.* **79**, 1134 (1983).

[10] L. D. Ziegler and B. Hudson, *J. Chem. Phys.* **79**, 1197 (1983).

[11] L. D. Ziegler and B. Hudson, *J. Phys. Chem.* **88**, 1110 (1984).

[12] L. D. Ziegler, P. B. Kelly, and B. Hudson, *J. Chem. Phys.* **81**, 6399 (1984).

[13] R. A. Desiderio, D. P. Gerrity, and B. Hudson, *Chem. Phys. Lett.* **115**, 29 (1985).

[14] R. R. Chadwick, D. P. Gerrity, and B. Hudson, *Chem. Phys. Lett.* **115**, 24 (1985).

[15] D. P. Gerrity, L. D. Ziegler, P. B. Kelly, R. A. Desiderio, and B. Hudson, *J. Chem. Phys.* **83**, 3209 (1985).

[16] L. D. Ziegler, B. Hudson, D. P. Strommen, and W. L. Peticolas, *Biopolymers* **23**, 2067 (1984).

[17] W. L. Kubasek, B. Hudson, and W. L. Peticolas, *Proc. Natl. Acad. Sci. U.S.A.* **82**, 2369 (1985).

[18] L. C. Mayne, L. D. Ziegler, and B. Hudson, *J. Phys. Chem.* **89**, 3395 (1985).

[19] L. C. Mayne and B. Hudson, *J. Am. Chem. Soc.,* submitted.

[20] B. Hudson, P. B. Kelly, L. D. Ziegler, R. A. Desiderio, D. P. Gerrity, W. Hess, and R. Bates, *in* "Advances in Laser Spectroscopy" (B. A. Garetz and J. R. Lombardi, eds.), Vol. 3, p. 1. Wiley, New York, 1986.

[21] B. Hudson, *Spectroscopy* **1**, 22 (1986).

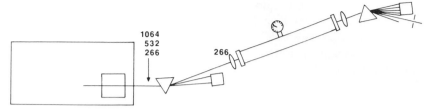

Fig. 1. A schematic illustration from above of the laser apparatus used to generate ultraviolet radiation for resonance Raman studies. The Nd:Yag laser is a Quanta-Ray DCR-2 30 Hertz model. The harmonic generating crystals are contained in a temperature-controlled metal housing (HG-1 with added heating tape and temperature controller; the current Quanta-Ray commercial model is an HG-2). The harmonics are separated with a 60 degree prism. The selected beam is sent to the lens of the hydrogen Raman shifting cell. The other beams, especially the 1064 nm IR beam, are sent to beam stops. The length of our Raman shifting cell is 0.75 m. The input and output lenses have focal lengths of 50 cm. The emerging Raman shifted beams are dispersed with a prism and one beam is selected for Raman excitation. All of the optical components are supracil quartz.

nents, including a model for the peptide bond, N-methylacetamide,[18] histidine, and some proline-containing peptides.[19] The qualitative interpretation of these spectra is well developed since most of the bands have been previously identified as to the types of vibrational motion involved. The quantitative understanding of the intensities of these bands is still in its infancy because of the complexity of the excited states of these species. However, the principles are well established and will be illustrated.

Experimental Methods

Laser Technology

A schematic illustration of the laser system used for these ultraviolet resonance Raman experiments is presented in Fig. 1. The basic component of this system is a commercial Nd:YAG laser (Quanta-Ray DCR-2). This device produces pulsed radiation at 1064 nm with an average power of roughly 6–7 W. The pulse duration is ~8 nsec and the repetition rate is typically 30 Hz. This corresponds to an energy per pulse of about 200 mJ and a peak power at the pulse maximum of 20–30 MW. When focused, this laser is easily able to generate electric fields that are sufficient to cause dielectric breakdown of air. These high peak powers result in very efficient conversion to shorter wavelengths using nonlinear optical crystals. In our applications we use KDP (potassium dihydrogen phosphate) or KD*P (potassium dideutero phosphate) crystals. A commercially available combination of these crystals is used to generate harmonics of the fundamental 1064 nm infrared radiation at 532, 355, and 266 nm. The average power levels at these wavelengths are typically 3.0 W, 1.2 W, and 600 mW. This corresponds to 20 mJ/pulse at 266 nm. Similar Nd:YAG lasers with an amplifier stage can produce 266 nm pulse energies of 100 mJ at 10 Hz.

The higher harmonic at 213 nm (1064 nm/5) can be generated by summing the 266 nm radiation with the fundamental. This is achieved with 90° phase matching using a special temperature controlled KDP crystal held at −40°.[22] The power generated at 213 nm is limited by temporary damage to the KDP crystal if the full power of the YAG laser is used. This damage is apparently due to the generation of absorbing f-centers from iron impurities in these crystals. Illumination at high intensity results in a brown discoloration of the crystal which attenuates the 213 and 266 nm radiation. Warming the crystal to room temperature removes this discoloration. The maximum output of 213 nm radiation is roughly 10 mW average power. Recent advances in nonlinear optical materials promise to eliminate this limitation on the intensity of 213 nm radiation.

The radiation produced at 532, 355 and 266 nm is sufficient to generate stimulated Raman scattering in molecular hydrogen gas, its isotopic variants and other gases. Higher pulse energies at 213 nm obtainable with new nonlinear crystals should be sufficient to permit similar shifting of this radiation. The result of the stimulated Raman shifting process is a series of collimated, powerful beams of radiation with frequencies shifted from the incident radiation by the vibrational frequency of the gas. For molecular hydrogen this frequency is 4155 cm^{-1}. The new radiation is displaced in both the Stokes (lower frequency) and anti-Stokes (higher frequency) directions relative to the incident radiation. This stimulated Raman shifting process is the key to the generation of a wide variety of frequencies with sufficient power for resonance Raman scattering well into the vacuum ultraviolet region. The stimulated Raman process can be understood by analogy with the stimulated emission process occurring in laser action. In the Raman case, the incident radiation results in normal spontaneous Raman scattering producing, primarily, Stokes-shifted output with no preferential propagation direction. For polyatomic molecules, there will be several shifted frequencies. Some of this secondary radiation travels in the direction of propagation of the incident beam. This radiation is able to enhance the spontaneous Raman process because it provides radiation at the correct frequency to stimulate the downward step. For a polyatomic molecule, one or perhaps two of all of the modes are more efficient in this

[22] G. H. Lesch, J. C. Johnson, and G. A. Massey, *IEEE J. Quantum Electron.* **12**, 83 (1976).

stimulated process and begin to dominate the intensity. The resulting stimulation process results in amplification of the Raman-shifted radiation traveling in-phase with the excitation beam. For molecular hydrogen and 266 nm excitation radiation, this will result in a new beam at 299 nm, 4155 cm^{-1} lower in energy. The two beams can now interact with each other via four-wave mixing processes. This nonlinear interaction results in the production of radiation at the frequency

$$\omega_3 = 2\omega_1 - \omega_2$$

Either of the two propagating frequencies, 266 or 299 nm, may act as ω_1 with the other being ω_2. The result is the generation of two new frequencies with wavelengths of 240 and 341 nm. These beams will also be colinear with the pump beam in the limit of a nondispersive Raman medium (with constant refractive index). Repetition of this process to produce further Stokes and anti-Stokes shifted beams is surprisingly efficient. Molecular hydrogen is a particularly good Raman shifting medium because of its large Raman cross section and its low refractive index dispersion. The important point about this process is that it has a threshold energy level. Below a certain energy per pulse, no shifted radiation is produced. Near threshold, the output is very unstable so that there are large fluctuations in the intensity of the shifted beams. At very high power levels the process tends to saturate because of thermal refractive effects in the medium.

A list of the wavelengths that can be generated using hydrogen as the shifting medium and the 532, 355, and 266 nm Nd:YAG harmonics is given in the table. Raman spectra have been obtained with wavelengths as short as 184 nm. The main point is that this one, simple method is able to generate a wide variety of wavelengths. The density of these wavelengths is probably sufficient for coverage of most biopolymer Raman excitations, i.e., sharp resonances falling between these values are not expected. Other media can be used to generate alternate frequencies but molecular hydrogen has the lowest threshold and the largest Raman shift.

The Raman shifting cell used for this process is commercially available (Quanta-Ray) or can be readily fabricated. The cell should be 1/2 to 1 m in length and fitted with thick (1/4 in.) Suprasil quartz or calcium fluoride windows. The window holder flanges should be lined with a gasket on the outside of the window to permit compression in response to the internal pressure. A valve and a gauge complete the cell. The optimum pressure of hydrogen gas depends on the particular Raman-shifted line of interest. Lower pressures (5–10 atm) are used for anti-Stokes scattering to minimize beam walk-off due to dispersion. A prism is used to select the particular Nd:YAG harmonic (532, 355, or 266 nm) that is to be used for Raman shifting. This beam is focused into the Raman shifter with a long

STIMULATED RAMAN LINES FROM H2

Wavelength	Yag harmonic	Number of quanta of H2 vibration
266.04	4th	0
252.67	2nd	5th
245.97	3rd	3rd
239.55	4th	1st
228.66	2nd	6th
223.13	3rd	4th
217.86	4th	2nd
212.83	5th	0
208.81	2nd	7th
204.19	3rd	5th
199.76	4th	3rd
192.13	2nd	8th
188.21	3rd	6th
184.45	4th	4th

focal length lens. The emerging beams are collimated with a matching long focal length lens and then dispersed with another prism. A particular beam is selected and directed to the sample.

One point worth mentioning is the inadvertant generation of stimulated rotational Raman scattering with this apparatus. The radiation produced by the KDP harmonic generating crystals is linearly polarized. Linear polarized radiation will only generate vibrationally shifted stimulated Raman scattering. Circularly polarized light also excites rotational and vibrational–rotational stimulated Raman radiation with frequencies quite close to the vibrationally shifted band selected for the resonance Raman excitation. The presence of crystallinity or strain-induced birefringence in the entrance window of the stimulated Raman shifting cell can induce ellipticity in the excitation beam and thus generate these rotational components.

An alternative procedure for generation of ultraviolet radiation is the use of a doubled dye laser. This has the advantage that continuously tunable radiation is available. It has the disadvantage that large changes in the excitation frequency require changing the dye. There is also a problem of dye degradation with prolonged use. This is particularly a problem for the blue coumarin dyes which lase in the 440 to 500 nm region. The shortest wavelength that can be generated by direct doubling of dye laser output is 217 nm. A modification of this procedure uses the summation of the 1064 nm radiation with the doubled output of efficient red dye laser

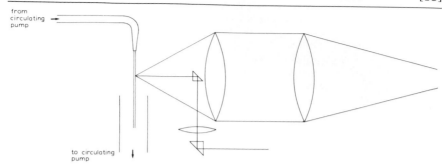

FIG. 2. A schematic illustration from the side of the sample geometry showing the excitation and collection arrangement. The liquid sample stream is directed downward into a collection tube. The excitation beam is directed upward and then back by prisms and the scattered radiation is collected by large quartz lenses. The use of a visibly fluorescent sample facilitates alignment of the collection optics.

radiation. An apparatus for ultraviolet resonance Raman scattering based on this approach has been described.[23]

The most efficient method for generation of tunable far ultraviolet radiation is to combine the two processes described above. The idea is to generate sufficient power in the doubled dye laser radiation that the pulse energy is above the stimulated Raman threshold. Doubled dye laser or dye laser plus 1064 nm radiation at, say, 220 nm is shifted to 202 nm by one anti-Stokes conversion in molecular hydrogen. Devices of this sort based on excimer lasers are commercially available. The use of Nd:YAG lasers for this procedure requires an amplifier stage following the oscillator.

Sample and Scattering Geometry

The radiation produced by the methods described above is directed to the sample using quartz prisms. (Most inexpensive mirrors will not withstand continued illumination with this UV radiation.) A final lens is used to focus the radiation on the sample. Backscattering geometry is used for most of our experiments. This configuration is obtained by using a very small Suprasil prism mounted on a post in the center of a large quartz lens which serves as the primary collection element (see Fig. 2). This geometry is used because backscattering is relatively insensitive to the value of the optical density of the sample at the particular excitation wavelength.

Our earlier experiments used a Spex 1401 half-meter double monochromator. More recently we have used a McPherson 1-m vacuum UV

[23] S. A. Asher, C. R. Johnson, and J. Murtaugh, *Rev. Sci. Inst.* **54**, 1657 (1983).

single monochromator. This device has a single curved grating as its only optical component, resulting in very high throughput in the ultraviolet region. A fairly large monochromator is needed for work in the far ultraviolet because of the increased number of wavenumbers/nm in this region. The vibrational Raman bands of biopolymers are usually somewhat over 5 cm^{-1} wide. This is a reasonable number to use for adequate resolution if no significant features are to be missed. At 500 nm this bandwidth of 5 cm^{-1} corresponds to a monochromator bandpass of 0.13 nm. At 200 nm the corresponding bandpass is 0.02 nm. Our experiments so far have dealt with homogeneous, nonscattering solutions. For these samples, stray light rejection with a single monochromator appears to be adequate for Raman shifts greater than about 200 cm^{-1}. Scattering samples require double or triple monochromators.

Another important consideration in UV Raman scattering is order sorting. If the excitation wavelength is, say, 210 nm then the 4000 cm^{-1} interval spanning the fundamental region of the Raman spectrum extends to 230 nm. If this region is covered in second order to increase resolution, the monochromator will also pass 420 to 460 nm light in first order. This is in the tail of the strong fluorescence of many proteins. This fluorescence will overwhelm the Raman signal. There are no inexpensive high efficiency filters for the far UV region. Our solution to this problem is to use a "solar blind" photomultiplier which only responds to light in the 200 to 300 nm region. Even with this detector, however, second order operation is not possible when there is intense fluorescence. The use of this detector in first order reduces stray fluorescence and room light signals.

Optical multichannel detection offers considerable benefits with respect to rate of data acquisition and rejection of laser pulse to pulse fluctuations as a source of noise. However, presently available detectors suffer somewhat in terms of resolution. Because of "bleed-through" between the detector elements, the narrowest lines are about three channels wide. This corresponds to a slit width of 75 μm. With a (high efficiency) 600 groove/mm grating, a one-meter monochromator has a dispersion in first order of about 1.6 nm/mm slit width. This means that the spectral bandwidth is about 0.12 nm for these multichannel devices. At 200 nm this corresponds to about 33 cm^{-1}. Use of a 1200 groove/mm grating reduces this to 16 cm^{-1}. This is probably adequate for most applications but often spectra at higher resolution will be needed in order to see if additional features can be resolved.

The pulsed output of the photomultiplier is processed with a box car amplifier (Stanford Research Systems) triggered by a signal from the laser. In most cases, 30 or 100 pulses are averaged corresponding to a time constant of 1 or 3 sec. The analog output of the box car amplifier is sent both to a strip chart recorder and to the ADC input of a minicomputer. A reference channel is not used in most cases.

The samples used for studies of biopolymers are circulating streams of aqueous solutions. A small dye laser pump circulates a 5 to 50 ml sample producing a downward directed column of liquid. The ultraviolet radiation strikes the front surface of the stream. This arrangement is used because it avoids all windows. We have found that Suprasil quartz and all other materials tested become fluorescent under prolonged illumination by focused UV radiation. Also, there seems to be a significant increase in sample damage at the window-solution interface. The other advantage of this sampling method is that it continuously exchanges the sample, minimizing the effects of degradation. The same arrangement can be used in a "once through" geometry if sample damage is a significant problem. The disadvantages of this sample system are the relatively large volume, the exposure of the sample to the atmosphere, and the relative difficulty of accurate temperature control. It is expected that these restrictions can be overcome by the use of a shielding layer of temperature controlled, humidified inert gas. In recent work, we have developed a flat-surface guided flow device in which a stream ~2 mm thick with exposed surfaces 5.7 mm wide is created by flow between two parallel glass plates. The flat surface of the stream permits reduction of stray light by directing the reflected incident radiation in a direction off the optic axis.

The concentration used in most of our work on small peptides has been 1–10 mM. Proteins have been examined at 100 μM and it is possible to use concentrations as low as 10 μM with spectral averaging and probably 1 μM using multichannel detection.

Water is a very poor Raman scatterer and has excellent UV transmission and therefore makes a good solvent for these studies. Phosphate, sulfate, or cacodylate can be used as buffers and internal frequency standards. The scattering from sodium sulfate in aqueous solution has been shown to provide a good calibrated intensity standard.[24] Additional frequency standards are provided by the excitation frequency (zero wavenumber shift) and by a small amount of the adjacent stimulated Raman shifted line (at 4155 cm^{-1} for hydrogen). Lines corresponding to atmospheric oxygen (at 1556 cm^{-1}) and nitrogen at (2330 cm^{-1}) are also often observed in spectra. Frequency calibration can also be provided by spectrometer wavelength standardization during each scan. Our method[21] uses a chopped low pressure mercury lamp synchronized with the Nd:YAG laser but out of phase. A second box car amplifier collects this calibration scan.

[24] J. M. Dudik, C. R. Johnson, and S. A. Asher, *J. Chem. Phys.* **82,** 1732 (1985).

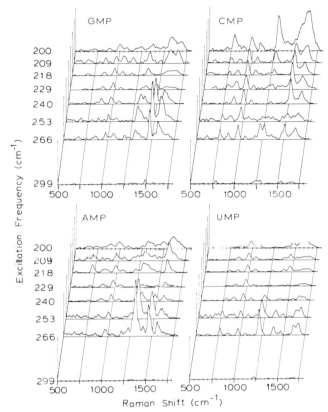

FIG. 3. The resonance Raman spectra of GMP, CMP, AMP, and UMP obtained with the excitation wavelengths indicated at the left of each panel. The concentration is 1 mM in each case except for the 299 nm spectra where a concentration of 10 mM was used. The spectra have been scaled to a constant value for the 994 cm^{-1} band of phosphate present at 1 M concentration. pH 7.0.[17]

It is unfortunate that both of the materials usually used for protein denaturation, guanidine and urea, are very strong UV Raman scatterers. This means that proteins will have to be denatured using other solvent systems such as extremes of temperature or pH or, perhaps, lithium or other salts. This has yet to be investigated.

Illustrative Examples

Nucleic Acid Bases

Raman spectra are presented in Figs. 3–5 for dilute (1 mM) aqueous solutions of the four ribonucleotides AMP, GMP, UMP, and CMP ob-

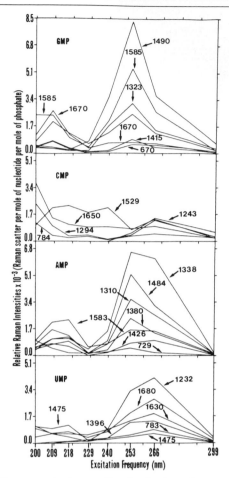

FIG. 4. Raman excitation profiles for the four ribonucleotides.[17]

tained with laser excitation at 299, 266, 253, 240, 229, 218, 209, and 200 nm.[16,17] These spectra were obtained in 1 M potassium phosphate at pH 7 (Fig. 3) or at the indicated pHs (Fig. 5). In most cases, 3–5 scans each of about 40 min duration have been averaged. Distinct evidence of strong, selective resonance enhancement is obtained in that the spectra have very different appearance at different excitation wavelengths. These spectra are presented on the same relative intensity scale by using the phosphate band at 994 cm^{-1} as an internal reference. Excitation profiles have been constructed for the strongest bands (Fig. 4). The excitation spectra for many of the vibrational bands are dominated by a peak corresponding to the lowest energy electronic transition near 260 nm. Smaller peaks are

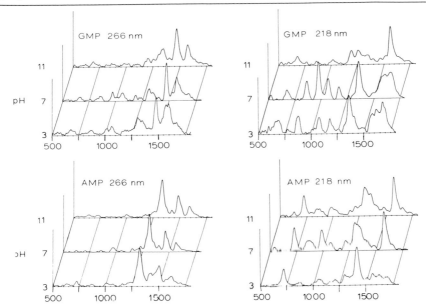

FIG. 5. The resonance Raman spectra of AMP and GMP obtained with 266 and 218 nm radiation. The pHs of the solutions are indicated.[17]

seen for higher energy electronic transitions. For some modes, the resonance enhancement is dominated by the higher energy transitions. A full description of the resonance Raman profiles of the nucleic acids will have to include several excited electronic states.[17,25] Conversely, these excitation profiles can be used as a method for sorting out overlapping electronic transitions in these complicated chromophores.

The scattering from the different bases is quite different in intensity with the strongest bands decreasing roughly in the order GMP > AMP > UMP > CMP. However, differential enhancement is possible by the proper choice of excitation wavelength. For example, between 220 and 230 nm the scattering of the 1529 cm^{-1} band of CMP is stronger than any of the bands of GMP. Similarly, the relative contribution of bands that overlap because of low resolution or intrinsic bandwidth, such as the 1585 cm^{-1} band of GMP and the 1583 cm^{-1} band of AMP, can be sorted out by comparison of the spectra obtained at two different wavelengths.

These spectra are most easily obtained using 266 nm radiation because this is near the peak of the excitation profile for the strongest modes and because of the relatively high average power of the radiation at this wave-

[25] P. Callis, *Ann. Rev. Phys. Chem.* **34,** 329 (1984).

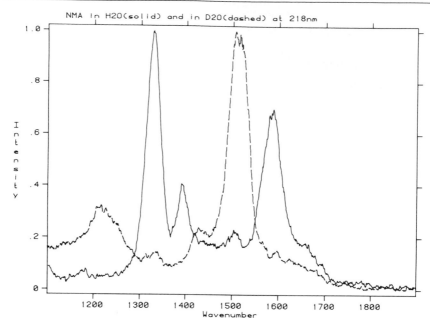

FIG. 6. The resonance Raman spectra of *N*-methylacetamide in water (solid curve) and deuterium oxide (dashed curve) obtained with excitation at 218 nm. The concentration was 10 m*M*.

length. On the other hand, two examples are given in Fig. 5 of cases where ionic species can be easily distinguished using far UV excitation (218 nm) but the spectra of these species are indistinguishable with 266 nm excitation. This demonstrates the utility of far UV resonance Raman spectroscopy for obtaining structural information.

Simple Peptides

Ultraviolet resonance Raman spectra of the simple peptide compound *N*-methylacetamide are shown in Figs. 6–8.[18] Three isotopic species are shown: the normal proto species, the N-deutero form, and the N(15)-deutero-C(13) form. The in-plane vibrations of the peptide band, amide I, II, III, etc. are modified on deuterium substitution. The modes of the deuterated form are designated amide I', II', etc. These spectra are of interest in several respects. The first is the essential absence of the C=O stretch (amide I) near 1680 cm^{-1}. This is one of the stronger bands in off-resonance spectra. It is believed that the apparent absence of this band in these spectra is due, at least in part, to a wide distribution of frequencies for this vibration due to a variety of hydrogen bonded structures. The major evidence for this is that this band is quite strong and sharp in

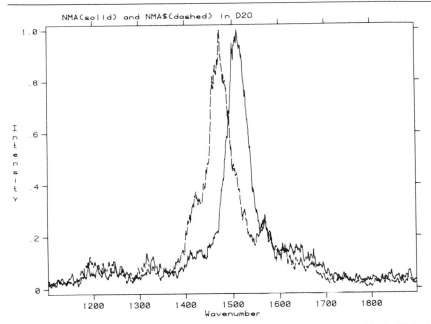

FIG. 7. The resonance Raman spectra of C(12), N(14) *N*-methylacetamide (solid line) in deuterium oxide compared to that for the isotopically substituted species with carbon-13 and nitrogen-15 in the peptide bond (also in deuterium oxide).

acetonitrile solution. The amide I band is also seen in protein resonance Raman spectra.

The second feature of interest is the high intensity of the amide II and II′ bands (Fig. 6). This result was anticipated in preresonance studies of this compound using longer wavelengths where some enhancement of the amide II band was observed.[26,27] In the deuterated form, the amide II′ band is by far the strongest component of the far UV spectrum. This band is very strong in infrared spectra but very weak in Raman spectra obtained with visible excitation. The enhancement with ultraviolet excitation can be understood in terms of the geometry changes associated with excitation of the peptide chromophore to its $\pi-\pi^*$ state. The resonance Raman spectra of these low symmetry species are dominated by those modes that are along the geometry change associated with the excitation. In this case the analysis can be put on a quantitative basis because of the outcome of the comparison of the spectra for normal and isotopically

[26] I. Harada, Y. Sugawa, H. Matsuura, and T. Shimanouchi, *J. Raman Spectrosc.* **4**, 91 (1975).
[27] Y. Sugawara, I. Harada, H. Matsuura, and T. Shimanouchi, *Biopolymers* **17**, 1405 (1978).

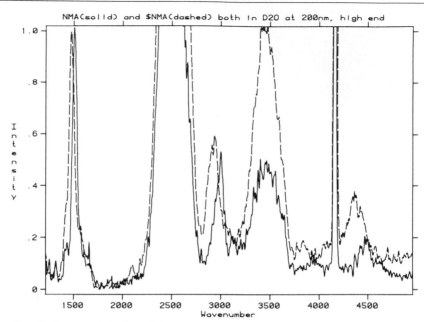

NMA(solid) and $NMA(dashed) both in D2O at 200nm, high end

FIG. 8. An expanded view of the resonance Raman spectrum of the same *N*-methylace-tamide species of Fig. 7 obtained with 200 nm excitation. The harmonics of the amide II' vibration are indicated as are the solvent bands. The strong sharp line at 4155 cm^{-1} is the next lower stimulated Raman line at 218 nm.

substituted species shown in Fig. 7. The shift upon replacement of the amide carbon and nitrogen by higher mass isotopes is roughly 47 cm^{-1}. The shift expected if the motion of this mode were a pure C—N stretch would be 55 cm^{-1}. This observation, in combination with the intensity observations, indicates that the displacement of the amide group on excitation to its $\pi-\pi^*$ excited state is predominantly along the C—N bond. This seems to be in disagreement with simple molecular orbital calculations. The assignment of the $\pi-\pi^*$ of this peptide as the dominant state responsible for resonance enhancement has recently been confirmed by determination of the excitation profiles for the strongest bands.[28]

The replacement of an amide deuteron by a hydrogen splits the intensity of the amide II' band between the amide II and III transitions (Fig. 6). Experiments performed in a 50–50 mixture of H_2O and D_2O show that the sum of the intensities of these two bands is equal to the intensity of the original amide II' band.[18] This is easily explained by the hypothesis

[28] J. M. Dudik, C. R. Johnson, and S. A. Asher, *J. Phys. Chem.* **89,** 3805 (1985).

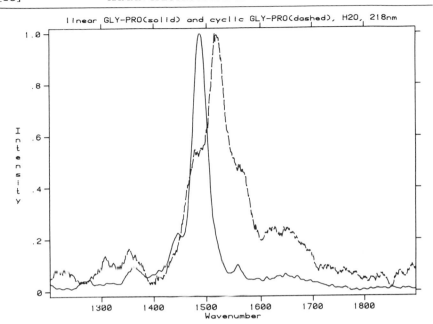

FIG. 9. Resonance Raman spectra of N-Gly-Pro-COOH (solid line) and cyclic-Gly-Pro (dashed line) in water obtained with 218 nm excitation.

that in the proto form the C—N stretching motion is spread between these two modes because of mixing with the in-plane bending motion of the C—N—H group. In the deutero form this motion occurs at lower frequency and is not coupled to the C—N stretch.

The spectrum of N-methylacetamide observed with 200 nm excitation (Fig. 8) exhibits a characteristic feature of resonance Raman spectra in the appearance of overtone bands. This is due to the dominance of the Raman scattering process by vibronic levels well up in the excited electronic state having large transition amplitudes to highly excited levels of the ground state.

The resonance Raman spectrum of the X-proline peptide bond obtained with ultraviolet radiation is similar to that of deuterated N-methylacetamide with one strong band near 1500 cm^{-1} (Fig. 9).[19] This is the expected result in that the absence of an amide proton eliminates coupling of the C—N and C—N—H motions in the same fashion as deuterium substitution. In this linear dipeptide the strong band appears at 1485 cm^{-1}. Restriction of the X-proline peptide bond into a cis configuration raises the dominant peak to a value of about 1515 cm^{-1}. Although further studies are needed, it appears that this method may be capable of detecting the cis

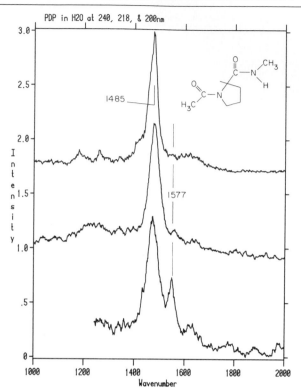

FIG. 10. Resonance Raman spectra of *N*-acetylprolinemethylamide obtained with excitation at 240, 218, and 200 nm (top to bottom). The band at 1485 cm^{-1} is due to the *N*-acetylproline linkage (see Fig. 9). The band at 1577 cm^{-1} is due to the prolinemethylamide linkage. This band is preferentially enhanced at shorter wavelengths. A study of the individual species acetylproline and prolyglycine shows that the optimum relative excitation of X-Pro/Pro-Y occurs at about 230 nm.

isomer of the X-proline peptide bond. In order to be useful in studies of complex structures such as proteins, it will be necessary to enhance the signal from the X-proline bond relative to that from other peptide linkages. It has been found[19] that excitation near 230 nm results in a maximum relative scattering of X-proline to other linkages of about a factor of 30. This is due to the relatively red shifted absorption of the X-proline linkage. An example of this behavior is shown in Fig. 10.

Several of the amino acid side chains have been investigated using this new method. Histidine is one of the most interesting in terms of protein structural studies and enzymatic mechanism. The spectra of Fig. 11 show that deuterium substitution and the state of ionization of this group have a considerable influence on the form of these spectra.

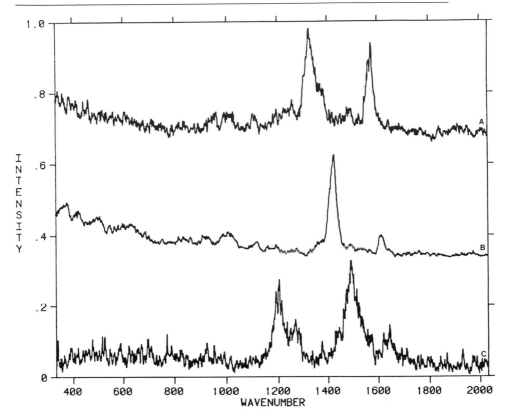

FIG. 11. Resonance Raman spectra of histidine obtained with 213 nm excitation in (A) D20 at pH 11.8, (B) D20 at pH 4.4, and (C) H20 at pH 4.0.

Summary

Ultraviolet resonance Raman scattering promises to be a useful technique for investigating the structure, refolding, and isotope exchange behavior of proteins and nucleic acids. Protein–nucleic acid interactions may be particularly amenable to this new method. These preliminary results and many others not reported here have demonstrated that these spectra are quite strongly enhanced, are often distinct from those obtained with visible excitation, are very sensitive to isotopic substitution and conformation, and, in many cases, are sensitive to the detailed wavelength used for excitation. This last observation should prove useful in sorting out complex overlapping bands in proteins by providing a check on any proposed assignment in the form of confirmatory intensity changes.

Ultraviolet resonance Raman spectroscopy is also clearly a useful technique for probing the geometries of excited electronic states of these species. It is also likely that this method can be useful in sorting out the complex pattern of electronic excited states of biopolymers. This is particularly needed if other well developed optical methods, such as circular dichroism, are to be put on a firm theoretical foundation.

The limits of sensitivity of UV Raman scattering have yet to be determined. Considerable improvements in the quality of spectra have resulted over the past year due primarily to changes in the laser hardware, the use of a high throughput monochromator and signal collection and processing methods. The use of multichannel detection will probably permit studies of micromolar concentration protein solutions. Advances in sample handling methods are clearly needed and reliable internal intensity standards at shorter wavelengths would be useful. Clearly, however, laser technology is no longer the limiting factor in such studies. Methods are now available that permit extension of this technique to wavelengths as short as 150 nm.

Acknowledgments

This work was supported by NIH Grant GM32323, NSF Grant PCM8308529, and NIH predoctoral training Grant GM07759-05. We thank Drs. Lawrence Ziegler and Daniel Gerrity who performed some of the early experiments on peptide components.

[16] Resonance Raman Studies of Ligand Binding

By Nai-Teng Yu

Introduction

A complete elucidation of the mechanism of function of a biological metal center requires knowledge of the exact nature of the metal–ligand bonds and their dependence on the protein environment. For understanding the cooperativity in hemoglobin (Hb) it is essential to know the effect of quaternary structure on the iron–ligand bond strength in both ligated and unligated states[1,2] in order to determine if the free energy of

[1] M. F. Perutz, *Annu. Rev. Biochem.* **48**, 327 (1979).
[2] D. L. Rousseau and M. R. Ondrias, *Annu. Rev. Biophys. Bioeng.* **12**, 357 (1983).

cooperativity is localized in these bonds[1,3] or distributed over many chemical bonds in the heme macrocycle, as well as the entire protein.[4]

Perhaps the most direct information on the nature of the iron–ligand bonds in hemoproteins has come from resonance Raman spectroscopy. Many iron–ligand stretching and bending vibrations have now been detected with Soret or charge-transfer excitation.[5-16] These vibrational frequencies may be employed to obtain force constants, bond lengths, and bond energy (or bond strength). The variations of the iron–ligand bond strength with protein conformation will then allow one to test the validity of some hypotheses pertaining to the mechanisms of protein control of heme reactivity: the proximal base tension effect,[17-20] the proximal imidazole deprotonation/H-bonding effect,[21-25] the distal side steric effects,[26-28] and the π-donor/acceptor interactions between the heme macrocycle and aromatic amino acids.[29-31]

The power of resonance Raman spectroscopy for ligand binding stud-

[3] J. M. Baldwin and C. Chothia, *J. Mol. Biol.* **129,** 175 (1979).

[4] J. J. Hopfield, *J. Mol. Biol.* **77,** 207 (1973).

[5] M. Tsubaki, R. B. Srivastava, and N.-T. Yu, *Biochemistry* **21,** 1132 (1982).

[6] N.-T. Yu, E. A. Kerr, B. Ward, and C. K. Chang, *Biochemistry* **22,** 4534 (1983).

[7] E. A. Kerr, H. C. Mackin, and N.-T. Yu, *Biochemistry* **22,** 4373 (1983).

[8] K. Nagai, T. Kitagawa, and H. Morimoto, *J. Mol. Biol.* **136,** 271 (1980).

[9] H. Hori and T. Kitagawa, *J. Am. Chem. Soc.* **102,** 3608 (1980).

[10] P. M. Champion, B. R. Stallard, G. C. Wagner, and I. C. Gunsalus, *J. Am. Chem. Soc.* **104,** 5469 (1982).

[11] B. Benko and N.-T. Yu, *Proc. Natl. Acad. Sci. U.S.A.* **80,** 7042 (1983).

[12] N.-T. Yu, B. Benko, E. A. Kerr, and K. Gersonde, *Proc. Natl. Acad. Sci. U.S.A.* **81,** 5106 (1984).

[13] S. A. Asher, this series, Vol. 76, p. 371, and references therein.

[14] R. H. Felton and N.-T. Yu, *in* "The Porphyrins" (D. Dolphin, ed.), Vol. III, Chap. VIII, Academic Press, New York, 1978, and references therein.

[15] H. C. Mackin, E. A. Kerr, and N.-T. Yu, *in* "Physical Methods in Heterocyclic Chemistry" (R. R. Gupta, ed.), Chap. 7. Wiley, New York, 1983, and references therein.

[16] T. G. Spiro, *in* "Iron Porphyrins" (A. B. P. Lever and H. B. Gray, eds.), Part II, p. 89, Addison Wesley, Reading, Massachusetts, 1983, and references therein.

[17] M. F. Perutz, E. J. Heidner, J. E. Lander, J. G. Bettlestone, C. Ho, and F. Slade, *Biochemistry* **13,** 2187 (1974).

[18] J. L. Hoard, *in* "Hemes and Hemoproteins" (B. Chance, R. W. Estabrook, and T. Yonetani, eds.), p. 9. Academic Press, New York, 1966.

[19] M. Rougee and D. Brault, *Biochemistry* **14,** 4100 (1975).

[20] J. Geibel, J. Cannon, D. Campbell, and T. G. Traylor, *J. Am. Chem. Soc.* **100,** 3575 (1978).

[21] H. B. Dunford, *Physiol. Veg.* **12,** 13 (1974).

[22] J. Peisach, *Ann. N.Y. Acad. Sci.* **244,** 187 (1975).

[23] M. Morrison and R. Schonbaum, *Annu. Rev. Biochem.* **45,** 861 (1976).

[24] T. Mincey and T. G. Taylor, *J. Am. Chem. Soc.* **101,** 765 (1979).

ies lies in its ability to detect the metal–ligand vibrations in both model complexes and hemoproteins. The structural data from model complexes as determined by X-ray diffraction can then be employed to deduce the ligand structure in hemoproteins, where the X-ray crystallographic analysis is less accurate. Infrared spectroscopy is capable of detecting the metal–ligand vibrations in model compounds in crystals or in nonaqueous solution; yet it is difficult, if not impossible, to obtain similar information in hemoproteins because of strong water absorption and the interference due to other vibrational modes from the protein matrix. The sensitivity and selectivity of resonance Raman techniques permit the detection of metal–ligand vibrations in aqueous solution as low as $\sim 10^{-5}$ M, compared to the $\sim 10^{-3}$ M concentration used in infrared studies of internal ligand vibrations such as $\nu(C\!-\!O)$,[32–35] $\nu(O\!-\!O)$,[36–39] $\nu(N\!-\!O)$,[40] $\nu(N\!=\!N\!=\!N)$[41,42] and $\nu(C\!\equiv\!N)$.[41,43]

[25] J. C. Schwartz, M. A. Stanford, J. N. Moy, B. M. Hoffman, and J. S. Valentine, *J. Am. Chem. Soc.* **101**, 3396 (1979).

[26] J. P. Collman, J. I. Brauman, K. M. Doxsee, T. R. Halbert, and K. S. Suslick, *Proc. Natl. Acad. Sci. U.S.A.* **75**, 564 (1978).

[27] P. W. Tucker, S. E. V. Phillips, M. F. Perutz, R. A. Houtchens, and W. S. Caughey, *Proc. Natl. Acad. Sci. U.S.A.* **76**, 1076 (1978).

[28] D. K. White, J. B. Cannon, and T. G. Traylor, *J. Am. Chem. Soc.* **101**, 2443 (1979).

[29] E. H. Abbott and P. A. Rafson, *J. Am. Chem. Soc.* **96**, 7378 (1974).

[30] J. A. Shelnutt, D. L. Rousseau, J. K. Dethmers, and E. Margoliash, *Proc. Natl. Acad. Sci. U.S.A.* **76**, 3865 (1979).

[31] M. A. Stanford, J. C. Swartz, T. E. Phillips, and B. M. Hoffman, *J. Am. Chem. Soc.* **102**, 4492 (1980).

[32] J. O. Alben and W. S. Caughey, *Biochemistry* **7**, 174 (1968).

[33] W. S. Caughey, J. O. Alben, S. McCoy, S. H. Boyers, S. Characke, and P. Hathaway, *Biochemistry* **8**, 59 (1969).

[34] P. W. Tucker, S. E. V. Phillips, M. F. Perutz, R. A. Houtchens, and W. S. Caughey, *Proc. Natl. Acad. Sci. U.S.A.* **76**, 1076 (1978).

[35] M. W. Makinen, R. A. Houtchens, and W. S. Caughey, *Proc. Natl. Acad. Sci. U.S.A.* **76**, 6042 (1979).

[36] C. H. Barlow, J. C. Maxwell, W. J. Wallace, and W. S. Caughey, *Biochem. Biophys. Res. Commun.* **55**, 91 (1973).

[37] J. C. Maxwell, J. A. Volpe, C. H. Barlow, and W. S. Caughey, *Biochem. Biophys. Res. Commun.* **58**, 166 (1974).

[38] J. C. Maxwell and W. S. Caughey, *Biochem. Biophys. Res. Commun.* **60**, 1309 (1974).

[39] W. S. Caughey, C. H. Barlow, J. C. Maxwell, J. A. Volpe, and W. J. Wallace, *Ann. N.Y. Acad. Sci.* **244**, 1 (1975).

[40] J. C. Maxwell and W. S. Caughey, *Biochemistry* **15**, 388 (1976).

[41] S. McCoy and W. S. Caughey, *Biochemistry* **9**, 2387 (1970).

[42] J. O. Alben and L. Y. Fager, *Biochemistry* **11**, 842 (1972).

[43] S. Yashikawa and W. S. Caughey, in "Biochemical and Clinical Aspects of Oxygen" (W. S. Caughey, ed.), p. 311. Academic Press, New York, 1979.

Resonance Raman studies of iron–axial ligand vibrations in hemoproteins have been greatly facilitated by the advent of highly sensitive multichannel detectors.[44] Experiments involving photodissociable ligands (e.g., CO, O_2, and NO) or unstable complexes need enhanced capability so that high quality RR spectra may be obtained by the use of low laser power and short data accumulation times. The availability of a broad range of excitation frequencies in the UV, visible, and IR regions make possible the systematic search for the charge-transfer resonance[10,45–53] needed to observe the ligand stretching vibrations. Excitations in the Soret region have been proven[5–9,11,12,54–60] to be fruitful in observing the ligand-associated vibrations presumably because many charge-transfer transitions underlie the Soret band[45–48] or the ligand vibrations are in resonance with the $\pi \rightarrow \pi^*$ Soret transition.[5,54]

This chapter will focus on recent advances in resonance Raman spectroscopy of ligand binding (exogenous and endogenous) to hemoproteins, as well as heme model complexes. Some aspects of Raman spectroscopic techniques essential for ligand binding studies will be presented first. Important bases for vibrational data analysis and interpretation will be treated next. The exogenous ligands discussed here include carbon monoxide, cyanide, nitric oxide, dioxygen fluoride, and azide; endogenous ligands include histidine, tryosine, and cysteine. Resonance Raman stud-

[44] N.-T. Yu and R. B. Srivastava, *J. Raman Spectrosc.* **9**, 166 (1980).
[45] N.-T. Yu and M. Tsubaki, *Biochemistry* **19**, 4647 (1980).
[46] M. Tsubaki, R. B. Srivastava, and N.-T. Yu, *Biochemistry* **20**, 946 (1981).
[47] M. Tsubaki and N.-T. Yu, *Proc. Natl. Acad. Sci. U.S.A.* **78**, 3581 (1981).
[48] H. C. Mackin, M. Tsubaki, and N.-T. Yu, *Biophys. J.* **41**, 349 (1983).
[49] S. A. Asher and T. M. Schuster, *Biochemistry* **18**, 5377 (1979).
[50] S. A. Asher and T. M. Schuster, *Biochemistry* **20**, 1866 (1981).
[51] S. A. Asher, L. E. Vickery, T. M. Schuster, and K. Sauer, *Biochemistry* **16**, 5849 (1977).
[52] K. Nagai, T. Kagimoto, A. Hayashi, E. Taketa, and T. Kitagawa, *Biochemistry* **22**, 1305 (1983).
[53] P. G. Wright, P. Stein, J. M. Burke, and T. G. Spiro, *J. Am. Chem. Soc.* **101**, 3531 (1979).
[54] M. A. Walters and T. G. Spiro, *Biochemistry* **21**, 6989 (1982).
[55] T. Kitagawa, M. R. Ondrias, D. L. Rousseau, M. Ikeda-Saito, and T. Yonetani, *Nature (London)* **298**, 869 (1982).
[56] A. Desbois, M. Lutz, and R. Banerjee, *Biochemistry* **18**, 1510 (1979).
[57] D. L. Rousseau, M. R. Ondrias, G. N. LaMar, S. B. Kong, and K. M. Smith, *J. Biol. Chem.* **258**, 1740 (1983).
[58] M. R. Ondrias, D. L. Rousseau, and S. R. Simon, *J. Biol. Chem.* **258**, 5638 (1983).
[59] M. R. Ondrias, D. L. Rousseau, T. Kitagawa, M. Ikeda-Saito, T. Inubushi, and T. Yonetani, *J. Biol. Chem.* **257**, 8766 (1982).
[60] R. S. Armstrong, M. J. Irwin, and P. E. Wright, *J. Am. Chem. Soc.* **104**, 626 (1982).

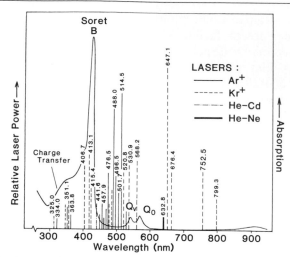

FIG. 1. Excitation wavelengths available from CW ion lasers (Ar⁺, Kr⁺, He–Cd, and He–Ne). The Soret (β) and visible (Q_0, Q_v) absorption bands of oxyhemoglobin are also shown.

ies of ligand binding to copper proteins,[61a] hemerythrin,[61b] and non-heme–iron dioxygenases[61c] have been reviewed elsewhere.

Experimental Techniques

Continuous-wave (CW) ion lasers are ideal excitation light sources for resonance Raman studies of ligand binding. In general, photodissociation or photochemical damage is more severe with pulse lasers, which also likely produce undesirable nonlinear effects. The four most commonly used ion lasers (Ar⁺, Kr⁺, He–Cd, and He–Ne) provide excitation lines ranging in wavelength from 320 to 800 nm (the entire region of Soret and visible absorption of the heme) (see Fig. 1). For continuous wavelength coverage, argon- and krypton-ion lasers are usually employed to pump a dye laser.

The multichannel Raman system, particularly suited for the detection of weak Raman signals under low laser power conditions, is shown in Fig.

[61a] T. M. Loehr and Joann Sanders-Loehr, *in* "Copper Proteins and Copper Enzymes" (R. Lontie, ed.), Vol. 1, p. 115. CRC Press, Boca Raton, Florida, 1984.

[61b] D. M. Kurtz, Jr., D. F. Shriver, and I. M. Klotz, *Coord. Chem. Rev.* **24**, 145 (1977); D. F. Shriver, *Adv. Inorg. Chem.* **2**, 117 (1980).

[61c] T. M. Loehr and W. E. Keyes, *in* "Oxidases and Related Redox Systems" (T. E. King, H. S. Mason, and M. Morrison, eds.), p. 463. Pergamon, Oxford, 1982.

2. The intensified SIT (silicon intensified target) shown in Fig. 3 has an enhanced detection capability,[44] especially in the 400 nm region.

To minimize photodissociation or the effect of heating, the sample-containing Raman cell is normally rotated at a speed of ~1000 rpm. Sample spinning may be difficult at cryogenic temperatures. An apparatus suitable for resonance Raman measurements with rotating samples at temperatures as low as 77 K has recently appeared[62] (see Fig. 4).

Figure 5 shows the experimental arrangements for measuring the depolarization properties of the sample. In a typical 90° scattering geometry, measurements of $I_\parallel$ (parallel to the incident polarization) and $I_\perp$ (perpendicular to the incident polarization) spectra are accomplished by simply rotating a polarizer between the collecting optics and the scrambler in front of the entrance slit of the spectrometer. Four types of Raman lines may be classified according to their polarization ratios (ρ), as defined by $I_\perp/I_\parallel$: polarized (p), depolarized (dp), inversely polarized (ip), and anomalously polarized (ap).

Basis for Spectral Interpretations

Vibrational Frequencies and Force Constants

It is not straightforward to deduce force constants from observed vibrational frequencies. In general, normal coordinate analysis is required to find a set of values for stretching and bending force constants consistent with a set of observable frequencies. However, the problem is simplified if characteristic group frequencies occur in molecules. To a first approximation, the group can be isolated from the rest of the molecule and coupling with the motion of the neighboring atoms neglected. The group gives rise to a set of characteristic frequencies when small numbers of atoms are tightly bound together, particularly if some of the atoms are much lighter than others and also if the group is bound less tightly to the rest of the molecule. For example, the imidazole ring of the proximal histidine is coordinated to the heme iron in myoglobins and hemoglobins. The chemical bonds of the ring are much stronger than that between the ring and the iron; the imidazole ring vibrations are not appreciably coupled with the stretching vibration of the $Fe-N_\varepsilon(His)$ bond. The $Fe-N_\varepsilon(His)$ bond stretching mode may be approximated by a diatomic model with suitably adjusted "reduced mass." Characteristic group vibrations also occur in the M–AB system, where M denotes transition metal and

---

[62] M. Braiman and R. Mathies, this series, Vol. 88, p. 648.

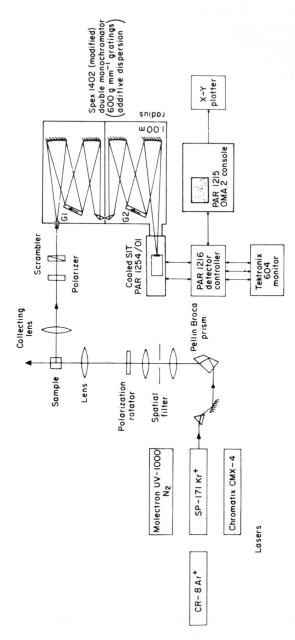

Fɪɢ. 2. Schematic diagram of a multichannel Raman system. From *Journal of Raman Spectroscopy*, N.-T. Yu and R. B. Srivastava,[44] Copyright (1980). Reprinted by permission of John Wiley & Sons, Ltd.

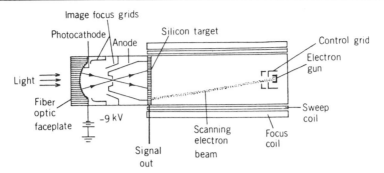

FIG. 3. Intensified vidicon detector (Silicon Intensified Target; Princeton Applied Research Model 1254).

AB diatomic ligand. This triatomic moiety is relatively isolated as far as their vibrations are concerned. Furthermore, the force constant between A and B is much greater than that between M and AB; the M–AB stretching mode, in fact, can be treated as a diatomic vibration; the reduced mass depends on the geometry.

The correlation between the stretching frequency of a chemical bond and its force constant depends on the nature of the potential energy curve. In general, the potential energy curve can be represented by a Maclaurin series expansion:

$$U(q) = U(0) + \left(\frac{dU}{dq}\right)_{q=0} q + \frac{1}{2!}\left(\frac{d^2U}{dq^2}\right)_{q=0} q^2 + \frac{1}{3!}\left(\frac{d^3U}{dq^3}\right)_{q=0} q^3 + \cdots \quad (1)$$

where $q = r - r_e$, a deviation from the equilibrium bond length. Since $U(0)$ is taken as zero and $(dU/dq)_{q=0} = 0$ at the minimum, the expression for the potential energy may be written as

$$U(q) = \frac{1}{2}\left(\frac{d^2U}{dq^2}\right)_{q=0} q^2 + \frac{1}{6}\left(\frac{d^3U}{dq^3}\right)_{q=0} q^3 + \cdots \quad (2)$$

Alternatively, one may write the potential function as

$$2U(r) = F_2(r - r_e)^2 + F_3(r - r_e)^3 + F_4(r - r_e)^4 + \cdots$$

where F_2, F_3, and F_4 are quadratic, cubic, and quartic force constant, respectively. A harmonic potential function is obtained when terms higher than quadratic are neglected. The force constant most frequently referred to is the quadratic one (F_2) which is the second derivative of the potential function evaluated at the potential minimum. With a harmonic

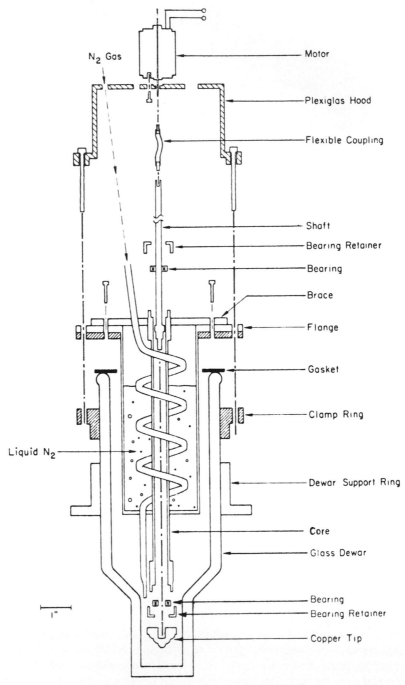

N₂ Gas

Motor

Plexiglas Hood

Flexible Coupling

Shaft

Bearing Retainer

Bearing

Brace

Flange

Gasket

Clamp Ring

Liquid N₂

Dewar Support Ring

Core

Glass Dewar

Bearing

Bearing Retainer

Copper Tip

1"

FIG. 4. Raman sample rotating device for cryogenic temperatures (down to 77 K). From Braiman and Mathies,[62] with permission.

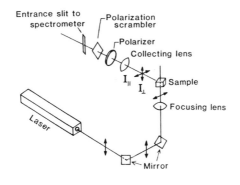

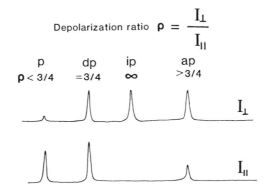

FIG. 5. Experimental configuration for the measurement of depolarization ratio.

potential function, the stretching frequency $\tilde{\nu}$ (in cm^{-1}) is related to force constant and reduced mass by

$$\tilde{\nu} = \frac{1}{2\pi c} \sqrt{\frac{k}{\mu}} \tag{3}$$

where $k \equiv F_2$, and $\mu \equiv m_1 m_2/(m_1 + m_2)$. Figure 6 shows two different potential functions. As one readily sees, the harmonic potential is a good approximation only in the bottom of the potential well. For most cases, we are concerned with the vibrational transition from $v = 0$ to $v = 1$; the harmonic potential has been frequently assumed.

Badger's Rule: Relation between Force Constant and Bond Length

There are many empirical formulas which describe the relationship between harmonic bond stretching force constant and equilibrium bond

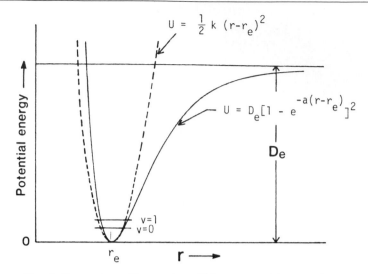

FIG. 6. Comparison of harmonic and Morse potential energy functions.

length. One of the simplest is Badger's rule,[63a]

$$F_2 = 1.86(r_e - d_{ij})^{-3} \tag{4}$$

where the constant d_{ij} has a fixed value for bonds between atoms from rows i and j of the periodic table. Later, Herschbach and Laurie[63b] modified it to give

$$r_e = d_{ij} + (a_{ij} - d_{ij})F_2^{-1/3} \tag{5}$$

where the parameters a_{ij} and d_{ij} are constants for atoms from rows i and j. Since diatomic ligands such as CO, NO, O_2, and CN^- consist of atoms from $i = 1$ and $j = 1$, the a_{ij} and d_{ij} values have been given[63b] as 1.91 and 0.68 Å, respectively. Thus, Badger's rule for these ligands are

$$r_e = 0.68 + 1.23\ k^{-1/3} \tag{6}$$

Combination of Eqs. (3) and (6) gives the following relationships for the variations of stretching frequency, force constant, and bond length for the ligands CO, NO, O_2, and CN^-:

$$\frac{\hat{v}^2}{\hat{v}'^2} = \frac{k}{k'} = \left(\frac{r'_e - 0.68}{r_e - 0.68}\right)^3 \tag{7}$$

[63a] R. M. Badger, *J. Chem. Phys.* **2**, 128 (1934).
[63b] D. R. Herschbach and V. W. Laurie, *J. Chem. Phys.* **35**, 458 (1961).

The correlations between equilibrium bond length and cubic (or quartic) force constant has also been given by Herschbach and Laurie[63b] but will not be discussed here.

Correlation between Force Constant and Bond Energy

If the potential is strictly harmonic, there will be no relation between force constant and bond energy. The bond energy is usually taken as the bond dissociation energy. The greater the energy required to break a chemical bond, the greater the bond strength. The determination of bond energy from the bond stretching force constant requires knowledge of the potential function. The most commonly used potential function for this purpose[8] may be the Morse potential.

$$U(r) = D_e\{1 - \exp[-a(r - r_e)]\}^2 \tag{8}$$

The second derivative of this potential evaluated at $r = r_e$ is $2D_e a^2$, which is equal to the harmonic force constant k. In other words, the force constant k is proportional to D_e, which may be taken as the bond dissociation energy. Thus, one may write the following relationship:

$$\frac{\tilde{\nu}}{\tilde{\nu}'} = \left(\frac{k}{k'}\right)^{1/2} = \left(\frac{D_e}{D_e'}\right)^{1/2} \tag{9}$$

Dependence of Metal–Ligand Stretching Frequency on Bending Angle

Assuming that the M–A–B grouping is linear and the A–B force constant is much greater than the M–A force constant, the stretching vibration of the M–AB bond may be treated as a diatomic molecule with $m_1 =$ mass of M and $m_2 =$ mass of AB. Direct application of Eq. (3) gives a relation between the stretching frequency and the force constant. There is no coupling between stretching and bending; the greater the stretching frequency, the greater the M–AB bond strength. However, for a bent M–A–B grouping, the reduced mass decreases relative to the linear case; the ν(M–AB) stretching frequency should increase even though there are no changes in force constants. In order to correlate the observed stretching frequency with the bond strength one needs to know the stretching force constant for the M–AB bond because the force constant and bond strength are directly related on the basis of Morse potential function.

To provide a specific example which shows the sensitivity of the ν(M–AB) stretching frequency on the $\angle$M–A–B angle, Yu et al.[12] have obtained the dependence of the Fe–CO stretching frequency on the

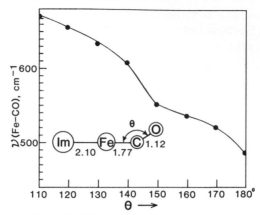

FIG. 7. Dependence of the ν(Fe–CO) stretching frequency on <Fe–C–O angle in carbonmonoxy hemes.

$\angle$Fe–C–O angle (θ) in carbonmonoxy hemes (see Fig. 7). With no changes in force constants, an increase in θ from 110 to 180° causes a decrease of 184 cm^{-1} in the ν(Fe^{II}–CO) frequency from 670 to 486 cm^{-1}.

Estimation of Bond Angle from Isotope Shifts

The ν(M–AB) frequency depends not only on the $\angle$M–A–B angle, but also on other factors such as polarity, H bonding, trans effect, etc. It may not be a good indicator of the M–A–B geometry. However, data for isotopic shifts for the two stretching frequencies may be employed to estimate the M–A–B angle, as shown by Yu et al.[6] in the case of carbonmonoxy hemes. With the assumption that the M–A–B bending force constant is much smaller than the stretching force constant, the expressions for the two stretching vibrations (ν_1 for M–AB and ν_2 for A–B) may be written

$$\lambda_{1,2} = \frac{1}{2}\left(\frac{k_1}{\mu_1} + \frac{k_2}{\mu_2}\right) \pm \frac{1}{2}\left[\left(\frac{k_1}{\mu_1} + \frac{k_2}{\mu_2}\right)^2 - 4k_1k_2\left(\frac{1}{M^2} + \frac{\sin^2\theta}{m_2^2}\right)\right]^{1/2} \quad (10)$$

where $\nu_1 = 4\pi^2(c\nu_1)^2$, $\nu_2 = 4\pi^2(c\nu_2)^2$, c = speed of light, k_1 and k_2 = bond stretching force constants for M–A and A–B, respectively, $\mu_1 = m_1m_2/(m_1 + m_2)$, $\mu_2 = m_2m_3/(m_2 + m_3)$, $M^2 = m_1m_2m_3/(m_1 + m_2 + m_3)$, and m_1, m_2, m_3 are masses for M, A, and B, respectively. The upper plus (+) sign is for λ_2 and the lower minus (−) for λ_1. For different isotopes, λ_1 and λ_2 (hence ν_1 and ν_2) shift to λ_{1i} and λ_{2i} (ν_{1i} and ν_{2i}), respectively, whereas the force constants k_1 and k_2 are unchanged. For isotope substitution at the A atom, m_2 becomes m_{2i}, and M^2 becomes $M_i^2 = m_1m_{2i}m_3/(m_1 + m_{2i} + m_3)$.

If the isotope label occurs in the B atom, m_3 becomes m_{3i}, and M^2 becomes $M_i^2 = m_1 m_2 m_{3i}/(m_1 + m_2 + m_{3i})$. The following expression may be obtained from Eq. (10):

$$\frac{\lambda_{1i}\lambda_{2i}}{\lambda_1\lambda_2} = \left(\frac{1}{M_i^2} + \frac{\sin^2\theta}{m_{2i}^2}\right)\bigg/\left(\frac{1}{M^2} + \frac{\sin^2\theta}{m_2^2}\right) \tag{11}$$

upon rearrangement, one obtains

$$\sin^2\theta = \frac{1/M_i^2 - W/M^2}{W/m_2^2 - 1/m_{2i}^2} \tag{12}$$

where

$$W = \frac{\lambda_{1i}\lambda_{2i}}{\lambda_1\lambda_2} = \left(\frac{\nu_{1i}\nu_{2i}}{\nu_1\nu_2}\right)^2 \tag{13}$$

Therefore, it is possible to estimate the $\angle$M–A–B angle (θ) from isotopic shifts involving two stretching frequencies. Calibration of the equation against data from a model compound with known θ is necessary because of the assumptions made. In the case of carbonmonoxy hemes, Yu et al.[6] found that the ν_{1i} values should be corrected by 2 cm^{-1}, i.e., $W = [(\nu_{1i} - 2)\nu_{2i}/(\nu_1\nu_2)]^2$. The application of Eq. (12) is expected to give a reasonable estimate of the $\angle$M–A–B angle (θ), provided that the angle θ is not smaller than 160°.

Ligand Binding to Metalloporphyrins and Hemoproteins

Bonding Interactions and Bonding Geometry

To facilitate the discussion on diatomic ligand binding to hemes, the bond properties of CO, CN$^-$, NO, and O$_2$, and their electronic configurations are compared in Table I. The anti-bonding π^* orbitals of CO and CN$^-$ are unoccupied, whereas those of NO and O$_2$ have one and two electrons, respectively. The occupancy of the π^* orbitals decreases the bond order but increases the bond length; the bond length increases in the order CO < CN$^-$ < NO < O$_2$. It is of interest to note that the shorter the bond length the greater the force constant, the bond energy and the bond stretching frequency. Badger's rule for the variation of force constant and equilibrium bond length for these diatomic ligands has been given by Herschbach and Laurie[63b] as $r_e = 0.68 + 1.23k^{-1/3}$.

There are mainly two types of bonding between transition metal and diatomic ligand, i.e., σ- and π-bonding as depicted in Fig. 8. First, there is σ-bonding involving the forward donation of lone-pair electrons from the

TABLE I
ELECTRONIC CONFIGURATIONS AND BOND PROPERTIES
OF DIATOMIC LIGANDS

Ligand (A–B)	CO	CN⁻	NO	O₂
σ^*	⬡	⬡	⬡	⬡
π^*	⬡ ⬡	⬡ ⬡	⭘ ⬡	⭘ ⭘
σ	⬇⬆	⬇⬆		
π	⬇⬆ ⬇⬆	⬇⬆ ⬇⬆	⬇⬆ ⬇⬆ ⬇⬆	⬇⬆ ⬇⬆ π ⬇⬆ σ

$$[\sigma(1s)]^2[\sigma^*(1s)]^2[\sigma(2s)]^2[\sigma^*(2s)]^2$$

	CO	CN⁻	NO	O₂
Bond length[a] (r_e, Å)	1.128	1.14	1.150	1.207
Force const. (k, mdyne/Å)	18.6[b]	16.5[c]	15.5[b]	11.4[b]
Bond energy[a] (kcal/mol)	256		162	118
ν(A–B) (cm⁻¹)	2145[d]	2083[e]	1877[d]	1555[d]
Badger's rule[f]		$r_e = 0.68 + 1.23\ k^{-1/3}$		

[a] H. B. Gray, "Electrons and Chemical Bonding," pp. 65, 83. Benjamin, New York, 1964.
[b] G. C. Pimentel and R. D. Spratley, "Understanding Chemistry," p. 575. Holden-Day, San Francisco, 1971.
[c] This value was calculated from the observed ν(C–N) frequency at 2083 cm⁻¹ (in H₂O).
[d] G. Herzberg, "Molecular Spectra and Molecular Structure. I. Spectra of Diatomic Molecules," p. 62. van Nostrand, Princeton, N.J., 1950.
[e] N.-T. Yu, A. Lanir, and M. M. Werber, *J. Raman Spectrosc.* **11**, 150 (1981).
[f] D. R. Herschbach and V. W. Laurie, *J. Chem. Phys.* **35**, 458 (1961).

ligand to the d_{z^2} orbital of the metal; it is especially favorable when the M–A–B linkage is linear. The second type of σ-bonding involves interactions between one of the ligand (π^*) orbitals and the metal (d_z^2) orbital. It is stabilized when there is occupancy by electron(s) in either the π^* or the d_{z^2} orbital. The distribution of electron density in this σ-bonding between the metal and a diatomic ligand depends to a great extent on the relative energies of the metal d_{z^2} and the ligand π^* level. Since the energy of the π^* level decreases in the order CO > NO > O₂, one would expect the fraction of σ-bonding electron density on the ligand to increase in the

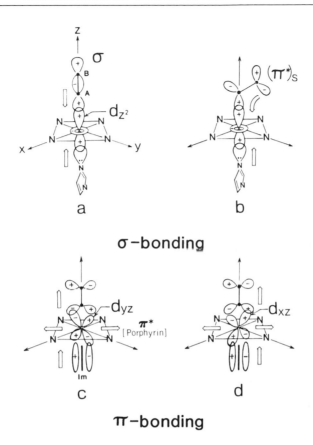

σ-bonding

π-bonding

FIG. 8. Schematic representation of σ- and π-bonding.

order O_2 > NO > CO if one considers a given metal with the same oxidation and spin state.

In π-bonding there is a back donation of electron from the metal $d_\pi(d_{xz}, d_{yz})$ to the ligand (π*) orbital. This π-back bonding enhances the metal–AB bond strength. However, the A–B bond is weakened because of the electron population in the antibonding π* orbital of the (A–B) ligand. In other words, the π-bonding causes the shortening of the metal–AB bond and a concomitant lengthening of the A–B bond. When the trans base is a strong π-donor, the donation of electrons to the d_π (metal) orbital should favor the formation of a stronger π-bond between the metal and a diatomic ligand. It should be noted that the π* (porphyrin) orbital competes for the same d_π(metal) electrons. Thus, the π electron density can be transferred among the base, the metal, the metal–ligand bond, and the π^* orbitals of diatomic ligand and porphyrin macrocycle.

The geometry of the M–A–B linkage is determined by the nature of σ- and π-bonding between the metal and the AB ligand. It is generally observed that when the number of electrons in the metal d and ligand π^* orbitals is greater than six, a bent M–A–B geometry is expected. Figure 9 displays both linear and bent geometries whose structural parameters are known from X-ray diffraction studies. Nitric oxide is a linear ligand in the [Mn^{II} + NO] system but a bent ligand when it binds to Fe^{II} heme. The salient features of the bonding interactions and bonding geometry[64–70] may be summarized as follows: (1) The d_π–p_π bonding interaction is maximized for linear M–AB geometry; (2) the ligand π^* double degeneracy is lifted when the M–A–B linkage has a bent geometry and one π^* orbital (symmetric with respect to the M–A–B plane) is favorable for σ bonding with the d_z^2 (metal) orbital (see Fig. 8b), whereas the other π^* orbital participates in π-bonding with the d_π(metal) orbital; (3) the angle of bending is determined by the relative importance of the (d_{z^2}–π^*) σ-bonding, d_{z^2}–lone pair interaction, and π back-bonding.

Exogenous Ligands

Carbon Monoxide. Carbon monoxide binds to simple ferrous porphyrins in a linear and perpendicular fashion with very high affinity ($P_{1/2} \sim 2 \times 10^{-4}$ Torr).[71] However, its binding affinity to hemoproteins is greatly reduced, by at least two orders of magnitude (e.g., in the case of sperm whale myoglobin).[71] This decreased affinity of hemoproteins for CO has been attributed[72–74] to the steric hindrance by the distal residues in the

[64] D. M. P. Mingos, *Inorg. Chem.* **12,** 1209 (1973).

[65] V. L. Goedken, J. Molin-Case, and Y.-A. Whang, *J. Chem. Soc. Chem. Commun.* 337 (1973).

[66] B. B. Wayland, J. V. Minkiewicz, and M. E. Abd-Elmageed, *J. Am. Chem. Soc.* **96,** 2795 (1974).

[67] W. R. Scheidt, Y. J. Lee, W. Luangdilok, K. J. Haller, K. Auzai, and K. Hatano, *Inorg. Chem.* **22,** 1516 (1983).

[68] S. M. Peng and J. A. Ibers, *J. Am. Chem. Soc.* **98,** 8032 (1976).

[69] P. L. Piciulo, G. Rupprecht, and W. R. Scheidt, *J. Am. Chem. Soc.* **96,** 5293 (1974).

[70] G. B. Jameson, G. A. Rodley, W. T. Robinson, R. R. Gagne, C. A. Reed, and J. P. Collman, *Inorg. Chem.* **17,** 850 (1978).

[71] T. G. Traylor, *Acc. Chem. Res.* **14,** 102 (1981).

[72] W. S. Caughey, *Ann. N.Y. Acad. Sci.* **174,** 148 (1970).

[73] M. F. Perutz, *Br. Med. Bull.* **32,** 195 (1976).

[74] J. P. Collman, J. I. Brauman, T. R. Halbert, and K. S. Suslick, *Proc. Natl. Acad. Sci. U.S.A.* **73,** 3333 (1976).

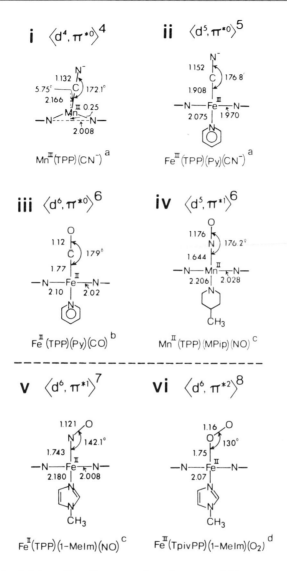

FIG. 9. Structural data and bonding geometry of some model heme complexes from X-ray diffraction. A bent geometry is observed when the number of electrons in the metal d and ligand π^* orbitals is greater than six. a, From Ref. 67; b, from Ref. 68; c, from Ref. 69; d, from Ref. 70. TPP, Tetraphenylporphyrin; Py, pyridine; Mpip, 4-methylpiperidine.

heme pocket. X-Ray[75-78] and neutron diffraction[79] studies of carbon-monoxy Hbs and Mbs indicate a distorted Fe–C–O linkage with respect to the porphyrin ring. The oxygen atom of the carbonyl is definitely displaced off the heme axis, but the location of the carbon atom is uncertain. Norvell et al.[79] interpreted the neutron diffraction data of Mb · CO in favor of a bent configuration of the CO ligand with the Fe–C–O angle at 135°. On the other hand, Baldwin[78] assumed a tilted CO geometry for Hb · CO with an Fe–C–O angle of 180°. In carbonmonoxy erythrocruorin Steigemann and Weber[77] determined the Fe–C–O angle as 161° and the Fe–C bond distance as 2.4 Å, which is quite different from the 1.80 Å value in Hb · CO found by Baldwin.[78]

Resonance Raman spectroscopy is well suited for resolving questions concerning the nature of the Fe–C–O distortion in carbonmonoxy hemoproteins. The three normal vibrations associated with the Fe–C–O linkage in both model complexes and hemoproteins are readily detectable with Soret excitation.[5-7] The iron–carbon stretching vibration (Fe–CO) was first identified at 507 cm^{-1} (Hb · CO), 512 cm^{-1} (Mb · CO), and 508 cm^{-1} (Carp Hb · CO) by Tsubaki et al.[5] The Fe–C–O bending vibration was detected[5] at ~577 cm^{-1}, which is clearly distinguishable from the ν(Fe–CO) vibration via its isotopic shift pattern. In Fig. 10 are presented the resonance Raman spectra of carbonmonoxy Mb with various ligand isotopes ($^{12}C^{16}O$, $^{13}C^{16}O$, $^{12}C^{18}O$, and $^{13}C^{18}O$). In order of increasing ligand mass the Raman line at 512 cm^{-1} displays a monotonous frequency decrease, to 509 cm^{-1} ($^{13}C^{16}O$), 504 cm^{-1} ($^{12}C^{18}O$), and 487 cm^{-1} ($^{13}C^{18}O$). In contrast, the 577 cm^{-1} line exhibits a decrease–increase–decrease "zigzag" behavior, expected of a bending mode in the Fe–C–O system, where the iron is much heavier and the masses for carbon and oxygen are comparable. As indicated in Fig. 11, for the Fe–C–O bending mode, the amplitude of vibration of the bound carbon is far greater than that of the terminal oxygen, since the moments of oscillation of these two atoms around the iron atom must approximately cancel. Thus, the effects of isotopic substitution at the terminal oxygen upon the Fe–C–O bending frequency would be small compared to those at the bound carbon. The assignments of the 512 cm^{-1} line to the Fe–CO stretching and the 577 cm^{-1} line to the Fe–C–O bending vibration are unambiguous in this case. The isotopic shift from 512 cm^{-1} ($^{12}C^{16}O$) to 504 cm^{-1} ($^{12}C^{18}O$) is approxi-

[75] E. J. Heidner, R. C. Ladner, and M. F. Perutz, J. Mol. Biol. 104, 707 (1976).
[76] E. A. Padlan and W. E. Love, J. Biol. Chem. 249, 4067 (1975).
[77] W. Steigemann and E. Weber, J. Mol. Biol. 127, 309 (1979).
[78] J. M. Baldwin, J. Mol. Biol. 136, 103 (1980).
[79] J. C. Norvell, A. C. Nunes, and B. P. Schoenborn, Science 190, 568 (1975).

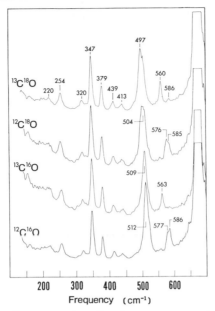

FIG. 10. Resonance Raman spectra of carbonmonoxy myoglobin with various ligand isotopes ($^{12}C^{16}O$, $^{13}C^{16}O$, $^{12}C^{18}O$, and $^{13}C^{18}O$). $\lambda_{exc} = 406.7$ nm; pH [reprinted with permission from Tsubaki et al.[5] (1982). Copyright 1982 American Chemical Society].

mately twice that from 512 cm^{-1} ($^{12}C^{16}O$) to 509 cm^{-1} ($^{13}C^{16}O$), indicating a nearly linear Fe–C–O linkage. In fact, the application of Eq. (12) results in an Fe–C–O angle of 175 ± 5°, in disagreement with the 135° value determined by neutron diffraction in crystals.[79] Thus it appears that the primary mechanism for the Fe–C–O distortion in Mb · CO (in solution) is the tilting of the Fe–C bond from the heme normal.

The ν(Fe–CO) stretching vibration in carbonmonoxy leghemoglobin (Lb · CO) was detected at 505 cm^{-1} by Armstrong et al.[60] and by Rousseau

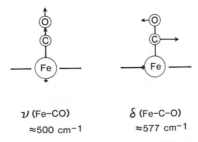

FIG. 11. Schematic representation of ν(Fe–CO) and δ(FE–C–O) vibrations.

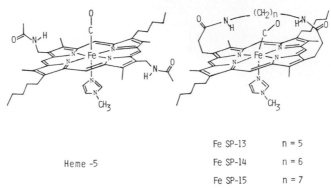

Fe SP-13	n = 5
Fe SP-14	n = 6
Fe SP-15	n = 7

Heme -5

FIG. 12. Chemical structures of unstrapped (heme-5) and strapped hemes. Reprinted with permission from Yu et al.[6] (1983). Copyright 1983 American Chemical Society.

et al.[57] Both groups observed a line at ~580 cm^{-1}, but did not mention if it was from the δ(Fe–C–O) bending mode. It is interesting to note that the ν(Fe–CO) mode is at the same frequency in Hb · CO and in Lb · CO (505 ± 1 cm^{-1}, according to Rousseau et al.[57]) even though these two globins are unrelated and the CO binding affinities are quite different. It was concluded[57] that the Fe–CO bond energy is unrelated to the ligand affinity. Furthermore, any significant changes on the iron–histidine bond should affect the Fe–CO bond[5]; the similarity in v(Fe–CO) between Hb · CO and Lb · CO suggests similar proximal environments.

Since the ligand vibrational frequencies can be affected by numerous factors, a systematic study on the effects of each factor employing model compounds is essential in order to establish reliable structural correlations. Yu et al.[6] reported a study of the distal steric affect on the Fe–CO stretching, Fe–C–O bending, and C–O stretching vibrations. They employed four synthetic hemes: a simple iron porphyrin (heme-5, according to Ward et al.[80]) without groups to hinder the CO binding and three "strapped hemes" with a 13-, 14-, or 15-atom hydrocarbon strap across the CO binding site (see Fig. 12). The base used was N-methylimidazole. In Fe^{II}(heme-5)(N-MeIm)CO complex, the ν(FE–CO) frequency was detected at 495 cm^{-1} (Fig. 13) which is lower than those observed in hemoproteins. The isotopic shift for the ($^{12}C^{16}O \rightarrow {}^{12}C^{18}O$) substitution is twice that for the ($^{12}C^{16}O \rightarrow {}^{13}C^{16}O$) substitution indicating a nearly linear Fe–C–O geometry, as expected.[68] While the ν(Fe–CO) mode at 495 cm^{-1} is quite strong, the δ(Fe–C–O) bending vibration was not detectable because of its intrinsically weak intensity when the CO ligand binds to the

[80] B. Ward, C.-B. Wang, and C. K. Chang, J. Am. Chem. Soc. 103, 5236 (1981).

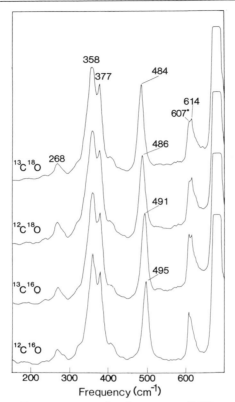

FIG. 13. Resonance Raman spectra of carbonmonoxy-Fe^{II}(heme-5)(N-MeIm) showing the absence of a δ(Fe–C–O) bending mode at ~575 cm^{-1}. Solvent: benzene with 10% methylene chloride. λ_{exc} = 406.7 nm. Reprinted with permission from Yu et al.[6] (1983). Copyright 1983 American Chemical Society.

heme in a linear and perpendicular fashion. The δ(Fe–C–O) bending mode becomes resonance enhanced when the Fe–C–O linkage is distorted by the strap, as demonstrated in the Fe^{II}(SP-14)(N-MeIm)CO complex (Fig. 14). The intensity of δ(Fe–C–O) relative to that of ν(Fe–CO) increases with a decrease in the strap length, as indicated in the Fe^{II}(SP-13)(N-MeIm)CO complex (Fig. 15).

Since it is known[80] that the shorter the strap length, the lower the binding affinity and hence the greater the steric hindrance or CO distortion, it was suggested[6] that the intensity ratio between δ(Fe–C–O) and ν(Fe–CO) may be a good indicator of the Fe–C–O distortion (i.e., primarily tilting). It is interesting that increasing the steric hindrance (by decreasing the strap length) reduces the CO binding affinity, but increases the ν(Fe–CO) stretching frequency: heme-5, 495 cm^{-1}; FeSP-15, 509

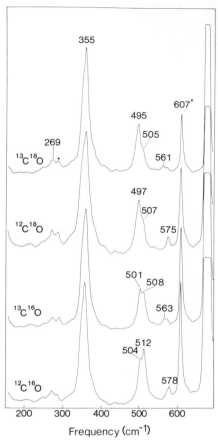

FIG. 14. Resonance Raman spectra of carbonmonoxy-Fe^{II}(SP-14)(N-MeIm) excited at 406.7 nm. Solvent: benzene with 10% methylene chloride. Reprinted with permission from Yu *et al.*[6] (1983). Copyright 1983 American Chemical Society.

cm^{-1}; FeSP-14, 512 cm^{-1}; FeSP-13, 514 cm^{-1}. The doubling of the ν(Fe–CO) vibration (Figs. 14 and 15) was due to the presence of a second species which has the CO bound to the unhindered side and the base to the strap side.[6] In addition, Yu *et al.*[6] reported the ν(C–O) stretching frequencies: heme-5, 1954 cm^{-1}; FeSP-15, 1945 cm^{-1}; FeSP-14, 1939 cm^{-1}; FeSP-13, 1932 cm^{-1}. The greater the steric hindrance, the lower the ν(C–O) stretching frequency. In other words, the ν(C–O) frequency decreases (1954 → 1945 → 1939 → 1932 cm^{-1}) with the decreasing CO binding affinity, in contrast with the ν(Fe–CO) stretching frequency. In essence, an increase in ν(Fe–CO) is accompanied by a decrease in ν(C–

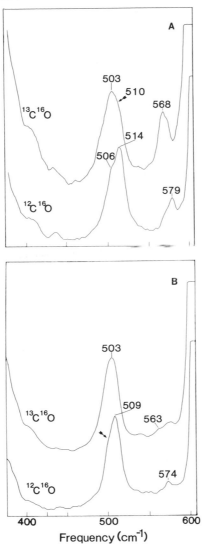

FIG. 15. Resonance Raman spectra of carbonmonoxy-Fe^{II} (SP-13)(N-MeIm) (A) and carbonmonoxy Fe^{II} (SP-15)(N-MeIm) (B); λ_{exc} = 406.7 nm. Reprinted with permission from Yu *et al.*[6] (1983). Copyright 1983 American Chemical Society.

O). In comparing Mb·CO and Hb·CO, the ν(Fe–CO) frequency in Mb·CO (~512 cm^{-1}) is higher than that in Hb·CO (507 cm^{-1}), whereas the ν(C–O) frequency is lower in Mb·CO (~1944 cm^{-1}) than in Hb·CO (1951 cm^{-1}).

Previously Tucker *et al.*[27] assumed that in hemoproteins the sp^2 (distal histidine) → π^*(CO) donation would weaken the C–O bond [hence decreasing the ν(C–O) stretching frequency] and oppose the d_π(Fe) → π^*(CO) donation; the Fe–C bond should be weakened.[27] Thus, one might expect a decrease in ν(Fe–CO) frequency as the ν(C–O) frequency decreases. In CO complexes of simple ferrous porphyrins, there is no sp^2 → π^* donation and the ν(C–O) frequency should appear higher as actually observed. One would also expect the ν(Fe–CO) frequency to be much higher in simple ferrous hemes than in hemoproteins, which was not observed. The main conclusion to be drawn from these model compound studies[6] is that the increase in ν(Fe–CO) frequency and the decrease in ν(C–O) frequency can be caused by nonbonding distal steric hindrance without involving the sp^2 donation. The lowering of the ν(C–O) stretching frequency in hemoproteins should not be attributed solely to the sp^2 donation. In carbonmonoxy erythrocruorin (insect hemoglobins from *Chironomus thummi thummi*), there is no sp^2 donation because of lack of a distal histidine[77] and the ν(Fe–CO) and ν(C–O) frequencies appear at 500 and 1961 cm^{-1}, respectively.[12] The ν(Fe–CO) frequencies in other simple ferrous hemes were found at 486 cm^{-1} in Fe^{II}(TPP)(*N*-MeIm)(CO),[7] 489 cm^{-1} in Fe^{II}(TpivPP)(*N*-MeIm)(CO),[7] and 496 cm^{-1} in Fe^{II}(OEP)(*N*-MeIm)(CO).[7]

The tilting or bending of the CO ligand lowers the CO(π^*) orbital energy and enhances its overlap with both the π and π^* orbitals of the porphyrin. In the ground state, the overlap facilitates the electron donation from a pyrrole ring to the CO ligand, decreasing the C–O bond order. However, the CO tilting decreases the π(Fe) → π^*(CO) donation, resulting in an increase in the CO bond order. The observed decrease in ν(C–O) frequency with the Fe–CO distortion indicates that the former effect is greater than the latter. The mechanism by which the Fe–C–O bending mode is resonance enhanced upon Soret excitation may reside in charge transfer from the porphyrin (π^*) to the CO(π^*) orbital. The electron density is high at the pyrrole nitrogens in the $e_g(\pi^*)$ orbital; the charge transfer is facilitated when the Fe–CO linkage is distorted. A significant electron population of the CO(π^*) orbital upon Soret excitation (porphyrin π → π^* transition) may cause the Fe–C–O linkage to bend in precisely the way caused by the Fe–C–O bending vibration. Since there is a plane of symmetry for a tilted and slightly bent Fe–C–O linkage, the ν(Fe–C–O) in-plane bending is totally symmetric, giving rise to a polarized Raman line at ~577 cm^{-1}. Its intensity is believed to derive from Albrecht's A term or Franck-Condon scattering mechanism.[14]

An inverse relationship between binding affinity and the strength of the iron–carbon bond was found by Kerr *et al.*[7] Intuitively it would appear that the greater the binding affinity, the stronger the iron–carbon bond strength. Unfortunately, this is just not the case in metalloporphyrins and hemoproteins. Kerr et al.[7] measured the ν(Fe–CO) frequencies in the CO complexes of iron[II] "picket fence" porphyrin, Fe[II]TpivPP, with trans ligands of different strength: N-methylimidazole (N-MeIm), 1,2-dimethylimidazole (1,2-Me$_2$Im), pyridine (Py), and tetrahydrofuran (THF). While the CO binding affinity of Fe[II]TpivPP(1,2-Me$_2$Im) is ~400 times lower than that of Fe[II]TpivPP(N-MeIm),[81] the ν(Fe–CO) frequency for the former (at 496 cm^{-1}) is higher than that for the latter (at 489 cm^{-1}).[7] Comparison of the CO binding to Fe[II]TpivPP(THF) and Fe[II]TpivPP(N-MeIm) reveals a similar relationship; the ν(Fe–CO) frequency (at 527 cm^{-1}) in Fe[II](TpivPP)(THF)(CO) is 38 cm^{-1} higher than that in Fe[II](TpivPP)(N-MeIm)(CO),[7] but the CO binding affinity is much lower for the THF complex.[82] In fact, X-ray crystallographic data of Fe[II](deutero)(THF)(CO)[83] revealed an unusually short Fe–C bond length (1.706 Å) compared to the 1.77 Å value found in Fe[II](TPP)(Py)(CO).[68] Since there is little difference in ν(Fe–CO) frequency between Fe[II](TPP)(Py)(CO) (at 486 cm^{-1})[12] and Fe[II]-(TpivPP)(N-MeIm)(CO) (at 498 cm^{-1}),[7] the Fe–C bond length in Fe[II]-(TpivPP)(N-MeIm)(CO) must be very close to 1.77 Å. By analogy, the Fe[II](TpivPP)(THF)(CO) complex must have the Fe–C bond distance of ~1.71 Å.[7,12]

It is important to realize that the ligand binding affinity in heme systems is *not* directly correlated with the strength of the Fe–ligand bond. To explain the seemingly paradoxical results, one must realize that there is an extensive charge redistribution upon ligand binding throughout the σ- and π-bonding network involving the axial ligands and the porphyrin macrocycle. Thermodynamically, the binding constant K, a ratio of k_{on} and k_{off}, is related to the free energy difference ($\Delta G°$) between final (ligated) and initial (unligated) states by $K = \exp[-(G' - G)/RT]$, where G' and G are the free energies of the final and initial states, respectively. In considering the free energies of ligated and unligated hemes, one should include the contributions from all the chemical bonds which are affected by charge redistribution upon ligand binding. Figure 16 shows the free energy diagram for the CO binding to Fe[II] "picket fence" porphyrin with hindered (1,2-Me$_2$Im) and unhindered (N-MeIm) bases. The unligated Fe[II](TpivPP)(1,2-Me$_2$Im) + CO has a higher free energy state because the

[81] J. P. Collman, J. I. Brauman, and K. M. Doxsee, *Proc. Natl. Acad. Sci. U.S.A.* **76**, 6035 (1979).

[82] M. Rougee and D. Brault, *Biochemistry* **14**, 4100 (1975).

[83] W. R. Scheidt, K. J. Haller, M. Fons, T. Mashiko, and C. A. Reed, *Biochemistry* **20**, 3635 (1981).

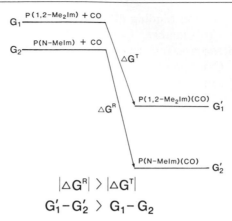

$$|\Delta G^R| > |\Delta G^T|$$

$$G_1' - G_2' > G_1 - G_2$$

FIG. 16. Comparison of free energy difference between CO-ligated and unligated hemes with hindered and unhindered axial bases. From Kerr et al.,[7] with permission.

Fe–N_ε(Im) bond is weaker; a weaker complex is less stable, and should have a higher free energy. ΔG^T and ΔG^R are the free energy changes associated with the CO binding. Because of the higher binding constant (~400 times) for the (N-MeIm) complex, $|\Delta G^R|$ is greater than $|\Delta G^T|$ by ~3.5 kcal mol^{-1}. Consequently, the difference between G_1' and G_2' must be greater than that between G_1 and G_2, by ~3.5 kcal mol^{-1}. This is readily seen from the following rearrangements: 3.5 kcal mol^{-1} ≃ $(G_2 - G_2') - (G_1 - G_1') = (G_1' - G_2') - (G_1 - G_2)$.

Thus, the ligated Fe^{II}(TpivPP)(1,2-Me$_2$Im)(CO) complex has a higher free energy than the ligated Fe^{II}(TpivPP)(N-MeIm)(CO) complex. If the formation of the Fe–C bond were an isolated event in the binding process, involving no changes in other parts of the complex, the Fe–C bond in the low affinity (1,2-Me$_2$Im) complex should be weaker than that in the high affinity (N-MeIm) complex. Since the reverse relation was observed, there must be some energy compensation by weakening the chemical bonds such as C–O, Fe–N_ε(Im), and those of the heme macrocycle. It is known that the stronger the Fe–C bond, the weaker the C–O bond.[7] The C–O bond in the low-affinity (1,2-Me$_2$Im) complex must be weaker than that in the high-affinity (N-MeIm) complex.[83a] Kerr et al.[7] also found that tension in a trans Fe–N_ε(Im) bond increases the Fe–C bond strength. In other words, the stronger the Fe–C bond, the weaker the Fe–N_ε(Im)

[83a] The ν(C–O) frequencies for Fe^{II}(TpivPP)(1,2-Me$_2$Im)(CO) and Fe^{II}(TpivPP)(N-MeIm)(CO) in benzene appear at 1962 and 1966 cm^{-1}, respectively (E. A. Kerr and N.-T. Yu, unpublished results).

bond. In general, there is a shortening of the $Fe-N_\varepsilon(Im)$ bond in going from unligated (high-spin, 5-coordinated) to ligated state (low-spin, 6-coordinated). It is conceivable that the amount of shortening (hence strengthening) may be smaller for the low-affinity (1,2-Me_2Im) complex because of the persistence of a tension on the $Fe-N_\varepsilon(Im)$ bond even in the ligated state. Additional sources of energy compensation may come from the heme macrocycle. However, the effect as reflected in ring mode frequency differences between low- and high-affinity complexes may be quite small[2] because of its much greater capacity for absorbing the free energy. An extra free energy of a few kcal mol^{-1} or less is only a very small fraction of the total bond energy of the heme macrocycle (3.5×10^3 kcal mol^{-1} assuming an average 125 kcal mol^{-1} per bond for 28 bonds).

The low-affinity (1,2-Me_2Im) complex in its unligated state may be a good model for the T-state deoxyHb. Since the 2-methyl group of 1,2-Me_2Im provides restraint to the motion of the axial base toward the porphyrin plane upon ligation, the tension on the $Fe-N_\varepsilon(Im)$ bond still persists in the ligated state. This results in different $\nu(Fe-CO)$ frequencies at 496 and 489 cm^{-1} for the low- and high-affinity ligated complexes. However, in comparing the CO complexes of human Hb (Kansas) and HbA, Tsubaki et al.[5] and Rousseau and Ondrias[2] found no difference in the $\nu(Fe-CO)$ stretching mode. This implies that there is no significant tension on the $Fe-N_\varepsilon$ (proximal histidine) bond in the T-state carbonmonoxyHb (Kansas) (+IHP). Thus the low-affinity (1,2-Me_2Im) complex is *not* a good model for the T-state ligated Hb. The role that a globin plays in the storage of free energy has not been readily modeled in synthetic hemes.

Nitric Oxide. Nitric oxide, like cyanide ion, binds to both ferric and ferrous hemoproteins.[84–89] The NO affinity to ferrous deoxyHb is greater than that of CO by a factor of 1500.[90] The binding of NO to the ferric forms of Mb and Hb is also strong, but the complex is unstable and undergoes slow autoreduction to form ferrous nitrosyl complexes.[84–89] The ferric HRP·NO complex, however, is quite stable for at least 10 hr at room temperature without noticeable autoreduction.[88] Although ferrous nitro-

[84] D. Keilin and E. F. Hartree, *Nature (London)* **139**, 548 (1937).

[85] A. Ehrenberg and T. W. Szczepkowski, *Acta Chem. Scand.* **14**, 1684 (1960).

[86] H. Kon, *Biochem. Biophys. Res. Commun.* **35**, 423 (1960).

[87] J. C. W. Chien, *J. Am. Chem. Soc.* **91**, 2166 (1969).

[88] T. Yonetani, H. Yamamoto, J. E. Erman, J. S. Leigh, Jr., and G. H. Reed, *J. Biol. Chem.* **247**, 2446 (1972).

[89] D. H. O'Keefe, R. E. Ebel, and J. A. Peterson, *J. Biol. Chem.* **253**, 3509 (1978).

[90] Q. H. Gibson and F. J. W. Roughton, *J. Physiol. (London)* **136**, 507 (1957).

syl complexes have been extensively studied by X-ray diffraction,[91-94] infrared,[94,95] visible absorption,[96-98] EPR,[88,89,92,97,98] EXAFS,[99] and resonance Raman,[100-108] very little is known about the nature of the bonding interactions between Fe^{III} and NO in hemoproteins and heme model complexes, presumably because of the intrinsic tendency toward spontaneous autoreduction,[84-89] diamagnetic properties (EPR silent),[85] and the high quantum yield of ligand photodissociation which has made resonance Raman studies difficult.

The [Fe^{III} + NO] system is isoelectronic with the [Fe^{II} + CO] system, having a total of six electrons in the metal d and ligand π^* orbitals; a linear Fe^{III}–NO linkage is expected in the absence of a distal steric effect (see section on "Bonding Interactions and Bonding Geometry"). In view of recent assignments of the Fe^{II}–CO stretching and Fe^{II}–C–O bending vibrations, it is of interest to see if the Fe^{III}–NO stretching and Fe^{III}–N–O bending modes are observable under Soret excitation. Furthermore, a comparison of these vibrational frequencies between the two systems should provide insight into the nature of these iron–axial ligand bonds in hemoproteins.

With excitation at 406.7 nm, Benko and Yu[11] have detected the Fe^{III}–NO bond stretching frequencies at 595 cm^{-1} (ferric Mb · NO), 594 cm^{-1} (ferric HbA · NO), and 604 cm^{-1} (ferric HRP · NO). The Fe^{III}–N–O bend-

[91] J. F. Deatherage and K. Moffat, *J. Mol. Biol.* **134,** 401 (1979).
[92] W. R. Scheidt and M. E. Frisse, *J. Am. Chem. Soc.* **97,** 17 (1975).
[93] W. R. Scheidt and P. L. Piciulo, *J. Am. Chem. Soc.* **98,** 1913 (1976).
[94] W. R. Scheidt, A. C. Brinegar, E. B. Ferro, and J. F. Kirmer, *J. Am. Chem. Soc.* **99,** 7315 (1977).
[95] J. C. Maxwell and W. S. Caughey, *Biochemistry* **15,** 388 (1976).
[96] M. F. Perutz, J. V. Kilmartin, K. Nagai, A. Szabo, and S. R. Simon, *Biochemistry* **15,** 378 (1976).
[97] E. Trittelvitz, H. Sick, and K. Gersonde, *Eur. J. Biochem.* **31,** 578 (1972).
[98] H. Hori, M. Ikeda-Saito, and T. Yonetani, *J. Biol. Chem.* **256,** 7849 (1981).
[99] R. G. Shulman, P. Eisenberger, S. Simon, S. Ogawa, and A. Mayer, *Biophys. J.* **37,** 91a (Abstr.) (1982).
[100] G. Chottard and D. Mansuy, *Biochem. Biophys. Res. Commun.* **77,** 1333 (1977).
[101] A. Debois, M. Lutz, and R. Banerjee, *Biochemistry* **18,** 1510 (1979).
[102] A. Szabo and L. D. Barron, *J. Am. Chem. Soc.* **97,** 660 (1975).
[103] K. Nagai, C. Welborn, D. Dolphin, and T. Kitagawa, *Biochemistry* **19,** 4755 (1980).
[104] D. M. Scholler, M.-Y. R. Wang, and B. M. Hoffman, *J. Biol. Chem.* **254,** 4072 (1979).
[105] J. D. Stong, J. M. Burke, P. Daly, P. Wright, and T. G. Spiro, *J. Am. Chem. Soc.* **102,** 5815 (1980).
[106] A. Debois, M. Lutz, and R. Banerjee, *Biochim. Biophys. Acta* **671,** 184 (1981).
[107] M. Tsubaki and N.-T. Yu, *Biochemistry* **21,** 1140 (1982).
[108] M. A. Walters and T. G. Spiro, *Biochemistry* **21,** 6989 (1982).

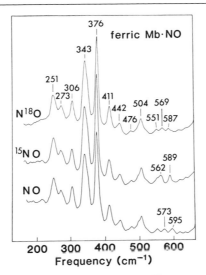

FIG. 17. Resonance Raman spectra of ferric Mb · NO. λ_{exc} = 406.7 nm; pH 7.2. From Benko and Yu,[11] with permission.

ing vibrations are located at 573 cm^{-1} (ferric Mb · NO) and 574 cm^{-1} (ferric HRP · NO), which are very similar to the Fe^{II}–C–O bending frequencies at ~578 cm^{-1} in (carbonmonoxy) Mb and HbA. The assignments were based on ^{15}NO and N^{18}O isotope shifts, which permit distinction between iron–ligand stretching and bending vibrations. Resonance Raman spectra of ferric Mb · NO with various ligand isotopes are presented in Fig. 17. It is interesting to note that while the Fe^{III}–NO and Fe^{II}–C–O bending frequencies are very similar the Fe^{III}–N–O and Fe^{II}–CO stretching frequencies differ by ~90 cm^{-1}. If both Fe^{III}–NO and Fe^{II}–CO linkages are indeed linear and the small distortion by the protein is similar, the stretching frequencies at ~600 cm^{-1} for the Fe^{III}–NO bond and at ~500 cm^{-1} for the Fe^{II}–CO bond may be employed to obtain the relative bond-stretching force constants. It was estimated[11] that the force constant for the Fe^{III}–NO bond is ~1.5 times greater than that for the Fe^{II}–CO bond. Thus it appears that the Fe^{III}–NO bond is stronger (or shorter) than the Fe^{II}–CO bond.

The origin of the greater bond strength for the Fe^{III}–NO bond compared to the Fe^{II}–CO bond was suggested[11] to be due to the participation in π-bonding of the unpaired electron in the π^*(NO) orbital. Upon binding, the unpaired π^*(NO) electron may couple antiferromagnetically to the unpaired d_π(Fe) electron to form a strong π-bond between the Fe^{III} and NO ligand, in addition to the σ-bond.

TABLE II
DIFFERENCES IN LIGAND STRETCHING
FREQUENCY BETWEEN FREE AND BOUND STATES

Ligand binding	$\nu(A-B)(cm^{-1})$		
	Free[a]	Bound	Δ
$O_2 \rightarrow$ deoxyHb	1555	1125[b]	−430
NO $\rightarrow$ deoxyHb	1877	1622[c]	−255
CO $\rightarrow$ deoxyHb	2145	1950[d]	−195
NO $\rightarrow$ ferric HRP	1877	1899[e]	+22

[a] From Table I.
[b] From Tsubaki and Yu.[47]
[c] From Stong et al.[105]
[d] From Tsubaki et al.[5]
[e] Converted from $\nu(^{15}N-O)$, Maxwell and Caughey.[40]

As previously discussed the $Fe^{II}-C-O$ distortion by the distal steric effect induces the RR enhancement of the $Fe^{II}-C-O$ bending mode with Soret excitation. The enhancement of the $Fe^{III}-N-O$ bending mode in the ferric nitrosyl complexes of Mb and HRP may also indicate the $Fe^{III}-N-O$ distortion in these hemoproteins. Comparison of isotopic shifts in the Fe–ligand stretching frequencies between ferric Mb · NO (Fig. 17) and ferrous Mb · CO (Fig. 10) reveals noteworthy differences, which have implications on the ligand geometry. The $N^{18}O$ shift (8 cm^{-1}) is not twice as large as the ^{15}NO shift (6 cm^{-1}), whereas the $C^{18}O$ shift (8 cm^{-1}) is slightly greater than two times the ^{13}CO shift (3 cm^{-1}). This may be taken as evidence that the $\angle Fe^{III}-N-O$ angle in ferric Mb · NO is smaller than the $\angle Fe^{II}-C-O$ angle in ferrous Mb · CO. The quantitative estimation of the $\angle Fe^{III}-N-O$ angle employing Eq. (12) must await the detection of $\nu(N-O)$ frequencies for two different isotopes. At present, the only piece of information available in the literature is the IR-detected $\nu(^{15}N-O)$ stretching frequency at 1865 cm^{-1} in ferric HRP · NO, a frequency 22 cm^{-1} higher than the free gas value (1866 cm^{-1}).[40] In contrast, the $\nu(C-O)$ and $\nu(N-O)$ stretching frequencies decrease 195 and 255 cm^{-1}, respectively, upon binding to ferrous hemes (see Table II).

Recently, Walters and Spiro[54] assigned two isotope-sensitive lines at 596 and 573 cm^{-1} as the $Fe^{II}-NO$ stretching and $Fe^{II}-N-O$ bending vibrations in the low-pH form (pH 5.8) of ferrous Mb · NO, which was thought to contain a penta-coordinated ferrous nitrosyl heme. Mackin et al.[109]

[109] H. C. Mackin, B. Benko, N.-T. Yu, and K. Gersonde, FEBS Lett. 158, 199 (1983).

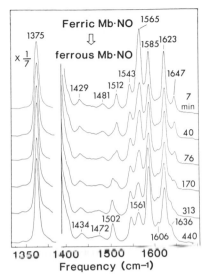

FIG. 18. The spontaneous autoreduction of ferric Mb · NO to ferrous Mb · NO as monitored by resonance Raman spectra in the 1350–1800 cm^{-1} region. pH 6.0; λ_{exc} = 406.7 nm. From Benko and Yu.[11]

found that Walters and Spiro's sample was a mixture of ferric and ferrous Mb · NO, and that, in agreement with earlier EPR studies,[97] there was no conversion of hexa- to penta-coordinated ferrous Mb · NO from pH 8.4 to 5.8. Mackin *et al.*[109] further demonstrated that there was no Soret-excited enhancement of the 596 and 573 cm^{-1} lines in a penta-coordinated ferrous Mb · NO prepared by interactions with sodium dodecyl sulfate. It was concluded[110] that Walter and Spiro's isotope-sensitive lines at 596 and 573 cm^{-1} are due to the ferric Mb · NO complex.

Time-dependent resonance Raman spectra in the conversion of ferric to ferrous Mb · NO have been reported[11] for the 1400–1700 cm^{-1} region. As shown in Fig. 18, the most striking spectral changes occur at 1512 and 1647 cm^{-1}, which shift to 1502 and 1636 cm^{-1}, respectively, upon autoreduction. The final spectrum taken at 440 min after the addition of NO to metMb is identical to the spectrum of ferrous Mb · NO[107] prepared by sodium dithionite reduction before nitric oxide was introduced. Unlike the conversion of low spin Fe^{III} to low spin Fe^{II} in the cyanide complex of HRP in which the π-electron density indicator at 1375 cm^{-1} shifts to 1362 cm^{-1}, the addition of one electron to the ferric Mb · NO complex does not alter the frequency at 1378 cm^{-1} (within ±1 cm^{-1}). Apparently, the differ-

[110] J. C. W. Chien and L. C. Dickinson, *J. Biol. Chem.* **252**, 1331 (1977).

ence in charge between the two oxidation states is accommodated by the complex without affecting the π^* electron density of the porphyrin macrocycle.

In the low-frequency (200–700 cm^{-1}) region, the autoreduction results in the appearance of a new ring mode at 357 cm^{-1} and an ^{15}NO isotope-sensitive line at 554 cm^{-1}. Previously, Chottard and Mansuy[100] in their studies of ferrous HbA · NO assigned a line at ~550 cm^{-1} as the Fe^{II}–NO stretching mode. This assignment has been generally accepted without question.[2,13,16,56,105–108] However, Benko and Yu[11] found that this vibration at 554 cm^{-1} in ferrous Mb · NO is insensitive to N^{18}O and suggested that it should be assigned as the δ(Fe^{II}–N–O) bending mode.

The validity of the assignment of the 554 cm^{-1} line as either the ν(Fe^{II}–NO) stretching or ν(Fe^{II}–N–O) bending vibration depends on the $\angle$Fe^{II}–N–O angle (θ). X-Ray structural data of Fe^{II}(TPP)(N-MeIm)(NO)[93] reveal a value of θ of 142°. In horse nitrosyl hemoglobin X-ray data[91] also indicate a value of 145° for both the α- and β-subunits. The Fe^{II}–NO bond length in both the model complex and horse Hb · NO (α- and β-subunits) is 1.74 Å. If θ is 142° or greater, the isotopic shift pattern is more consistent with the assignment of a δ(Fe^{II}–N–O) bending mode. However, if θ is close to 90° the ν(Fe^{II}–NO) stretching frequency would be insensitive to the terminal ^{18}O substitution, and the isotopic shift pattern alone may not be good enough to distinguish these two vibrational modes. EPR studies of [^{15}N]nitrosylhemoglobin Kansas by Chien and Dickinson[108] indicated a large difference in the θ value between the α-subunit (167°) and the β-subunit (105°). In ferrous Mb · NO[110,111] and Hb · NO,[112] the values of θ were also estimated from EPR data as ~110°. Resonance Raman studies of nitrosyl Hb (Kansas) should be of importance in testing the validity of EPR estimates.

Ferrous nitrosyl Hb in the R-structure has four hexacoordinated hemes, whereas in the T-structure, as induced by inositol hexaphosphate (IHP), two of the four Hb chains (i.e., the α chains) become penta-coordinated.[95,96,103] Stong et al.[105] reported the detection of the ν(Fe^{II}–NO) stretching vibration at 592 cm^{-1} from penta-coordinated α-subunits, which is very close to the ν(Fe^{III}–NO) stretching vibration in ferric Hb · NO (594 cm^{-1}) and ferric Mb · NO (595 cm^{-1}),[11] presumably hexacoordinated. Tsubaki and Yu[107] detected no enhancement of the isotope-sensitive line at ~592 cm^{-1} in the same (Hb · NO + IHP) system upon Soret excitation at 406.7 nm. Mackin et al.[109] also reported no enhancement of the ν(Fe^{II}–NO) stretching mode in penta-coordinated ferrous

[111] L. C. Dickinson and J. C. W. Chien, J. Am. Chem. Soc. 93, 5035 (1971).
[112] J. C. W. Chien, J. Chem. Phys. 51, 4220 (1969).

Mb · NO. It is possible that the $\nu(Fe^{II}-NO)$ stretching mode from penta-coordinated ferrous nitrosyl hemes is resonance enhanced via a charge-transfer mechanism near the 454.5 nm excitation wavelength used by Stong et al.[105] Further studies, especially excitation profile measurements near 454.5 nm, should help to resolve the question of whether a penta-coordinated nitrosyl heme does have the $\nu(Fe^{II}-NO)$ stretching vibration at 592 cm^{-1}.

Dioxygen. The nature of dioxygen binding to hemoproteins in general, and hemoglobin in particular, has been a matter of keen and continuing interest to both chemists and biochemists.[113-128] Ever since the first magnetic susceptibility measurement of oxyhemoglobin by Pauling and Coryell[113] in the 1930s, there has been considerable controversy over the ligand geometry[113,114] and electronic configuration of the iron.[113,115-121,123] X-Ray structural determinations on oxy "picket fence" complex,[124] oxymyoglobin,[127] and oxyhemoglobin[128] have now established that the Fe–O$_2$ moiety has an end-on geometry. However, there are still questions regarding the detailed geometry, in particular the Fe–O bond distances and the Fe–O–O angles.[77,127,128] A large variation in the Fe–O–O angle from 115° (Mb · O$_2$)[127] to 170° (oxyerythrocruorin)[77] does not seem likely in view of the small difference (~5 cm^{-1})[129] in the Fe–O–O bending vibration[11] (or

[113] L. Pauling and C. D. Coryell, *Proc. Natl. Acad. Sci. U.S.A.* **22**, 210 (1936); L. Pauling, *Nature (London)* **203**, 182 (1964); L. Pauling, *Proc. Natl. Acad. Sci. U.S.A.* **74**, 2612 (1977).

[114] J. S. Griffith, *Proc. R. Soc. Ser. A* **235**, 23 (1956).

[115] J. Weiss, *Nature (London)* **203**, 83 (1964).

[116] G. Lang and W. Marshall, *J. Mol. Biol.* **18**, 358 (1966).

[117] J. D. Wittenberg, B. A. Wittenberg, J. Peisach, and W. E. Blumberg, *Proc. Natl. Acad. Sci. U.S.A.* **67**, 1846 (1970).

[118] A. S. Koster, *J. Chem. Phys.* **56**, 3161 (1972).

[119] M. Cerdonio, A. Conqiu-Castellano, L. Calabrese, S. Morante, B. Pispisa, and S. Vitale, *Proc. Natl. Acad. Sci. U.S.A.* **75**, 4916 (1978).

[120] J. S. Philo, U. Dreyer, and T. M. Schuster, *Biophys. J.* **41**, 414a (Abstr.) (1983).

[121] C. H. Barlow, J. C. Maxwell, W. J. Wallace, and W. S. Caughey, *Biochem. Biophys. Res. Commun.* **58**, 166 (1973).

[122] H. Brunner, *Naturwissenschaften* **61**, 129 (1974).

[123] B. H. Huynh, D. A. Case, and M. Karplus, *J. Am. Chem. Soc.* **99**, 6103 (1977); D. A. Case, B. H. Huynh, and M. Karplus, *J. Am. Chem. Soc.* **101**, 4433 (1979).

[124] J. P. Collman, R. R. Gagne, C. A. Reed, W. T. Robinson, and G. A. Rodley, *Proc. Natl. Acad. Sci. U.S.A.* **71**, 1326 (1974); J. P. Collman, J. I. Brauman, T. R. Halbert, and K. S. Suslick, *Proc. Natl. Acad. Sci. U.S.A.* **73**, 3333 (1976).

[125] C. A. Reed and S. K. Cheung, *Proc. Natl. Acad. Sci. U.S.A.* **74**, 1780 (1977).

[126] C. K. Chang and T. G. Traylor, *J. Am. Chem. Soc.* **95**, 8477 (1973).

[127] S. E. V. Phillips, *J. Mol. Biol.* **142**, 531 (1980).

[128] B. Shaanan, *Nature (London)* **296**, 683 (1982).

[129] B. Benko, N.-T. Yu, and K. Gersonde (unpublished results).

the Fe–O_2 stretching mode)[122] at ~570 cm^{-1} between these two proteins. Another controversy is centered around the electronic configuration of the iron and the electronic charge on the bound dioxygen. Both Hb·O_2 and Hb·CO have been believed to be diamagnetic; the iron is in the ferrous low spin d^6 (S = 0) state and the oxygen changes from its normal S = 1 state to an S = 0 state upon binding. However, some spectroscopic evidence supports a configuration for Hb·O_2 with the iron in a ferric low spin state and the oxygen bound as superoxide (O_2^-).[115-121] It was then postulated that an antiferromagnetic coupling exists between the two unpaired spins (one on the iron and the other on the oxygen) to give a diamagnetic ground state.[117] Indeed, Cerdonio et al.[119] have reported paramagnetism of the heme in human Hb·O_2 at room temperature. However, Philo et al.[120] were unable to reproduce the data of Cerdonio et al.[119]

Resonance Raman evidence in favor of the Fe^{3+}–O_2^- formulation has been presented by Yamamoto et al.[130] and Spiro and Strekas.[131] Their argument was primarily based on the similarities in porphyrin ring vibrations at ~1377 (band I), ~1564 (band III), and ~1640 cm^{-1} (band V) between Hb·O_2 and low spin ferric species such as metHb cyanide. These Raman frequencies, particularly band I, reflect the electron density in the antibonding π^* orbital of the porphyrin macrocycle.[2,16] A decrease in the band I frequency from ~1355 cm^{-1} in deoxyHb to ~1377 cm^{-1} in oxyHb may be attributed to decreased electron density in the $e_g(\pi^*)$ orbital, with resultant increase in the π-bond order. Since it is not a direct measure of the charge on the iron, Raman evidence for the ferric state of the heme iron in oxyHb is not compelling. The notion of bound dioxygen as a superoxide (O_2^-) has been derived from the bound O–O stretching frequency[38,47,48] at 1100–1165 cm^{-1} in model compounds and hemoproteins, which happens to fall into the range characteristic of a superoxide ion found in molecules like NaO_2. The decrease in ν(O–O) from 1555 cm^{-1} (free O_2) to ~1100–1165 cm^{-1} is primarily caused by the backdonation of d_π(Fe) electrons into the $\pi^*(O_2)$ antibonding orbital, causing reduction in the O–O bond order. It does not necessarily mean that there is a net negative charge on the bound O_2. Because of the forward σ donation from O_2 to the iron (3 d_{z^2}, 4s, and $4p_z$) orbitals, there may be little net charge transfer to oxygen, as indicated by extended SCF-CI Pariser–Parr Pople (PPP) calculations.[123]

[130] T. Yamamoto, G. Palmer, D. Gill, I. T. Salmeen, and L. Rimai, *J. Biol. Chem.* **248**, 5211 (1973).
[131] T. G. Spiro and T. C. Strekas, *J. Am. Chem. Soc.* **96**, 338 (1974).

Since the reduction in the AB ligand bond order is primarily caused by the back-donation of electrons into the $\pi^*(AB)$ antibonding orbital, the decrease in $\nu(A-B)$ frequency upon binding may give information regarding the extent of electron donation. This decrease is ~430 cm^{-1} for O_2 (1155 $\rightarrow$ 1125 cm^{-1}),[47] 255 cm^{-1} for NO (1877 $\rightarrow$ 1622 cm^{-1}),[107] and ~195 cm^{-1} for CO (2145 $\rightarrow$ 1950 cm^{-1})[5] (see Table II). Here the free gas values were taken from Table I and the bound $\nu(A-B)$ frequencies are those of Hb/Mb complexes given in the cited references. Thus, the amount of $d_\pi(Fe) \rightarrow \pi^*(AB)$ back-donation decreases in the following order: $O_2 > NO > CO$, in reverse order for the energy of the π^* level.[132,133]

The $Fe^{II}-O_2$ moiety gives rise to three vibrational modes: $\nu(Fe^{II}-O_2)$ stretching, $\delta(Fe^{II}-O-O)$ bending, and $\nu(O-O)$ stretching. The $\nu(Fe^{II}-O_2)$ stretching vibration in oxyhemoglobin was first assigned by Brunner[122] to a Raman line at 567 cm^{-1} (excited at 488.0 nm) which shifts to 540 cm^{-1} upon $^{16}O_2 \rightarrow {}^{18}O_2$ isotope substitution. The same vibration has been detected at 577 cm^{-1} (oxymyoglobin),[56] 576 cm^{-1} (oxyleghemoglobin),[134] and 568 cm^{-1} [$Fe^{II}(TpivPP)(N-MeIm)O_2$].[135] While this assignment has been widely accepted, it is now being questioned by Benko and Yu.[11] On the basis of the insensitivity of this frequency to the isotopic label at the terminal oxygen atom it was suggested[11] as the $\delta(Fe^{II}-O-O)$ bending vibration. So far, the $\nu(O-O)$ stretching vibration of oxy iron hemes has not been directly detected by resonance Raman spectroscopy. However, the intensity increase of a porphyrin ring mode at 1125 cm^{-1} upon $^{18}O_2$ substitution (Mb $\cdot$ O_2) has been used[47] to argue that the $\nu(O-O)$ stretching mode must be very close to this frequency where a resonance interaction occurs between the $\nu(^{16}O-{}^{16}O)$ and the 1125 cm^{-1} porphyrin mode.

Direct observation of the $\nu(O-O)$ stretching vibration in hemoproteins via resonance Raman enhancement is possible if the heme iron is replaced by cobalt.[47] With excitation of 406.7 nm, Tsubaki and Yu[47] detected a strong isotope-sensitive line at 1137 cm^{-1} in oxyCoMb and oxyCoHbA, which is different from the 1105 cm^{-1} frequency identified by Maxwell and Caughey as the $\nu(O-O)$ stretching vibration in the infrared spectrum of oxyCoHbA (containing Co-deuteroporphyrin IX).[38] Resonance Raman spectra of oxyCoHbA with $^{16}O_2$ and $^{18}O_2$ are presented in Fig. 19, which

[132] R. F. Fenske and R. L. DeKock, *Inorg. Chem.* **9**, 1053 (1970).
[133] D. W. Turner and D. P. May, *J. Chem. Phys.* **45**, 471 (1966).
[134] M. J. Irwin, R. S. Armstrong, and P. E. Wright, *FEBS Lett.* **133**, 239 (1981).
[135] J. M. Burke, J. R. Kincaid, S. Peters, R. R. Gagne, J. P. Collman, and T. G. Spiro, *J. Am. Chem. Soc.* **100**, 6083 (1978).

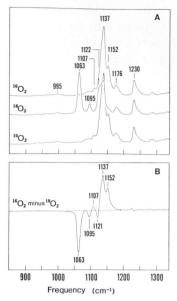

FIG. 19. Resonance Raman spectra of oxyCoHbA (A) and difference spectrum (CoHbA $^{16}O_2$ minus CoHbA $^{18}O_2$) (B) in the 900 to 1300 cm^{-1} region. From Tsubaki and Yu,[47] with permission.

shows clearly two additional isotope-sensitive lines at 1107 and 1152 cm^{-1}. The one at 1107 cm^{-1} presumably corresponds to the infrared band at 1105 cm^{-1}. In the $^{18}O_2$ spectrum there are two new lines at 1063 and 1095 cm^{-1}, which were assigned to the $\nu(^{18}O-^{18}O)$ stretching frequency from two different conformers.[47] The two frequencies at 1107 and 1137 cm^{-1} in the $^{16}O_2$ spectrum arise from resonance interaction between a $\nu(^{16}O-^{16}O)$ mode at ~1121 cm^{-1} and an accidentally degenerate porphyrin ring mode at 1121 cm^{-1}. The third frequency at 1152 cm^{-1} ($^{16}O_2$ spectrum) is the $\nu(^{16}O-^{16}O)$ mode from a second conformer, whose $\nu(^{18}O-^{18}O)$ frequency is at 1095 cm^{-1}. The 1152 cm^{-1} component is very weak in oxy-CoMb[47] and is absent in oxyCo^{II} mesoporphyrin IX-substituted myoglobin.[48] Thus the $\nu(O-O)$ mode at ~1121 cm^{-1} is the major conformer and that at 1152 cm^{-1} a minor one. In protein-free model complexes such as CO^{II}(TpivPP)(N-MeIm)(O$_2$),[48] there is only a single $\nu(O-O)$ frequency at 1153 cm^{-1}. The lowering of the $\nu(O-O)$ frequency for the major conformer in oxyCoMb and oxyCoHb may be caused by some specific interactions between bound oxygen and distal residues. Previously, EPR and functional studies on various cobalt-substituted Hbs and Mbs have already suggested the presence of an oxygen–histidine hydrogen bond.[136–138] X-ray (oxyCoMb[139] and oxyFeHb[128]) and neutron (oxy-FeMb[140]) diffraction studies also revealed the existence of such a hydrogen bond. Kitagawa *et al.*[141] demonstrated the sensitivity of the $\nu(O–O)$ frequency in both oxyCoMb and oxyCoHb to the replacement of H_2O by D_2O. It shifts from 1130 (H_2O) to 1138 cm^{-1} (D_2O) in oxyCoHb. The results imply that the bound oxygen interacts with an adjacent exchangeable proton, presumably from the distal histidine.[141]

The fact that the $\nu(O–O)$ vibration in oxyCo-hemes, but not in oxyFe-hemes, can be resonance enhanced upon Soret excitation suggests the presence of a charge-transfer (CT) transition underlying the intense Soret band. The contribution from coupling with the porphyrin $\pi \to \pi^*$ transition is insignificant as judged from its absence upon Q-band excitation. In contrast, the $\nu(Fe^{II}–CO)$ stretching at ~ 500 cm^{-1} and $\delta(Fe–N–O)$ bending at 554 cm^{-1} can be resonance enhanced in both the Soret and Q-band regions.[5,108] The CT transition responsible for the $\nu(O–O)$ enhancement was suggested as the type $\pi^*(u_g^{*}O_2/d_{xz}) \to \sigma^*(d_{z^2}/Co/\pi_g^*)$ (see Fig. 20 for a simplified molecular orbital energy diagram). With Co^{II} as the metal, there are nine electrons to populate these orbitals; the $\pi^*(d_{xz}/\pi_g^*)$ antibonding orbital is half occupied ($S = \frac{1}{2}$). Electron spin resonance studies[142] indicated that the unpaired electron resides in an orbital predominantly localized on dioxygen. This unpaired electron can be promoted upon excitation to the antibonding σ^* molecular orbital involving primarily $d_{z^2}(Co)$ and $\pi_g^*(O_2)$. This electronic displacement would cause elongation of the Co–O bond and contraction of the O–O bond in the excited state. In general, the vibrational modes which resemble the distortions in the excited state are likely to be resonance enhanced.[143] Indeed, the $\nu(Co–O_2)$

[136] M. Ikeda-Saito, T. Iizuka, H. Yamamoto, F. J. Kayne, and T. Yonetani, *J. Biol. Chem.* **252**, 4882 (1977).

[137] M. Ikeda-Saito, M. Brunori, and T. Yonetani, *Biochim. Biophys. Acta* **533**, 173 (1978).

[138] M. Ikeda-Saito, H. Hori, T. Inbushi, and T. Yonetani, *J. Biol. Chem.* **256**, 10267 (1981).

[139] G. A. Petsko, D. Rose, D. Tsernoglou, M. Ikeda-Saito, and T. Yonetani, *in* "Frontiers of Biological Energetics" (P. L. Dutton, J. S. Leigh, and A. Scarpa, eds.), p. 1011. Academic Press, New York, 1978.

[140] S. V. E. Phillips and B. P. Schoenborn, *Nature (London)* **292**, 81 (1981).

[141] T. Kitagawa, M. R. Ondrias, D. L. Rousseau, M. Ikeda-Saito, and T. Yonetani, *Nature (London)* **298**, 869 (1982).

[142] B. M. Hoffman, D. L. Diemente, and F. Basolo, *J. Am. Chem. Soc.* **92**, 61 (1970); D. Getz, E. Melemud, B. C. Silver, and Z. Dori, *J. Am. Chem. Soc.* **97**, 3846 (1975); R. K. Gupta, A. S. Mildvan, T. Yonetani, and T. S. Srivastava, *Biochem. Biophys. Res. Commun.* **67**, 1005 (1975).

[143] A. Y. Hirakawa and M. Tsuboi, *Science* **188**, 359 (1975); M. Tsuboi and A. Y. Hirakawa, *J. Raman Spectrosc.* **5**, 75 (1976).

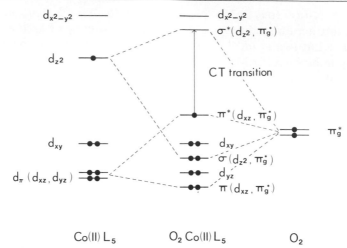

FIG. 20. Schematic energy level diagram for molecular orbitals in the $O_2Co^{II}L_5$ system. The charge-transfer (CT) transition from π^* (d_{xz}, π_g^*) to π^* (d_{z^2}, π_g^*) is responsible for the resonance enhancement of ν(O–O) stretching vibration. From Tsubaki and Yu,[47] with permission.

stretching mode in oxyCoMb and oxyCoHb have also been observed[47] with excitation at 406.7 nm.

The existence of the Co^{II}–O_2 CT transition in the Soret region near 400 nm was suggested earlier by Nozawa et al.[144] in the O_2 adducts of Co^{II} mesoporphyrin IX dimethyl ester, and by Wayland et al.[145] in Co^{II}(TPP)O_2. Recent observations of both ν(O–O) and ν(Co–O_2) in Co^{II}(TPP)(Py)O_2 with excitation at 457.9 nm by Bajdor and Nakamoto[146] also support the charge-transfer mechanism for the enhancement of ν(O–O) and ν(Co–O_2) in oxyCo-hemes. The Co^{II}–O_2 charge-transfer bands in Co^{II} Schiff base complexes are located in the 500–600 nm region.[147]

The evidence for the existence of two conformers in oxyFeMb/HbA comes from a combination of resonance Raman and infrared data. The intensity increase at 1125 cm^{-1} in the RR spectrum of oxyFeMb upon $^{16}O_2$ → $^{18}O_2$ substitution[47] indicates the presence of a ν(O–O) frequency at

[144] T. Nozawa, M. Hatano, H. Yamamoto, and T. Kwan, *Bioinorg. Chem.* **5**, 267 (1973).
[145] B. B. Wayland, J. V. Minkiewitz, and M. E. Abd-Elmageed, *J. Am. Chem. Soc.* **96**, 2795 (1974).
[146] K. Bajdor and K. Nakamoto, *J. Am. Chem. Soc.* **105**, 678 (1983).
[147] K. Nakamoto, M. Suzuki, T. Ishiguro, M. Kozuka, Y. Nishida, and S. Kida, *Inorg. Chem.* **19**, 2822 (1980); M. Suzuki, T. Ishiguro, M. Kozuka, and K. Nakamoto, *Inorg. Chem.* **20**, 1993 (1981); K. Nakamoto, Y. Nonaka, T. Ishiguro, M. W. Urban, M. Suzuki, Kozuka, Y. Nishida, and S. Kida, *J. Am. Chem. Soc.* **104**, 3386 (1982).

~ 1125 cm^{-1}, which is accidentally degenerate with a porphyrin ring mode at 1125 cm^{-1} (vibrational perturbation). On the other hand, infrared data indicate the existence of a second $\nu(O-O)$ at 1148 cm^{-1} (oxyFeMb)[148] and 1156 cm^{-1} (oxyHbA).[149] The second (or minor) conformer in oxyCo and oxyFeMb/HbA presumably does not have a hydrogen bond between bound oxygen and the distal histidine because its $\nu(O-O)$ at ~ 1150 cm^{-1} is similar to the values found in Co^{II}(TpivPP)(N-MeIm)O$_2$ at 1153 cm^{-1} and Fe^{II}(TpivPP)(N-MeIm)(O$_2$)[150] at 1160 cm^{-1}.

The similarity in $\nu(O-O)$ between Co^{II}(TpivPP)(N-MeIm)O$_2$ and Fe^{II}(TpivPP)(N-MeIm)O$_2$ suggests that the Co^{II}-O$_2$ and Fe^{II}-O$_2$ bonds may also be similar. Since Burke et $al.$[135] and Hori and Kitagawa[9] have already assigned the $\nu(Fe-O_2)$ stretching vibration at 568 cm^{-1} in Fe^{II}(TpivPP)(N-MeIm)O$_2$, it came as a surprise that the $\nu(Co-O_2)$ stretching vibration was detected[48] at 516 cm^{-1} in Co^{II}(TpivPP)(N-MeIm)O$_2$. Earlier, Maxwell and Caughey,[151] on the basis of the similarity in the $\nu(O-O)$ vibration between iron and cobalt oxyHb, also argued in favor of similar Fe-O$_2$ and Co-O$_2$ bonds. Since the $\angle$Co^{II}-O-O and $\angle$Fe^{II}-O-O angles are similar ($\sim 130°$),[138,152] the 52 cm^{-1} difference between $\nu(Fe-O_2)$ and $\nu(Co-O_2)$ cannot be accounted for by different degrees of mixing in the normal mode or by the metal mass difference, which would shift the frequency by ~ 4 cm^{-1}. Thus, the frequency difference of 48 cm^{-1} must be attributed to the much weaker Co-O$_2$ bond, compared to the Fe-O$_2$ bond. However, it is rather peculiar that the $\nu(Fe-O_2)$ and $\nu(Co-O_2)$ frequencies can differ by as much as ~ 50 cm^{-1} when the $\nu(O-O)$ frequencies are nearly the same. The difficulties may be removed by assuming that the 568 cm^{-1} line in Fe^{II}(TpivPP)(N-MeIm)O$_2$ is the δ(Fe^{II}-O-O) bending mode, as suggested by Benko and Yu[11] for the 567 cm^{-1} line in oxyHb. It is reasonable for the δ(Fe^{II}-O-O) bending mode to appear at ~ 570 cm^{-1} because the δ(Fe^{II}-C-O) bending and δ(Fe^{II}-N-O) bending vibrations have been definitely located at ~ 577 and ~ 554 cm^{-1}, respectively. A search for the $\nu(Fe-O_2)$ stretching frequencies in oxyFe-hemes can now

148 W. S. Caughey, H. Shimada, M. P. Tucker, S. Kawanishi, S. Yoshikawa, and L. J. Young, in "Oxygenase and Oxygen Metabolism" (M. Nozaki, S. Yamamoto, Y. Ishimura, M. J. Coon, L. Ernster, and R. W. Estabrook, eds.). Academic Press, New York, 1983.

149 J. O. Alben, G. H. Bare, and P. P. Moh, in "Biochemical and Clinical Aspects of Hemoglobin Abnormalities" (W. S. Caughey, ed.), p. 607. Academic Press, New York, 1978.

150 J. P. Collman, J. I. Brauman, T. R. Halbert, and K. S. Suslick, Proc. Natl. Acad. Sci. U.S.A. 73, 3333 (1976).

151 J. C. Maxwell and W. S. Caughey, Biochem. Biophys. Res. Commun. 60, 1309 (1974).

152 G. B. Jameson, G. A. Rodley, W. T. Robinson, R. R. Gagne, C. A. Reed, and J. P. Collman, Inorg. Chem. 17, 850 (1978).

be made. Direct excitation into the $Fe^{II}-O_2$ charge-transfer transition at $\sim$325 nm[153] may bring out both the $\nu(Fe-O_2)$ and $\nu(O-O)$ vibrations.

The trans (proximal) base influence on the strength of the $Co-O_2$ bond has been investigated by Mackin et al.,[48] who found that the $\nu(Co-O_2)$ at 527 cm^{-1} for the low affinity $Co^{II}(TpivPP)(1,2-Me_2Im)(O_2)$ complex is 11 cm^{-1} higher than the 516 cm^{-1} value for the high affinity $Co^{II}(TpivPP)(N-MeIm)(O_2)$ complex. The O_2 binding affinity is lowered by a factor of 6.4 on going from the unstrained (N-MeIm) complex to the (1,2-Me$_2$Im) complex.[26] In other words, tension on the trans-imidazole bond increases the $Co-O_2$ bond strength, but decreases the O_2 binding affinity. A similar relation was found with carbon monoxide binding to Fe-hemes. However, in oxyFe-hemes, tension on the trans-imidazole bond decreases both the O_2 binding affinity[154] and the $Fe-O_2$ bond strength.[155] The $Fe-O_2$ bond length is longer in $Fe^{II}(TpivPP)(2-MeIm)O_2$ (i.e., 1.90 Å)[155] than in $Fe^{II}(TpivPP)(N-MeIm)O_2$ (1.75 Å);[152] the binding affinity is lower for the (2-MeIm) complex by three orders of magnitude.[154] Nevertheless, resonance Raman data exhibit only an 8–11 cm^{-1} difference in the $\nu(Fe-O_2)$ stretching frequency between the two complexes, which is too small to account for the 0.15 Å bond length. As pointed out by Walters et al.,[154] Badger's rule predicts a difference of only $\sim$0.01 Å for a 10 cm^{-1} decrease in $\nu(Fe-O_2)$. The discrepancy may be caused by the incorrect assignment of the 567 cm^{-1} line as the $\nu(Fe-O_2)$ stretching mode.[135] It may be more appropriate to assign it as the $\delta(Fe-O-O)$ bending vibration.[11]

The difference in the trans effect between oxy Fe-hemes and carbonmonoxy Fe-hemes may be rationalized in terms of the relative σ and π donating character of the imidazole ligands. In σ-bonding, the electron donations from both sides of the heme compete for the same $Fe(d_{z^2})$ orbital. A change from N-MeIm to 1,2-Me$_2$Im decreases the σ donation from the proximal side, and hence increases the σ-bonding for both the $Fe-O_2$ and $Fe-CO$ bonds. In π-bonding, the interaction between $d_\pi(Fe)$ and $\pi^*(O_2)$ is stronger than that between $d_\pi(Fe)$ and $\pi^*(CO)$. It is well known that O_2 depends more strongly upon the π basicity of the proximal base than does CO.[126] Thus, the substantially stretched $(Fe-N_\varepsilon)$ bond of the (1,2-Me$_2$Im) complexes decreases the imidazole to iron π donation,

[153] W. A. Eaton, L. K. Hanson, P. J. Stephens, J. C. Sutherland, and J. B. R. Dunn, J. Am. Chem. Soc. 100, 4991 (1978).

[154] M. A. Walters, T. G. Spiro, K. S. Suslick, and J. P. Collman, J. Am. Chem. Soc. 102, 6857 (1980).

[155] G. B. Jameson, F. S. Molinaro, J. A. Ibers, J. P. Collman, J. I. Brauman, E. Rose, and K. S. Suslick, J. Am. Chem. Soc. 100, 6769 (1978); 102, 3224 (1980).

which decreases the strength of the $d_\pi(Fe)-\pi^*(O_2)$ interaction to a greater extent relative to the $d_\pi(Fe)-\pi^*(CO)$ interaction. With O_2 binding, the weakening of the $Fe-O_2$ bond strength in the constrained complex indicates a large decrease in π bonding between Fe and O_2 which more than counterbalances the increase in σ-bonding. With CO binding, the increase in Fe–CO σ-bonding more than compensates the decrease in π-bonding.

Without a proximal base the O_2 binding affinity for $Co^{II}TPP$ is weak, but its complex can be stabilized at low temperature. Kozuka and Nakamoto[156] were able to detect the $\nu(Co-O_2)$ and $\nu(O-O)$ stretching vibrations at 345 and 1278 cm^{-1}, respectively, in the $Co^{II}(TPP)O_2/Ar$ matrix by infrared spectroscopy. The $\nu(Co-O_2)$ at 345 cm^{-1} is unusually low compared to 520 cm^{-1} in the hexa-coordinated $Co^{II}(TPP)(Py)O_2$ complex.[146] Here, a weak $Co-O_2$ bond corresponds to weak O_2 binding. On the other hand, the $\nu(O-O)$ frequency at 1278 cm^{-1} is high, relative to the 1144 cm^{-1} value in $Co^{II}(TPP)(Py)O_2$.[48,146] It appears that the higher the $\nu(O-O)$ frequency, the lower the $\nu(Co-O_2)$ frequency. It is of interest to note that the order of $\nu(O-O)$ for "base-free" adducts is $Co^{II}(TPP)O_2$ (1278 cm^{-1})[156] > $Fe^{II}(TPP)O_2$ (1195/1106 cm^{-1})[157] > $Mn^{II}(TPP)O_2$ (983 cm^{-1}).[158]

Cyanide. Cyanide is one of the strongest ligands that bind to ferric hemoproteins. For example, in the case of human ferric Hb (at pH 7.0, ionic strength 0.05 M and 20°) the binding constant is $8 \times 10^8\ M^{-1}$, which amounts to a free energy change of 12 kcal/mol.[159] The additional factor which favors the formation of the iron–ligand bond is the electrostatic attraction between CN^- and ferric iron. Therefore, one would expect a very strong (hence short) $Fe^{III}-CN^-$ bond. Quite contrarily, a recent X-ray crystallographic analysis of $Fe^{III}(TPP)(Py)(CN^-)$ reveals an unusually long $Fe^{III}-CN^-$ bond (1.908 Å),[160] which is 0.14 Å longer than the $Fe^{II}-CO$ bond (1.77 Å) in $Fe^{II}(TPP)(Py)(CO)$.[161] The $\nu(Fe^{III}-CN^-)$ stretching vibration identified at 452 cm^{-1} in $Fe^{III}(TPP)(Py)(CN^-)$[162] is 34 cm^{-1} lower than the $\nu(Fe^{II}-CO)$ stretching frequency at 486 cm^{-1} in $Fe^{II}(TPP)(Py)(CO)$.[12] Even the CN^- mass is 2 amu lighter than the CO mass. The $\nu(Fe^{III}-CN^-)$

---

[156] M. Kozuka and K. Nakamoto, *J. Am. Chem. Soc.* **103**, 2162 (1981).

[157] K. Nakamoto, T. Watanabe, T. Ama, and M. W. Urban, *J. Am. Chem. Soc.* **104**, 3744 (1982).

[158] M. W. Urban, K. Nakamoto, and F. Basolo, *Inorg. Chem.* **21**, 3406 (1982).

[159] E. Antonini and M. Brunori, "Hemoglobin and Myoglobin in Their Reactions with Ligands," p. 278. North-Holland Publ., Amsterdam, 1971.

[160] W. R. Scheidt, Y. J. Lee, W. Luangdilok, K. J. Haller, K. Auzi, and K. Hatano, *Inorg. Chem.* **22**, 1516 (1983).

[161] S. M. Peng and J. A. Ibers, *J. Am. Chem. Soc.* **98**, 8032 (1976).

[162] T. Tanaka and N.-T. Yu (unpublished results).

vibration in human cyanometHb was detected at 452 cm^{-1} by Henry,[163] and at 455 cm^{-1} by Tsubaki and Yu (unpublished results, cited in Ref. 5). More recently, Yu et al.[12] employed two different isotopes ($^{13}CN^-$ and $C^{15}N^-$) to distinguish the $\nu(Fe^{III}-CN^-)$ stretching (at 453 cm^{-1}) from the $\delta(Fe^{III}-C-N^-)$ bending (at 410 cm^{-1}) in the cyanide complex of erythrocruorin (monomeric insect hemoglobin III from *Chironomus thummi thummi*). The $\nu(Fe^{III}-CN^-)$ frequencies from hemoproteins are indeed very similar to that in $Fe^{III}(TPP)(Py)(CN^-)$, indicating that the $Fe^{III}-CN^-$ bond length in human Hb and insect Hb derivatives must be very close to 1.91 Å, a value much shorter than the 2.2 Å value estimated by Steigeman and Weber[77] from X-ray diffraction analysis of cyanometerythrocruorin. This estimate, together with the estimate of the $Fe^{II}-CO$ bond length (2.40 Å)[75] in carbonmonoxyerythrocruorin, indicates substantial errors in the X-ray analysis of Steigeman and Weber.

The ligand pockets in hemoproteins hinder binding of cyanide and carbon monoxide in their preferred linear axial coordination modes and force them to assume a distorted geometry. X-Ray crystallographic studies[75,164-167] show that the N-atom of cyanide is definitely displaced off the heme normal, but the poorly defined C-atom position does not permit a distinction between a bent and a tilted (or kinked) geometry. However, Yu et al.[12] found that the $\nu(Fe^{III}-^{13}CN^-)$ and $\nu(Fe^{III}-C^{15}N^-)$ frequencies are identical at 450 cm^{-1}, an indication that the FeCN linkage is essentially linear. Therefore, it was concluded[12] that the distortion in the FeCN group of hemoproteins must predominantly result from a tilting mechanism, as suggested by Deatherage et al.[166]

To account for the large free energy decrease upon cyanide binding to ferric hemes, one must consider contributions from the formation of the iron-cyanide bond and the associated bond strengthening/weakening in various parts of the complexes. If the globin conformation differs between liganded and unliganded states, its contribution must also be included. In comparing cyanide and fluoride binding to aquametHb (or aquametMb), one may neglect the contribution from the globin because the large difference in binding affinity between F^- and CN^- is not unique to Hb or Mb. First, there is creation of a chemical bond between the heme iron and the

[163] E. R. Henry, Ph.D. thesis, Princeton University, Princeton, NJ (1980).
[164] P. A. Bretscher, *Nature (London)* **219,** 606 (1968).
[165] W. A. Hendrickson and W. E. Love, *Nature (London) New Biol.* **232,** 197 (1971).
[166] J. F. Deatherage, R. S. Loe, C. M. Anderson, and K. Moffatt, *J. Mol. Biol.* **104,** 687 (1976).
[167] T. L. Poulos, S. T. Freer, R. A. Alden, N. H. Xuong, S. L. Edwards, R. C. Hamlin, and J. Kraut, *J. Biol. Chem.* **253,** 3730 (1978).

cyanide ligand which is associated with a decrease in free energy. The cyanide ligand induces a high to low spin transition which shortens numerous chemical bonds such as Fe–histidine, C–N, Fe–N (pyrrole), and those in the heme macrocycle. The ν(C–N) stretching frequency increases from 2083 cm^{-1} (KCN in aqueous solution)[168] to 2127 cm^{-1} (cyanometMb).[169,170] The frequency changes in the porphyrin ring modes in the 1400–1700 cm^{-1} region (characteristic of the bond order in the heme macrocycle) are also striking. In going from aquametHb to cyanometHb the frequency increases are[14,171,172]: 1610 → 1642 cm^{-1} (band V), 1561 → 1588 cm^{-1} (band IV), and 1481 → 1508 cm^{-1} (band II). This implies a large decrease in free energy contributed by the porphyrin ring. However, in going from aquametHb to fluorometHb, there are frequency decreases: 1610 → 1608 cm^{-1} (band V), 1561 → 1556 cm^{-1} (band IV), and 1481 → 1482 cm^{-1} (band II). It is known that the free energy decrease associated with fluoride binding to human aquametHb is only 2.5 kcal/mol,[159] a value 9.5 kcal/mol less than that for cyanide binding.

Cyanide also binds to ferrohemes, but the binding affinity is much lower. The resulting complexes such as ferromyoglobin–cyanide and ferroperoxidase–cyanide are readily photodissociable (quantum yield ~1)[172,173] presumably because [Fe^{II}–CN$^-$] is isoelectronic with the [Fe^{II}CO] system. Attempts to observe the Fe^{II}–CN$^-$ vibrations so far have been unsuccessful.

Fluoride. The binding of fluoride to ferric Hb/Mb results in the formation of fully high spin complexes.[174,175] Addition of IHP to metHb · F$^-$ induces the conversion of the protein from the R to the T form.[176] Concomitant with this structural change is a 20-Å red-shift of the ~600 nm charge-transfer band. X-Ray diffraction analysis of metHb · F$^-$ by Deatherage *et al.*[177] revealed a heterogeneity in the fluoride environment; there may be an equilibrium between two species, those with and without hydrogen bonding to water.

[168] N.-T. Yu, A. Lanir, and M. M. Weber, *J. Raman Spectrosc.* **11**, 150 (1981).

[169] S. McCoy and W. S. Caughey, *Biochemistry* **9**, 2387 (1970).

[170] W. S. Caughey, *Adv. Inorg. Biochem.* **2**, 95 (1980).

[171] T. G. Spiro, J. D. Stong, and P. Stein, *J. Am. Chem. Soc.* **101**, 2648 (1979).

[172] D. Keilin and E. F. Hartree, *Biochem. J.* **61**, 153 (1955).

[173] B. M. Hoffman and Q. H. Gibson, *Proc. Natl. Acad. Sci. U.S.A.* **75**, 21 (1978).

[174] J. G. Bettlestone and P. George, *Biochemistry* **3**, 707 (1964).

[175] M. F. Perutz, E. J. Hidner, J. E. Ladner, J. G. Beetlestone, C. Ho, and E. F. Slade, *Biochemistry* **13**, 2187 (1974).

[176] G. Fermi and M. F. Perutz, *J. Mol. Biol.* **114**, 421 (1977).

[177] J. F. Deatherage, R. S. Loe, C. M. Anderson, and K. Moffat, *J. Mol. Biol.* **104**, 687 (1976).

Upon excitation in the charge-transfer region of metHb · F⁻, Asher *et al.*[51] detected a selectively enhanced doublet at 468 and 439 cm⁻¹, which were assigned to the $\nu(Fe^{III}-F^-)$ stretching vibrations, corresponding to the unperturbed and H-bonded species, respectively. Similar doublets have also been detected[50] at 461 and 422 cm⁻¹ in metMb · F⁻, at 466 and 441 cm⁻¹ in met (α-subunit) · F⁻, and at 471 and 444 cm⁻¹ in met (β-subunit) · F⁻. These frequencies are much lower than the $\nu(Fe^{III}-F^-)$ stretch at 600 cm⁻¹ in ferric octaethylporphyrin fluoride[178] and at 580 cm⁻¹ in the fluoride complex of ferric mesoporphyrin IX dimethyl ester.[179] Asher *et al.*[51] argue that the environment of the iron in metHb · F⁻ is quite different from that in fluoride complexes of metalloporphyrins. In penta-coordinated fluoride complexes, the iron lies ~0.45 Å out-of-plane toward the fluoride ligand,[180] whereas in metHb · F⁻ the iron lies 0.3 Å out-of-plane toward the proximal-histidine on the opposite side of the fluoride ion.[181] Thus, the iron is displaced about 0.75 Å in metHb · F⁻ compared to ferric porphyrin fluoride, resulting in a ~120 cm⁻¹ decrease in frequency of the $\nu(Fe^{III}-F^-)$ stretch in metHb · F⁻.

The original assignment of the doublet to the $\nu(Fe^{III}-F^-)$ stretch in metHb · F⁻ by Asher *et al.*[51] was based on indirect evidence, without isotope shift data. However, it was supported by the work of Desbois *et al.*,[56] who detected an ($^{56}Fe \rightarrow {}^{54}Fe$) isotope-sensitive line at 462 cm⁻¹ in metMb · F⁻ upon Soret excitation. It now appears that the assignment of Asher *et al.*[51] is correct.

Asher *et al.*[182] demonstrated the pH sensitivity of the $\nu(Fe^{III}-F^-)$ stretching frequency. By lowering the pH from 7.0 to ~5.0, the $\nu(Fe^{III}-F^-)$ stretch at 461 (metMb · F⁻) and 468 cm⁻¹ (metHb · F⁻) shifts to 399 and 407 cm⁻¹. This was correctly interpreted as resulting from protonation of the distal histidine and the formation of a hydrogen bond to the fluoride ligand. The stronger the hydrogen bond, the lower the $\nu(Fe^{III}-F^-)$ stretching frequency. The relationship between $\nu(Fe^{III}-F^-)$ frequency and three types of fluoride environment is depicted in Fig. 21.

An unusually low $\nu(Fe^{III}-F^-)$ frequency was detected (in the present author's laboratory) at 385 cm⁻¹ in the fluoride complex of horseradish peroxidase (see Fig. 22), where strong hydrogen bonding between the

[178] J. Kincaid and K. Nakamoto, *Spectrosc. Lett.* **9,** 19 (1976).

[179] T. G. Spiro and J. M. Burke, *J. Am. Chem. Soc.* **98,** 5482 (1976).

[180] J. L. Hood, *in* "Porphyrins and Metalloporphyrins" (K. M. Smith, ed.). Elsevier, New York, 1975.

[181] J. F. Deatherage, R. S. Loe, and K. Moffat, *J. Mol. Biol.* **104,** 723 (1976); M. F. Perutz, A. R. Fersht, S. R. Simon, and G. C. K. Roberts, *Biochemistry* **13,** 2174 (1974).

[182] S. A. Asher, M. L. Adams, and T. M. Schuster, *Biochemistry* **20,** 3339 (1981).

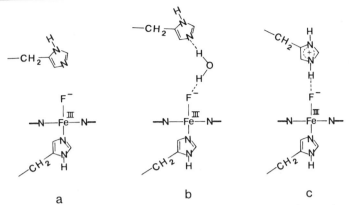

FIG. 21. Variations of $\nu(Fe^{III}-F^-)$ stretching frequency with the microenvironment: (a) no hydrogen bond, 460–470; (b) hydrogen bond with H_2O, 420–440 cm^{-1}; (c) strong hydrogen bond with a protonated distal histidine, 380–410 cm^{-1}. After Asher *et al.*,[182] with modifications.

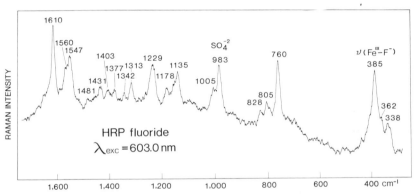

FIG. 22. Resonance Raman spectrum of HRP fluoride excited at 603.0 nm (CMX 4 xenon flashlamp-pumped dye laser; 25 pulses per second). First presentation here, not yet reported elsewhere.

distal histidine and F^- is expected.[183] It is generally recognized[183–187] that the weak acid HF is kinetically a more significant inhibitor than its conjugate base F^-. The ionization of HF ($HF \rightleftarrows H^+ + F^-$) at the active site is prerequisite to the ligation of F^- by the ferric ion.[183] The distal histidine

183 G. R. Schonbaum, R. A. Houtchens, and W. S. Caughey, *in* "Biochemical and Clinical Aspects of Oxygen" (W. S. Caughey, ed.), p. 195. Academic Press, New York, 1979.
184 B. Chance, *J. Biol. Chem.* **194,** 483 (1952).
185 G. R. Schonbaum, *J. Biol. Chem.* **248,** 502 (1973).
186 J. E. Erman, *Biochemistry* **13,** 34 (1974).
187 D. Dolman, G. A. Newell, M. D. Thurlow, and H. B. Dunford, *Can. J. Biochem.* **53,** 495 (1975).

accepts the released proton to form an imidazolium ion which then hydrogen bonds to the fluoride.

Asher et al.[51] and Rousseau et al.[2] examined the effects of IHP-induced R → T structural transition on the resonance Raman spectrum of metHb · F$^-$. Both groups formed only small frequency shifts (less than 1 cm^{-1}) in certain porphyrin ring modes. However, no detectable shifts occur at the Fe–F stretching doublet.[50,51] It appears that the 20-Å absorption red-shift induced by IHP does not result from a change in the Fe–F bond. The effect of the R → T conversion in metHb · F$^-$ is to perturb heme macrocyclic structure without significantly altering the heme out-of-plane iron distance or the Fe–F bond length.

An important energy compensation between the heme macrocycle and the Fe–F bond was noted by Asher et al.[50] The macrocyclic modes at 760, 1429, 1550, and 1611 cm^{-1} in α^{III}F are higher in frequency than the corresponding ones (758, 1427, 1547, and 1609 cm^{-1}) in β^{III}F, whereas the α-subunit $\nu(Fe^{III}–F^-)$ doublet at 466 and 441 cm^{-1} appears 5 cm^{-1} lower than the β-subunit doublet.

It is interesting to note that the $\nu(Fe^{III}–F^-)$ stretch in metMb · F$^-$ can be resonance enhanced in both the charge-transfer (~600 nm)[50] and Soret regions (~442 nm).[56,188] The $\nu(Fe^{III}–F^-)$ stretch in ferric leghemoglobin fluoride was detected at 470 and 446 cm^{-1} upon excitation at 441.6 nm. The resonance Raman spectrum of ferric HRP fluoride reported by Teraoka et al.[189] contains an unusually intense line at 385 cm^{-1}, which has not been assigned, but may correspond to the $\nu(Fe^{III}–F^-)$ stretching vibration at 385 cm^{-1} selectively enhanced via charge-transfer excitation (Fig. 22). In contrast, the spectrum of hog intestinal peroxidase fluoride[190] excited at 441.6 nm exhibits no Raman lines assignable to the $\nu(Fe^{III}–F^-)$ stretch.

Azide. Unlike cyanide and fluoride, the binding of azide to hemoproteins results in the formation of both high spin and low spin complexes at room temperature.[191] In metMb · N$_3^-$ and metHb · N$_3^-$, the energy separation between the high spin ($S = 5/2$) and low spin ($S = 1/2$) states is comparable to the thermal energy at room temperature so that the spin equilibrium is temperature dependent. The equilibrium constants are different between myoglobin and hemoglobin derivatives. McCoy and Caughey[41] identified the bound azide antisymmetric stretch at 2023 and

[188] A. Debois and M. Lutz, in "Raman Spectroscopy: Linear and Nonlinear" (J. Lascombe and P. V. Huong, eds.), p. 737. Wiley, New York, 1982.
[189] J. Teraoka, T. Ogura, and T. Kitagawa, *J. Am. Chem. Soc.* **104,** 7354 (1982).
[190] S. Kimura, I. Yamazaki, and T. Kitagawa, *Biochemistry* **20,** 4632 (1981).
[191] T. Iizuka and M. Kotani, *Biochim. Biophys. Acta* **154,** 417 (1968); **181,** 275 (1969); **194,** 351 (1969).

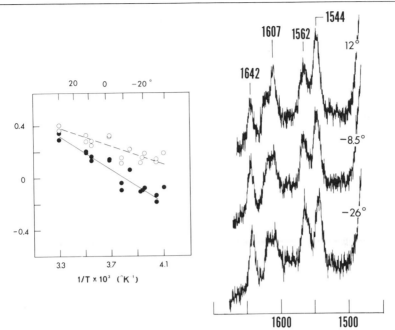

FIG. 23. Effect of temperature on the resonance Raman spectrum of metmyoglobin azide in the 1500–1700 cm^{-1} region. The logarithms of peak height intensity ratios, log (I_{1607}/I_{1642}) (●) and log (I_{1544}/I_{1642}) (○) are plotted against $1/T$. λ_{exc} = 647.1 nm. Reprinted with permission from Tsubaki et al.[46] (1981). Copyright 1981 American Chemical Society.

2046 cm^{-1} in metMb · N$_3^-$, and at 2026 and 2048 cm^{-1} in metHb · N$_3^-$. They assigned the lower frequencies to low spin and higher frequencies to high spin. Indeed, the relative intensities of the two infrared bands exhibited a temperature-dependent behavior expected from a thermal spin equilibrium.

The resonance Raman spectrum of metMb · N$_3^-$ in the porphyrin ring mode (1500–1700 cm^{-1}) region displays two sets of lines from both high spin and low spin species. As shown in Fig. 23, the two high spin lines appear at 1607 and 1544 cm^{-1}, and the low spin line at 1642 cm^{-1}. By decreasing temperature from 12 to −26°, the intensities of the high spin lines decrease relative to that at 1642 cm^{-1}.

Metmyoglobin azide (or methemoglobin azide) has an absorption band at ~640 nm which was assigned to an x,y-polarized charge-transfer from the high spin species by Eaton and Hochrasser.[192] Upon freezing to 77 K, metMb · N$_3^-$ is converted completely to the low spin form; yet its absorption spectrum still exhibits a weak absorption at ~650 nm, which

[192] W. A. Eaton and R. M. Hochrasser, J. Chem. Phys. **49**, 985 (1968).

was suggested[192] as a z-polarized charge-transfer transition from the low spin species. With excitation at 638.3 nm, Asher *et al.*[51] detected a selectively enhanced Raman line at 413 cm^{-1} in metMb · N$_3^-$ and assigned it to the ν(Fe^{III}–N$_3^-$) stretching vibration (without isotope shift data). Later, Desbois *et al.*[101] observed a line at 570 cm^{-1} (excited at 441.6 nm) which was assigned by them to the ν(Fe^{III}–N$_3^-$) stretch on the basis of a 16 cm^{-1} ^{15}N$_3$ isotope shift. The 157 cm^{-1} difference between the two frequencies is indeed quite large. However, Asher and Schuster[49] suggested that the 413 cm^{-1} line is from high spin and the 570 cm^{-1} mode from the low spin form, but both were still the ν(Fe^{III}–N$_3^-$) stretch.

To test the validity of these assignments, Tsubaki *et al.*[193] studied the isotopic shift of the 413 cm^{-1} line and its temperature effect, and compared the depolarization ratios of the 413 and 570 cm^{-1} lines. With 647.1 nm excitation they detected a strong line at 411 cm^{-1} in metMb · N$_3^-$ which shifts only 6 cm^{-1} upon ^{15}N$_3$ isotope substitution, compared to the 16 cm^{-1} observed for the 570 cm^{-1} line by Desbois *et al.*[101] The 411 cm^{-1} line is polarized, whereas the 570 cm^{-1} line is depolarized. Furthermore, the intensity of the 411 cm^{-1} line increases with decreasing temperature, behavior not expected if the 411 cm^{-1} mode were due to high spin species because the high spin absorption at 640 nm decreases with decreasing temperature. On the basis of these new experimental findings and normal coordinate calculations, Tsubaki *et al.*[193] concluded that the ~411 cm^{-1} line is the ν(Fe^{III}–N$_3^-$) stretch of *low spin* species, and that the depolarized line at ~570 cm^{-1} is the out-of-plane internal azide bending mode, also of low spin species. Here, the plane is defined by Fe and N=N=N with the $\angle$Fe–N$_1$–N$_2$ angle being 110°.

Tsubaki *et al.*[193] have discussed four types of charge-transfer which may be responsible for the enhancement of various Fe–ligand and internal ligand modes: (1) porphyrin (π) → high spin Fe(d_π), (2) azide (n) → low spin iron (d_{z^2}), (3) azide (π) → low spin iron (d_{z^2}), and (4) azide (π) → porphyrin (π^*) (high spin). The lack of enhancement of the internal azide modes and the detection of the 411 cm^{-1} ν(Fe^{III}–N$_3^-$) stretch upon excitation in the vicinity of 600 nm have led Tsubaki *et al.*[193] to propose that the z-polarized charge-transfer band at ~650 nm is of the azide (n) → low spin iron (d_{z^2}) type or porphyrin (π) → iron d_{z^2} type. On the other hand, upon Soret excitation (at 406.7 nm) not only the ν(Fe^{III}–N$_3^-$) stretch but also the nontotally symmetric internal azide modes (i.e., 573 cm^{-1} out-of-plane bending and 2024 cm^{-1} antisymmetric stretch) from the low spin species have been detected. This is a good evidence that there is a low spin charge-transfer transition underlying the Soret band, which may be as-

[193] M. Tsubaki, R. B. Srivastava, and N.-T. Yu, *Biochemistry* **20**, 946 (1981).

signed as azide $(\pi) \rightarrow$ low spin iron (d_{z^2}). It was pointed out[193] that the resonance enhancement of nontotally symmetric azide modes requires mixing (vibronic coupling) between the proposed charge-transfer state and the Soret state via a Herzberg–Teller scattering mechanism.[14] Nontotally symmetric azide modes are not resonance enhanced in the cytochrome c–heme octapeptide azide complex upon Soret excitation. It was suggested[193] that the similar charge-transfer band may have been shifted away and that the Soret band alone is not capable of enhancing these azide internal vibrations.

Resonance Raman studies of azide binding to manganese-substituted myoglobin ($Mn^{III}Mb$) were reported by Yu and Tsubaki.[194] Nontotally symmetric bound azide vibrations were observed at 650 (bending) and 2039 cm^{-1} (antisymmetric stretch) upon excitation at $\sim$400–460 nm (band V). The assignments of these two vibrations were based on the excellent agreement of their $^{15}N_3^-$ isotope shifts (22 and 70 cm^{-1}) with the calculated values (22 and 69 cm^{-1}), the depolarized nature and their similarities to the corresponding vibrations in inorganic azide compounds. The azide internal bending mode at 650 cm^{-1} in $Mn^{III}Mb \cdot N_3^-$ is 80 cm^{-1} higher than that in $Fe^{III}Mb \cdot N_3^-$. No $\nu(Mn^{III}-N_3^-)$ stretching vibration can be detected under the conditions that the azide internal modes were observed. This has led Yu and Tsubaki[194] to suggest the charge-transfer transition as azide $(\pi) \rightarrow$ porphyrin (π^*) rather than azide $(\pi) \rightarrow Mn(d_{z^2})$ or azide $(n) \rightarrow$ $Mn(d_{z^2})$. Recently, Spiro[16] suggested that a charge-transfer transition of azide $(\pi) \rightarrow Mn(d_\pi)$ character is also consistent with the lack of $\nu(Mn^{III}-N_3^-)$ enhancement.

The ligation properties of manganese^{III} and iron^{III} hemes are quite different. The cyanide ion, which is a very strong ligand for Fe^{III} hemes, does not bind to manganese III myoglobin. A weaker ligand such as azide does bind. The interactions between the π orbitals of azide and the porphyrin ring may play an important role in azide binding to $Mn^{III}Mb$.[194]

Endogenous Ligands

Histidine. Most hemoproteins contain an iron–histidine linkage; the variations of the $Fe-N_\varepsilon(His)$ bond strength had been implicated in the protein regulation of heme reactivity.[1] Recent identification of the $\nu(Fe^{II}-N_\varepsilon)$ stretching mode[8,9,195] at $\sim$220 cm^{-1} in deoxyHb and Mb represents an important advance in studies of hemoproteins; it is now possible to monitor directly the strength of the iron–histidine bond. The suggestion of the

[194] N.-T. Yu and M. Tsubaki, *Biochemistry* **19**, 4647 (1980).
[195] J. Kincaid, P. Stein, and T. G. Spiro, *Proc. Natl. Acad. Sci. U.S.A.* **76**, 4156 (1979).

mode as essentially involving the Fe–N (pyrrole) bonds[196] has not been substantiated by others. In Table III are listed the $\nu(Fe^{II}-N_\varepsilon)$ frequencies (193–244 cm^{-1}) in high spin ferrous hemoproteins and model compounds. Included also in Table III are the $\nu(Fe^{III}-N_\varepsilon)$ frequencies for the ferric hemoproteins such as aquametMb, ferric HRP, and ferric HRP fluoride, first identified by Teraoka and Kitagawa.[197]

The decrease in $\nu(Fe^{II}-N_\varepsilon)$ in the R → T transition of deoxyHbA was interpreted by Nagai and Kitagawa[198] as resulting from mechanical tension (hence weakening) of the Fe–N_ε(His) bond. This is supported by the evidence that the $\nu(Fe^{II}-N_\varepsilon)$ frequencies in model compounds with sterically hindered bases (2-MeIm or 1,2-Me$_2$Im) are generally lower than those with unhindered bases (N-MeIm or Im). However, Stein et al.[199] provided an alternative interpretation involving hydrogen bonding of the N_δ proton of the proximal imidazole.[200] Based on the notion that in hemoglobin the N_δ proton may participate in H-bonding with a carbonyl oxygen of the peptide backbone,[201] they suggested that the H-bond is weakened in the T state. The weakening of the N_δ H-bond is expected to decrease the electron-donating power of N_ε and to weaken the Fe–N_ε(His) bond. In other words, a stronger N_δH-bond (the extreme case being deprotonation) should strengthen the Fe–N_ε(His) bond. They cited two pieces of evidence in support of their interpretation. First, the $\nu(Fe^{II}-N_\varepsilon)$ frequency at 220 cm^{-1} for Fe^{II}(PP)(2-MeIm) in water is higher than the 205 cm^{-1} value for Fe^{II}(PPDME)(2-MeIm) in benzene. However, the $\nu(Fe^{II}-N_\varepsilon)$ mode appears at 195 cm^{-1} for Fe^{II}(PP)(1,2-Me$_2$Im) in H$_2$O and Fe^{II}(PPDME)(1,2-Me$_2$Im) in benzene. For 2-MeIm, the difference in $\nu(Fe^{II}-N_\varepsilon)$ was attributed to H-bonding of the N_δH proton to solvent H$_2$O and a weakening of the H-bond in benzene. In (1,2-Me$_2$Im) complexes, H-bonding is absent because the N_δ proton is replaced by a methyl group. Second, they demonstrated that when Fe^{II}(OEP)(2-MeIm) in dimethylformamide is deprotonated, the $\nu(Fe^{II}-N_\varepsilon)$ frequency increases from 212 to 239 cm^{-1}.

[196] A. Desbois and M. Lutz, Biochim. Biophys. Acta 671, 168 (1981); A. Desbois, M. Momenteau, B. Loock, and M. Lutz, Spectrosc. Lett. 14, 257 (1981).

[197] J. Teraoka and T. Kitagawa, J. Biol. Chem. 256, 3969 (1981).

[198] K. Nagai and T. Kitagawa, Proc. Natl. Acad. Sci. U.S.A. 77, 2033 (1980).

[199] P. Stein, M. Mitchell, and T. G. Spiro, J. Am. Chem. Soc. 102, 7795 (1980).

[200] F. A. Walker, M.-W. Lo, and M. T. Ree, J. Am. Chem. Soc. 98, 5552 (1976); J. D. Satterlee, G. N. LaMar, and J. S. Frye, J. Am. Chem. Soc. 98, 7275 (1976); J. Peisach and W. B. Mims, Biochemistry 16, 2795 (1977); J. S. Valentine, R. P. Sheridan, L. C. Allen, and P. Kahn, Proc. Natl. Acad. Sci. U.S.A. 76, 1009 (1979); M. A. Stanford, J. C. Swartz, T. E. Phillips, and B. M. Hoffman, J. Am. Chem. Soc. 102, 4492 (1980).

[201] R. C. Ladner, E. J. Heidner, and M. F. Perutz, J. Mol. Biol. 114, 385 (1977).

TABLE III
THE Fe-IMIDAZOLE STRETCHING FREQUENCIES

Compounds	$\nu(Fe^{II}-N_{\varepsilon})(cm^{-1})$	References
$Fe^{II}(PP)(1,2-Me_2Im)$, 0.1% CTAB	193	204
$Fe^{II}(PocPiv)(1,2-Me_2Im)/CH_2Cl_2$	193	a
$Fe^{II}(PP)(1,2-Me_2Im)$, 0.01% CTAB	199	204
$Fe^{II}(TpivPP)(1,2-Me_2Im)/CH_2Cl_2$	200	9
$Fe^{II}(PPDME)(2-MeIm)/CH_2Cl_2$	200	197
$Fe^{II}(PPDME)(2-MeIm)/DMF$	205	197
$Fe^{II}(PP)(2-MeIm)$, 0.1% CTAB	206	204
$Fe^{II}(OEP)(2-MeIm)/DMF$	208	197
$Fe^{II}(PP)(2-MeIm)/SDS$ micelle	208	197
$Fe^{II}(PP)(2-MeIm)/TX-100$ micelle	208	197
$Fe^{II}(TpivPP)(2-MeIm)/CH_2Cl_2$	209	9
$Fe^{II}(PocPiv)(2-MeIm)/CH_2Cl_2$	215	a
$Fe^{II}(PP)(2-MeIm)/H_2O$	219	197
$Fe^{II}(PP)(2-MeIm)$, 0.001% CTAB	220	204
$Fe^{II}(TpivPP)(1-MeIm)/CH_2Cl_2$	225	9
$Fe^{II}(PocPiv)(1-MeIm)/CH_2Cl_2$	227	a
$Fe^{II}(PPDME)(2-MeIm^-)/Me_2SO_4$	233	197
$Fe^{II}(OEP)(2-MeIm^-)/DMF$	233	197
	239	199
$Fe^{II}(PocPiv)(Im)/CH_2Cl_2$	236	a
DeoxyMb (sperm whale)	220	197
Deoxy insect Hb, CTT III at pH 6.0, 9.6	218	12
DeoxyHbA (human)	216	8
Deoxy NES des-Arg HbA	222	8
Deoxy leghemoglobin	224	57,212
Deoxy hybrid Hb,$\alpha(Co)_2\beta(Fe)_2$	218	b
Deoxy hybrid Hb,$\alpha(Fe)_2\beta(Co)_2$	201,212	b
Ferro intestinal peroxidase, pH 8.3	254	216
Ferro HRP-C at pH 8	241	197
Ferro HRP-C at pH 5	244	197
Ferro lactoperoxidase at pH 7.4	248	216
AquametMb at pH 7.8	248	197
Ferric HRP at pH 9.6	282	197
Ferric HRP fluoride	267	197
HRP compound I	248	218
Ferric lactoperoxidase	261	216
HRP compound II	287	215

a J. P. Collman, J. I. Brauman, T. J. Collins, B. L. Iverson, G. Lang, R. B. Patterson, J. L. Sessler, and M. A. Walters, *J. Am. Chem. Soc.* **105**, 3038 (1983).

b M. R. Ondrias, D. L. Rousseau, T. Kitagawa, M. Ikeda-Saito, T. Inubushi, and T. Yonetani, *J. Biol. Chem.* **257**, 8766 (1982).

Since a decrease in $\nu(Fe^{II}-N_\varepsilon)$ can be caused by either mechanical tension or the electronic effect induced by the weakening of the $N_\delta H$-bonding, resonance Raman data alone are not sufficient to determine the causative factors which lower the $\nu(Fe^{II}-N_\varepsilon)$ frequency in the T-state deoxyHb. However, Nagai et al.[202] have shown that it is possible to distinguish the two factors by a combination of resonance Raman and NMR data. If the effect is primarily caused by a change in $N_\delta H$-bonding, both $\nu(Fe^{II}-N_\varepsilon)$ and the $N_\delta H$ contact shift change in opposite directions. On the other hand, a simple change in the iron–histidine tension should produce changes in the same direction for both the $N_\delta H$ contact shift and $\nu(Fe^{II}-N_\varepsilon)$. In reduced horseradish peroxidase, an increase in the $N_\delta H$ contact shift[202] is accompanied by a decrease in $\nu(Fe-N_\varepsilon)$,[202] indicating a dominant influence of $N_\delta H$-bonding on iron–histidine bonding. In contrast, NMR and resonance Raman data of R- and T-state deoxyHb[203] argue strongly against a primary contribution of the electronic effect induced by altered $N_\delta H$-bonding to a backbone carbonyl, as suggested by Stein et al.[199]

Resonance Raman studies of deoxyHbs (chemically modified, mutant, and in the presence of allosteric effectors) have led to the discovery of an inverse linear correlation[204] between the frequencies of the porphyrin ring mode at ~ 1357 cm^{-1} (sensitive to π-electron density)[16,205] and the $\nu(Fe^{II}-N_\varepsilon)$ stretch at ~ 217 cm^{-1}, for a majority of the globins examined. As shown in Fig. 24, an increase in the $\nu(Fe^{II}-N_\varepsilon)$ frequency corresponds to a decrease in the porphyrin ring frequency. To the extent that this correlation is valid, the changes in iron–histidine bonding are transmitted to the porphyrin π^* electron density via $Fe(d_\pi)$–porphyrin (π^*) backbonding. Nagai and Kitagawa[198] and Perutz[206] have already proposed that globin-induced changes in the iron–histidine bond can give rise to electron-density changes in the porphyrin macrocycle. Ondrias et al.[204] suggested that the reverse is possible, i.e., variations in the heme electron density caused by changes in the porphyrin environment could also alter the iron–histidine bond. In terms of free energy, it appears that the loss in the iron–histidine bond is either completely or partly compensated by the gain in the heme macrocycle.

[202] K. Nagai, G. N. LaMar, T. Jue, and H. F. Bunn, Biochemistry 21, 842 (1982).
[203] J. Teroaka and T. Kitagawa, Biochem. Biophys. Res. Commun. 93, 694 (1980).
[204] M. R. Ondrias, D. L. Rousseau, J. A. Shelnutt, and S. R. Simon, Biochemistry 21, 3428 (1982).
[205] J. A. Shelnutt, D. L. Rousseau, J. M. Friedman, and S. R. Simon, Proc. Natl. Acad. Sci. U.S.A. 76, 4409 (1979).
[206] M. E. Perutz, Proc. R. Soc. London, Ser. B 208, 135 (1980).

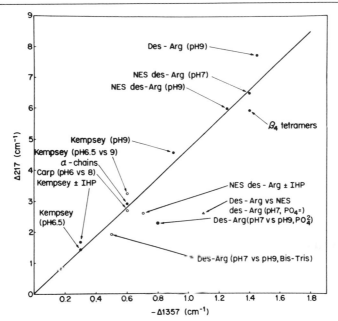

FIG. 24. A correlation between differences in the 1357 cm^{-1} line and the Fe–His stretching frequency. Reprinted with permission from Ondrias *et al.*[204] (1982). Copyright 1982 American Chemical Society.

Friedman and co-workers[207] examined the time dependence of the two modes (at ~1357 and ~217 cm^{-1}) following photolysis of HbCO derivatives and found that the changes in frequency of these two modes were correlated from 10 nsec to several microseconds. The frequency differences in these modes, referenced to the corresponding modes in deoxy-HbA, agree with the correlation in Fig. 24.

Metastable species of deoxyhemoglobin (Hb*) generated by photolysis of Hb·CO at cryogenic temperature (80 K) have been studied by Ondrias *et al.*,[208] who found an unusually high $\nu(Fe^{II}-N_\varepsilon)$ stretching frequency at 240 cm^{-1}, compared to the ~216 and ~222 cm^{-1} values in the T-state and R-state deoxyHb, respectively. At 10 K this frequency is even higher (at 244 cm^{-1}).[209] The porphyrin ring modes in the 200–450 cm^{-1} region are very similar between the spectrum of Hb* resulting from photolyzed Hb·CO at 80 K (or 10 K) and that obtained at room temperature

[207] J. M. Friedman and K. B. Lyons, *Nature (London)* **284**, 570 (1980); J. M. Friedman, D. L. Rousseau, M. R. Ondrias, and R. A. Stepnoski, *Science* **218**, 1244 (1982).
[208] M. R. Ondrias, J. M. Friedman, and D. L. Rousseau, *Science* **220**, 615 (1983).
[209] M. R. Ondrias, D. L. Rousseau, and S. R. Simon, *J. Biol. Chem.* **258**, 5638 (1983).

10 nsec after photolysis. They also resemble deoxyHb spectra except for the mode at ~341 cm^{-1}. It is likely that photolysis of Hb · CO results in a complete breakage of the Fe–CO bond, although the CO molecule may be still retained within the heme pocket, analogous to the photolysis of Mb · CO and cytochrome oxidase.[210] On the other hand, the EXAFS by Chance et al.[211] revealed only a small increase (less than 0.05 Å) in the Fe–CO bond upon photolysis of Mb · CO at 4 K. An increase of 0.05 Å in the Fe–CO bond length should lower the ν(Fe–CO) stretching frequency by ~30 cm^{-1}. Since the metastable Hb* species at 4 K may be viewed as an intermediate between deoxyHb and Hb · CO, it may be possible to enhance both the ν(Fe^{II}–N$_\varepsilon$) and ν(Fe^{II}–CO) stretching vibrations. So far, attempts to detect the ν(Fe^{II}–CO) stretch in Hb* have been unsuccessful.

It is possible that the origin of leghemoglobin's high ligand affinities is an axial compression of the proximal histidine, the coverse of the Perutz trigger mechanism[73] of hemoglobin cooperativity. Resonance Raman data of Irwin et al.[212] and Armstrong et al.[213] on deoxyleghemoglobin (Lb) may be employed as evidence for an expanded porphyrin core[214] and an increased Fe–N (pyrrole) bond strength (compared to Mb). However, Rousseau et al.[57] reexamined resonance Raman spectra of deoxyleghemoglobin and found some important discrepancies. The core size marker line is at nearly the same frequency in Lb (at 1554 cm^{-1}) as in HbA (1555 cm^{-1}), in contrast to the results of Armstrong et al.[213] who reported a large difference in this mode. Comparison of Lb and HbA (or Mb) does indicate that the ν(Fe^{II}–N$_\varepsilon$) frequency is higher in Lb (at 224 cm^{-1}) than in HbA (216 cm^{-1}) or Mb (220 cm^{-1}). Rousseau et al.[57] found that major differences between deoxyHbA and Lb result from conformational changes in the vinyl groups of the heme. They believed that the difference in the vinyl π-conjugation into the porphyrin ring could make a significant contribution to the difference in ligand binding affinity for Lb and HbA.

The ν(Fe^{II}–N$_\varepsilon$) stretching frequency at 244 cm^{-1} for ferro HRP[197] is unusually high compared to those of deoxyMb (220 cm^{-1}) and deoxyHbA (216 cm^{-1}). This line exhibits a pH-dependent frequency shift (~3 to 5

[210] J. O. Alben, F. P. Moh, F. G. Fiamingo, and R. A. Altschuld, *Proc. Natl. Acad. Sci. U.S.A.* **78**, 234 (1981); J. O. Alben, F. P. Fiamingo, and R. A. Altschuld, VII *Int. Biophys. Congr., 7th, Pan-Am. Biochem. Congr., 3rd, Mexico City, Aug. 23–28* (1981).

[211] B. Chance, R. Fischetti, and L. Powers, *Biochemistry* **22**, 3820 (1983).

[212] M. J. Irwin, R. S. Armstrong, and P. E. Wright, *FEBS Lett.* **133**, 239 (1981).

[213] R. S. Armstrong, M. J. Irwin, and P. E. Wright, *Biochem. Biophys. Res. Commun.* **95**, 682 (1980).

[214] L. D. Spaulding, C. C. Chang, N.-T. Yu, and R. H. Felton, *J. Am. Chem. Soc.* **97**, 2517 (1975).

cm^{-1}) toward lower frequency at higher pH with a midpoint pH value of 7 for isoenzyme C and 5.5 for isoenzyme A$_2$. In general, the deprotonation of the proximal histidine is expected to raise the $\nu(Fe^{II}-N_\varepsilon)$ stretching frequency.[197,199] The frequency decrease in $\nu(Fe^{II}-N_\varepsilon)$ of ferro HRP at alkaline pH suggests that the proximal histidine is not deprotonated. It has been suggested[197] that the proximal histidine in ferro HRP is strongly hydrogen bonded at both sides of the transition, and an appreciable strain is exerted on the Fe-N$_\varepsilon$ bond on the alkaline side due to a globin conformational change effect upon distal histidine ionization.

The $\nu(Fe^{III}-N_\varepsilon)$ stretching frequency at 274 cm^{-1} for ferric HRP is also unusually high compared to that of aquametMb (at 248 cm^{-1}).[197] This mode also shows a pH-dependent frequency shift around pH 11. The highest $\nu(Fe^{III}-N_\varepsilon)$ frequency was found at 287 cm^{-1} in HRP compound II, which shifts to 248 cm^{-1} in HRP compound I.[215] The assignments of these vibrations are only tentative because they have not been confirmed by isotope shift data.

Differences between animal and plant peroxidases were found by Kitagawa *et al.*[216] The $\nu(Fe^{III}-N_\varepsilon)$ stretch appears at 248 cm^{-1} in ferro lactoperoxidase and at 254 cm^{-1} in ferro intestinal peroxidase. These frequencies also exhibit pH dependence.

Tyrosine. There are several hemoproteins which have an endogenous tyrosine coordinated in the axial position to the ferric heme. These include catalase,[217] HbM Boston [His-E7(58)$\alpha \rightarrow$ Tyr],[218] and HbM Iwate [His-F8(87)$\alpha \rightarrow$ Tyr].[219] Protein crystallographic studies show that in beef liver catalase[217] tyrosine-357 occupies the fifth coordination site, with an Fe^{III}-O (phenolate) bond length of 1.5 Å; in HbM Boston[218] the distal Tyr(E7) of the abnormal α-subunit binds to the ferric heme without a trans proximal Fe-His(F8) bond; and in HbM Iwate[219] both distal His(E7) and proximal Tyr(F8) bind to the heme iron in the abnormal α-subunit. Other mutant human hemoglobins such as HbM Emory [His-E7(63)$\beta \rightarrow$ Tyr] and HbM Hyde Park [His-F8(92)$\beta \rightarrow$ Tyr] also have a tyrosine ligand at either the distal or proximal position in the β-subunit, but there are no X-ray structural data to confirm its coordination to the heme iron.

[215] J. Teraoka, T. Ogura, and T. Kitagawa, *J. Am. Chem. Soc.* **104,** 7354 (1982).
[216] T. Kitagawa, S. Hashimoto, J. Teraoka, S. Nakamura, H. Yajima, and T. Hosoya, *Biochemistry* **22,** 2788 (1983).
[217] T. J. Reid, III, M. R. N. Murthy, A. Sicignano, N. Tanaka, W. D. L. Musick, and M. G. Rossman, *Proc. Natl. Acad. Sci. U.S.A.* **78,** 4767 (1981).
[218] D. D. Pulsinelli, M. F. Perutz, and R. L. Nagel, *Proc. Natl. Acad. Sci. U.S.A.* **70,** 3870 (1973).
[219] J. Greer, *J. Mol. Biol.* **59,** 107 (1971).

Nagai *et al.*[220] were the first to demonstrate that the ring vibrations of a heme-bound phenolate and the $Fe^{III}-O$ (phenolate) stretching mode can be resonance enhanced via a charge-transfer transition. The phenolate ring vibrations were identified at ~ 1606 and ~ 1505 cm^{-1}. The $Fe^{III}-O$ stretching vibration was tenatively assigned to the 603 cm^{-1} line in HbM Boston and the 589 cm^{-1} line in HbM Iwate. The relatively high frequency for this stretching mode may be attributed to the unusually short (hence strong) $Fe^{III}-O$ bond length (1.5 Å) as revealed in the structure of beef liver catalase. The CO stretching vibration of phenolate was also tentatively identified at 1278 cm^{-1} (HbM Boston) and 1308 cm^{-1} (HbM Iwate). The excitation profiles of the two modes at ~ 600 and ~ 1300 cm^{-1} suggest the existence of a charge-transfer band around 475–510 nm. Comparison of these two modes between HbM Boston and HbM Iwate reveals that the lower the $\nu(Fe^{III}-O)$ frequency, the higher the $\nu(C-O)$ frequency. A greater donation of electrons from phenolate to the heme iron would strengthen the $Fe^{III}-O^-$ bond, but weaken the phenolate CO bond.

Nagai *et al.*[220] have also carried out resonance Raman studies of a Fe^{III}-phenolate containing model compound, $Fe^{III}(PPDBE)(OC_6H_4NO_2)$, where PPDBE = protoporphyrin IX dibutyl ester. Although the visible absorption and EPR spectra are similar to those of HbM Boston and HbM Iwate,[221] it exhibits neither the phenolate internal modes nor the $Fe^{III}-O^-$ stretching vibration upon excitation between 441 and 515 nm. It was suggested[220] that the phenolate in the model compound is perpendicular to the heme plane, but becomes tilted in HbM Boston and HbM Iwate because of some restriction by the protein. The tilted geometry may allow direct interaction between the π orbitals of phenolate and that of the porphyrin, providing an avenue for the charge-transfer enhancement mechanism. The differences in the C–O and $Fe^{III}-O^-$ modes between HbM Boston and HbM Iwate may then be attributed to a difference in the tilting angle of the phenolate with respect to the heme plane.

Resonance enhancement of phenolate vibrations in non-heme iron–tyrosine proteins upon excitation in the CT band has been demonstrated for serum transferrin,[222–224] and protocatechuate-3,4-dioxygenases.[225–229]

[220] K. Nagai, T. Kagimoto, A. Hayashi, F. Taketa, and T. Kitagawa, *Biochemistry* 22, 1305 (1983).
[221] E. W. Ainscough, A. W. Addison, D. Dolphin, and B. R. James, *J. Am. Chem. Soc.* 100, 7585 (1978).
[222] Y. Tomimatsu, S. Kint, and J. R. Scherer, *Biochem. Biophys. Res. Commun.* 54, 1067 (1973).
[223] Y. Tomimatsu, S. Kint, and J. R. Scherer, *Biochemistry* 15, 4918 (1976).
[224] B. P. Gaber, V. Miskowski, and T. G. Spiro, *J. Am. Chem. Soc.* 96, 6868 (1974).
[225] Y. Tatsuno, Y. Saeki, M. Iwaki, T. Yagi, M. Nozaki, T. Kitagawa, and S. Otsuka, *J. Am. Chem. Soc.* 100, 4614 (1978).

Cysteine. The carbon monoxide complexes of ferrous cytochrome *P*-450 and chloroperoxidase exhibit a characteristic Soret absorption band at ~450 nm.[230] The fact that this unique Soret band position can be reproduced in thiolate-ligated CO complexes of iron porphyrins, is suggestive of the presence of an Fe–S bond in these two hemoproteins.[231-234] Other indirect evidence for the Fe–S bond includes evidence from EPR,[235,236] MCD,[237] and EXAFS[238] studies.

The search for the ν(Fe–S) stretching vibration in cytochrome *P*-450 derivatives was initiated in 1974 in the author's laboratory. With excitation at 514.5 nm the low-frequency resonance Raman spectrum of the cytochrome *P*-450–camphor complex (ferric, high spin) exhibited a very unique vibrational mode at 351 cm^{-1}, which was first suggested by Felton and Yu[14] as the Fe^{III}–S$^-$ bond stretching vibration (see Fig. 25). On the other hand, Champion and Gunsalus[239] reported on a selectively enhanced Raman line at 351 cm^{-1} in the same complex excited at 363.8 mm, which was assigned by them to a porphyrin ring mode involving Fe–N (pyrrole) stretching. More recently, Champion *et al.*[10] employed isotopically enriched (^{54}Fe, ^{34}S) cytochrome *P*-450–camphor samples to confirm the earlier suggestion by Felton and Yu.[14] They found that there was a -4.9 ± 0.3 cm^{-1} downshift upon ^{32}S $\rightarrow$ ^{34}S isotope substitution and a $+2.5 \pm 0.2$ cm^{-1} unshift upon ^{56}Fe $\rightarrow$ ^{54}Fe isotope substitution (Fig. 26). It now appears that there are two charge-transfer states (one at ~365 nm and the other at 540 nm) responsible for the enhancement of the ν(Fe–S) stretch at 351 cm^{-1} in the cytochrome *P*-450–camphor complex. These

[226] W. E. Keyes, T. M. Loehr, and M. L. Taylor, *Biochem. Biophys. Res. Commun.* **83**, 941 (1978).

[227] R. H. Felton, L. D. Cheung, R. S. Phillips, and S. W. May, *Biochem. Biophys. Res. Commun.* **85**, 844 (1978).

[228] L. Que, Jr. and R. H. Heistand, II, *J. Am. Chem. Soc.* **101**, 2219 (1979).

[229] C. Bull, D. P. Ballou, and I. Salmeen, *Biochem. Biophys. Res. Commun.* **87**, 836 (1979).

[230] R. Sato and T. Omura, "Cytochrome P-450," p. 192. Academic Press, New York, 1978.

[231] S. C. Tang, S. Koch, G. C. Papaefthymiou, S. Foner, R. B. Frankel, J. C. Ibers, and R. H. Holm, *J. Am. Chem. Soc.* **98**, 2414 (1976).

[232] C. K. Chang and D. Dolphin, *Proc. Natl. Acad. Sci. U.S.A.* **73**, 3338 (1976).

[233] J. P. Collman and T. N. Sorrell, *J. Am. Chem. Soc.* **97**, 5948 (1975).

[234] S. P. Cramer, J. H. Dawson, K. Hodgson, and L. P. Hager, *J. Am. Chem. Soc.* **100**, 7282 (1978).

[235] K. Murakami and H. S. Mason, *J. Biol. Chem.* **242**, 1102 (1967).

[236] C. R. E. Jefcoate and J. L. Gaylor, *Biochemistry* **8**, 3464 (1969).

[237] J. H. Dawson, R. H. Holm, J. R. Trudell, G. Barth, R. E. Linder, E. Bunnenberg, C. Djerassi, and S. C. Tang, *J. Am. Chem. Soc.* **98**, 3707 (1976).

[238] S. Cramer, J. Dawson, K. Hodgson, and L. P. Hager, *J. Am. Chem. Soc.* **100**, 7282 (1978).

[239] P. M. Champion and I. C. Gunsalus, *J. Am. Chem. Soc.* **99**, 2000 (1977).

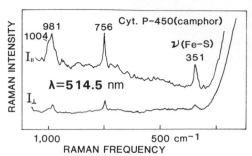

FIG. 25. Resonance Raman spectra ($I_\parallel$ and $I_\perp$) of cytochrome P-450. Camphor complex excited at 514.5 nm. The ν(Fe–S) stretching mode is the most pronounced line in the 100–700 cm^{-1} region.

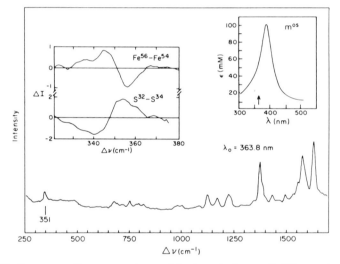

FIG. 26. Resonance Raman spectrum of ferric cytochrome P-450·camphor complex excited at 363.8 nm. Reprinted with permission from Champion et al.[10] (1982). Copyright 1982 American Chemical Society.

two charge-transfer bands may correspond to the single-crystal Z-polarized absorption maxima at ~330 and 567 nm.[240]

The excitation profile of the ν(Fe–S) stretch at 351 cm^{-1} in the cytochrome P-450–camphor complex has been obtained.[241] The 40-data point profile between 488 and 600 nm clearly displays a maximum at 538 nm.

[240] L. K. Hanson, S. G. Sligar, and I. C. Gunsalus, *Croat. Chem. Acta* **49**, 237 (1977).
[241] L.-S. Duncan Cheung, Ph.D. thesis, Georgia Institute of Technology, 1978.

The 351 cm^{-1} excitation profile in the 300–400 nm region was constructed by Dr. P. M. Champion and co-workers (private communication). The information provided by resonance Raman spectroscopy may be considered the most direct and independent evidence for the existence of an Fe–S bond in the P-450–camphor complex. The isotope shifts of the ν(Fe–S) stretching frequency have been employed to estimate the Fe–S–C angle as 125–145° in P-450–camphor, a value somewhat larger than the Fe–S–C bond angle of 100.4° reported for a synthetic P-450 model complex.[231]

Acknowledgment

Work from the author's laboratory described here was supported by Grant GM 18894 from the National Institutes of Health.

Section III

Macromolecular Conformational Stability and Transitions

[17] Calculation of Electrostatic Interactions in Proteins

By JAMES B. MATTHEW and FRANK R. N. GURD

Introduction

The classical treatment of hydrogen ion titration curves of proteins regards the molecule as an impenetrable sphere on which amino acid residues are grouped into classes of intrinsically identical sites with their charges uniformly distributed over the surface.[1] The elegant simplicity of this frequently used model is offset by its inability to yield electrostatic information about individual groups, their specific roles, and their interactions.

In the more realistic discrete-charge electrostatic theory,[2,3] the amino acid groups are point charges positioned at fixed sites on the surface of the protein or are buried a short distance within the interior of the molecule which is assumed to be a continuous medium of low dielectric constant. The theory was successfully tested on a variety of model compounds. A full treatment of the hydrogen ion titration curve for tetrameric human hemoglobin was carried out by Orttung.[4] While previous treatments had required burial of the charges up to 1 Å into the low dielectric medium, Orttung found that in order to obtain proper fit of the data it was necessary to place all charges at the surface of the dielectric boundary. A much more efficient iterative algorithm was developed by Tanford and Roxby[5] who applied it in analyzing the hydrogen ion titration curve of lysozyme. All charges were buried at a uniform depth of 0.4 Å to give a calculation that agreed with the experimental titration curve.

In an attempt to overcome the uncertainty over the burial parameter and to allow for the irregular surface of a real protein, Shire et al.[6] introduced a modification into the Tanford model whereby, for each individual group, the magnitude of electrostatic intramolecular interaction was reduced in direct proportion to the extent of the group's exposure to the

[1] K. Linderstrøm-Lang, C. R. Trav. Lab. Carlsberg 15, 70 (1924).
[2] C. Tanford and J. G. Kirkwood, J. Am. Chem. Soc. 79, 5333 (1957).
[3] C. Tanford, J. Am. Chem. Soc. 79, 5340 (1957).
[4] W. H. Orttung, Biochemistry 9, 2394 (1970).
[5] C. Tanford and R. Roxby, Biochemistry 11, 2192 (1972).
[6] S. Shire, G. I. H. Hanania, and F. R. N. Gurd, Biochemistry 13, 2967 (1974).

solvent. The degree of exposure of each group is measured by its solvent accessibility parameter.[7] Subsequently, it has been shown that the discrete-charge model can be fruitfully employed to study several aspects of electrostatic effects in proteins.

To date, the static-accessibility-modified theory has been successfully applied to a variety of proteins that include sperm whale myoglobin and 11 species variations,[8,9] oxy- and to deoxyhemoglobins and their interactions involving hydrogen ions, chloride ions, carbamino adducts, and organic phosphate polyanions,[10–13] BPTI[14] and ribonuclease.[15] The intent of these studies has been to test the versatility and predictive power of the algorithm by treating a variety of proteins. The mathematical detail,[10] the sensitivity to variations in model parameters,[10,16] and the effects of charge array dissymmetries with respect to charge conformation and static solvent accessibility[11] have been examined.

The generality of this approach is illustrated by the fact that all computations are based on the same consistent set of intrinsic pK values[8,16] with the appropriate solvent accessibility parameter obtained from the known atomic coordinates.[17]

Computational Methods

Calculation of pK_i

In these computations it is assumed that for each pH and ionic strength the unique protein charge array confers an electrostatic potential at site i causing the apparent pK to deviate from pK_{int}. The influence of the summed electrostatic fields at a given site i, controlling its effective

[7] B. K. Lee and F. M. Richards, *J. Mol. Biol.* **55**, 379 (1971).

[8] L. H. Botelho, S. H. Friend, J. B. Matthew, L. D. Lehman, G. I. H. Hanania, and F. R. N. Gurd, *Biochemistry* **17**, 5197 (1978).

[9] S. H. Friend and F. R. N. Gurd, *Biochemistry* **18**, 4612 (1979).

[10] J. B. Matthew, G. I. H. Hanania, and F. R. N. Gurd, *Biochemistry* **18**, 1919 (1979).

[11] J. B. Matthew, G. I. H. Hanania, and F. R. N. Gurd, *Biochemistry* **18**, 1928 (1979).

[12] J. B. Matthew, G. I. H. Hanania, and F. R. N. Gurd, *Biochemistry* **20**, 571 (1981).

[13] J. B. Matthew, S. H. Friend, and F. R. N. Gurd, *in* "Hemoglobin and Oxygen Binding" (C. Ho, ed.), p. 231. Elsevier, New York, 1982.

[14] K. L. March, D. G. Maskalick, R. D. England, S. H. Friend, and F. R. N. Gurd, *Biochemistry* **21**, 5241 (1982).

[15] J. B. Matthew and F. M. Richards, *Biochemistry* **21**, 4989 (1982).

[16] J. B. Matthew, S. H. Friend, L. H. Botelho, L. D. Lehman, G. I. H. Hanania, and F. R. N. Gurd, *Biochem. Biophys. Res. Commun.* **81**, 416 (1978).

[17] J. B. Matthew, G. I. H. Hanania, and F. R. N. Gurd, *Biochem. Biophys. Res. Commun.* **81**, 410 (1978).

pK_i, is derived by an iterative scheme expressed as

$$pK_i = (pK_{int,i}) - \frac{1}{2.303RT} \sum_{j \neq i} W'_{ij} Z_j$$

$$= (pK_{int,i}) + \sum_{j \neq i} \Delta pK_{ij} Z_j \tag{1}$$

where $(pK_{int,i})$ is the intrinsic ionization constant of the group i, the value it would have in the absence of any influence from other charged sites on the protein, and W'_{ij} are the static-accessibility modified Tanford–Kirkwood work factors. The iteration is continued at each pH until all pK_i and Z_i values stabilize.

The pK_i of almost every group is found to vary with pH, and it is convenient to define the parameter $pK_{1/2}$ for a group, this being the pH at which a particular group is half-titrated. In the absence of electrostatic interaction $\sum_{j \neq i} \Delta pK_{ij}$ goes to zero, the pK_{int} equals $pK_{1/2}$, and the variations of pK_i with pH disappear. In the reference state for all of these calculations, the groups are maximally accessible to solvent and have no electrostatic interactions with each other. This hypothetical state can be approached for many proteins under conditions of complete denaturation. If the random configuration has a high charge density, however, e.g., poly(DL-lysine), the assumption that equates the random coil to the reference state $(\Delta pK_{ij=0})$ would not be valid. Note also that in an intact protein if the pK_i of a particular group happens to equal $pK_{int,i}$, this does not necessarily imply a lack of electrostatic interactions but may be a reflection of strong but cancelling interactions with neighboring groups.

Intrinsic pK Values. The intrinsic proton dissociation constants for acidic and basic groups of proteins are based on measured pK values for model compounds. Following previous work[8–16] we distinguish two categories of proton binding sites. The first class comprises all groups with normal pK_{int} values: terminal carboxyl, 3.60; Asp, 4.00; Glu, 4.50; Cys SH, 9.0; Tyr, 10.00; Lys, 10.40; and Arg, 12.00. Values for terminal residues are specific to the amino acids, 7.00–8.00. For glutamic acid, aspartic acid, histidine, and arginine where two atoms exist for possible charge placement, we choose the most solvent-accessible atomic coordinate, unless dictated otherwise by the possibility of ion pairing. For histidine the chemical character of N^τ and N^π differs[8] such that pK_{int} is taken as 6.60 or 6.00, respectively, depending on which locus is more exposed to solvent and untrammeled by internal interactions. In practice the choice is sometimes difficult, in which case the value of 6.60 could provide a rule of thumb.

The second class is that of "masked" groups. These residues are generally charge sites which are buried in their neutral form well within the protein interior. This class is encountered in ribonuclease only in the case of Tyr-97 which would already be in neutral form under all but the most basic conditions,[15] but widely and consistently among histidine residues in myoglobins[8] and hemoglobins.[10]

Titratable groups that are hydrogen bonded to components of the protein surface fall into a third category. The practice of altering the pK_{int} values of these groups by ± 0.5 is an ad hoc formalism based on chemical experience which is difficult to test for its necessity in the protein setting. Such groups are relatively rare on present indications, usually restricted to carboxylic acids or arginine, and significantly affect the predictions only at pH extremes. Where present they most likely signal a particular functional adaptation, and particular attention should be given to obtaining accurate experimental comparisons.

The argument is often made that the partitioning of a titratable group from the high dielectric environment of the solvent to the protein–solvent interface will result in a new intrinsic proton affinity shifted in favor of the neutral species. This effect is usually discussed in the context of an isolated ion's "self energy." The fundamental question is whether partitioning of the charge site at the protein–solvent dielectric interface is sufficient to account for the observed protein pK values, or are these effects small by comparison to the through space interactions with other charge sites. Recently, it was shown that there is no correlation between a groups fractional burial at the solvent–protein boundary and the sign or magnitude of its variation from the intrinsic pK value.[17a] Cationic and anionic acids both have positive and negative deviations from their solvent intrinsic pK values whose magnitude is unrelated to their fractional burial. A shift in intrinsic proton affinity is likely for sites that are totally removed from solvent by protein steric constraints (the class of sites referred to above as "masked"), however, for partially exposed groups the charge site interactions as modulated by the solvent accessibility at the protein-solvent dielectric interface appear to be dominant.

Electrostatic Work Factors. At the core of these calculations is the electrostatic interaction term, W_{ij}. Following Tanford and Kirkwood[2] the free energy of interaction, W_{ij}, is calculated for a pair of unit charges placed on the surface of an appropriate low dielectric sphere for all separation, r_{ij}. The details of this calculation are given elsewhere.[2,9]

Figure 1 illustrates the direct application of this calculation to a sphere of radius 18 Å.[15] The variation of W_{ij} with separation distance r_{ij} is cast in

[17a] J. B. Matthew, *Ann. Rev. Biophys. Chem.* **14**, 387 (1985).

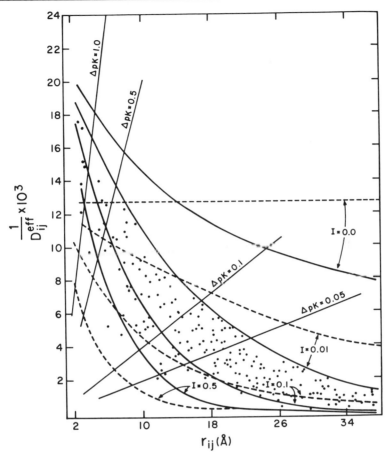

FIG. 1. Energy of interaction for pairs of unit charges placed on a sphere of radius 18 Å, with an ion exclusion radius of 20 Å. The charges are separated by a distance r_{ij}. The bulk dielectric constants of the external medium and the interior of the sphere are 78.5 and 4.0, respectively. The energy is presented as the reciprocal of the effective dielectric constant, D_{ij}^{eff}, for the particular charge pair. This parameter was calculated from the Tanford–Kirkwood procedure [Eq. (2)] for ionic strengths, 0.0, 0.01, 0.10, and 0.50, shown as solid lines, and from the Debye–Hückel screening relationship [Eq. (3)] at the same ionic strengths (broken lines). A sample of the parameters used in the calculations for RNase S at ionic strength 0.01 is shown as solid circles. These values were derived from the Tanford–Kirkwood relations as modified by the incorporation of the mean static accessibility factor as indicated in Eq. (4). The values of $(D_{ij}^{eff})^{-1}$ and r_{ij} which lead to four particular changes in ionization behavior are shown by the contours of constant ΔpK. (From Matthew and Richards.[15])

terms of an effective dielectric constant by using the Bjerrum formalism:

$$W_{ij}r_{ij} = \frac{1}{D_{ij}^{\text{eff}}} \qquad (2)$$

The reduction in free energy of interaction with increasing charge separation at finite ionic strength, i.e., the increase in the effective dielectric constant, is based on a Poisson–Boltzmann distribution of counterions around the charge sites i and j at the dielectric interface. The broken curves show the effective dielectric constant values calculated for Debye–Hückel screening for small ions in solution, following Hill[18] as

$$D_{ij}^{\text{eff}} = D_{H_2O}e^{\kappa r_{ij}} \qquad (3)$$

where κ is the usual Debye–Hückel parameter proportional to the square root of the ionic strength.

At all distances greater than 14 Å, and at ionic strengths greater than 0.01 M for distances greater than 8 Å, the effective dielectric constant is greater than that of water, indicating the dominance of the external dielectric. Only at short distances, e.g., $r_{ij} = 3$–10 Å, where ΔpK_{ij} values are greater than 0.3, are the W_{ij} values noticeably sensitive to the presence of the low dielectric region. For two charges separated by 4 Å the effective dielectric value for their interaction is 54.3 at $I = 0.01$ M. If the internal dielectric constant had been chosen as 10 rather than 4, the effective dielectric value would rise only to 57.0. It is particularly noteworthy that the experimentally observed pK values for groups involved in salt bridges, where the maximum shift in pK is about 1.5, correspond to an effective dielectric value of 50 at a separation of 3.0 Å.[13]

The Tanford–Kirkwood algorithm for calculating W_{ij} as a function of r_{ij} allows one to investigate the effect of sphere curvature and the effect of displacing one or both charges inside the dielectric boundary. However, for an approximate calculation of interaction energy, i.e., ΔpK_{ij}, for two surface charges Eq. (3) with D solvent equal to 50, will give a crude, working estimate.

Incorporation of the Solvent-Accessibility Parameter. In the modification of the Tanford–Kirkwood treatment, solvent accessibility (SA), specific for each group, has been incorporated into the calculation of W_{ij} as follows:

$$W'_{ij} = W_{ij}(1 - \overline{SA}_{ij}) \qquad (4)$$

where $\overline{SA}_{ij}$ is the average accessibility of sites i and j.[13-15]

[18] T. L. Hill, *J. Phys. Chem.* **60,** 253 (1956).

If a good set of atomic coordinates is available, solvent accessibilities are calculated for each titratable group, to account for the unique dielectric environment of individual groups at the protein–solvent interface.

The electrostatic energies of interaction W'_{ij} for the pairwise interactions of all charged sites i and j on ribonuclease are plotted as points in Fig. 1.[15] The values calculated by the Tanford–Kirkwood formalism at $I = 0.01$ when reduced by $(1 - \overline{SA}_{ij})$ are distributed between the curve for $I = 0.01$ as an upper limit and a curve equivalent to $I = 0.5$ as a lower limit. In the same way, the W'_{ij} values for ribonuclease at $I = 0.1$ (not shown) fall between the coulombic screening curves of $I = 0.1$ and 0.5.

Physical Interpretation of the Accessibility Modified T–K Formalism. Different accessibilities of the various groups reflect the ability of the protein to restrict both solvent interactions and the effective sequestering of counterions. When $\overline{SA}_{ij}$ exceeds 0.95, the interaction energy between the two charge sites is negligible, and neither one perturbs the pK value of the other. For lower values of $\overline{SA}_{ij}$ the protein plays the role of sterically restricting the access of solvent and mobile counterions to the high local field of the charge sites: $\Sigma\psi > kT$. At an average accessibility of 0.05 or less, the protein is assumed to restrict the counterions to the ionic exclusion radius where the Poisson–Boltzmann ionic strength approximation is valid, i.e., $\Sigma\psi < kT$. Hence, the charge sites at low accessibility are allowed to interact as calculated by the Tanford–Kirkwood formalism. The linear relation between W'_{ij} and $\overline{SA}_{ij}$ approximates the $1/r$ dependence of the local field around a point charge. The use of the $\overline{SA}_{ij}$ factor in reducing electrostatic free energy portrays a higher effective coulombic shielding for solvent-exposed sites.[6] This shielding, due to higher local ionic strength, can be interpreted as a higher effective local dielectric constant.[13,15]

Computation of Solvent Accessibility

Following Lee and Richards[7] the "accessible surface area" is calculated as the area around an atom in $\mathring{A}^2$ over which the center of the solvent (i.e., water) molecule can be placed, enabling it to maintain van der Waals contact without penetrating any other atom in the protein. The influence of protein structure on the solvent accessibility of individual charge sites is obtained by normalization to the accessibility of the same atom in a model tripeptide where accessibilities are assumed to attain their maximum value indicative of the fully hydrated site.

Procedure. The method of Lee and Richards[7] treats the protein structure as a set of interlocking spheres of appropriate van der Waals radii. This continuous structure is sliced by a series of parallel planes with predetermined spacing at 0.25 Å. The intersection of each sphere with a given plane appears as a circle, and the outermost trace of these circles becomes the envelope of a van der Waals surface of the protein section in that plane. For calculation, a sphere is centered at each atomic site in the coordinate space and is assigned a radius equal to the sum of radii for the atom and the solvent (1.40 Å for water). The surface section is the locus of the center of a solvent molecule as it rolls along the protein molecule, and any part of the atom's surface that lies on the envelope is regarded as accessible. The atom's solvent accessibility is thus proportional to the length of the exposed arcs, summed over each of the sections which pass through the atom.

The present computational method follows essentially that of Lee and Richards except that an initial scan of distances is first made to eliminate all atoms beyond 6.40 Å from the selected atom, this being the largest sum of two radii each including water. Radii were taken as follows: 3.10 Å for main chain alpha carbon atoms, 2.92 Å for main chain carbonyl oxygens, 2.95 Å for main chain amide nitrogens, and 3.20 Å for all other atoms.[7,19] At 0.25 Å spacing, 25 sections can be cut parallel to the X–Y plane through the principal atom. Within each section, atoms intersected by the plane give arcs of various radii. For the principal atom, the length of the arc in that plane is obtained by determining the fraction of the circumference that does not lie within the effective radius of a neighboring atom. The accessible surface area (A) is obtained by summation over the 25 slices:

$$A = \sum_i [0.25(R)(L_i)/(R^2 - Z_i^2)^{1/2}] \tag{5}$$

where R is the effective radius, L_i is the length of an arc in a given section i, and Z_i is the perpendicular distance from the center of the atomic sphere to the planar section i, all values being in Å. The accessibility of the atom is then defined as the accessible surface area in A, normalized and expressed as percentage:

$$\text{Accessibility} = 100(A)/4\pi R^2 \tag{6}$$

A clearer appraisal of the influence of protein structure on the solvent accessibility of individual atoms or groups is obtained by normalization to the corresponding accessibility in a model tripeptide where, it may be assumed, accessibilities attain their maximum values. The "fractional static accessibility" (SA) is expressed as:

$$SA = (A \text{ in protein})/(A \text{ in tripeptide}) \tag{7}$$

[19] A. Bondi, *J. Phys. Chem.* **68**, 441 (1964).

Accessible surface areas have been calculated for the two beta trans configurations of Ala-X-Ala peptides, where X is an amino acid residue.[7] We have observed that accessibility values intermediate between those for the two peptide structures are obtained if one considers the tripeptide Ala-X-Ala coordinates as taken from helical regions. These fractional accessibility values vary from 0.01, indicating virtually no van der Waals contact with water in the static structure, to 1.0, inferring no protein structural restraints on solvation. Each protein charge site known to be titratable in the native structure has some solvent contact; thus they are all at the protein solvent interface.

Effects of Side Chain Orientation. Because of the role that the calculated accessibility plays in these computations, the variability due to the quality of the crystallographic data must be considered. Even in a refined high resolution crystal structure, side chain placement of surface charge residues is often completely arbitrary. To address this issue Table I lists the accessibility results for two sets of sperm whale ferrimyoglobin coordinates.[20,21]

On the whole, the results were similar but the magnitudes were significantly different for many surface residues. Some examples are shown in Table I of changes in atomic coordinates from Watson[20] to Takano[21] that are variable, and in some cases quite large. As a consequence, the numbers of neighboring atoms within 6.40 Å of the principal atom differ considerably, and also those within a single radius of 3.20 Å. Accessibility differences, listed in the last column of Table I, can therefore be traced to the repositioning of the principal atom or of its neighboring atoms. A comparison of all charged residue accessibilities between the two data sets indicates that about half of the accessibility values differ by 20% from the previously published data, and in some cases the difference exceeds 50%. It is clear that computations of surface areas and accessibilities for protein atoms need to be interpreted in the context of the refinement procedure and possible poetic license exercised.

Examples and Extensions

pH Dependence of j Site Contributions to pK_i

At a given pH, the unique protein charge distribution confers an effective pK_i value on each titratable site i. pK_i is determined by the interplay of the entire protein charge complement. The coulombic influence on site

[20] H. C. Watson, *Prog. Stereochem.* **4**, 299 (1969).
[21] T. Takano, *J. Mol. Biol.* **110**, 537 (1977).

TABLE I

COMPARISON OF ACCESSIBILITY DIFFERENCES FOR SOME RESIDUES IN SPERM WHALE MYOGLOBIN AS DERIVED FROM WATSON[20] AND TAKANO[21] X-RAY COORDINATE

Residue	Helix	Atom	Ref.	X-Ray coordinates			Number of atoms at distance of		[Eq. (6)] Accessibility	Change in accessibility (T–W)
				X	Y	Z	≤6.4 Å	3.2 Å		
1 Val	1NA	N	W	−2.90	17.60	15.50	24	4	24.8	+13.8
			T	−4.42	16.67	16.94	15	4	38.6	
12 His	10A	NE2	W	15.00	9.20	24.30	22	6	8.4	+7.4
			T	14.40	9.23	24.95	19	4	15.8	
16 Lys	14A	NZ	W	17.00	8.60	23.30	27	3	9.2	−8.6
			T	17.25	8.70	22.18	38	2	0.6	
45 Arg	3CD	NH1	W	24.10	31.80	6.30	41	4	12.2	−3.6
			T	24.28	31.80	6.03	44	4	8.6	
145 Lys	22H	NZ	W	−1.70	34.30	5.40	28	7	1.7	+26.0
			T	−2.17	28.01	7.78	31	3	27.7	
148 Glu	25H	OE1	W	−3.20	35.10	4.20	25	4	26.4	+15.1
			T	−3.67	35.05	3.91	15	3	41.5	

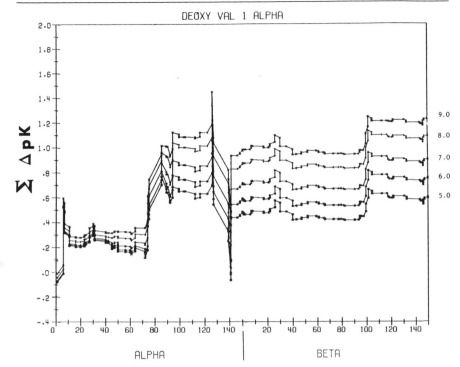

FIG. 2. Summation plots of intramolecular electrostatic contributions, ΔpK_{ij}, defined in Eq. (1) to the value of pK_i of the amino acid residue Val-1α in deoxyhemoglobin tetramer, computed at $I = 0.01\ M$ and at pH 5.0, 6.0, 7.0, 8.0, and 9.0 (curves 1–5, respectively). The sums are cumulative, showing the residue sequence numbers of the groups in their α and β chains. A contribution attributed to a particular site represents the summed effect of that site from each dimer. In the case of Val-1α, being very near the hemoglobin tetramer symmetry axis, the effects are nearly equal between dimers. The net values at the extreme right represent the electrostatic potential acting on the deoxyhemoglobin Val-1α site at the given pH.[10]

i exerted by each one of the other sites j is proportional to the charge Z_j and to the extent of their burial within the protein molecule ($1 - \overline{SA}_{ij}$), but it is inversely proportional to intercharge distance r_{ij}.

Figures 2 and 3 show the final iterative calculation of pK_i at five pH values for two sites in deoxyhemoglobin tetramer, Val-1α and His-117β, respectively.[10] The electrostatic contributions to pK_i are summed according to the formalism of Eq. (1). The individual ΔpK_{ij} contributions are summed cumulatively from left to right along the x axis following the j site residue number in its respective chain, so that at the extreme right the level corresponds to the total electrostatic contribution to the effective magnitude of pK_i for site i at the given pH. Every step in each figure

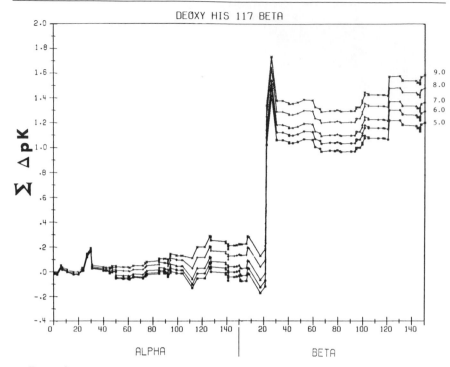

FIG. 3. Summation plots, corresponding to those in Fig. 2 and under the same conditions, showing electrostatic contributions to the value of pK_i of the amino acid residue His-117β. The sharp rises at Glu-22β and Glu-26β illustrate the dominating effect of close interaction with these two negative sites, yielding the extremely high pK_i value of 8.2 for His-117β at pH 9.0. In contrast to Val-1α the His-117β site is off the dyad axis, and the effects are primarily the result of one dimer site, the second being generally at a greater r_{ij}.[10]

represents the summed effects of the given residue in both of the pairs of α or β chains.

The charge Z_i is positive for residues Val-1α and His-117β, the examples shown in the figures. It follows from Eq. (1) that a rise in the level reflects the influence of negatively charged sites (Z_j is negative), while a drop in level corresponds to pK_i depression under the influence of positively charged sites. Nearly constant levels in the plots represent sections of the chain sequence whose residues are either nontitratable, neutral at the pH, or such that the combination of the effects of Z_j, $\overline{SA}_{ij}$ and r_{ij} tends to balance out.

Figure 2 also shows that, at $I = 0.01\ M$ and over the range of pH 5–9, the computed electrostatic effect on Val-1α totals from 0.6 to 1.2 which, based on $pK_{int} = 7.00$, yields an effective pK_i which varies from 7.6 at pH 5 to 8.2 at pH 9. The midtitration value, $pK_{1/2}$, is 8.1. On the whole, it is

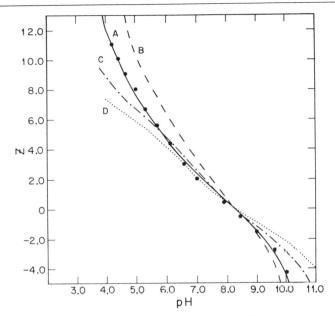

FIG. 4. Computed curves for Mb, 0.2 mM, $I = 0.01$ M, 25°. Curve A, according to the present treatment; curve B, with SA taken always as 1; curve C, with SA taken always as 0; curve D, with all groups placed 1 Å inside the low dielectric medium.[16]

seen that there are a large number of small contributions from other groups determining the resultant pK_i value for Val-1α.

In contrast, Fig. 3 shows that the electrostatic effect on His-117β is dominated by very strong interaction with two negative groups, Glu-22β and Glu-26β. A pK_i rise of 1.2–1.6 occurs, yielding the extremely high pK_i values of 7.8 at pH 5 and 8.2 at pH 9 (based on $pK_{int} = 6.60$). Here again, lesser effects from more distant groups are not negligible.[10]

Titration Curves

The theoretical hydrogen ion titration curve of a protein is formed by summing electrostatic contributions from every charged site in the molecule. Figure 4 shows experimental points for the titration of sperm whale myoglobin at 25° and ionic strength 0.01 M, and four theoretical titration curves for the same conditions.[16] The continuous curve (A) is predicted from the present model and falls in close agreement with the observed points. The steepest curve (B) uses the assumptions of the present model except that SA is taken as 1 in all cases, which has the effect of obliterating the electrostatic contributions. The curve (C) that falls first on the other side of the experimental points is computed with the assumption that SA is zero, in effect treating all charged groups as inaccessible to solvent. The least shapely curve (D) involves the same assumption of zero SA with the depth parameter d of Tanford and Kirkwood set equal to 1.0 Å, in keeping with the practice of the original treatment.[3,5]

Figure 5 shows the computed net protein charge, Z, varying with pH from pH 5 to 10 at 25°, at three ionic strength levels, 0.0, 0.01, and 0.1 M (broken curves, top to bottom, respectively) for deoxyhemoglobin.[10] The points are experimental taken from Rollema et al.[22]

The effect of ionic strength on the titration curve brings out an uncommon feature, in that the crossover region is near the acid end at pH 5. Human deoxyhemoglobin A has its isoionic point near pH 7.4, and it is around that region that the net protein charge would normally be expected to be nearly independent of ionic strength.[23] Since the monomeric hemoprotein myoglobin shows classical titration behavior in this respect,[6] the question arises as to whether this special feature in hemoglobin may be a consequence of its quaternary structure or of its charge configuration.

Simulated Titration Curves

Various charge distributions can be simulated by using the same model and formalism that are applied to proteins.[11] In a typical computation, a uniform (symmetric) charge distribution was achieved by placing 14 point charges on the vertices of a 24-faced regular polyhedron inscribed within a low dielectric sphere of radius of 10.0 Å. The sites were assigned pK_{int} values to simulate amino acid residues: 4.00 for acidic groups, 10.40 or 12.00 for basic groups, and 6.60 for intermediate histidine-like groups. A mean static accessibility fraction, SA, of 0.50 was used for many of the trials; varying this value for all point charges together does not alter the overall form of the results, but separate variation for specific sites produces profound effects as described below. Nonuniform charge distributions were simulated by pairing oppositely charged groups at an ion-pair distance of 3.0 Å. Interactions were also modulated by reducing the solvent accessibilities of ion-paired groups to 0.20. The dielectric constant used for the internal region of the protein was $D_{int} = 4.0$, and that for the surrounding medium was $D_{ext} = 78.5$ at 25°.[2,9,15] Since the sphere had a radius of 10.0 Å, the ion exclusion radius was taken to be 12.0 Å. Calculations of effective pK_i values, and charges Z_i, as functions of pH and ionic strength, were then made.[11] As noted above, the classical expectation for protein titration behavior is that in the pH region near the isoelectric point

[22] H. S. Rollema, S. H. de Bruin, L. H. M. Janssen, and G. A. J. van Os, *J. Biol. Chem.* **250**, 1333 (1975).

[23] C. Tanford, *Adv. Protein Chem.* **17**, 69 (1962).

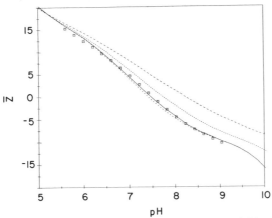

FIG. 5. The hydrogen ion titration curve of human deoxyhemoglobin showing the variation of net protein charge, $\bar{Z}$, with pH at 25°. Experimental data are from Rollema *et al.*[22] The
computed curves are at three levels of ionic strength: 0.0, 0.01, and 0.10 *M* (from top to
bottom, all broken curves). The full curve is the computed titration curve, protons bound per
tetramer, when the effect on hydrogen ion equilibria of Cl⁻ ion binding at the Val-1α–Arg-
141α site is included.[10]

there should be a minimal dependence on ionic strength. This behavior is
predicted by application of Debye–Hückel theory to the smeared charge
model of globular proteins.[1,24] As Fig. 6 shows,[11] this is not the case for
hemoglobin, although myoglobin and cytochrome *c* do indeed follow the
classical pattern.[6,16]

Figure 7 compares the simulated titration curves for two hypothetical
polyelectrolyte models chosen to mimic the charge distribution and *SA*
values characteristic of hemoglobin α chain and myoglobin.[11] The paired
charge model is shown by curves 1 and 2 for ionic strengths of 0.00 and
0.10 *M*, respectively, and the uniform distribution model is shown by
curves 3 and 4 for the corresponding ionic strength values. The model
consists of 14 point charges placed on a sphere of radius 10.0 Å. The
charges are assigned to three types: six carboxylic acids of $pK_{int} = 4.00$;
four imidazoles, $pK_{int} = 6.00$; and four amines, $pK_{int} = 10.00$. The paired-
charge, nonclassical pattern of curves 1 and 2 was achieved by placing the
carboxylic and amine charges at the vertices of an inscribed regular polygon but with the imidazoles located with maximum symmetry on the
spherical surface 3.0 Å from carboxylate points. The classical pattern of
curves 3 and 4 placed all 14 charges at the vertices of the inscribed
polygon which increased the distance between each imidazole point

24 P. Debye and E. Hückel, *Phys. Z.* **24**, 305 (1923).

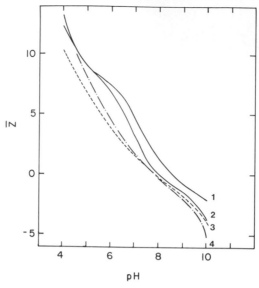

FIG. 6. Comparison of ionic strength effects on titration curves, net protein charge $\bar{Z}$ vs pH at 25°, for human ferrohemoglobin α-chain monomer (full curves, 1 and 2) and sperm whale ferromyoglobin (broken curves, 3 and 4). Curves 1 and 3 apply at $I = 0.0\ M$ and curves 2 and 4 at $I = 0.10\ M$.[11]

charge and the nearest carboxylate point charge to approximately 6.5 Å. The nonpaired charges were assigned SA values of 0.50, while the SA values of members of charge pairs were taken as 0.20.[11]

The broken curves 3 and 4 in Fig. 7 for the uniform geometry of point charges show, classically, a minimum sensitivity to ionic strength near $Z = 0$, whereas the nonuniform geometry represented in the solid curves 1 and 2 produces a nonclassical minimum sensitivity to ionic strength well displaced from the region of $Z = 0$. The characteristics of these curves, therefore, mimic the effects seen for the hemoglobin α chain and myoglobin in Fig. 6 for correspondingly designated curves.

If we compare the predicted ionic strength variation of the individual titration curves for a number of amino acid residues in hemoglobin α chain with their equivalent groups in myoglobin, the same phenomenon is observed, namely, classical behavior by individual groups in myoglobin but not in hemoglobin α chain. Figure 8 shows the computed behavior for the Val-1 group in both proteins.[11] As expected, the variation of ionic strength dependence of pK_i with pH is strongest at the lowest ionic strength, $I = 0\ M$, and is weakest at the highest, $I = 0.1\ M$, in both cases. However, the three Val-1 curves intersect near pH 8.4 in myoglobin

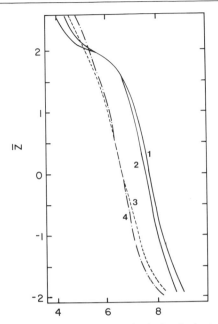

FIG. 7. Simulated titration curves for a hypothetical polyelectrolyte model having 14 point charges placed on a sphere of radius 10.0 Å. The charges were assigned to three types: six carboxylic acids, $pK_{int} = 4.00$; four imidazoles, $pK_{int} = 6.60$; and four amines, $pK_{int} = 10.00$. Curves 1 and 2 correspond to a paired-charge distribution, at 0 and 0.10 M ionic strength, respectively. Curves 3 and 4 correspond to a uniform charge distribution, at 0.0 and 0.10 M ionic strength, respectively.[11]

(curves 4, 5, and 6), whereas the corresponding curves in hemoglobin α chain (curves 1, 2, and 3) meet near pH 4.

Other groups follow the pattern. It appears that the fundamental difference in ionic strength behavior between hemoglobin and myoglobin may lie not so much in the oligomeric protein structure or in the chemical nature of the charged groups themselves as in their unique distribution.

Treatment of Specific Ion Binding Sites

The presence of bound anions such as chloride and inorganic or organic phosphates requires special consideration. These ions contribute to coulombic influences and may themselves represent titratable proton sites. They also have the ability to diffuse away to relieve or enhance interactions rather than shifting protonation state. In early work[11,12] we relied on chemical or crystallographic data for specific ion site location.

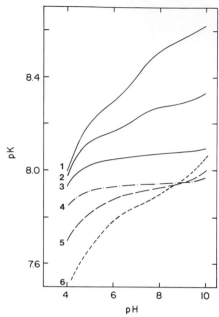

FIG. 8. Comparison of ionic strength effect on the electrostatically calculated pK_i values for a single amino acid residue, the amino-terminal valine group, in human ferrohemoglobin A α-chain monomer (solid curves, 1, 2, and 3) and sperm whale ferromyoglobin (broken curves, 4, 5, and 6). The pH variation at 25° is shown at three levels of ionic strength: 0.0 M, curves 1 and 6; 0.01 M, curves 2 and 5; and 0.1 M, curves 3 and 4.[11]

The coordinates also can be chosen by locating any area of high potential at the protein solvent interface.[15] The thermodynamics of ion binding are treated as follows. The term $\Sigma \Delta p K_{iq}$ is calculated for the ion, q. This value is used to define directly the association constant, calculated in the iterative procedure at various pH values to yield the pH dependence of ion association. For example, a site whose potential is calculated to be $5kT$ at zero ionic strength, $3kT$ at 0.01, and $2kT$ at 0.1 would have monovalent anion affinities of 150, 20, and 10 $1/M$, respectively. It should be noted that no side chain rearrangement has been allowed in response to binding, so that chelation effects, if any, are omitted, and the calculated magnitudes of interaction may be underestimates. When the anion is itself a titratable species such as inorganic phosphate, the calculation for the site includes terms in the iteration reflecting this additional ionizable group. Based on the calculated protein charge array and the known protein surface topography a procedure has been developed for the identification of specific ion binding sites at the protein surface based on electrostatic criteria.[15] In particular, the active site of ribonuclease can be

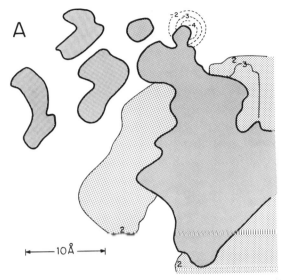

FIG. 9. Calculated electrostatic field for a cross section through the active site of ribonuclease S. The section is perpendicular to the y axis at position $y = -6$. The calculations were made on a square grid 1 Å on a side. The heavily shaded areas are inside the van der Waals envelope of the protein. The field contour interval is $1kT$. Positive values are given by the solid contour and negative values by dashed contours. Positive field contours of $2kT$ and above are lightly shaded: (A) pH 8.0, anion site(s) unoccupied; (B) pH 6.0, anion site(s) unoccupied; (C) pH 6.0, phosphate in anion site A2. These contours were drawn for an effective ionic strength of zero.[15]

identified as a general anion binding site and thus allows an evaluation of the treatment in the context of the measured active-site histidine proton affinities.

Active Site in Ribonuclease: An Anion Binding Site. At the isoionic pH (9.2), net charge equals zero, and the measured number of associated anions or cations is zero. Saroff and Carroll[25] have shown the number of chloride ions bound per mole of RNase A to be zero at pH 9.6, 0.50 at pH 6.6, and 2.0 at pH 4.5 (total chloride concentrations of 0.012, 0.016, and 0.085 M, respectively). Plots of the electrostatic potential at the protein surface at several ionic strengths show two sites with exceptionally positive potential at pH 6.0. The number of solvent-accessible, high potential sites increases to at least five by pH 4.0. One of the two positive centers identified at pH 6.0 is the active-site cluster.

The calculated electrostatic potential,[15] expressed in units of kT, in the active site of ribonuclease S at pH values of 8 and 6 is shown in parts A

[25] H. S. Saroff and W. R. Carroll, *J. Biol. Chem.* **237**, 3384 (1962).

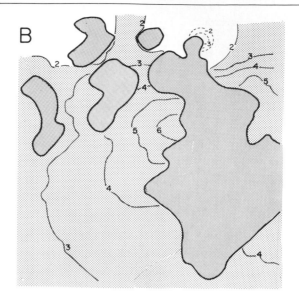

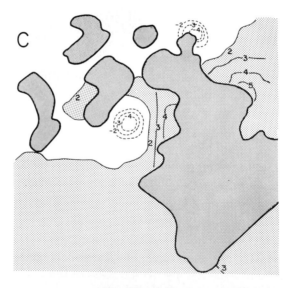

FIG. 9B and C.

TABLE II

CALCULATED $pK_{1/2}$ VALUES IN THE PRESENCE OF BOUND MONO-
OR DIVALENT ANION[15 a-c]

	Unoccupied[d]	Cl⁻ position A1	Cl⁻ position A2	PO_4^{2-}/HPO_4^-	
				Position A1	Position A2
Histidine-12	6.38	6.86	6.80	7.20	7.52
Histidine-119	6.34	6.64	7.02	6.82	7.24

[a] Cohen et al.[26]

[b] Griffin et al.[27]

[c] Positions A1 and A2 for bound anion placement are described in Ref. 15.

[d] All the $pK_{1/2}$ values were calculated assuming a Debye screening of 0.10; the explicit exclusion of a site-bound ion is given as a reference for comparison with the site-bound cases.

and B of Fig. 9, respectively. The pH-dependent development of the anion binding site is evident. At pH 8, where the two active-site histidines are essentially unprotonated (i.e., $pK_{1/2}$ values of ~6.3), the active-site potential is $(2-3)kT$ compared to $(5-6)kT$ at pH 6. Figure 9C illustrates the long-range effect of binding a divalent anion in this site at pH 6.0. The effect of including two negative charges is damped by an increase in protonation of the active-site histidines.

Electrostatic Effects of Anion Binding at the Active Site on Proton Uptake and Release. Since the net charge of a phosphate ligand is influenced by and also influences the protonation of its binding domain, we can calculate net proton uptake or release only by considering both components. The observed shifts in apparent proton affinities for the active-site histidine residues in response to the binding of several anionic ligands are tabulated in Ref. 15. With increasing anionic charge on the ligand, the observed $pK_{1/2}$ values are seen to increase from ~6.0 to 8.0.

Table II[26,27] gives the computed effective $pK_{1/2}$ values for histidine-12 and histidine-119 when full occupancy ligands are simulated at two alternative positions in the active site.[15] The values computed for the unoccupied site are given in column one for reference. These predicted effects due to ligand binding should be viewed as underestimates as no side chain rearrangements have been allowed to maximize coulombic interactions. At interactive distances of less than 5 Å, a movement to more realistic chelate geometry involving a reduction in distance of 1 Å can easily increase the effects listed in Table II by 20%.

[26] J. S. Cohen, J. H. Griffin, and A. N. Schechter, *J. Biol. Chem.* **248,** 4305 (1973).

[27] J. H. Griffin, A. N. Schechter, and J. S. Cohen, *Ann. N.Y. Acad. Sci.* **222,** 693 (1973).

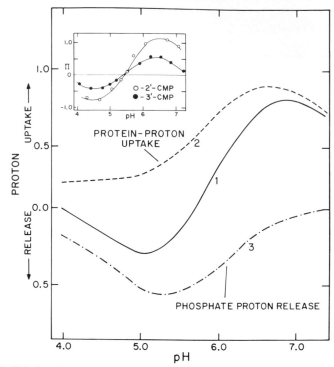

FIG. 10. Calculated proton uptake or release induced by full occupancy of anion site A1 by phosphate at $I = 0.01$ as a function of pH (solid curve). This calculated uptake or release is resolved into the dashed curve: protein–proton uptake (curve 2) and phosphate ion proton release (curve 3). The inset shows the observed uptake or release for 2′-CMP (○) and 3′-CMP (●).[15] Data are from Hummel and Witzel.[28]

     The net release or uptake of protons attributable to ligand binding to the active site of RNase S can be computed by using the increased pK_i values of the protein calculated at a given pH in the presence of ligand and the calculated proton release by the ligand. In Fig. 10 the solid curve shows the calculated net proton uptake or release from RNase S in response to inorganic phosphate association. This net proton flux is further resolved into protein uptake and phosphate release. While the absolute magnitude of these calculated contributions will vary with ion placement, ionic strength, and site occupancy, it is significant that the pH dependence and magnitude of proton flux closely follow the experimental values of Hummel and Witzel[28] over the entire pH range. The inset in Fig. 10

[28] J. P. Hummel and H. Witzel, *J. Biol. Chem.* **241**, 1023 (1966).

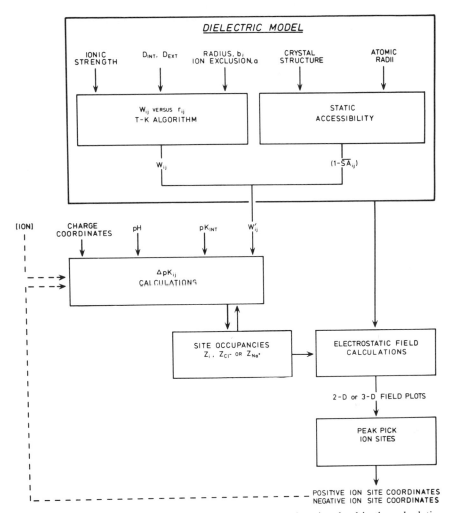

FIG. 11. This figure describes the flow of the computations involved in the calculations that have been presented in the text. All the calculations stem from the computations that take place within the dielectric model box. Two separate computations take place at this level. (1) Computations of the $\overline{SA}_{ij}$ terms. The input for these computations consists of the X-ray crystallographic coordinates of the protein being studied, and the atomic radii of the various atoms in the protein molecule. (2) Computations of W_{ij} vs r_{ij} table. For a given protein, a radius b and an ion exclusion radius a are defined. Using D_{int}, D_{ext}, and the Tanford–Kirkwood algorithm, a numerical table of W_{ij} values is calculated as described in the text. At every chosen ionic strength, a W_{ij} value corresponds with one r_{ij}. The charge coordinates, pK_{int} of various amino acid side chains, and the pH range over which the calculations are performed, are entered at the next stage of the calculations. The charge coordinates are used to calculate r_{ij} for all pairs of charges. Each r_{ij} corresponds to a W_{ij} value in the previously calculated W_{ij} table. The W_{ij} values that are selected from the table are modulated by the previously calculated $\overline{SA}_{ij}$ term as described in the text. The resulting W'_{ij} values are the core of the further calculations that are described in the text.

shows the proton uptake or release observed when saturating quantities of 2'-CMP or 3'-CMP are added to RNase A.

Conclusion

The present electrostatic treatment is successful. By use of the static accessibility modification of the Tanford–Kirkwood model, a simple and efficient computational procedure yields quantitative and qualitative results in agreement with experimental data. This is particularly satisfying since non-coulombic effects and arbitrary curve fitting were excluded. The introduction of a solvent accessibility parameter apparently corrects the model adequately so that the smooth protein-solvent boundary of the model is able to mimic the real protein surface. The assumption that the structure of the protein in solution throughout the pH range of interest is essentially that of the crystal seems to hold reasonably well. Minor perturbations may of course occur in solution and could be expected to alter charge configurations. Specific chloride ion binding is an example that has been taken into account above. Figure 11 outlines the computational flow of electrostatic interactions as laid out in the text.

The working electrostatic model for the protein molecule adapts easily to a variety of problems and enables one to evaluate electrostatic contributions to such diverse phenomena as the binding of charged ligands, amino acid substitutions of site-directed mutagens, and any known tertiary or quaternary structural change.[29]

Acknowledgment

The assistance of J. E. Beecher Matthew in the preparation of this manuscript is greatly appreciated.

[29] J. B. Matthew, M. A. Flanagan, B. Garcia-Moreno E., K. L. March, S. J. Shire, and F. R. N. Gurd, *C.R.C. Crit. Rev. Biochem.* **18,** 91 (1985).

[18] Stabilization and Destabilization of Protein Structure by Charge Interactions

By JAMES B. MATTHEW and FRANK R. N. GURD

Introduction

The overall free energy of protein stabilization is often taken as the sum of contributions

$$\Delta G = \sum_{1}^{N} (\Delta G_{i,el} + \Delta G_{i,h} + \Delta G_{i,conf} + \Delta G_{i,vw}$$

$$+ \Delta G_{i,hb}) + \Delta G_{buried} + \Delta G_{s\text{-}s} \tag{1}$$

where the summation over all residues refers, respectively, to electrostatic, hydrophobic, conformational, van der Waals, and hydrogen bonding contributions. It is variations in the electrostatic component which give rise to the pH and ionic strength-dependent properties of proteins. The conformation transition of sperm whale myoglobin in acid solution is an example of a poised process in which variation of pH and ionic strength clearly show the importance of electrostatic interactions for stability of the native structure. Figure 1 shows the effect of pH on the percentage of the native structure present at four different values of ionic strength. The two measures of the native and denatured states were absorbance at 409 nm[1] and molar ellipticity at 222 nm.[2] These variables detect, respectively, the loss of the characteristic environment of the heme[3] and the loss of helical folding of the polypeptide backbone elements.[4] The changes in these two variables are clearly concurrent, indicating a cooperative process involving the whole protein molecule.

Figure 1 shows that the dependence on ionic strength is marked, indicating that the native state is destabilized relative to the denatured state as the electrostatic interactions are lessened with increasing ionic strength. Note that the variation in the effect is most marked at the lower ionic

[1] E. Breslow and F. R. N. Gurd, *J. Biol. Chem.* **237,** 371 (1962).
[2] D. Puett, *J. Biol. Chem.* **248,** 4623 (1973).
[3] H. Theorell and A. Ehrenberg, *Acta Chem. Scand.* **5,** 371 (1951).
[4] E. Breslow, E. S. Beychok, K. D. Hardman, and F. R. N. Gurd, *J. Biol. Chem.* **240,** 304 (1965).

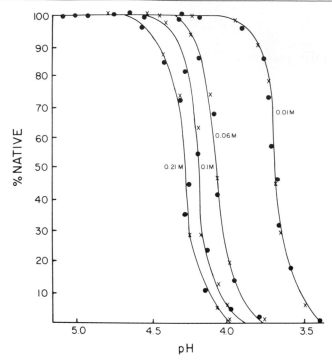

FIG. 1. Dependence of unfolding transition of sperm whale ferrimyoglobin on pH. The ordinate shows the percentage of the native form found for the substantially two-state process (see text). The curves show results for ionic strength values of 0.21, 0.10, 0.06, and 0.01 M, respectively: (●) 409 nm; (×) 222 nm.[8]

strengths, as expected from the modified Debye–Hückel treatment.[5,6] Breslow and Gurd[1] have shown that the proton titration curve for the denatured state is much less dependent on ionic strength than that for the native state. It follows that intramolecular electrostatic effects are important for the stability of native structure. Furthermore, since the increase in ionic strength reduces the absolute magnitude of the electrostatic interactions, it follows that the net effect of these terms is to contribute to the stability of the native structure under the conditions of the experiments reported in Fig. 1.

The electrostatic contribution to protein stability may be estimated from the change in summed electrostatic free energy terms for all charge

[5] C. Tanford and J. G. Kirkwood, *J. Am. Chem. Soc.* **79,** 5333 (1957).
[6] S. J. Shire, G. I. H. Hanania, and F. R. N. Gurd, *Biochemistry* **13,** 2967 (1974).

sites, $\Sigma\Delta G_{i,el}$, according to the modified Tanford–Kirkwood theory.[7,8] If the electrostatic free energy of placing a pair of unit charges i and j on the protein is W_{ij}, by summing W_{ij} terms for all pairwise interactions, the summed electrostatic free energy with respect to the reference state is obtained (see this volume [17]). In the reference state for all these calculations, the groups are maximally accessible to solvent and have no electrostatic interaction with each other. This hypothetical state can be approached for many proteins under conditions of complete denaturation. If the random configuration has a high charge density, e.g., poly(DL-lysine), the assumption that equates the random coil to the reference state would not be valid.

Protein molecules are unique polyelectrolytes in that they carry their own partial complement of counterions which exhibit charge cancellations at their characteristic isoionic point, and "tight" binding in the form of intramolecular salt bridges. It is expected that when the protein's own charge array is sufficiently asymmetric, by either composition, geometry, or both, the surface electrostatic potential in the absence of steric hindrance will dictate specific ion binding sites. Any calculation of protein stabilization as a function of pH and ionic strength must include the phenomenon of "specific" ion binding where applicable. Except where specific evidence to the contrary can be established such calculations assume that the crystal structure is valid over the pH and ionic strength range of the computation.

Counterions bound at specific sites are distinguished from other associated counterions by the inclusion of their interaction energy contributing to the electrostatic stability of the macromolecule. When the charge geometry on a macromolecule is such that a counterion intervenes between surface ions of like charge, then the ion–counterion force vectors enhance the conformational stability of the macromolecule by acting as structural struts. These specific ion binding sites at the molecule surface are identified using the charge distribution and surface topography of the macromolecules[9,10] (see this volume [17, Fig. 9]). Coordinates chosen as specific ion site locations correspond to the areas of high potential at the molecule–solvent interface. Often, the electrostatic energy, $\varepsilon\Psi$, for a small ion in the vicinity of a highly charged polymer is calculated to be several kT. The potential at a given site and thus the binding constant

[7] J. B. Matthew, S. H. Friend, and F. R. N. Gurd, *Biochemistry* **20**, 571 (1981).
[8] S. H. Friend and F. R. N. Gurd, *Biochemistry* **18**, 4612 (1979).
[9] J. B. Matthew and F. M. Richards, *Biochemistry* **21**, 4989 (1982).
[10] J. B. Matthew, P. C. Weber, F. R. Salemme, and F. M. Richards, *Nature (London)* **301**, 169 (1983).

varies with pH, ionic strength, and proximity/occupancy of the other ion sites.

Computational Methods

Electrostatic Computation of Stability

Free energies of interaction of the ionizable groups have been calculated (see this volume [17], and Ref. 11) according to the procedures described in this volume [17]. The electrostatic work factor, the interaction energy between two formal changes, i and j, at a spherical interface between two media of different dielectric constant, W_{ij}', is equal to W_{ij}, computed from the Tanford–Kirkwood treatment,[12] multiplied by the factor $(1 - \overline{SA}_{ij})$ where $\overline{SA}$ is the mean fractional static solvent accessibility. Computed pK_i values are sometimes expressed as $pK_{1/2}$, those values applying at the pH of half-titration. The charge borne on a given site, Z_i or Z_j, varies between -1 and 0 or 0 and $+1$ to allow for fractional saturation. The overall electrostatic free energy contribution is represented as

$$\Delta G_{el} = \sum_{i=j}^{n} \sum_{j<i} W_{ij}(1 - \overline{SA}_{ij})Z_iZ_j \tag{2}$$

where Z_i and Z_j are the fractional charges at the sites i and j at the pH in question, and n is the total number of potentially charged groups in the protein.

Treatment of Specific Ion Binding Sites

The presence of bound anions such as chloride and inorganic or organic phosphates require special consideration. These ions contribute to coulombic influences and may themselves represent titratable proton sites. They also have the ability to diffuse away to relieve or enhance interactions rather than shifting protonation state. The thermodynamics of ion binding are treated as follows.[9,13] Specific ion site coordinates are determined by locating the area of high potential at the protein–solvent interface. The term ΔG_{el} is calculated for the protein with and without bound full occupancy, nontitratable anion. The electrostatic work to

[11] M. A. Flanagan, B. Garcia-Moreno, S. H. Friend, R. J. Feldmann, H. Scouloudi, and F. R. N. Gurd, *Biochemistry* **22**, 6027 (1983).

[12] C. Tanford and R. Roxby, *Biochemistry* **11**, 2192 (1972).

[13] J. B. Matthew and F. M. Richards, *J. Biol. Chem.* **258**, 3039 (1983).

place a diffusible ion, q, at the solvent interface can be expressed as

$$\Delta G_q^{+/-} = \sum_{i=1}^{N} W_{iq}(1 - \overline{SA}_{iq})Z_i(+/-1) \tag{3}$$

The calculated static accessibility for surface bound ions usually varies from 0.48 to 0.55. For a permanent anion such as chloride, the difference between these two values for full or no occupancy is assumed to give the binding constant. This binding constant is calculated in the iterative procedure at various pH values to yield the pH dependence of the anion association. If the ion affinity for a high potential site is defined as

$$K_k = \exp(\Delta G_q^{+/-}/RT) \tag{4}$$

where K_k is the binding constant for $q^{+/-}$ at the kth site, then the fractional occupancy of each site, k, is given by the binding constant for that site and the concentration of $q^{+/-}$:

$$Z_k = (K_k[q^{+/-}])/(1.0 + K_k[q^{+/-}]) \tag{5}$$

Because k, the number of ion sites, can be greater than one, the calculation is carried out in an iterative manner using Eq. (2). The ith and jth sites then include the cation sites k and for the kth site Z_k (the ion occupancy from the previous step) is substituted for Z_i or Z_j. The iterative procedure is carried out until the change in the sum of Z_k between iterative steps is less than 10^{-3}. When the anion is itself a titratable species such as inorganic phosphate, the calculation for the fully occupied site includes an intrinsic pK value in the iteration reflecting this additional ionizable group. The actual procedure for HPO_4^{2-} placed a potential negative charge on two of the four tetrahedral oxygen atoms.[9] One of these was assigned the first pK of phosphoric acid and the other the second. They were treated as independent sites in the calculation (with their own cross term omitted), and the net protonation was calculated from the final occupancies. The calculation was not sensitive to the choice of which oxygen was used because they are so close together (1.9 Å).

Examples and Discussion

Effects of pH and Ionic Strength on Protein Stability

   Overall Stabilization of Tertiary Structure. Summation of all the computed electrostatic interactions felt by individual residues in sperm whale myoglobin according to W_{ij} values yields the overall electrostatic free

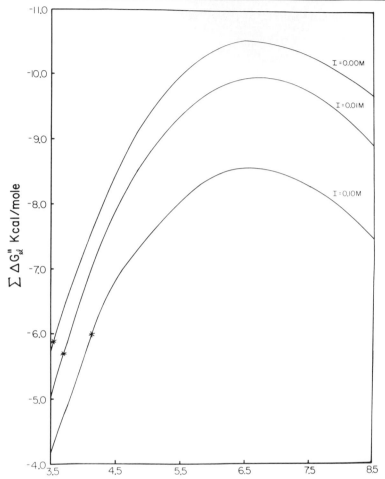

FIG. 2. Dependence of summed electrostatic free energy terms, in kilocalories per mole, between individual charged sites as a function of pH at ionic strength values of 0.00, 0.01, and 0.10 M. The asterisked point on each curve at low pH represents the midpoint pH for the unfolding transition, extrapolated or interpolated from Fig. 1, at the particular ionic strength.[8]

energy contribution to the stability of the native structure, ΔG_{el}, according to Eq. (2).

Figure 2 shows the results of the calculation made for the three conditions of ionic strength, 0.00, 0.01, and 0.10 M.[8] Here ΔG_{el}, expressed in kilocalories per mole, is plotted as the ordinate against pH as the abscissa. Over the range of pH involved, the free energy change resulting from the full constellation of charge point interactions is negative, indicating substantial stabilization ascribable to this source. As would be expected for a net stabilizing effect, the result of increasing the ionic strength is to reduce the stabilization. Experimentally, this is what is generally found (see Fig. 1). Increasing the ionic strength from 0.00 to 0.10 M decreases the stabilization rather generally by approximately 20%. The relative effect of ionic strength on W_{ij} is distance dependent, and the overall ionic strength effect observed here fits roughly with the effect to be found when the charged groups are on the average 6 Å apart and average values of SA are taken into account.[14]

The maximum stabilization at each ionic strength in Fig. 2 occurs near pH 6.5, well below the isoionic point near pH 8.30. Substantially the same pH profile was obtained by Acampora and Hermans[15] when they plotted the midpoints of thermal denaturation transitions against pH. This result is significantly different from what is predicted for a uniform charge distribution of a similar magnitude of net charge according to the smeared-charge model of Linderstrøm-Lang.[16]

The maximum stabilization falls in a range, pH 5.5–7.5, where histidine residues are the prominent titrating class. The breadth of the maximum reflects in part the fact that the majority of the titratable histidine residues has a wide range of $pK_{1/2}$ values complicated by a considerable range of SA values. The net electrostatic field sensed by histidine residues of higher $pK_{1/2}$ is negative whereas those of low $pK_{1/2}$ sense predominantly positive environments. Protonation of the first set with their higher $pK_{1/2}$ values generally promotes preferential stabilization in the relatively negative environments and produces the rise in overall stabilization seen in Fig. 2. Conversely, protonation of histidine residues in the relatively positive environments occurs at the lower pH values, and it is the protonation of these low pK histidines below pH 6.5 that deleteriously affects the summed electrostatic free energy values.

Since the electrostatic interactions depend not only on the state of charge of a given group under consideration but also on those of all other groups in the molecule, the stabilization or destabilization sensed by each group varies over the entire pH range and must be defined separately for each pH value.

The distinction between the pH of the maximum stability at pH 6.5 and that of the isoionic point at pH 8.3 is thus related to the properties of

[14] J. B. Matthew, G. I. H. Hanania, and F. R. N. Gurd, *Biochemistry* **18,** 1919 (1979).
[15] G. Acampora and J. Hermans, *J. Am. Chem. Soc.* **89,** 1543 (1967).
[16] K. Linderstrøm-Lang, *C. R. Trav. Lab. Carlsberg* **15,** 70 (1924).

the charge array determined by both geometrical location and solvent accessibility.[8,17]

The points on each curve in Fig. 2 correspond to the pH midpoint for the acid-denaturation transition (Fig. 1). These values were estimated by interpolation and extrapolation of the denaturation results and are entered as asterisks.[17a] They indicate that in each case the midpoint of the acid transition occurs under conditions in which the net electrostatic stabilization of the native structure amounts to 5.7–6.0 kcal/mol.

Contribution of Individual Sites to Electrostatic Stability of a Protein. To evaluate the contributions of specific amino acid residues, the individual sites' electrostatic stability can be plotted as a function of pH or ionic strength. In Fig. 3A, B, and C the individual site contributions of sperm whale myoglobin are given for each acidic side chain, each titratable histidine residue, and each strongly basic arginine and lysine residue. The computations were made for 25° and an ionic strength of 0.01 M. An overall summation of all sites in Fig. 3 will correspond to twice the value given in Fig. 2 due to the double counting of all pairwise interactions. Note the remarkable variety of patterns illustrated by the various sites, implying in a vivid way the vast range of functional adaptation that electrostatic stabilization may provide.[8]

Minor Role of a "Trigger Group." It is common practice to identify an ionizing group such as His-64 with a "trigger" role if, as observed here, its $pK_{1/2}$ corresponds to the midpoint pH of a conformational transition.

In Fig. 3B it may be seen that the slope at the midpoint of the change in $\Delta G''_{el}$ with pH, $\Delta G''_{el}/\Delta pH$, is indeed steep for His-64. However, it can be estimated from the individual curves that His-64 contributes only 11% of the net slope; Arg-31 and Lys-42 also show especially rapid changes, for example. The overall transition will combine $\Delta G''_{el}/\Delta pH$ terms to produce the narrow, cooperative transition profile (Fig. 1).[8] Conversely, the analysis of the relation between titration of certain individual residues and any pH-dependent conformational changes outside of the unfolding region may prove especially fruitful.

Ion Binding and pH-Dependent Electrostatic Free Energy

The calculated contributions of electrostatic free energy to the stability of the native structure for ribonuclease S at ionic strengths of 0.01 and 0.15 [Eq. (5)] are shown in Fig. 4B, curves 1 and 2, respectively. At all but

---

[17] J. B. Matthew, G. I. H. Hanania, and F. R. N. Gurd, *Biochemistry* **18**, 1928 (1979).

[17a] The results in Fig. 1 are substantially unchanged by varying the electrolyte between KCl, sodium acetate, and sodium citrate, which indicates the absence of specific ion effects. Potassium perchlorate, on the other hand, appears to have a specific destabilizing effect.[8]

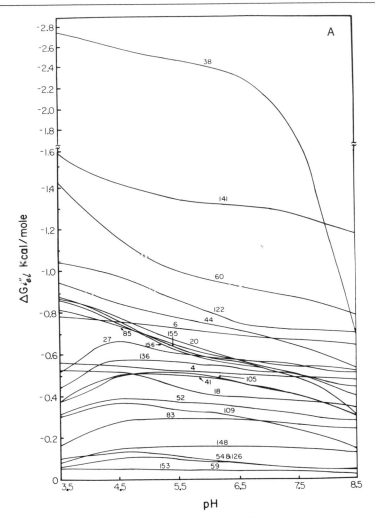

FIG. 3. (A) Computed free energy contributions, $\Delta G''_{i,\mathrm{el}}$, in kilocalories per mole, for acidic groups over the pH range from 3.5 to 8.5. The COOH terminal is denoted by 153. Residues 20, 27, 44, 60, 122, 126, and 141 are aspartic acid. All others are glutamic acid. Note that the scale is condensed beyond -1.6 kcal/mol in the ordinate. The ionic strength is 0.01 M and the temperature is 25°. (B) Corresponding plot for histidine residues. Note that the negative ordinate is condensed. (C) Corresponding plot for basic residues. The NH_2 terminal is denoted by 1. Residues 31, 45, 118, and 139 are arginine. The iron atom is denoted by 154. All others are lysine.[8]

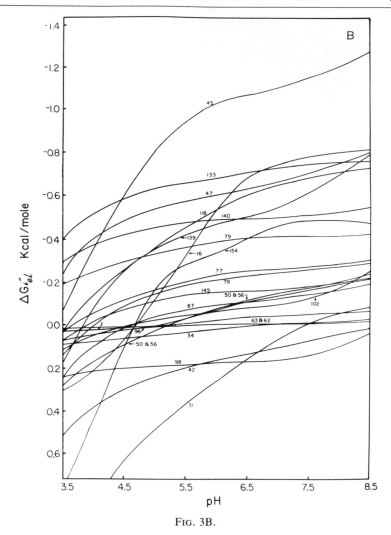

FIG. 3B.

the most acidic pH values the free energy is negative, indicating substantial stabilization. The pH-dependent free energy of stabilization calculated at 0.10 ionic strength is not shown but lies between curves 1 and 2 through the entire pH range. Measurements of the transition temperature with changing pH at constant ionic strength (I = 0.16) lead to a pH-dependent component for the transition free energy whose maximum value is 5.4 kcal/mol at pH 5–7. A very similar pH dependence of protein

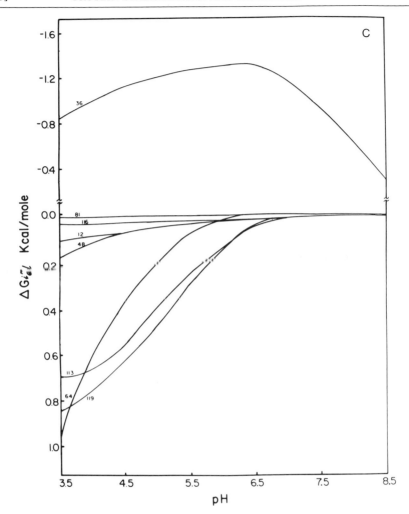

FIG. 3C. (See legend on p. 445.)

stability observed by Hermans and Scheraga[18] for RNase A is included in Fig. 4. For the pH range of 4–7, the observed and predicted values at $I = 0.15$ are within 1 kcal, but the acid limb of the calculated curve drops more steeply than the experimental one. This is consistent with the fact that the calculation did not allow for the known uptake of bound chloride

[18] J. Hermans and H. A. Scheraga, *J. Am. Chem. Soc.* **83,** 3283 (1961).

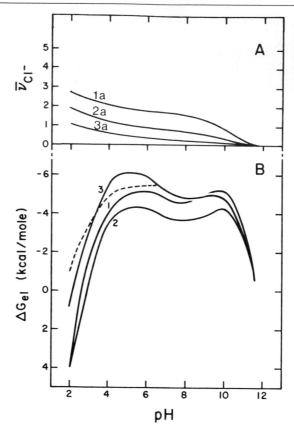

FIG. 4. (A) Sum of the predicted chloride ion uptake, $\bar{\nu}$, for the five anionic binding sites of RNase S as a function of pH. The calculated chloride uptake is shown for ionic strengths 0.15, 0.04, and 0.01 (curves 1a, 2a, and 3a, respectively). The concentration of free chloride ion is assumed equal to the ionic strength. (B) Summed electrostatic free energy of interaction between charged sites, ΔG_{el} [Eq. (5)] expressed in kilocalories per mole as a function of pH for ribonuclease S. Curve 1 corresponds to ionic strength 0.01 in the absence of specific ion uptake. Curves 2 and 3 are calculated at 0.15 ionic strength in the absence and presence of specific anion uptake, respectively, as a function of pH (see A).[9] The dashed line represents an experimental determination of the electrostatic component of protein stability for RNase A ($I = 0.16$, 25°).[18]

ion and the resultant stabilization. Saroff and Carroll[19] have shown that the number of chloride ions bound per mole of RNase A is zero at pH 9.6, 0.50 at pH 6.6, and 2.0 at pH 4.5, with total chloride concentrations of 0.012, 0.016, and 0.085 M, respectively.

    Plots of the electrostatic potential at the protein surface[9] at several

[19] H. A. Saroff and W. R. Carroll, *J. Biol. Chem.* **237,** 3384 (1962).

ionic strengths show two sites with exceptionally positive potential at pH 6.0. The number of solvent-accessible, high-potential sites increases to at least five[19a] by pH 4.0. Curve 3 shows the pH-dependent stabilization ($I = 0.15$) when the calculated partial occupancies of the five anion sites identified at the solvent interface for RNase S are included in the calculation. With this inclusion, the difference between the experimental and calculated stabilities at $I = 0.15$ is less than 0.5 kcal at pH 6.0, and the difference in the acid limb is reduced by 70%.

Figure 4A shows the predicted chloride uptake by RNase S at ionic strengths of 0.15, 0.04, and 0.01 (curves 1a, 2a, and 3a). These values represent the monovalent anion site occupancies summed over the five high potential sites. The chloride binding data of Saroff and Carroll[19] agree with the predicted anion uptake between pH 9.6 and 4.5 and its variation as a function of ionic strength. The data below pH 4.0[20] were obtained at total chloride concentrations less than 0.04, yet the measurements indicated that RNase A binds more than three chloride ions. The calculations below pH 4.0 at the same ionic strength predict half that amount. Because the calculations are based on the crystal structure, it is not surprising that they do not accurately predict the marked increase in chloride affinity of ribonuclease in the pH range 2–4. The ion binding studies were carried out at 25° where the midpoint for the acid-induced conformational transition is approximately pH 2.0.[18] At low pH, where repulsive interactions begin to dominate the charge array stability, conformational displacements or rearrangements which relieve unfavorable interactions or facilitate anion uptake would be expected.[21]

Experimental determinations of the thermal transition of RNase A have been measured at two pH values as a function of ionic strength.[18] At pH 6.5 the transition temperature for RNase A varies between 62 and 63° for the ionic strength range 0.01–0.50. The calculated stabilities (curves 1–3) at pH 6.5 show that two opposing effects lead to this insensitivity to increasing ionic strength. The increased coulombic screening at increased ionic strength (curve 2) is offset by the stabilization of specific ion binding (curve 3). At a pH of 0.7 where the net charge on the protein has more than tripled to 18+, the transition temperature varies from 26.6° ($I = 0.16$) to 31.5° ($I = 0.32$). This increase in stability with ionic strength could be

---

[19a] Five anion sites were identified by locating areas of high positive potential outside the protein van der Waals envelope. For illustration, we give the distance in Å from each site to three titratable groups in the structure: site 1, 4.0 to Lys-31, 6.0 to Lys-37, and 6.6 to Tyr-92; site 2, 4.0 to Lys-7, 5.1 to Arg-10, and 6.9 to Lys-41; site 3, 5.0 to His-12, 8.6 to Lys-41, and His-119; site 4, 4.0 to Lys-104, 5.6 to His-105, and 6.3 to Val-124.

[20] G. I. Loeb and H. A. Saroff, *Biochemistry* **3**, 1819 (1964).

[21] V. Glushko, P. J. Lawson, and F. R. N. Gurd, *J. Biol. Chem.* **247**, 3176 (1972).

explained in terms of preferential coulombic screening of the repulsive forces but more likely reflects the stability imparted by increased bound chloride.

Electrostatic Free Energy of Anion Binding at the Active Site

In the case of azide anion binding to ferrimyoglobin the dependence on ionic strength and pH may be defined in terms of the summed individual interaction terms of all other charged sites with the heme iron coordination site, terms that are present in the reactant protein but obliterated in the product. Correlations with experiment are very satisfactory and self-consistent, and point up the importance of taking the whole interactive discrete-charge lattice into account.[22,23] In this case the coordination complex formation is the primary source of stabilization.

The same general formulation has been applied to ribonuclease.[9] The additional electrostatic stabilizations of the overall protein charge array introduced by anion binding can be cast as pH-dependent association constants, $\log K_{anion} = \Delta G_{el}/(2.3RT)$. These pH-dependent association constants are plotted in Fig. 5 for a monovalent ion such as chloride and for inorganic phosphate at ionic strengths of 0.01 and 0.10. Figure 5 also shows the experimental association constants[24] for inorganic phosphate, 2'-CMP, 3'-CMP, and 3'-UMP. The predicted pH dependence for inorganic phosphate accurately follows the observed pH dependence, and the overall predicted magnitude of the electrostatic component of phosphate binding is 75–80% of the experimental value. (As discussed above, the magnitude of the predicted effect can increase at least 20% in response to specific placement or side chain rearrangement.) The broken lines drawn through the nucleotide data correspond to the predicted pH dependence of inorganic phosphate, demonstrating that these anions follow the pH-dependent behavior observed and predicted for inorganic phosphate. The pH-dependent component of 3'-UMP and 3'-CMP must be about 2.5 kcal/mol while the specific interactions of 2'-CMP must account for 4.0 kcal/mol. A portion of the difference in the 2'- and 3'-nucleotide binding energy can be attributed to the different electrostatic field at the position of 2'- and 3'-phosphate groups. The nature of such a shift is shown by curves 2 and 2' in Fig. 5.

Nucleotides such as 2'-CMP or 3'-CMP in binding to RNase S present the additional complexity of substantial nonelectrostatic components to

[22] S. H. Friend, K. L. March, G. I. H. Hanania, and F. R. N. Gurd, *Biochemistry* **19**, 3039 (1980).
[23] K. L. March, Ph.D. thesis, Indiana University, 1983.
[24] D. G. Anderson, G. E. Hammes, and F. G. Walz, Jr., *Biochemistry* **7**, 1637 (1968).

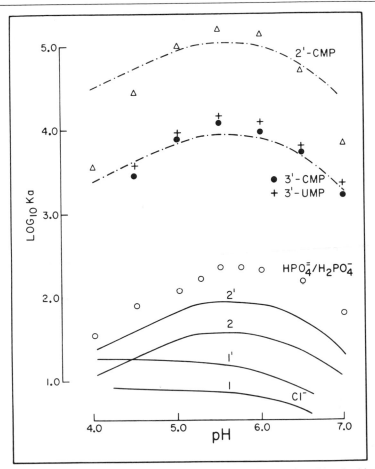

FIG. 5. pH-dependent additional electrostatic stabilization produced by the binding of univalent and divalent anions in the RNase S active site (position A1) expressed as pH-dependent association constants. Curves 1 and 1' are for a univalent anion such as chloride at ionic strengths of 0.10 and 0.01, while curves 2 and 2' are calculated for the titratable phosphate ion. Data are those of Anderson $et\ al.$[24] determined spectrophotometrically for the pH-dependent binding of inorganic phosphate (○), 3'-CMP (●), 3'-UMP (+), and 2'-CMP (△). The broken lines drawn through the nucleotide data are the calculated pH-dependent phosphate curves translated up to demonstrate the common pH dependence of these phosphate compounds. (The protonation of the pyrimidine ring nitrogen in the acid region has not been taken into account in these calculations but would lead to a decrease in binding for CMP at pH values near 4.0.)[9]

binding energy (cf. azidomyoglobin[22,23]). In these cases the anionic portion of the ligand may have to compromise the charge placement that maximizes electrostatic interaction in order to accommodate the steric restraints inferred by hydrogen bonding or nonpolar interactions. In these cases the possible variation in the pH-dependent component of binding is evaluated by testing the sensitivity to charge placement within the ion site.

Electrostatic Stabilization of Protein Association

Electrostatic free energy contributions to assembly processes are computed by subtracting the summed contributions of the reactants from that of the product. In the case of the hemoglobins such analyses can be particularly revealing expressed in terms of either monomer to tetramer[25] or dimer to tetramer[26] association. The latter more physiological step for either the deoxy or oxytetramer assembly involves the juxtaposition of substantial contact surfaces that were previously exposed to solvent. The contacts are somewhat different for the deoxy and oxy systems, offering the opportunity to sample the application of similar principles to two distinct cases. Each assembly process is unique in terms of r_{ij} and $\overline{SA}_{ij}$ values between sites in different subunits and in $\overline{SA}_{ij}$ values between sites in a given subunit in which the SA_i or SA_j is altered. The increased stability conferred in either case is thus partly based on stabilizing interactions between the dimer pairs and partly on stabilizing interactions sensed within a given dimer. The two assembly systems also differ in binding affinity for anions.[25,26] Electrostatic interactions play a more marked role in the case of the deoxy assembly process according to both the theoretical and experimental evidence.[26] Since the overall equilibria of the assembly processes can be measured experimentally,[27] the fractional contribution of electrostatic and nonelectrostatic processes can be computed for normal and various abnormal hemoglobins.[28]

Whereas electrostatic stabilization accounts for approximately half of the energy of dimer–tetramer assembly in hemoglobin, its role is marginal in the combination of trypsin with bovine pancreatic trypsin inhibitor, BPTI.[23] The combination of these two highly cationic proteins is directed and stabilized by a pattern of interaction that exploits, among other things, low SA values for certain anionic groups and generally high SA values for cationic groups. For the docking of cytochrome c with flavo-

[25] S. H. Friend, J. B. Matthew, and F. R. N. Gurd, *Biochemistry* **20**, 580 (1981).
[26] M. A. Flanagan, G. K. Ackers, J. B. Matthew, G. I. H. Hanania, and F. R. N. Gurd, *Biochemistry* **20**, 7439 (1981).
[27] A. H. Chu, B. W. Turner, and G. K. Ackers, *Biochemistry* **22**, 604 (1983).
[28] M. Crowl-Powers, Ph.D. thesis, Indiana University, 1984.

doxin more highly asymmetric fields direct the combination that is stabilized significantly by ionic interactions between lysine residues on cytochrome c and acidic residues of flavodoxin.[10] The electrostatic potential around two DNA binding proteins, catabolite activator protein (CAP) and the CRO repressor protein, is also constructed to assist in their proper association with DNA.[29,30] In fact, complementary electrostatic interactions between negatively charged B-DNA and positively charged Cro repressor protein have been shown to substantially contribute to the formation energy of sequence specific and nonspecific Cro–DNA complexes.[31,32] It should be noted that the formation of charge clusters on protein faces to enhance association rates and binding constants is electrostatically unfavorable. These charge arrays contribute to the isolated protein's overall stability because the remaining charge array interactions are overwhelmingly favorable.

Conclusion

The electrostatic contribution to the stability of a protein is a part of the puzzle involving various bonded and nonbonded interactions [Eq. (1)]. In cases where both the electrostatic component and the overall process can be, respectively, computed and measured, it is feasible in principle to dissect out the relative contributions of the other terms. This point has been reached for the assembly of hemoglobin and numerous variants which will illustrate the relative importance of various nonelectrostatic terms.[26–28] It has also been reached with the estimation[9] of the pH-independent component of 3'-CMP, 3'-UMP, and 2'-CMP interactions with the active site in ribonuclease (Fig. 5).

Each protein studied shows a specific pattern of stabilization or destabilization resulting from a unique set of long-range, overlapping coulombic fields associated with each charge site. The individual contributions and hence the overall stabilization are affected as the array is modified by changes in pH, ionic strength, binding of charge components, or alteration of solvent exposure.

Acknowledgment

The assistance of J. E. Beecher Matthew in the preparation of this manuscript is greatly appreciated.

[29] D. H. Ohlendorf, Y. Takeda, W. F. Anderson, and B. W. Matthews, *J. Biomol. Struct. Dyn.* **1,** 553 (1983).

[30] T. A. Steitz, I. T. Weber, and J. B. Matthew, *Cold Spring Harbor Symp. Quant. Biol.* **47,** 419 (1982).

[31] D. H. Ohlendorf and J. B. Matthew, *Adv. Biophys. (Japan)* **20,** 137 (1985).

[32] J. B. Matthew and D. H. Ohlendorf, *J. Biol. Chem.* **260,** 5860 (1985).

[19] Measurements of Diffusion and Chemical Kinetics by Fluorescence Photobleaching Recovery and Fluorescence Correlation Spectroscopy

By N. O. PETERSEN and E. L. ELSON

Chemical kinetics and rates of lateral transport (diffusion or flow) can be measured by observing spontaneous fluctuations in reactant concentrations which occur even when the system rests in equilibrium. This can be accomplished by using fluorescence to monitor the numbers of molecules of specified types in a defined open region of the reaction system. The feasibility of this approach has been demonstrated, and the method, called fluorescence correlation spectroscopy (FCS) has found practical application. Because it is based on microscopic fluctuations FCS is an intrinsically statistical method, however, and so requires that the reaction system be stable during prolonged measuring intervals. For studies of less stable systems (e.g., living cells in culture) a method has been developed which is based on observation of the rate of relaxation of a local macroscopic concentration gradient. The gradient is produced by rapidly photolyzing a portion of the reactant molecules in a small open region of the system. This method, which we call fluorescence photobleaching recovery (FPR), yields results which are approximately equivalent to those obtained by FCS. Theoretical advantages of FCS are balanced by practical experimental advantages of FPR. The major applications of this combined technology up to now have been in studies of lateral diffusion in model membranes (FCS and FPR) and on animal cell surfaces (FPR).

Introduction

Fluorescence correlation spectroscopy (FCS) and fluorescence photobleaching recovery (FPR) [also termed fluorescence recovery after photobleaching (FRAP) or fluorescence microphotolysis (FM)] are two methods for measuring either rates of simple transport processes such as diffusion or convective flow or rates of transport coupled with chemical reactions in open systems. Although the two methods are closely related in experimental implementation and theoretical interpretation, they differ somewhat in their range of application and in the motivations for their developments.

There has been considerable theoretical and practical interest in extracting chemical rate constants and transport coefficients from observations of spontaneous fluctuations of the concentrations of the components of systems in thermodynamic equilibrium. This is interesting theoretically because it provides a demonstration of the fundamental postulate of Onsager that the regression back to equilibrium of spontaneous microscopic fluctuations and of macroscopic displacements from equilibrium (in the linear regime) are governed by the same phenomenological laws.[1] Measurements of fundamental phenomenological rates via spontaneous fluctuations have practical application for studies of systems which are not susceptible to or which can be damaged by the perturbations of state necessary to generate macroscopic displacements from equilibrium. There has also been particular interest in conformational fluctuations ("breathing" modes) of proteins and nucleic acids.

One spectroscopic approach to measuring phenomenological rates in equilibrium systems was provided by quasielastic light scattering (QLS). This approach has proved to be extremely successful in measurements of molecular transport processes such as diffusion and electrophoresis.[2] Despite considerable theoretical attention and some experimental work, however, QLS has not been generally successful in studying chemical kinetics. One problem is that the polarizability which determines the scattering strength of the reaction system is not a strong function of the progress of the reaction.[3] FCS overcomes this difficulty by using an optical property, fluorescence, which can be more sensitive than light scattering to chemical change. Although one motivation in the development of FCS was the desire to measure conformational fluctuations of proteins and nucleic acids, this has not yet been successfully accomplished. Nevertheless, applications of FCS to other kinds of problems have proved the potential utility of the method.[4,5]

FPR was originally developed as a way of measuring the lateral diffusion coefficients of protein and lipid components of biological membranes.[6-9] The rates of diffusion of molecules in liquid solution or in membranes report on their interactions with the solvent and other solute

[1] L. Onsager, *Phys. Rev.* **37**, 405 (1931).
[2] B. Berne and R. Pecora, "Dynamic Light Scattering." Wiley, New York, 1976.
[3] V. A. Bloomfield and J. A. Benbasat, *Macromolecules* **54**, 609 (1971).
[4] M. B. Weissman, *Annu. Rev. Phys. Chem.* **32**, 205 (1981)
[5] D. Magde, *Mol. Biol., Biochem. Biophys.* **24**, 43 (1977).
[6] R. Peters, J. Peters, K. H. Tews, and W. Bahr, *Biochim. Biophys. Acta* **367**, 282 (1974).
[7] M. Edidin, Y. Zagyanski, and T. J. Lardner, *Science* **191**, 466 (1976).
[8] K. Jacobson, E. Wu, and G. Poste, *Biochim. Biophys. Acta* **433**, 215 (1976).
[9] J. Schlessinger, D. E. Koppel, D. Axelrod, K. Jacobson, W. W. Webb, and E. L. Elson, *Proc. Natl. Acad. Sci. U.S.A.* **73**, 2409 (1976).

molecules. Biological processes are frequently controlled or regulated by the rate and extent of interaction among key components as in the binding of ligands to enzymes and the association of ligand–receptor complexes on membranes with each other and other membrane or cytoskeletal structures. In principle measurements of diffusion rates can reveal details of both the equilibrium and kinetic properties of these interactions.[10,11] This potential is just beginning to be explored as new techniques for measuring diffusion of specific components in complex mixtures emerge. Although FPR has mainly been used to measure diffusion on cells, its range of potential application is wider. Like FCS it can be used to measure chemical kinetics as well as transport. In this chapter we will stress the application of FCS and FPR to measure both diffusion coefficients and chemical rate constants in open systems, in which local concentration changes due to diffusion and reaction are coupled. These fluorescence techniques are specifically designed for (but not restricted to) measurements on small systems such as individual biological cells. Moreover, they permit observation of specific components of complex mixtures because of the spectroscopic selectivity of the fluorescent probes which can be used. Using one method or the other it is possible to study diffusion processes ranging from the fastest expected for proteins in aqueous solution to the slowest observed in paracrystalline membranes.

FPR and FCS have not previously been described in these volumes, although several reviews have appeared elsewhere.[12-17] This presentation will therefore emphasize the equipment design, the theoretical background, and general experimental protocols. Specific protocols, such as preparation of fluorescent probes, will not be described since they are dictated by the nature of the particular biological system.

Historically, fluorescence photobleaching experiments have been applied most extensively in studies of diffusion of membrane lipids and proteins in both model and cell membrane systems.[12,14-16] Nevertheless, both FPR and FCS have been used to study diffusion and chemical kinet-

[10] E. L. Elson and J. Reidler, *J. Supramol. Struct.* **12,** 481 (1979).

[11] D. Koppel, *J. Supramol. Struct.* **17,** 61 (1981).

[12] E. L. Elson and W. W. Webb, *Annu. Rev. Biophys. Bioeng.* **4,** 311 (1975).

[13] D. E. Koppel, D. Axelrod, J. Schlessinger, E. L. Elson, and W. W. Webb, *Biophys. J.* **16,** 1315 (1978).

[14] M. Edidin, *in* "Membrane Structure" (J. B. Finean and R. H. Michell, eds.), p. 37. Elsevier, Amsterdam, 1981.

[15] R. Peters, *Cell Biol. Int. Rep.* **5,** 733 (1981).

[16] N. O. Petersen, S. Felder, and E. L. Elson, *in* "Handbook of Immunology" (D. M. Weir and L. A. Herzenberg, eds.). Blackwell, Edinburgh, 1984.

[17] N. O. Petersen, *Can. J. Biochem. and Cell Biol.* **62,** 1158 (1984).

ics in solution[12,13,18-23] including measurements of intracellular diffusion of cytoplasmic components.[24,25] The areas of application illustrate both the versatility and the limitations of these techniques. FPR is particularly suited for diffusion measurements in the slower limit (10^{-17}–10^{-11} m^2 sec^{-1}, i.e., with characteristic times 0.1–10^5 sec) which is ideal for membrane studies whereas FCS is better suited for the faster diffusion (10^{-12}–10^{-9} m^2 sec^{-1}, i.e., with characteristic times 10^{-3}–1 sec) observed in solutions. These techniques then provide complementary measurements encompassing a dynamic range of 8 orders of magnitude in rate measurements. Moreover, each provides different ancillary information which is frequently as important as the dynamic measurement. Specifically, FPR experiments yield a measure of the proportion of the fluorescent molecules which are free to move at the timescale of the experiments, while FCS experiments give an estimate of the number-average density of fluorophores which can be interpreted in terms of the aggregation state of particles in solution or in membranes. Hence both FPR and FCS experiments can provide valuable molecular information about biological and biochemical systems. Throughout this chapter we shall therefore stress both the common and the distinguishing features of the techniques to clarify which approach is better suited to particular problems.

General Features of the Approaches

Optical measurements of spontaneous fluctuations of chemical concentrations can in principle be used to investigate the rates of chemical reactions and transport processes both in small systems such as living cells and in defined biochemical systems without imposing external perturbations. In practice living cells are usually not sufficiently stable to permit measurement of the microscopic fluctuations which occur in systems in equilibrium. Nor are the optical changes which result from spontaneous conformational fluctuations of biological macromolecules typically large enough to permit direct optical detection. It is therefore usually more feasible to obtain rate parameters for these kinds of systems by

[18] E. L. Elson and D. Magde, *Biopolymers* **13**, 1 (1974).
[19] D. Magde, E. L. Elson, and W. W. Webb, *Biopolymers* **13**, 29 (1974).
[20] D. Magde, W. W. Webb, and E. L. Elson, *Biopolymers* **17**, 361 (1978).
[21] C. Andries, W. Guedens, J. Clauwaert, and H. Geerts, *Biophys. J.* **43**, 345 (1983).
[22] R. D. Icenogle and E. L. Elson, *Biopolymers* **22**, 1919 (1983).
[23] R. D. Icenogle and E. L. Elson, *Biopolymers* **22**, 1949 (1983).
[24] J. W. Wojcieszyn, R. A. Schlegel, E.-S. Wu, and K. A. Jacobson, *Proc. Natl. Acad. Sci. U.S.A.* **78**, 4407 (1981).
[25] R. G. A. Safranyos and S. Caveney, *J. Cell Biol.* **100**, 736 (1985).

measuring the relaxation of the system back to equilibrium after it has been subjected to a macroscopic perturbation of state. For example, temperature jump methods have long been used to study the mechanisms of enzymatic catalysis and of macromolecular conformation changes. In an FPR experiment an intense pulse of light is used to generate by photolysis a spatial concentration gradient which can relax by diffusion or chemical reaction. This is analogous to the generation of a spatially uniform concentration perturbation which relaxes only by chemical reaction in a conventional temperature, electric field, or pressure jump measurement. Under appropriate constraints optical detection of spontaneous microscopic and induced macroscopic concentration changes can provide similar information about chemical kinetics and rates of transport.

Fluorescence is especially useful for measuring the concentrations of specific molecules in complex biological systems because of the sensitivity with which it can be detected, its chemical and spectroscopic selectivity, and the ease and rapidity of its measurement with minimal perturbation to the experimental system (under physiological conditions if necessary). Another advantage is the high spatial resolution which can be obtained using fluorescence microscopy which permits measurements in small ($\sim 1 \ \mu m$) subregions of biological cells.

The two closely related techniques, FCS and FPR, have been developed with these points in mind.[26] The two methods share a common simple conceptual basis. The numbers of molecules of specified types in a defined open observation volume are measured as functions of time. In a chemical reaction system the molecules of a certain reactant can appear or disappear either by being created or destroyed in the chemical reactions or by being transported (by diffusion or convective flow) into or out of the volume. Hence the rate of change of concentration is determined by the rate of transport (and the size of the observation region) and by the chemical reaction rates. Therefore by measuring rates of concentration change one can obtain both chemical rate constants and transport coefficients such as diffusion coefficients and flow velocities. Measurements of concentration changes in an *open* volume provide information simultaneously about both transport and chemical kinetics. The price of this advantage is an increase in the complexity of the theoretical apparatus needed to interpret the measurements. In the absence of chemical reactions it is relatively simple to interpret measurements on open volumes to obtain transport coefficients. On the other hand chemical kinetics uncomplicated by transport can in principle be obtained from analogous mea-

---

[26] E. L. Elson, J. Schlessinger, D. E. Koppel, D. Axelrod, and W. W. Webb, *in* "Membranes and Neoplasia: New Approaches and Strategies" (V. T. Marchesi, ed.), pp. 137. Liss, New York, 1976.

surements on small *closed* systems. In practice, however, photochemical interference and other problems make this kind of measurement very difficult.

Instrumentation

The objective of FPR and FCS experiments is to observe the temporal variation in fluorescence originating from a geometrically well-defined volume. Generally the observation volume is defined by the illumination volume. Some of the advantages of FPR and FCS measurements derive from the ability to study small systems using an epifluorescence microscope in conjunction with a laser as the light source for illumination. In this section we describe the equipment associated with spot-photobleaching because of its simplicity and versatility and because of its adaptability to both FPR and FCS experiments. Measurement systems using more complex arrangements for bleaching and detection have been developed and are especially useful for certain applications.[27-34] The overall design of a typical FPR/FCS instrument is illustrated in Figs. 1 and 2 and consists of an ion laser, an epifluorescence microscope, focusing and intensity modulation optics, a photomultiplier tube with photon counting electronics, accessory electronics for controlling shutters and protecting the photomultiplier tube, and computer interfacing for controlling the experiment and collecting the data. The only difference between the FPR and FCS equipment is the presence of optics that permit brief exposures to high illumination intensity in the former. There is of course also a difference in the subsequent treatment of the data.

Illumination Sources

Argon and krypton ion lasers are good sources of illumination because their emissions are in the visible region of the spectrum. Many biological systems absorb in the ultraviolet or near-ultraviolet (about 400 nm and below) but few absorb extensively in the visible region. This provides the opportunity of labeling selected components of the biological system with

[27] D. E. Koppel, *Biophys. J.* **28,** 281 (1979).
[28] B. A. Smith and H. M. McConnell, *Proc. Natl. Acad. Sci. U.S.A.* **75,** 2759 (1978).
[29] B. A. Smith, W. R. Clark, and H. M. McConnell, *Proc. Natl. Acad. Sci. U.S.A.* **76,** 5641 (1979).
[30] F. Lanni and B. R. Ware, *Rev. Sci. Instrum.* **53,** 905 (1982).
[31] J. Davoust, P. F. Devoux, and L. Leger, *EMBO J.* **1,** 1233 (1982).
[32] R. Peters, A. Brunger, and K. Schulten, *Proc. Natl. Acad. Sci. U.S.A.* **78,** 962 (1981).
[33] N. L. Thompson, T. P. Burghardt, and D. Axelrod, *Biophys. J.* **33,** 435 (1981).
[34] R. M. Weis, K. Balakrishnan, B. A. Smith, and H. M. McConnell, *J. Biol. Chem.* **257,** 6440 (1982).

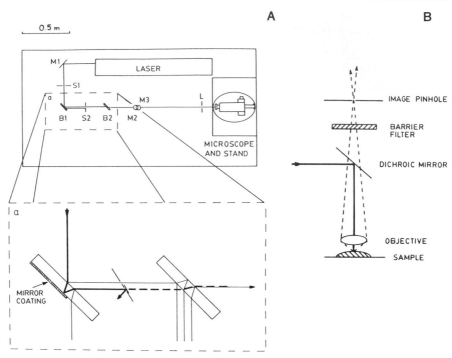

FIG. 1. Schematic representation of an FPR/FCS system. (A) The laser beam is reflected by a mirror (M1) onto a beam splitter (B1) coated with a silver coating (inset) to maximize the reflection at the back surface. The weaker beam (the monitor beam) is passed unimpeded to the second beam splitter (B2) where it is recombined with the stronger beam (the bleach beam) after two internal reflections. This beam splitting approach was first suggested by D. E. Koppel.[24] The bleach beam is blocked (except during the bleach) by a shutter (S2). The beam is aligned with the optical axis of the microscope by a pair of mirrors (M2 + M3) in a beam steerer. The single lens L provides the appropriate focusing prior to entry into the epifluorescence condenser of the microscope. The first shutter (S1) permits blocking of the entire beam between experiments. The equipment is contained on a 4′ × 8′ vibration isolation table and the optical components are covered with a Plexiglas housing to minimize dust accumulation. (B) In the microscope the laser beam is reflected by a dichroic mirror and focused by the objective onto a small part of the sample. The fluorescence is collected by the objective and after passing through the dichroic mirror and a barrier filter is focused on the aperture of an image plane pinhole. The photomultiplier is located above the pinhole.

fluorescent probes which do absorb in the visible region (450–600 nm) and measuring the dynamics of these components with little or no background contribution. The selectivity is important in complex mixtures and generally outweighs the disadvantages inherent in any labeling technique (e.g., perturbation by labeling, complexity of chemistry). In special cases ultraviolet illumination is desirable and can be obtained from the more power-

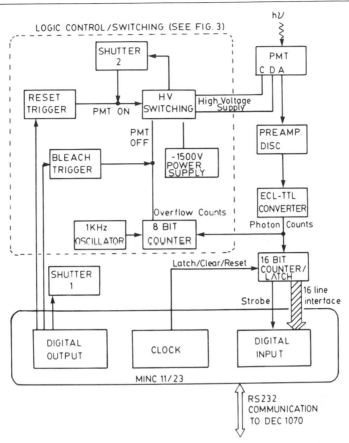

FIG. 2. Functional block diagram of an FPR/FCS system. Fluorescence photons from the illumination volume are imaged by the microscope onto a GaAs reflective photomultiplier tube (PMT; RCA CA31034A). The transient photon currents are amplified by the preamplifier/discriminator circuit (Lecroy MLV100 chip) producing ECL pulses. These are converted to TTL logic pulses which constitute the photon counts. A 16-bit counter with internal latches is interfaced with the Digital Input Module and the Clock Module of a DEC MINC 11/23 computer. The count intervals are controlled by the clock which provides a latch signal which is also used to clear and reset the counter. When the data are latched a strobe signal to the Digital Input Module initiates data transfer to the computer. The photon counts are monitored with an 8-bit counter which is reset with a 1 kHz oscillator. If the count rate exceeds 128 kHz the 8-bit counter overflows causing the high voltage switching circuit to reverse the bias of the PMT protecting the latter against excessive currents. The computer software controls shutter 1 (S1, Fig. 1) directly but controls the bleach shutter (S2, Fig. 1) through the high voltage switching circuit. A computer pulse (or a manual switch) initiates a bleach trigger which causes the high voltage switching circuit to reverse the bias of the PMT. When bleaching is completed the PMT is turned back on again and counting proceeds. The PMT can also be reset by software or a manual switch following an overflow condition.

ful (>4 W total power) versions of these lasers with special optics. Although other illumination sources are possible, the lasers are advantageous for several reasons: (1) they provide wavelength selectivity to excite different chromophores, (2) the power output can be controlled and is easily attenuated to the useful range (1 μW to 5 mW), (3) the beam is collimated so that the long pathlength necessary for focusing optics and shutters is not a technical problem, and (4) the transverse intensity distribution can readily be defined as a convenient two-dimensional Gaussian distribution.

Optical Components and Microscope

The optical components serve three functions, one of which is uniquely required for the FPR experiment: (1) to focus the laser beam to a desired size at the entry into the epifluorescence port of the microscope, thereby producing the required beam size at the specimen in the focal plane of the objective (typically 1–4 μm radius beams yield convenient diffusion times), (2) to attenuate with neutral density filters the laser power such that the illumination power on the sample is in the 1 μW range during observation and photon counting, and (3) to modulate the intensity of the illumination beam in the FPR experiment so that a brief (1–100 msec) photobleaching pulse of high intensity (1–10 mW) can be produced exactly coincident with the measuring beam.

Focusing can be achieved with a single lens as in Fig. 1, but if pathlength constraints require it, more lenses may be used. The position and focal length of the lens are determined by the microscope optics and the desired beam size. For example a 2-μm beam at the focal plane of a 40× objective on a Zeiss Universal microscope requires a 48-μm beam focused 22 mm in front of the first lens in the epicondenser assembly. The optics necessary to produce this may be calculated from the lens propagation formulas for Gaussian beams[35]:

$$\frac{1}{w_i^2} = \frac{1}{w_o^2 f^2}\left[(d_o - f)^2 - \left(\frac{\pi w_o^2}{\lambda}\right)^2\right] \tag{1}$$

$$d_i = f + (d_o - f)\left(\frac{w_i}{w_o}\right)^2 \tag{2}$$

where d_o is the distance of the lens from the object (the beam waist in the laser or the focal point of a prior lens), w_o is the beam radius at d_o, d_i is the

---

[35] H. Kogelnick and T. Li, *Appl. Opt.* **5,** 1550 (1966).

distance of the image from the lens, w_i is the beam radius at d_i, and f is the focal length of the lens. The propagation formula can be applied succesively to other lenses in the system. It is important to note that the focal properties (d_i and w_i) depend on the wavelength, λ, so the beam size at the sample will vary with wavelength. Therefore the lens position must be adjustable over at least a few centimeters to accommodate the range of laser wavelengths used in the experiments.

Attenuation of the laser beam is generally needed because the laser output is not stable at low power output (low laser current). The optical elements in the path and in the microscope typically attenuate the beam to about 20% of the laser output. Stable power can be obtained at 30–50 mW. Illumination power during photocounting should not exceed a few microwatts at the sample. Thus for FCS experiments a 1000-fold attenuation is needed. In FPR or FPR/FCS equipment part if not most of this attenuation is achieved with the internal reflection losses in the beam splitters. It is important to note that any optical element including neutral density filters will, if improperly aligned, shift the beam direction and beam focus. Consequently the position in the optical path for these elements is best chosen where they will introduce the least error.

Intensity modulation is important in the FPR experiments which require precisely timed, short pulses of high intensity illumination which in turn are spatially coincident with the low intensity monitoring beam. Early designs utilizing neutral density filters or Pockel cells caused optical alignment changes[36] or intensity profile distortions.[37] These have now been replaced by pairs of beam splitters in any of several geometries[38] one of which is shown in Fig. 1. Here we rely on the reflected beam at each glass–air interface being attenuated by >90%: the monitor beam is thus attenuated ~10,000× while the bleach beam is attenuated by ~3×. A fast shutter controls the duration of the bleach pulse. A second shutter before the beam splitters permits control of the duration of exposure of the sample to the monitor beam. This is particularly important in experiments with photolabile fluorescent probes.

The microscope can be any of a number of regular or inverted epifluorescence microscopes available provided that the frame is reasonably rigid, that the laser can be directed into the epifluorescence port and replace the standard arc light source, and that a photometer housing and a

[36] B. G. Barisas, *Biophys. J.* **29**, 545 (1980).
[37] L. S. Barak, E. A. Nothnagel, M. B. Schneider, and W. W. Webb, private communication (1980).
[38] D. E. Koppel, *Biophys. J.* **28**, 281 (1979).

photomultiplier tube can be attached. For optimal background elimination, it is desirable to have image-plane pinholes in the photometer housing which match the beam sizes employed in the experiments. It is useful to have an iris diaphragm mounted at each end of the epifluorescence unit on the optical axis of the microscope. These facilitate alignment of the laser beam along the optical axis thereby eliminating most optical abberations and distortions in the microscope.[13,39,40]

Photomultiplier and Photon Counting

The photomultiplier tube (PMT) should be as sensitive as possible in the wavelength region from 450 to 700 nm. Reflective type GaAs PMT designs are well suited for this application. In order to minimize dark counts cooling the PMT is advisable. We use a dry-ice-cooled PMT housing but thermoelectric devices which can cool to about 245 K are satisfactory and often more convenient.

The low fluorescence intensities encountered in the FPR and FCS measurements (<10,000 cps) necessitate photon counting and therefore require both preamplifier/discriminator and photon counter circuits. Commercial systems are available (e.g., from Princeton Applied Research or Ortec) but it is relatively straightforward to incorporate these functions in the other electronic circuitry which must be custom built anyway. Figures 2 and 3 illustrate one such system and this is discussed further below.

Data logging can be achieved with either a transient recorder, a signal averager with digital input, or a dedicated microprocessor capable of acquiring data in real time. The FCS experiment would benefit from a direct linkage to a correlator, but this is not essential if long data records can be stored and processed rapidly. For the FCS experiments it is important that data can be sampled at 10–100 kHz rates if rapid fluctuations corresponding to diffusion coefficients of about 10^{-9}–10^{-10} m^2 sec^{-1} are to be measured. The FPR experiments have somewhat less stringent requirements but need the capability of acquiring and storing 500–1000 data points at up to 10 kHz rates (10 points per msec). Signal averaging of repetitive bleach and recovery cycles is useful for improving the signal-to-noise (i.e., the precision) for samples of entirely mobile components (e.g., lipid probes in some membranes and freely diffusing solutes in solution). Fortunately, the most rapidly recovering systems with poorest signal-to-noise because of rapid data collection rates (and therefore high shot noise) are also the systems most amenable to signal averaging because they

[39] M. B. Schneider and W. W. Webb, *Appl. Opt.* **20**, 1382 (1981).
[40] N. O. Petersen and W. B. McConnaughey, *J. Supramol. Struct. Cell. Biochem.* **17**, 213.

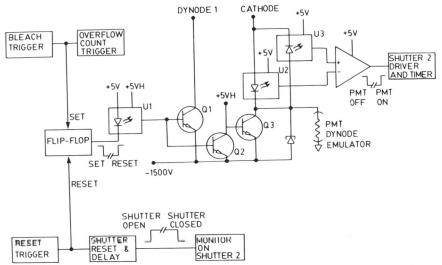

FIG. 3. Schematic representation of the logic and high voltage switching electronics. A set signal on the flip-flop activates optocoupler U1 turning transistors Q1 and Q2 on while turning Q3 off. This sets the first dynode voltage at -1500 V while the cathode voltage adjusts to -1350 V through the combination of a zeener diode and a resistance network which emulates the dynode chain of the PMT. In this reverse-biased condition the PMT is inactive (and protected) and U3 is active causing a low signal to trigger the shutter driver and timer for shutter 2 to open. This feature ensures that the bleach shutter does not open unless the PMT is reversed biased. When shutter 2 has closed, an optical sensor provides a reset signal to the flip-flop whereby U1, Q1 and Q2 turn off activating Q3 and U2 and providing the normal bias voltage of -1500 V to the cathode. The voltage on dynode 1 is then determined by the dynode chain resistance in the PMT. This switching circuit was originally designed and implemented by J. Bloom and W. W. Webb at Cornell University.

generally exhibit full recovery. Proteins in membranes are only partly mobile,[1-6] but the slow diffusion permits lower acquisition rates and therefore more precision in each data point so that signal averaging is less critical. In favorable cases, an analog signal from a photon counter can be displayed directly on a strip-chart recorder for analysis, but this precludes curve-fitting of the data and severely restricts the dynamic range of the FPR experiment.

Computer Interfacing and Control

Because of the improvements in data analysis achieved by curve fitting the FPR data, and because of the need to perform correlation analysis of the fluorescence data in FCS measurements, the data are invariably transferred to a computer at some stage. The computer interfacing depends on the level of control to be exercised by the computer and whether it is a dedicated or a timesharing system. A dedicated microprocessor offers the advantage that the data collection and instrument control can be automated but frequently with some reduction in the maximum acquisition rates. A remote computer offers the advantage that data processing is rapid and does not interfere with the data collection phase of the experiments, i.e., one researcher can process data without tying up the FPR/ FCS equipment. The optimal arrangement permits data collection and instrument control with a dedicated microprocessor which can transfer the data to a remote computer for further processing. The design outlined in Fig. 2 illustrates the latter approach.

In this system, data collection is controlled and timed by the microcomputer clock and data transfer occurs via a 16-bit parallel interface from the 16-bit counter to an input module. The instrument control, that is timing of the shutter opening and other electronic control signals, is provided by some of the lines on a 16-bit parallel interface from an output module to trigger circuits in the electronic control system. Once collected, the data are transferred from the microcomputer disc storage to the mainframe computer via an RS232 communications port. The major limitation in this approach is in FCS experiments which require long data sets to improve accuracy. Memory size limitations preclude collection of very long data records unless transferred to virtual memory or disc. This takes time and thus limits the rate of data acquisition. If one is willing to do intermediate processing, that is collect an intermediate length data record, correlate and then resume data collection, then this is not a serious problem. These problems can be avoided by using a dedicated correlator which can provide a continuously updated correlation function during the data acquisition period (supplied, e.g., by Langley-Ford, Nicomp, Milvern, or Honeywell).[41]

Control Electronics

The control electronics must serve at least two functions: to time the occurrence and duration of the bleach in FPR experiments and to protect the photomultiplier tube from the very high intensity of fluorescence during the bleach pulse. The block diagram in Figs. 2 and 3 illustrate an electronic control system which also provides interfacing to a microprocessor, control of the shutter blocking the whole beam (shutter 1) and which contains preamplifier/discriminator, photon counting, and PMT protection circuits as well as power supplies. Some details are presented in the figure captions and more are available on request.

[41] B. Chu, *Phys. Scripta* **19,** 458 (1979).

Theoretical Description of FPR and FCS

General Considerations

A fundamental distinction between methods based on macroscopic (FPR) and microscopic (FCS) concentration changes rests on the fact that the latter but not the former must, intrinsically, be analyzed in statistical terms. The relaxation of the concentration of a chemical component of the system, after initially having been displaced *macroscopically* from its equilibrium mean value, is governed deterministically by the conventional phenomenological chemical rate constants and transport coefficients. Therefore the values of these quantities can in principle be obtained to an accuracy limited only by the measuring accuracy from observations of the behavior of a single macroscopic relaxation. In contrast, the space-time behavior of the concentration when displaced by a spontaneous fluctuation *microscopically* from its mean value is related only probabilistically to the phenomenological coefficients. Measurement of a single fluctuation even with infinite accuracy would be insufficient to determine these coefficients accurately. Thus a statistical analysis is required to determine these coefficients from measurements averaged over many spontaneous fluctuations.

The first task in developing an analytical scheme is to relate the concentration of the fluorescent components in the system to the measured fluorescence intensity $i(t)$. Although most photobleaching measurements have been carried out to determine diffusion coefficients in systems with only a single fluorescent component, we shall develop the following for systems in which there can be n fluorescent components each of which can both diffuse and interchange with other components by chemical reactions. If the concentration of the jth fluorescent component at position r and t is $C_j(r,t)$, then the detected fluorescence at t due to all of the n fluorescent species will be[13,18]

$$i(t) = \sum_{j=1}^{n} Q_j \int d^3r I(r) C_j(r,t) \varepsilon(r) \tag{3}$$

where $I(r)$ is the excitation intensity profile, the Q_js are constants which account for the efficiencies of optical absorption and fluorescence emission, and $\varepsilon(r)$ is a measure of the efficiency of fluorescence detection of the experimental apparatus.[13,40]

The statistical analysis in FCS experiments is conveniently accomplished in terms of a correlation function,[7-9] $\phi_{jl}(r,r',\tau)$, which defines the average correlation of a concentration fluctuation of the jth component at r and t with a concentration fluctuation of the lth component at r' and a later time $t + \tau$. The concentration fluctuations can be expressed as $\delta C_j(r,t) = C_j(r,t) - \bar{C}_j$ where $\bar{C}_j$ is the equilibrium mean value of C_j. Then

$$\phi_{jl}(r,r',\tau) = \langle \delta C_j(r,t)\delta C_l(r,t + \tau)\rangle = \langle \delta C_j(r,0)\delta C_l(r',\tau)\rangle \qquad (4)$$

where $\langle \ldots \rangle$ denotes an average over many fluctuations (either as a time sequence or ensemble average). We have assumed that the system is stationary so that $\phi_{jl}(r,r',\tau)$ is independent of t.

Concentration fluctuations are detected experimentally as fluorescence fluctuations:[18,20]

$$\delta i(t) = \sum_j Q_j \int d^3r I(r)\varepsilon(r)\delta C_j(r,t) \qquad (5)$$

where $\delta i(t) = i(t) - i$. Then the measured behavior is embodied in a fluorescence fluctuation autocorrelation function, $G(\tau)$, which is defined and related to the ϕ_{jl} as follows:

$$G(\tau) = \langle \delta i(0)\delta i(\tau)\rangle$$

$$= \sum_{j,l} \int Q_j Q_l \int d^3r d^3r' I(r)I(r')\varepsilon(r)\varepsilon(r')\phi_{jl} \qquad (6)$$

Experimentally $G(\tau)$ can be considered to be defined as

$$G(\tau) = \frac{1}{T} \lim_{T \to \infty} \int_0^T \delta i(t)\delta i(t + \tau)\, dt \qquad (7)$$

The meaning of the fluctuation autocorrelation function has been discussed previously.[12] The relaxation of a system displaced macroscopically from equilibrium in FPR experiments may be represented simply as

$$\Delta i(\tau) = \sum_j Q_j \int d^3r I(r)\varepsilon(r)\Delta C_j(r,\tau) \qquad (8)$$

where "Δ" denotes a macroscopic departure from equilibrium.

Interpretation of the measurements demands that $G(\tau)$ and $\Delta i(\tau)$ be expressed in terms of chemical rate constants and transport coefficients. In the absence of chemical reactions the space-time evolution of concentration fluctuations may be simply calculated from the diffusion equation:

$$\frac{\partial \delta C_j(r,t)}{\partial t} = D_j \nabla^2 \delta C_j(r,t)$$

When, however, components of the system can be created or destroyed by chemical reactions, these provide an additional path for the relaxation of concentration fluctuations. Even for closed systems the analysis of chemical kinetics for multireaction systems can become complex.[42] Analysis of kinetics in an open reaction system is, however, still more complicated due to the coupling of diffusion and chemical reaction. As in any problem in chemical kinetics one must begin by postulating a reaction mechanism from which it is possible to derive the rate of change of the concentration of each of the components. We shall suppose that these rates of change can be described by a set of equations linear in the concentration of each of the components. Hence, the rate of change of the concentration of the jth component due only to chemical reaction is

$$\frac{d\delta C_j(r,t)}{dt} = \sum_l T_{jl}\delta C_l(r,t)$$

For a system of first-order reactions the matrix elements T_{jl} are made up of chemical rate constants. If there are higher order reactions, we shall suppose that the kinetic equations describing them can be linearized. Then the elements T_{jl} may also contain equilibrium reactant concentrations. This linearization is certainly legitimate for the very small spontaneous concentration fluctuations measured by FCS. Then the equations for the coupled diffusion–reaction system are

$$\frac{\partial\delta C_j(r,t)}{\partial t} = D_j\nabla^2\delta C_j(r,t) + \sum_l T_{jl}\delta C_l(r,t) \qquad (9)$$

We shall also suppose that the macroscopic concentration changes generated in FPR experiments are small enough to permit linearization if necessary. Hence Eqs. (9) will also hold for $\Delta C_j(r,t)$. Convective flow has been omitted from this development. It has been discussed previously for both fluorescence fluctuation and macroscopic relaxation experiments.[13,20,43]

In both fluctuation and macroscopic relaxation experiments the observation area is taken to be small compared to the total area of the reaction system. Therefore for both kinds of experiments Eqs. (9) may be solved subject to the boundary conditions:

$$\begin{matrix} \delta C_j(r,t) = 0 \\ \Delta C_j(r,t) = 0 \end{matrix} \quad \text{at } r = \infty$$

The initial conditions for the two kinds of experiments are different, however. FCS experiments are typically carried out in dilute solutions so

[42] G. H. Czerlinski, "Chemical Relaxation." Dekker, New York, 1966.
[43] D. Axelrod, D. E. Koppel, J. Schlessinger, E. L. Elson, and W. W. Webb, *Biophys. J.* **16**, 1055 (1976).

that we can assume ideal behavior. Then at any instant of time a fluctuation in the concentration of one component at some position r is independent of a fluctuation of the same component at any other position r' or of any other component at any position. Furthermore, the number of molecules of a given species in any volume will be governed by the Poisson distribution. This initial condition may, then, be expressed as

$$\langle \delta C_j(r,0)\delta C_l(r',0)\rangle = \bar{C}_j\delta_{jl}\delta(r - r') \tag{10}$$

In a macroscopic relaxation experiment the initial condition for Eqs. (9) [i.e., the initial concentration gradients $\Delta C_j(r,0)$] is determined by the means used to prepare the system in its initial nonequilibrium state. In FPR an intense pulse of focused light photolyzes irreversibly a portion of the fluorescent components in a small volume of the reaction system. Hence the $\Delta C_j(r,0)$ depend on the intensity of the bleaching pulse, $I'(r)$, its duration, T, and the kinetic mechanism of the irreversible photolysis reactions. It is simplest to assume an irreversible first-order reaction. (This can and should be verified experimentally in each application.) Thus

$$\frac{dC_j(r,t)}{dt} = -\alpha_j I'(r) C_j(r,t) \tag{11}$$

so that after exposure to the bleaching pulse from $t = -T$ to $t = 0$ the concentration of the jth component is reduced to

$$C_j(r,0) = \bar{C}_j \exp[-\alpha_j I'(r) T] \tag{12}$$

to yield the initial condition

$$\Delta C_j(r,0) = \bar{C}_j\{\exp[-\alpha_j I'(r) T] - 1\} \tag{13}$$

A simple approach to solving Eqs. (9) is to carry out a Fourier transformation to yield

$$\frac{dC_j(\nu,t)}{dt} = -\nu^2 D_j C_j(\nu,t) + \sum_l T_{jl} C_l(\nu,t) \tag{14}$$

where $C_j(\nu,t)$ is the Fourier transform of $\delta C_j(r,t)$ with ν the Fourier transform variable. The solutions of this set of equations are resolved into "normal modes" which combine the effects of the chemical reactions and the diffusion of each of the components. Associated with the sth normal mode there is a characteristic rate, $\lambda^{(s)}$, which is an eigenvalue of the matrix

$$M_{jl} = -\nu^2 D_j\delta_{jl} + T_{jl}$$

There are also amplitude factors associated with sth normal mode which are developed from $X_j^{(s)}$ the right eigenvectors of M_{jl} and from $(X^{-1})_j^{(s)}$, the inverse matrix of eigenvectors. Using the appropriate initial condition for the FCS experiment [Eq. (10)], this eventually yields the following expression for the photocurrent autocorrelation function:

$$G(\tau) = (8\pi^3)^{-1} \sum_{j,l} Q_j Q_l \bar{C}_j \int d^3v \phi(v) \sum_s A_{jl}^{(s)}(v) \exp[\lambda^{(s)}(v)\tau] \quad (15)$$

In this equation $A_{jl}^{(s)}(v) = X_j^{(s)}(X^{-1})_l^{(s)}$.

Since the eigenvectors are orthonormal,

$$\sum_s A_{jl}^{(s)} = \delta_{jl}$$

The function $(8\pi^3)^{-1}\phi(v)$ is simply the square magnitude of the Fourier transform of the beam profile detected, $I(r)\varepsilon(r)$. Details of this derivation have been published.[18]

For the photobleaching experiment use of the initial condition, Eq. (13), yields

$$\Delta i(\tau) = (8\pi^3)^{-1} \sum_j Q_j \int d^3v \sum_s \exp[\lambda^{(s)}(v)t] \sum_l A_{jl}^{(s)}(v) \bar{C}_l \psi_l(v) \quad (16)$$

where

$$\frac{\bar{C}_l}{8\pi^3} \psi_l(v)$$

is the product of the Fourier transform of $I(r)$ and the Fourier transform of $\Delta C_l(r,0)$.

As expected, there is a close similarity between Eqs. (15) and (16) which demonstrates the approximate equivalence of photobleaching and fluctuation measurements. This is more clearly seen when Eq. (16) is expressed for very small extents of bleaching. We note that

$$\phi(v) = \int d^3r \exp(ivr)I(r) \, \varepsilon(r) \int d^3r' \exp(-ivr')I(r') \, \varepsilon(r') \quad (17a)$$

and

$$\psi_l(v) = \int d^3r \exp(-ivr)I(r) \, \varepsilon(r) \int d^3r' \exp(ivr')$$

$$[\exp(-\alpha_l I'(r')T)] - 1] \quad (17b)$$

Then for $\alpha_l I'(r')T \ll 1$ and supposing that photobleaching and monitoring beam profiles are identical except for intensity, $I'(r) = AI(r)$, we obtain

$$\psi_l(\nu) = -\alpha_l TA\phi'(\nu) \tag{18a}$$

where $\phi'(\nu) = \phi(\nu)$ if $\varepsilon(r) = 1$. Hence

$$\Delta i(\tau) = \frac{-TA}{8\pi^3} \sum_{j,l} Q_j \alpha_l \bar{C}_l \int d^3\nu\phi'(\nu) \sum A_{jl}{}^{(s)} \exp(\lambda^{(s)}\tau) \tag{18b}$$

Therefore for low levels of photobleaching [and $\varepsilon(r) = 1$] $G(\tau)$ and $\Delta i(\tau)$ are very similar except in the relative weights of amplitude components. Apart from constants $G(\tau)$ and $\Delta i(\tau)$ differ mainly in that Q_l in the former is replaced by α_l in the latter. This near identity is to be expected on the basis of the equivalent behavior of linear macroscopic relaxations and appropriately averaged spontaneous microscopic fluctuations.[1]

Solutions in Special Cases

With the laser operating in TEM00 mode the spot photobleaching apparatus described in the previous section provides a transverse Gaussian illumination intensity profile. In this arrangement the photobleaching and monitoring beams differ only by the attenuation factor A when the optics are properly aligned and accordingly Eqs. (18a) and (18b) are valid.

In general it is important to recognize that the illumination volume is defined by both the transverse distribution and the intensity distribution along the direction of propagation (z) of the beam.[13,40] Thus

$$I(r) = \frac{2P}{\pi w^2(z)} e^{-[2(x^2+y^2)/w^2(z)]} \tag{19}$$

where $r = (x, y, z)$. In practice, the variation of $w(z)$ is small so that a cylindrical illumination volume provides a reasonable approximation. In that case the concentration fluctuations perpendicular to the direction of beam propagation are more important than those parallel to it. In membrane studies the fluorescence is confined to the plane of the membrane and the variation in $w(z)$ is relevant only if the membrane is not in the focal plane or there are multiple membranes.[40] For solution studies the light collection efficiency of the microscope optics determines contributions to the measured emission intensity of fluorophores off the focal plane. The point collection efficiency $\varepsilon(r)$ used in Eqs. (3) through (17) is defined by the convolution of the image with the transmission profile determined by the image plane aperture. The functional form of $\varepsilon(r)$ has been described,[13] but it has not yet been used to evaluate FCS correlation functions or FPR recovery curves measured in solution. [The original FCS experiments, which were carried out on solutions of ethidium bromide and DNA, evaded this problem by using a long focal length lens rather than a microscope to define the illumination profile. Then the profile was essentially constant along the direction of propagation of the beam through the thin sample, and so $\varepsilon(r)$ could be set equal to unity.[19]] An approximation has been used[23] in which the combined collection efficiency and illumination along the z-direction is assumed to be Gaussian such that

$$I(r)\,\upsilon(r) = \frac{2P}{\pi w^2}\, e^{-[2(x^2+y^2)/w^4]} e^{-[2z^2/w_z^2]} \tag{20}$$

where the transverse illumination radius w is assumed constant over the relevant depth of focus and where w_z represents the distance along the z-axis at which the intensity is e^{-2} of that at the plane of focus. For most applications of FCS and FPR to measurements in solution this approximation is likely to be adequate (especially if $w_z \gg w$) and particularly for measurements of relative rates of diffusion or chemical kinetics. However, *absolute* determinations of rate constants and transport coefficients are subject to systematic errors through the approximation of $\varepsilon(r)$. In membrane or surface work, the fluorescence is confined to two dimensions and consequently $\varepsilon(r)$ is unity. Then the solution to Eqs. (6) and (8) are particularly simple since

$$I(r)\,\varepsilon(r) = \frac{2P}{\pi w^2}\, e^{-[2(x^2+y^2)/w^2]} \tag{21}$$

where w is the beam size radius in the focal plane.

For the simplest ease where there are no chemical processes and only diffusing species, the correlation function for the FCS experiment is [18,19]

$$G(\tau) = \sum_j G_j(0)(1 + \tau/\tau_{D_j})^{-1} \tag{22}$$

where the characteristic diffusion times, τ_{D_j}, are related to the individual diffusion coefficients, D_j, by

$$\tau_{D_j} = \frac{w^2}{4D_j} \tag{23}$$

The factor 4 appears because of the two-dimensional nature of the problem. In solution measurements arranged so that $w_z \gg w$ in Eq. (21), the transport problem remains essentially a two-dimensional problem and τ_{D_j} has the same form.[44]

For the same problem the variation in time of the fluorescence intensity $i(t) = F(t)$ is given by a power series[16,43]

$$F(t) = \sum_j X_j[\phi_j F(-)f_j(t) + (1 - \phi_j)F(0)] \tag{24}$$

where the X_js are the mole fractions of each diffusing species, the ϕ_js represent the fraction of the jth species which is mobile on the experimental timescale, $F(-)$ is the absolute fluorescence prior to the photobleaching pulse, $F(0)$ is the absolute fluorescence immediately after the photobleaching pulse, and $f_j(t)$s are the relative fluorescence due to the mobile portion of each fluorophore and is

$$f_j(t) = \sum_{n=0}^{\infty} (-K_j)^n \{n![1 + n(1 + 2t/\tau_{D_j})]^{-1}\} \tag{25}$$

Here the K_js are parameters which measure the extent of bleaching as $K_j = \lambda_j I'T$ where λ_j is the first-order rate constant for photobleaching, I' is the intensity of the laser beam at the center of the beam in the focal plane during the photobleaching pulse which has a duration of T. The τ_{D_j}s in Eq. (25) are the same as those in Eqs. (22) and (23). Allowance for the possibility that a fraction of each species is immobile on the experimental time scale is motivated by observations of the diffusion of membrane proteins.[9]

In the limit of small bleaches Eqs. (15) and (18) illustrate the similarity between the FCS and FPR experiments. This is even clearer from Eqs. (24) and (25) in the limit of totally mobile species ($\phi_j \equiv 1$). Small bleaches implies $K \ll 1$ which reduces Eq. (25) to the first two terms ($n = 0,1$) so

$$f_j(t) = 1 - [\tfrac{1}{2}K(1 + t/\tau_{D_j})^{-1}] \tag{26}$$

Hence the functional form of the decay of the correlation function is the same as the recovery of fluorescence following a small bleach, except for the amplitude factors as discussed above.

[44] R. D. Icenogle, Ph.D. thesis, Cornell University (1981).

When there is only a single diffusing species, the results simplify further to commonly used equations[18,43]

$$G(\tau) = G(0)(1 + \tau/\tau_{D_j})^{-1} \tag{27}$$

and

$$f(t) = \sum_{n=0}^{\infty} (-K)^n \{n![1 + n(1 + 2t/\tau_{D_j})]^{-1}\} \tag{28}$$

where the latter in the limit of small bleaches is

$$f(t) = 1 - K/2(1 + t/\tau_{D_j})^{-1} \tag{29}$$

Similar results have been described for systems undergoing either uniform or laminar flow or one of these in combination with a diffusion process.[20,43] In all cases, the FPR results are the same as those of the FCS results in the limit of small bleaching only.

Measurements of Chemical Kinetic Processes

The use of FCS and FPR to measure chemical kinetics in open systems leads to greater analytical complexity than would be encountered in closed systems. Consider, for example, the simple second-order reaction:

$$A + B \underset{k_b}{\overset{k_f}{\rightleftharpoons}} C \tag{30}$$

where B is a fluorescent ligand which binds to a nonfluorescent molecule A to produce fluorescent complex C. We shall suppose that B and C, but not A, absorb the excitation radiation and are vulnerable to photolysis. In a closed system a single reaction rate parameter, R, describes the relaxation to equilibrium of the reaction when subjected to a sufficiently small but macroscopic perturbation. Then

$$-\Delta C_A(t) = -\Delta C_B(t) = \Delta C_C(t) = \Delta C_C(0) \exp(-RT) \tag{31}$$

where

$$R = k_f(\bar{C}_A + \bar{C}_B) + k_b \tag{32}$$

This is a consequence of the conservation of mass. In an open system, however, A, B, and C can change independently; mass is not conserved. Therefore, the fluctuation behavior must be described in terms of three kinetic "normal" modes. A detailed analysis of the reaction of Eq. (30) in an open volume has been presented.[18] A consequence of this complexity is that more detailed and more precise measurements must be obtained to characterize the open system than would be required for the simpler corresponding closed system.[22,23]

For the FPR experiment we could suppose that B and C are simply eliminated optically and chemically from the system by the photobleaching pulse. This would demand that the photochemical reaction be somehow propagated from B to the A molecule to which it is bound so that the complex C is eliminated as supposed. Then the initial conditions for a relaxation of this system would be

$$\Delta C_A(r,0) = 0 \tag{33}$$

$$\Delta C_j(r,0) = \bar{C}_j\left[\exp(-\alpha_j I'(r)T) - 1\right]; \qquad j = \text{B,C} \tag{34}$$

Several other possibilities may, however, be distinguished. For example, the photolysis pulse might destroy B whether free or bound with no effect on A whether free or bound. Then the effect of the pulse is to eliminate B and to convert C to A. Therefore for this example $\Delta C_B(r,0)$ and $\Delta C_C(r,0)$ are as listed above but $\Delta C_A(r,0) = -\Delta C_C(r,0)$. This difference in initial conditions does not influence the characteristic rates of the several relaxation modes of the two examples, but it does determine differences in their relative amplitudes. A detailed analysis of photobleaching experiments requires information about the kinetic mechanism of the photolysis reactions and their effects on the chemical reaction systems under study. If only the relaxation amplitude depends on the details of the photochemical response, a thorough characterization may not be necessary in some instances. Furthermore most of the required information might be obtained simply by observing the dependence of relaxation amplitudes on reactant concentrations. In complex systems, however, the expected fluorescence recovery curve may depend strongly on the mechanism of the bleaching reaction even to the extent that the results obtained by FCS and FPR appear quite different.[22,23] Therefore it seems to be a significant theoretical advantage of the fluctuation approach that it does not have to deal with these complications.

Comparison of FCS and FPR Measurements

A substantial advantage of the fluctuation approach is that, since the reaction system rests in thermodynamic equilibrium throughout the experiment, there is no need to define and characterize a process to prepare an initial nonequilibrium state. In some perturbation/relaxation experiments the response of the system to the perturbation can also provide useful information. For example, temperature- and pressure-jump experiments can yield information about reaction enthalpy and volume changes, respectively. In photobleaching experiments, however, information about the photochemical response of the system is needed to interpret the measurements but typically is not of intrinsic interest.

Balancing this theoretical advantage of FCS is a considerable practical disadvantage which results from the intrinsically statistical character of the fluctuation method. Individual fluctuations of concentration and fluorescence are small and difficult to measure. Even if the temporal behavior of a single fluctuation could be measured with high precision, however, it would still be necessary to average the behavior of many fluctuations to characterize the phenomenological chemical rate constants and transport coefficients with adequate precision. This is due to the stochastic character of spontaneous fluctuations.[1] A detailed analysis of the precision of fluorescence fluctuation experiments has been published.[45] As a result of this requirement for long periods of averaging the fluorescence fluctuation approach is applicable only to stable systems. Applications to simple solutions may be complicated by unwanted photobleaching which could occur to a significant extent even at low levels of excitation because of the extended duration of the measurements.[19] Applications to animal cell surface phenomena are further complicated by systematic motions of these cells such as ruffling, extension of microvilli, and locomotion. Local motions may generate fluorescence changes which can overwhelm the small fluorescence signals due to spontaneous concentration fluctuations. Longer range motions can remove the observation region entirely from view during the course of an extended period of data accumulation.

Conversely it is a substantial practical advantage of the photobleaching approach that it is not intrinsically statistical and does not require extended periods of data accumulation. Phenomenological rate constants and transport coefficients may be accurately determined from observation of a single macroscopic relaxation to the limits of precision of the fluorescence measurements. (Of course, measurements of repeated relaxations may be used to improve precision.) Therefore photobleaching experiments have proved useful in studies of cell surface mobility while the fluctuation approach is frequently impractical for studies of living cells.

Although there are no inherent limitations on the range of relaxation times which can be observed by FCS, the practical slower limit arising from intrinsic instability of the systems is at about 1 Hz fluctuation frequencies corresponding to a diffusion coefficient of about 10^{-12} m^2 sec^{-1} (assuming the beam radius is about 1 μm). Thus FCS experiments are easily applicable to solution studies of large molecules including polymers of large molar masses.

Many macromolecules have diffusion coefficients in this range and can

[45] D. E. Koppel, *Phys. Rev. A* **10**, 1938 (1974).

therefore be studied by either or both techniques to exploit their particular advantages. Recent studies of ethidium bromide binding to DNA illustrates the power of combining the techniques of FCS and FPR.[22,23]

It is evident that FCS or FPR measurements can provide equivalent *dynamic* information. The two techniques do however provide very different ancillary information arising from the difference in the amplitude factors. As indicated in Eq. (24), the extent of recovery of fluorescence after photolysis in an FPR experiment depends on the fraction of mobile molecules, ϕ. This parameter is unity for solution diffusion, but is typically less than that for proteins diffusing in cell membranes.[12–17] In many cases, changes in mobile fraction have provided information more interesting and useful than the diffusion measurement itself.[46,47] In contrast, the amplitude of the correlation measurements is sensitive to the immobile fraction only indirectly as it contributes to the mean fluorescence but not to the fluctuations. However, the $G_j(0)$ values in Eq. (22) depend on Q_j^2. As a consequence, the correlation function amplitudes are sensitive to the state of polymerization or aggregation of the molecular species observed.[20,45] This has been exploited in measurements of molecular masses of very large DNA molecules[48] and is being explored as a means of studying protein aggregation in cell membranes.[17]

General Experimental Considerations

FCS and FPR experiments were designed and developed to take advantage of the selectivity and sensitivity of fluorescence spectroscopy. The price for these advantages is the frequent necessity to use adventitious fluorescent labels and the resulting complication of interpretation due to possible perturbations of the system of these labels. In studies of diffusion or aggregation the probe perturbations are likely to be small. In studies of chemical kinetics, it is probable that the rates of reactions and the equilibrium constants may be affected by the probe. In such cases the choice of probe and/or the site of labeling may greatly affect the results and alternatives should be explored.

The particular probe employed depends principally on the biological problem under investigation. Since the equipment generally employs ion-lasers the fluorophore must absorb in the region 450–530 or 530–660 nm depending on the laser type. The probe must be reasonably photostable so that little photolysis occurs during the observation of the recovery or

[46] J. A. Reidler, P. M. Keller, E. L. Elson, and J. Lenard, *Biochemistry* **20**, 1345 (1981).
[47] D. C. Johnson, M. J. Schlesinger, and E. L. Elson, *Cell* **23**, 423 (1981).
[48] M. Weissman, H. Schindler, and G. Feher, *Proc. Natl. Acad. Sci. U.S.A.* **73**, 2776 (1976).

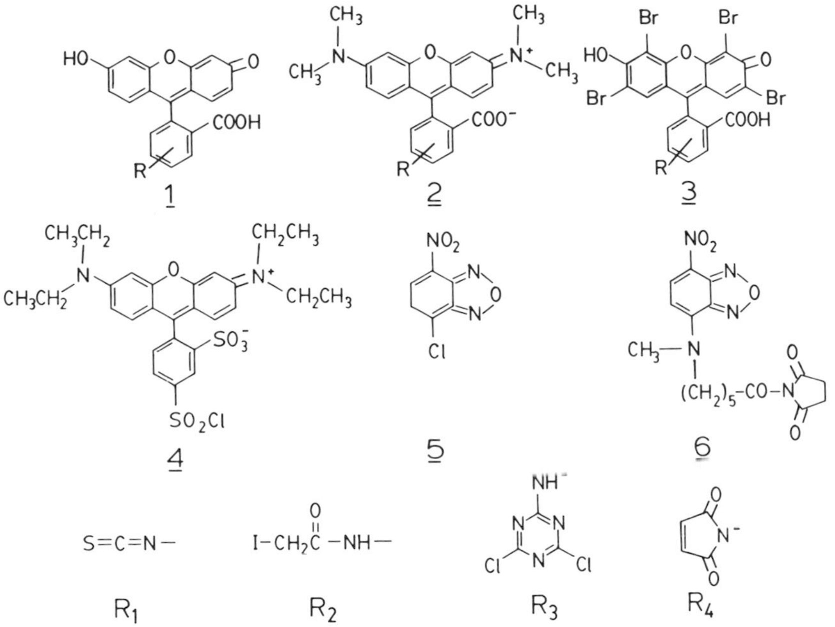

FIG. 4. Chemical structures of some commonly used fluorescent molecules and the reactive groups available with these. The names and spectroscopic characteristics are listed in the table.

fluctuations. Yet for FPR experiments it is necessary that the probe can be irreversibly photolysed with first-order kinetics so that the standard theoretical apparatus is applicable to the data analysis. Furthermore, it is desirable to maximize the fluorescence quantum yield to enhance the sensitivity. In most cases one also needs small, nonperturbing probes which are easily derivatized to react with appropriate functional groups on biologically interesting molecules under mild conditions (e.g., neutral pH and ambient temperatures).

A number of quite different probes satisfy most of these criteria and some of these are listed in the table along with their chemical structure shown in Fig. 4. Comprehensive bibliographies are available in the "Handbook of Fluorescent Probes" (published by Molecular Probes, Inc., 24750 Lawrence Road, Junction City, Oregon) and in "Fluorescent Probes" (G. S. Beddard and M. A. West, eds., published by Academic Press in 1981; see pages 184–235). A few selected references are provided in the table. There are naturally many variations on these themes leading, for example, to a host of related rhodamine derivatives with different chemical and spectroscopic properties. There are also many approaches

SOME COMMONLY USED FLUORESCENT MOLECULES

Fluorescent probe	λ_{ex}^{max} (nm)	λ_{em}^{max} (nm)	ϕ_f	Comments	Selected references
Fluorescein (1)	490	520	0.15–0.3	ϕ_f depends on degree of labeling because of self-quenching. Emission is pH sensitive. Photolabile	a,b,c,d
Tetramethylrhodamine (2)	550	580	0.1	Fairly hydrophobic and chemistry of labeling is harder in aqueous media. Extensive dialysis needed. Photostable	e,f,g
Eosine (3)	522	560 (690)	Very low phosphorescence	Good probe for rotational diffusion. Low fluorescence yield is less good for FPR/FCS measurements	h,i
Lissamine rhodamine (4)	540	570	Low	Easier to label with than 2 but lower ϕ_f. Not yet used extensively	c
Nitrobenzoxadiazole (NBD-Cl) (5)	340 (–Cl) 430 (–SR) 470 (–NHR) 480 (–NR$_2$)	— 520 530 530	0 — 0.05 0.05	Good probe to react with SH or amino groups. NBD-Cl is not fluorescent. Emission wavelength and yield are very solvent dependent. Photostable	j,k
NBD-methyl amino hexanoic acid-NHS (6)	480	535	0.05	Developed to react with hindered amines	l,m

Reagent	Comments
Isothiocyanate R$_1$	Reacts readily at slightly basic pH with free amines on, e.g., lysines. Typically react for several hours at room temperature. Can label in culture
Iodoacetamide R$_2$	Reacts well at neutral pH in aqueous media with –SH groups
Dichlorotriazinyl R$_3$	Reacts more rapidly than isothiocyanates at slightly acid pH with –SH, NH$_2$, and histidine. Can react with tyrosine
Maleimide R$_4$	Reacts readily and rapidly (<1 hr) with amino groups (similar to succinimide derivatives as in **6**)

[a] Elson and Reidler.[10]
[b] Safranyos and Caveney.[25]
[c] M. Wilchek, S. Spiegel, and Y. Spiegel, *Biochem. Biophys. Res. Commun.* **92**, 1215 (1980).
[d] M. Schindler, D. E. Koppel, and M. P. Sheetz, *Proc. Natl. Acad. Sci. U.S.A.* **77**, 1457 (1980).
[e] Edidin.[14]
[f] Peters.[15]
[g] D. Axelrod, *Proc. Natl. Acad. Sci. U.S.A.* **77**, 4823 (1980).
[h] D. E. Golan and W. Veatch, *Proc. Natl. Acad. Sci. U.S.A.* **77**, 2537 (1980).
[i] B. K. K. Fung and L. Stryer, *Biochemistry* **17**, 5241 (1978).
[j] B. Derzko and K. Jacobson, *Biochemistry* **19**, 6050 (1980).
[k] L. C. Cantley, Jr. and G. G. Hammes, *Biochemistry* **15**, 1 (1976).
[l] N. O. Petersen, *Spectrosc. Int. J.* **2**, 408 (1983).
[m] N. O. Petersen, *Can. J. Chem.* **63**, 77 (1985).

to introduce spacers for reaction with sterically hindered groups and two examples are listed in the table.

The transport processes or chemical kinetics which are observed can depend on which molecular components are labeled. For example in the model[18] where A is much larger than B [see Eq. (30)] so that the rates of diffusion of A and C are similar and smaller than that of B, the measured diffusion rate will depend on whether A or B is the fluorescent component. In the absence of fluorescence changes upon reaction the diffusion rate reflects the chemical kinetics only if B is the fluorescent component. On the other hand, labeling A and not B would permit studies of the transport of A without the complication of the chemical kinetics. Thus the choise of labeling site will determine the type of questions which can be answered.

Because of the possible perturbing effects of the derivatization with fluorescent probes, FCS and FPR measurements should be subjected to rigorous controls for background fluorescence, specificity of binding, purity of labeled compounds, and, when possible, biological activity of the derivatized molecule.

FCS experiments are frequently longer than FPR experiments and therefore place severe requirements on sample stability. For solution studies, the sample must be kept in suspension and to prevent evaporation of solvent in the small volumes typically employed, it is common to maintain the samples in small rectangular microtubes which are sealed at both ends. This also minimizes escape of gases such as CO_2 which can seriously alter the pH. For work with cells it is frequently impractical to use sealed or closed containers. The time one can work with a specimen is often restricted by the response of the cells to the changing environment of the medium. In order to minimize these changes and maximize the viability of the cells it is recommended that the experiments be done with the cells exposed to a defined culture medium (e.g., M199 from GIBCO) buffered with a nonvolatile buffer (e.g., 25–30 mM HEPES) and without serum. It is essential that this medium contains *no* Phenol Red (or other indicator dyes) since this contributes to the background fluorescence. In this way it is possible to maintain viable cells for an hour or more.

Data Analysis and Interpretation

The optimal approach to data analysis is to adapt a nonlinear least-squares fitting algorithm[16] to the complete theoretical solution to the fluctuation or recovery experiment. Several approaches have been discussed most recently in detail for the FPR experiment.[16] It is in principle straightforward to extend the curve fitting algorithms to any functional form. For example, Eqs. (27) and (28) are analyzed by very similar procedures in our laboratories.

The FCS experiments require efficient calculations of the correlation function $G(\tau)$. For large data sets, this is a serious computational problem if done by a conventional summation of time-lag products. Considerable timesavings can be had by employing modern digital fast Fourier transform procedures.[49,50]

The interpretation of FCS and FPR experiments must reflect the problem being studied but can be affected by limits of reproducibility, experimental artifacts, and approximations in theory. Detailed consideration of these problems have been presented[16] for the FPR experiments. Likewise the precision of the correlation experiments has been discussed.[45]

Conclusions

Methods developed for measuring chemical kinetics and transport via fluorescence detection of local concentration fluctuations have proved useful in their own right and have suggested possible applications for closely related photobleaching methods. Up to now the major application of this combined technology has been the measurement (mainly by FPR) of lateral diffusion on cell surfaces and model membranes. The conceptual approach of FCS has also been applied to measurements of rotational motion.[51,52] Up to now there has been less emphasis on FCS and FPR measurements in solutions. Nevertheless both methods have unique capabilities for studies of small fragile, heterogeneous, and dilute systems.

Studies of cell surface mobility have revealed that the lateral motion of several plasma membrane proteins is constrained by forces in addition to the viscosity of the lipid bilayer membrane matrix. Interactions between cell surface proteins and cytoskeletal structures are now being actively investigated using electron microscopic,[53] immunofluorescence,[54] and biochemical[55] methods. FPR and related methods have unique capabilities to supplement these "static" probes of structure with investigations of the dynamics of interactions of cell surface components. The develop-

[49] E. O. Brigham, "The Fast Fourier Transform." Prentice-Hall, New York, 1974.
[50] C. Chatfield, "The Analysis of Time Series: An Introduction." Chapman & Hall, London, 1980.
[51] M. Ehrenberg and R. Rigler, Chem. Phys. 4, 390 (1974).
[52] S. R. Aragon and R. Pecora, Biopolymers 14, 119 (1975).
[53] R. E. Webster, D. Henderson, M. Osborn, and K. Weber, Proc. Natl. Acad. Sci. U.S.A. 75, 5511 (1978).
[54] U. Groschel-Stewart, Int. Rev. Cytol. 65, 193 (1980).
[55] V. Bennet, J. Cell. Biochem. 18, 49 (1982).

ment of this approach is just now beginning. Although it is promising in principle, it is still too soon to decide whether sufficiently accurate measurements can be obtained over a wide enough range of conditions to yield the desired information.

It does not seem likely that FCS will be much used to study conformational fluctuations. Optical changes associated with conformational transitions are typically too small for successful application of this approach. Application of FPR to this kind of problem would require that particular conformational isomers be specifically sensitive to photolysis. This considerably narrows the choice of acceptable systems. A different approach would be to measure the relaxation of very small displacements from equilibrium generated by small perturbations of state variables such as temperature and pressure. Although those would be forced displacements, they could approach spontaneous fluctuations in magitude. Since the optical signals generated by such small changes in state are also very small, it is necessary to perturb and measure repetitively with signal averaging to achieve adequate precision. A method of this kind, based on small repetitive pressure jumps, has been developed.[56,57] Perturbation magnitudes are typically 10- to 100-fold smaller than in corresponding temperature- or pressure-jump experiments in which only one or a few transient relaxations are observed.

Acknowledgments

The development of FCS and FPR techniques has involved a large number of researchers in many laboratories across the world. This work was supported by NIH Grants GM 21661, GM 30299, and GM 27160 (to ELE), and NSERC, Canada Grants U0109 and E5837 as well as ADF (UWO) Grant 81-10 (to NOP).

[56] R. M. Clegg, E. L. Elson, and B. W. Maxfield, *Biopolymers* **14,** 883 (1975).
[57] H. R. Halvorson, *Biochemistry* **18,** 2480 (1979).

[20] Fluorescence Lifetimes with a Synchrotron Source

By STEVEN G. BOXER and RODNEY R. BUCKS

Synchrotron Radiation for Timing Experiments

Synchrotron radiation is a unique light source for use in the measurement of fluorescence phenomena. The light pulses generated by an electron storage ring cover all wavelengths of interest to spectroscopists working with biological molecules, have a high repetition rate, and are immune to shot-to-shot fluctuations in intensity. While mode-locked lasers can produce shorter pulses with greater pulse powers, they do not have the ease of excitation tunability of a synchrotron source. If photon-counting techniques are used, the low pulse power of synchrotron sources is not an important consideration. Visible light from the synchrotron had been used routinely by operators at the Stanford Synchrotron Radiation Laboratory (SSRL) for luminosity measurements and position adjustments. The notion of intercepting a fraction of this beam for lifetimes measurements was conceived by Drs. Ian Munro and Andrew Sabersky.[1]

The fluorescence lifetimes port at SSRL utilizes light pulses generated by electrons stored in the Stanford Positron Electron Asymmetric Ring (SPEAR). Although similar capabilities are available at other synchrotron sources, the authors' experience is limited to SSRL. The reader interested in other synchrotron sources should consult the abstracts of the 1983 Meeting of the American Society for Photobiology where these sources are compared.[2] As the characteristics, advantages, and problems associated with the measurement of fluorescence lifetimes at SSRL are typical, we will only discuss this facility.

The minimum repetition rate of the light pulses is 1.28 MHz, which corresponds to a single bunch of electrons circulating in the ring. This minimum repetition rate is determined by the ratio of the speed of light and the ring circumference. The total number of bunches of electrons which can be stored in SPEAR is 280; this value depends on the frequency of the accelerating rf field used to supply energy to the electrons as they circulate (358 MHz in SPEAR).

In practice, a pattern consisting of a small number of electron bunches (≤ 17) is stored in the ring. At any time this number is determined by a compromise among the needs of the many users of radiation emitted by the synchrotron source. The simplest pulse configuration occurs during runs when colliding beam experiments (electrons and positrons) are performed by high energy physicists. A single electron bunch (20–30 mA) circulates in the ring; this gives rise to light pulses which are separated by ~781 nsec. Due to the geometry of the optics the lifetimes port only receives light from the stored electrons, not the positrons. Another common mode of operation is the "timing" mode, where 4 or 5 single bunches of electrons are spaced roughly equally around the ring. The time between light pulses is 195 nsec (4 bunches) or 156 nsec (5 bunches). In order to

[1] I. H. Munro and A. P. Sabersky, *in* "Synchrotron Radiation" (H. Winick and S. Doniach, eds.), p. 323. Plenum, New York, 1980.
[2] *Abstr. Annu. Meet. Am. Soc. Photochem. Photobiol., 11th* **37,** 525–529 (1983).

obtain the highest energy radiation (X rays), highest intensity, and long beam storage time, a common mode of operation is four groups of four adjacent electron bunches, roughly equally spaced (4 × 4). Within the groups of four, the bunches of electrons are separated by 2.79 nsec. This mode is useless for fluorescence lifetimes as the pulse spacing within the groups of four is much too short. This can be circumvented by injecting a seventeenth electron bunch between two of the groups of four (4 × 4 + 1); the single bunch is isolated from any group of four by ~89 nsec. It has been found that this does not degrade overall performance. The electronics can be adjusted so that only fluorescent events caused by the light pulse from the single electron bunch are recorded, even though the detector sees events generated by each pulse. Of course, if the fluorescence lifetime of the sample is long (say longer than one-fifth of the interpulse spacing), some of the above configurations may not be suitable.

Under typical operating conditions of SPEAR (3.0 GeV, 5–10 mA in the timing electron bunch) the electron bunch and therefore the light pulse has a full width at half maximum of ~200 psec, although pulses as short as 55 psec are possible.[1] The pulse is roughly Gaussian in temporal profile, but is not perfectly so. Typically, there are about 10^4 photons per pulse (4 nm band width). Sabersky has studied the details of the pulse substructure using a streak camera.[3] Depending on the precise SPEAR operating conditions irregular pulse shapes and shape changes have been observed. These irregularities could affect the accuracy of measurements of very short fluorescence lifetimes; no systematic study has yet been undertaken.

The lifetimes port is located in a large freight container on top of the radiation shielding of the storage ring as shown in Fig. 1. The light is generated in one of the SPEAR bending magnets; it is reflected from a Be mirror and passes out of the beam vacuum chamber through a 1-cm Suprasil window. A front surface aluminized beam splitter (50–90%) reflects the light beam upward to the lifetimes port where an adjustable mirror directs the beam horizontally. The dimensions of the beam after reflecting from the mirror in the lifetimes port are about 1 × 5 cm (roughly rectangular). The optical components and ambient atmosphere determine the available wavelength range in the port (longer than 200 nm). With standard monochrometers, narrow bandwidth excitation can be obtained over the entire ultraviolet, visible, and infrared range. Experiments at higher energy (vacuum ultraviolet, soft X ray, etc.) require special high-vacuum systems and are available at other SSRL beam lines. The beam is highly linearly polarized in the vertical direction (Fig. 1).

[3] A. P. Sabersky and M. H. R. Donald, SLAC-PUB-2696, PEP-NOTE-350, February 1981.

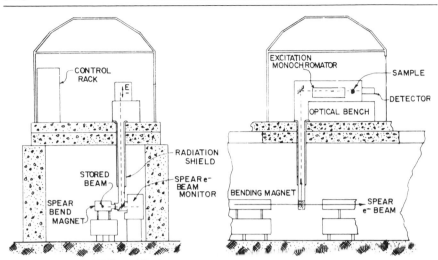

FIG. 1. An illustration of the fluorescence lifetimes port beamline at SSRL. The light beam is reflected from a Be mirror in the vacuum chamber, passes out through a Supracil window, and is reflected upward into the shack containing the sample and electronics.

Time-Correlated Single-Photon Counting

The theory of time-correlated single-photon counting has been discussed in many places and the interested reader is referred to the article by Yguerabide[4]; only a brief description of the important points will be given here. A fluorescent sample is irradiated with a short pulse of light; the amount of time which elapses between excitation and the detection of the first fluorescent photon is recorded. The data are a histogram of the number of fluorescent events which occurred at a given time after excitation and correspond to the fluorescence decay. Because of the recovery time and design of the electronics, one photon or less per light pulse is detected. This means that any fluorescent photons which strike the photocathode of the photomultiplier tube after the first photon will not be detected. If the accumulated data are to accurately represent the true fluorescence decay, it is necessary to avoid the situation where two or more fluorescent photons strike the photocathode per excitation pulse. Because only the first photon is detected, this would weight the decay toward shorter times. The simple solution is to count events with a low efficiency in terms of the number of fluorescent events detected compared to the number of light pulses which impinge on the sample. It has been shown that a detection rate of less than 2% of the sample excitation rate will lead to a measured decay which has less than 1% distortion relative to

[4] J. Yguerabide, this series, Vol. 25, p. 498.

the true decay.[4] Therefore, counting rates at SSRL are maintained lower than 25 kHz to avoid distortion of the data.

Instrumentation

The light from the horizontal mirror is focused onto the entrance slit of a monochromater which allows the selection of narrowband excitation. A number of grating sets are available for the monochrometer (e.g., 300, 750, and 1000 nm blaze) and the gratings are easily interchangeable (SPEX Doublemate). The slit width can also be adjusted. Upon exiting the monochrometer the light beam is directed onto the fluorescence sample, and fluorescence is detected at 90° to the direction of the excitation beam. Several photomultiplier tubes (PMTs) have been used, including Hamamatsu R1333 and RCA 8850 and 8852. The fluorescence is typically filtered with colored glass or interference filters to eliminate scattered light or it can pass through another monochrometer. The spread in the bunch of photoelectrons as it moves from the photocathode down the dynode chain (transit time jitter) typically broadens the observed pulse width from ~200 to ~600 psec.

A block diagram of the electronics is shown in Fig. 2; this is typical of photon-counting systems.[5] The heart of the system is the time-to-amplitude converter (TAC). The TAC accepts a start and stop pulse and produces an output voltage which is proportional to the time which elapses between these two pulses. At SSRL the start signal is initiated by a photon impinging on the photocathode of the PMT. The PMT tube output then goes to a constant fraction discriminator (CFD), and the CFD output is the start pulse for the TAC. It is necessary to use a constant fraction discriminator rather than a leading edge discriminator for this application because of the amplitude fluctuations of the PMT output pulse. The timing characteristics of the CFD are not dependent on pulse amplitude. The stop pulse originates in a signal from an rf electrode placed in the beam chamber which generates a signal as the electron bunch passes. The rf electrode signal is input to the Three Flavor Beam Trigger (TFBT). The TFBT is a discriminator which was built at the Stanford Linear Accelerator Center for use in timing applications on another storage ring. It is designed to respond to the fast, bipolar rf electrode pulse with little time walk as the pulse amplitude decays with time (due to inevitable degradation of the electron beam current). The output from the TFBT is input to a 100 MHz leading edge discriminator which produces the stop pulse.

Because there may be more than a single electron bunch in the storage ring (e.g., $4 \times 4 + 1$), it is necessary to reject data which are not due to the

[5] K. G. Spears, L. E. Cramer, and L. D. Hoffland, *Rev. Sci. Instrum.* **49**, 255 (1978).

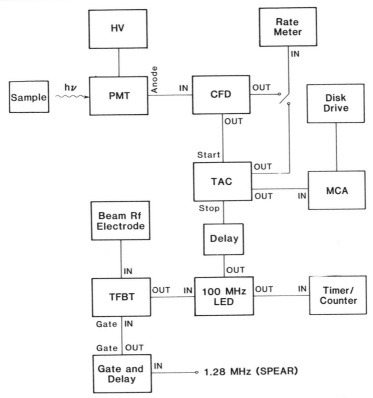

Fig. 2. Schematic diagram of the electronics used for time-correlated single-photon counting. CFD, Constant fraction discriminator; HV, high voltage; LED, leading edge discriminator; MCA, multichannel analyzer; PMT, photomultiplier tube; TAC, time-to-amplitude converter; TFBT, three flavor beam trigger.

single bunch. This is accomplished by gating the TFBT; it only produces an output pulse when the gate is activated. A 90-nsec gate pulse is produced by a gate and delay generator and supplied to the TFBT. A 1.28 MHz signal from the SPEAR master oscillator is used as the input to the gate and delay generator. The delay of the gate and the delay generator can be adjusted so that the TFBT is active for any 90 nsec within the 781-nsec cycle period of SPEAR. By adjusting this delay, the stop signal can be derived from the single bunch of electrons or any other of the electron bunches. The TAC only produces an output voltage pulse when it receives both a start and a stop signal. By gating the stop signal, fluorescent events which are detected by the PMT, but were caused by other than the desired electron bunch, can be ignored.

The TAC produces an output voltage which is proportional to the time interval between the start and stop pulses, and this is input to a Multi-Channel Analyzer (MCA), run in pulse height analysis mode. The MCA records the number of times that the TAC produced a given voltage by placing a count in channel n which corresponds to the voltage $n\Delta V$. ΔV is the voltage interval per channel of the MCA. The voltage is proportional to a time interval, and therefore the channel corresponds to an elapsed time between detection of a fluorescent photon (start) and the timing signal (stop). Because the timing signal is fixed in time with respect to the excitation pulse (both are derived from the circulating electron bunch) each channel of the MCA data contains the record of the number of fluorescent events detected at time $n\Delta t$ ($\Delta t \propto \Delta V$) after excitation. This is the fluorescence decay.

Although the pulse-to-pulse stability with a synchrotron source is extraordinary, the circulating electron current slowly decays over a period of hours. A separate photomultiplier has been provided which measures the exciting beam intensity as a function of time for data scaling during very long-term averaging.

In normal operation both a scattered light pulse profile (e.g., scattered light from coffee creamer) and a fluorescence decay are collected for each sample. Because the time response of the PMT varies with the wavelength of incident light, it is important to collect the pulse profile data at the wavelength that the sample emits. This is easily done by setting the excitation monochrometer at that wavelength (this calibration is easy to perform given the tunability of the light; this can be a source of considerable experimental difficulty with dye laser excitation). The temporal profile of synchrotron radiation is invariant with respect to wavelength, an important advantage of synchrotron radiation. The arrival time of each wavelength will vary slightly because of the intervening optics.

If the MCA is operated in multichannel scaling mode (MCS), time-resolved fluorescence excitation spectra can be obtained. This is possible because synchrotron radiation acts as a white light source, and the excitation can be scanned with a monochrometer. In MCS mode the MCA records the frequency of occurrence of events versus elapsed time. By using the upper level and lower level discriminators to set the limits of a voltage range (therefore time range) on the MCA, the MCA will record the number of events occurring as a function of time which have a voltage greater than the lower level setting and lower than the upper level setting. If the upper and lower level discriminators are set while the MCA is in pulse height analysis mode, one can see that a region of the decay can be selected which corresponds to events which occur at a given time with respect to excitation. If the MCA (in MCS mode) and excitation monochrometer are scanned simultaneously, the MCA records an excitation spectrum of events which occur during the selected time interval of the decay. The convenience of doing this type of experiment with a readily tunable source cannot be overstated.

A nonlinear least-squares routine is used to fit exponential decays to the fluorescence data. The method is known as the linearly constrained modified diagonal method and is based on a program originally written at the National Institutes of Health.[6,7] The program will fit a single exponential or the sum of exponential decays to the data. Many more or less sophisticated versions of software for deconvolution have been described; several are commercially available.

Performance

The major timing limitations in the experimental set-up are the response time and timing jitter of the PMT and time walk of the timing signal (stop pulse). The time jitter of the electronics is small (<50 psec) and this adds little to the observed pulse width, while the jitter in the PMT response broadens the measured pulse profile from the actual ~200 to ~650 psec. The addition of the TFBT solves the timing pulse walk problems. The rf electrode picks up the signal from the electron bunch ~3 m from the light source in the bending magnet and is directly beneath the lifetimes port. Because both the light pulse and rf electrode signals are generated directly from the beam, the timing pulse should not walk in time. The TFBT is insensitive to changes in signal amplitude as the beam decays, and produces a very stable timing signal. The 1.28 MHz master oscillator signal cannot be used as the timing pulse. The light pulse arrival time varies with respect to the 1.28 MHz signal as the beam decays and as the phase of the 358 MHz rf cavities are adjusted or drift.

The fluorescence lifetimes of several compounds with known lifetimes were used to test the capabilities of the facility. The measured fluorescence lifetimes of a chlorophyllide dimer in CH_2Cl_2 (100 ± 30 psec),[8] Rose Bengal in methanol (520 ± 50 psec),[5] and Rhodamine 6G in ethanol (3.8 ± 0.2 nsec)[9] all agree with the reported lifetimes within the quoted experimental error. There is evidence that the longitudinal structure of the electron bunch changes with time under certain operating conditions,[3]

[6] R. I. Schrager, *J. Assoc. Comp. Mach.* **17,** 446 (1970).
[7] A. Grinvald and I. Z. Steinberg, *Anal. Biochem.* **59,** 583 (1974).
[8] M. J. Pellin, M. R. Wasielewski, and K. J. Kaufmann, *J. Am. Chem. Soc.* **102,** 1868 (1980).
[9] D. R. Lutz, K. A. Nelson, C. R. Gochanour, and M. D. Fayer, *Chem. Phys.* **58,** 325 (1981).

and this will interfere with the measurement of fluorescent lifetimes which are comparable to the pulse width.

The high repetition rate of the excitation pulse at SPEAR allows the rapid accumulation of fluorescence decays with good signal-to-noise. Accumulation times of 5 min were typical for samples which had a fluorescence quantum yield of 0.3, absorption at the excitation wavelength of 0.05, and an excitation band pass of 5 nm. Low sample optical densities are required to prevent artificial lengthening of the fluorescence lifetime due to emission–reabsorption.

Several applications have been described in the literature. Monro, Pecht, and Stryer examined the fluorescence lifetimes and fluorescence anisotropy decay for a number of small proteins each containing a single tryptophan residue.[10] The combination of high intensity excitation light in the near ultraviolet (280 nm) and the highly polarized beam gave very good signal-to-noise and permitted a comparison of the degree of motional reorientation of tryptophan in these proteins. Our group has studied fluorescence lifetimes from a wide range of chlorophyll derivatives,[11,12] synthetic chlorophyll–protein complexes,[13] and photosynthetic reaction centers.[14] In the case of synthetic aggregates of chlorophyll-type chromophores, multiple decays are quite common. It is very important to distinguish decays due to impurities which may have a very high fluorescence quantum yield from real components of interest. The ability to record time-resolved excitation spectra has proved most useful in discriminating impurity fluorescence from the desired decay components. It is straightforward to scan the excitation monochrometer from 300 to 700 nm and compare the spectra giving rise to the fastest and slower decaying components.[12] Although such an experiment is possible in principal with a synchronously pumped dye laser system, it is a major undertaking, involving many different laser dyes and enormous variations in intensity over such a wide spectral range.

At the present time there have been relatively few applications of this technique to biological problems. This is undoubtedly because of the false impression that synchrotron experiments are exotic and inaccessible to the nonspecialist. The facility at SSRL is available for general use and is no more complex to operate than a commercial lifetimes apparatus. A

[10] I. Monro, I. Pecht, and L. Stryer, *Proc. Natl. Acad. Sci. U.S.A.* **76**, 56 (1979).
[11] R. R. Bucks and S. G. Boxer, *J. Am. Chem. Soc.* **104**, 340 (1982).
[12] R. R. Bucks, I. Fujita, T. L. Netzel, and S. G. Boxer, *J. Phys. Chem.* **86**, 1947 (1982).
[13] S. G. Boxer and K. A. Wright, *Biochemistry* **20**, 7546 (1981).
[14] C. E. D. Chidsey, Ph.D. thesis, Stanford University (1983).

new facility at the Brookhaven National Laboratory is likely to be even more flexible and equally useful for experiments in the ultraviolet region.

Acknowledgments

The Lifetimes Port at SSRL is supported in part by NIH Biotechnology Grant RR 01209-01. Work in the authors' lab was supported by NSF and GRI Grants PCM 8303776 and 82-260-0089 respectively. Steven G. Boxer is an A. P. Sloan and Camille and Henry Dreyfus Teacher-Scholar Fellow.

[21] Fluorescence Polarization at High Pressure

By ALEJANDRO A. PALADINI, JR.

Introduction

For many years pressure has been used as a key variable in gathering information on the structure of matter. About a hundred years ago the effect of pressure on living systems was reported for the first time by two French scientists, P. Regnard and A. Certes.[1] Since then it has become quite common to submit living organisms to pressure. The results obtained in those early efforts show that the general aim was to prove that pressure had an effect (usually negative) on the systems studied rather than its use as a thermodynamic variable.

The reader interested in a review of the literature dealing with high-pressure effects upon biological systems is directed to those recently published by Morild[2] and by Heremans.[3]

However, the assumptions, limitations, and consequences of the theory employed so far in the interpretation of experimental data can be found in a review by Weber and Drickamer.[4] Hydrostatic pressure is a scalar magnitude and as such it is expected to act without preferential direction on the media to which it is applied. Furthermore, the use of pressure permits the study of effects related exclusively to changes in volume. This is not generally possible when the temperature and/or chemical composition of the system has been selected as the disturbing

---

[1] P. Regnard, "Recherches Experimentales sur les Conditions Physiques de la Vie dans les Eaux." Librairie de L'Academie de Medicine, Paris, 1891.
[2] E. Morild, *Adv. Protein Chem.* **34,** 93 (1981).
[3] K. A. H. Heremans, *Annu. Rev. Biophys. Bioeng.* **11,** 1 (1982).
[4] G. Weber and H. G. Drickamer, *Q. Rev. Biophys.* **16,** (1983).

variable. It should also be noted that the study of pressure effects ought to be done in real time. Whatever the magnitude monitored, it must be studied while the pressure is applied to the system. The only exception would be when the changes occur at such low rates that meaningful determinations could be made before the system reverts back to equilibrium at atmospheric pressure. In general, one can expect to alter the chemical equilibrium of a system only if there is a net positive or negative volume difference between the products and reactants involved in it. This volume change will be affected by an increase in pressure, which will favor the form(s) with lower final volume (Le Chatelier).

One of the magnitudes that can be used to study the Brownian rotational motion of molecules in solution is the measurement of linear polarization of fluorescence, and a number of instruments which allow its determination have been described.[5–16] All but one of them were designed to study samples at atmospheric pressure, in which case one must rely on changes in the temperature and viscosity of the sample to characterize the rotational properties of the system under study. Alternatively, isothermal increase of hydrostatic pressure provides a direct method to increase the viscosity of the system without changing the composition or the total energy of the medium. Recently a high-pressure aparatus[17] was described to allow polarization measurements under pressures of 10^{-3} to 10 kbar. To introduce the necessary corrections requires an independent determination, at the beginning and at the end of the pressure run of the absolute value of the polarization at atmospheric pressure.

The physical requirements of a high-pressure bomb for fluorescence polarization determinations are in our view simplicity of construction, an internal volume as small as possible, pressure transmitting fluid transparent to the wavelengths of excitation and emission, total isolation of the sample from the hydrostatic fluid, and polarization scrambling due to the birefringence of the windows at high pressure kept at a minimum or

[5] G. Weber, *J. Opt. Soc. Am.* **46**, 962 (1956).
[6] J. Teissier, B. Valeur, and L. Monnerie, *J. Phys. E* **8**, 700 (1975).
[7] R. J. Kelly, W. B. Dandliker, and D. E. Williamson, *Anal. Chem.* **48**, 846 (1976).
[8] J. E. Wampler and R. J. Desa, *Anal. Chem.* **46**, 563 (1974).
[9] R. F. Chen and R. L. Bowman, *Science* **12**, 729 (1965).
[10] R. D. Spencer, F. B. Toledo, B. T. Williams, and N. L. Yoss, *Clin. Chem.* **19**, 838 (1973).
[11] D. A. Deranleau, *Anal. Biochem.* **16**, 438 (1966).
[12] G. Weber and B. Bablouzian, *J. Biol. Chem.* **241**, 2558 (1966).
[13] A. Bruck, E. Sahar, E. Agmon, and M. Shinitsky, *Appl. Opt.* **16**, 564 (1977).
[14] B. Witholt and L. Brand, *Rev. Sci. Instrum.* **29**, 1271 (1968).
[15] S. Ainsworth and E. Winter, *Appl. Opt.* **3**, 371 (1964).
[16] J. P. Bentz, J. P. Beyl, G. Beinert, and G. Weill, *Eur. Polym. J.* **11**, 711 (1975).
[17] G. S. Chryssomallis, H. G. Drickamer, and G. Weber, *J. Appl. Phys.* **46**, 3084 (1978).

accounted for by a correction factor. Additionally the polarization values obtained from the use of this type of bomb should be absolute, that is, independent of determinations by another instrument. We describe here a high-pressure bomb meeting these conditions and set out the corrections required by the depolarization due to birefringence of the windows. The high-pressure bomb is part of a photon-counting fluorescence polarization instrument. An analog instrument can also be used, the only limitation being its lower sensitivity. The four ports and the external dimensions of the high-pressure bomb described allow absorption and emission determinations with a great variety of commercial spectrophotometers and spectrofluorometers.

Description of the High-Pressure Bomb

In order to meet the experimental requirements for measuring fluorescence depolarization of protein solutions already stated the high-pressure bomb has to meet several others.

First of all, the bomb is supposed to be able to withstand pressures up to at least 3 kbars. Second, it should have four optical ports and one hydrostatic port to permit its pressurization. Third, its design should allow for easy loading and unloading of samples and prevent them from making contact with either the inner metallic bomb walls or the pressure-transmitting fluid. To all these requirements we have to add the obvious one of having the smallest internal volume possible in order to increase the efficiency of the pressure-delivery pump which has a total capacity of 10 ml. If the volume of the high-pressure bomb and the rest of the setup (high-pressure lines, valves, and gauge) is too large compared to the pump volume, refilling the pump will be required during experiments. This will reduce the efficiency of the system and jeopardize any time studies under pressure.

Figure 1 illustrates the high-pressure cell and Fig. 2 two of the four window plugs. All the high-pressure seals were made using Bridgman's principle of unsupported area.[18] The cell as well as the plugs were constructed with a special steel alloy called Vascomax 300, developed by Teledyne Vasco Company, Chicago. This material can be hardened after machining to achieve Rockwell #58 to #62, thus allowing the cell dimensions to be small in comparison with the dimensions required if standard steel were to be used instead of Vascomax.

In order to monitor the cell temperature, a conduit 2 mm in diameter and 23 mm depth is drilled in the bomb body. A thermocouple inserted in

[18] P. W. Bridgman, "The Physics of High Pressure." Bell, London, 1931.

FIG. 1. Side view of the high-pressure bomb. The square holes in the window plugs are wrench fittings machined on the conical port opening. The pressure inlet is at the top of the bomb.

it will therefore follow the temperature variations of the bomb and not of the sample inside it. To be able to ascertain the sample temperature, it is necessary to wait until the whole system, cell plus sample, reaches equilibrium. However, if it is imperative to monitor the sample temperature, the fourth port can be used to introduce the required electrical connection of a thermistor.

Several experiments proved that the bomb temperature does not differ significantly from the sample temperature, provided 5 to 10 min of stabilization is allowed after each temperature change.

An exploded view of one of the window plugs is shown in Fig. 3. It shows the plug body, the three sealing rings (2 made of Everdur and one of lead), the extractor ring, the fused quartz window, and the window cap. The three rings that form the plug packing have different functions in the high-pressure sealing. The lead one seals the cell during the low stages of pressurization, and the two Everdur ones do the same at higher pressures.

Careful machining of these sealing rings is required. Their dimensions are not given explicitly because they can be deduced from the dimension

FIG. 2. High-pressure window plug assembly. The right plug is shown completely assembled with quartz window, window cap, and sealing rings. Extractor ring is not shown. The mirror-polished surface can be appreciated in the left plug.

FIG. 3. Exploded view of a window plug. From bottom: plug body, first Everdur ring, lead ring, second Everdur ring, extractor ring, fused quartz window, and window cap.

of the high-pressure bomb window parts. Tolerance in their machining should be +0.000 and −0.005 of an inch.

The window plug must be torqued down at about 14 kg · m. Therefore an extractor ring must be present in the plug assembly to allow extraction of the packing in the event of packing replacement or window cleaning.

The windows are made of fused quartz of the following characteristics: T19 Suprasil 1; birefringence, 5 nm/cm; practically fluorescence free and having optically flat surfaces, with the following dimensions: diameter, 0.50 ± 0.05 in.; thickness, 0.30 ± 0.05 in. The seal between the window and the plug port is made by mirror-polishing the metal port surface upon which the window will rest. The pressure will make the seal by pressing both surfaces against each other. The window cap gives the window mechanical support when the cell is handled at atmospheric pressure. It is advisable to use a thin washer of Teflon between the window and the window cap to prevent scratches. The polishing of the plug surface is critical because any deviation from an optically flat surface will result in leaking under pressure. Figure 2 shows the plug surface against which the window rests. The seal between the window and the plug is considered acceptable only when the plug can be lifted by holding the window. A very useful check to employ during polishing is to observe the interference rings (also known as Newton rings) that appear when the window is placed over the polished steel surface. Only when these rings are all concentric with the port is a good seal obtained.

There is another critical requirement in machining and it relates to the surfaces of the plug and of the bomb upon which the packing rings rest. These surfaces must be polished very carefully because even the smallest (<0.0001 in.) scratch will result in a leak at high pressure. Since at this stage of construction the cell is already hardened, any scratches must be removed by a careful polishing process. Packing rings in this case will have to be constructed specifically for each port.

The only maintenance (once the bomb is leak free) consists in washing the cell interior regularly to prevent accumulation of foreign materials coming through the pressure line and/or dust collected during sample changes.

Top and side views of the bomb are shown in Figs. 4 and 5. Figure 6 shows one of the window plugs and the pressure inlet plug. For simplicity, some dimensions and tolerances were omitted from these figures.

In the interior of the bomb there is enough room to accommodate a fused quartz cylindrical cell of about 1.5 ml volume and 10 mm in diameter. This cell, which is shown in Fig. 7, is attached to a brass base which prevents movement with respect to the four window plugs. The stopper of this cell is made of a polyethylene tube with one of its ends closed by

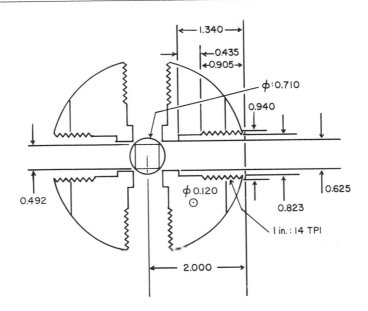

FIG. 4. Top view of the high-pressure bomb. All dimensions are in inches. Tolerances are omitted. In the critical places they are usually $+0.005$ and -0.000 in. From A. A. Paladini and G. Weber, *Rev. Sci. Instrum.* **52,** 419 (1981). Reprinted by permission of the publisher.

heating. The purpose of this tube is to allow pressure equalization between the hydrostatic medium (ethanol) and the sample inside the cell, preventing at the same time mixing of both liquids. Almost any solvent that remains transparent at the wavelength that will be used for excitation can be used. Experiments have to be performed to check the absence of leaks, to or from the inner cell. For example, a dye highly soluble in ethanol is placed in the cell and the bomb pressurized and depressurized several times. A leak is detected by the presence of the dye in the external ethanol where it is easily measured by spectrophotometry. This or any other suitable control experiment should be performed regularly to ensure proper isolation of the sample from the pressurizing fluid.

A cylindrical cuvette is not a drawback in the study of fluorescence depolarization. Being surrounded completely by alcohol its curved surface almost disappears due to the closeness of the indexes of refraction of quartz and ethanol.

Figure 8 shows the sample cell inside the pressure bomb. This diagram illustrates the fact that the aperture of the bomb is maximized by the conical shape of the plug openings; it also shows that the top plug is shaped so as to allow enough space for the inner cell top without unneces-

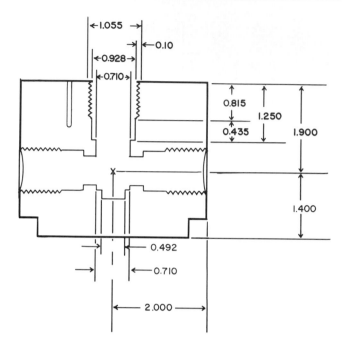

FIG. 5. Side view of the high-pressure bomb. Tolerances are omitted. In the critical places they were usually +0.005 and −0.000. From A. A. Paladini and G. Weber, *Rev. Sci. Instrum.* **52**, 419 (1981). Reprinted by permission of the publisher.

sarily increasing the bomb volume (about 10 ml). The bomb has four ports, one of them is used as the excitation port, two more ports (perpendicular to the first one) are used to observe the fluorescence emission, and the fourth permits exit of the excitation light, preventing it from bouncing back into the detection system by multiple reflections.

Figure 9 shows the high-pressure apparatus: the pump used is from High Pressure Equipment, Erie, Pennsylvania, and consists of a manually operated piston screw pump and is specially designed for use in applications where a liquid is to be compressed within a small volume in order to develop pressure. Disassembly and careful cleaning of the pump on arrival is recommended. These pumps are usually employed with oil and the slightest contamination of the lines and bomb with this material may severely reduce the transmission of the ultraviolet excitation light. At this time, it is advisable to add an extra Teflon seal to the standard set of five that comes with the pump to compensate for the lower viscosity of the ethanol as compared with oil.

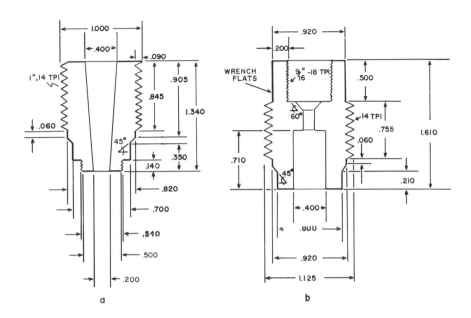

FIG. 6. Side view diagrams of the window port (a) and of the pressure inlet port (b). Dimensions in inches. Tolerances are omitted. In the critical places they were usually +0.005 and −0.000. From A. A. Paladini and G. Weber, *Rev. Sci. Instrum.* **52**, 419 (1981). Reprinted by permission of the publisher.

The gauge (Bourdon type) can be bought from Olson Engineering and Sales Corporation, Burlington, Chicago, and should have a range of operation that matches that of the pump.

The high-pressure line arrangement shown in Fig. 9 is such that the pump can easily be taken out of the bomb pressure line for refilling, emergency repairs during experiments, or just to release stress.

Provisions must be made to be able to connect the gauge either to the bomb or to the pump by means of two valves. In this way, pressure shocks to the bomb are avoided when, for example, pump refilling is necessary in the middle of a run.

Pipes, valves, and connectors can be bought from the same company that builds the pump. It is advisable to buy the coning and threading tools, which allow assembling of the high-pressure line at the laboratory bench.

Although the pressures involved in this type of research are relatively high, the whole procedure is reasonably safe mainly because the pressure is obtained by fluid compression. However, careful treatment of the lines

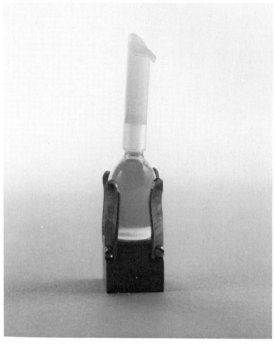

FIG. 7. Inner pressure cell and brass holder. The polyethylene tubing is about 25 mm long; the cylindrical quartz cell is 10 mm in diameter and 25 mm in height; the brass holder is a square block of 12 mm wide and 10 mm high.

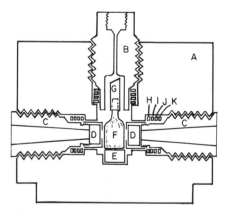

FIG. 8. Side view of high-pressure bomb (A) with the inner cell (F), its holder (E), blackened window ports (C), and the pressure inlet (B). Sealing rings I, J, K, and extractor H are shown making no contact with the walls and ports to facilitate visualization. D, quartz windows; G, polyethylene flexible stopper. From A. A. Paladini and G. Weber, *Rev. Sci. Instrum.* **52**, 419 (1981). Reprinted by permission of the publisher.

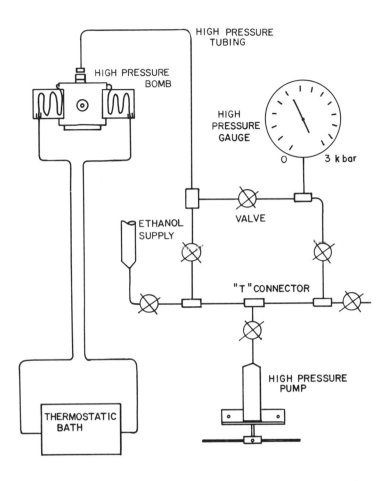

Fig. 9. Diagram of the high-pressure lines and temperature control.

and valves is advised. It should be pointed out finally that when bending lines at right angles a radius of less than 50 mm should not be used.

A thermostatic bath, Model LT-50 from Neslab Instruments, Portsmouth, New Hampshire, can be used to vary and hold the temperature of the bomb between +30 and −40° within less than half a degree of temperature variation. The thermostatic fluid, methanol, in our system, is circulated through four steel plates with copper tubing soldered to them. These plates are attached to the bomb body. The efficiency in introducing or removing heat is not high (it takes almost 1 hr to achieve a 10° change), but a faster and therefore more efficient system requires a larger bomb size to accommodate the fluid tubing within the bomb body. A larger bomb size would not fit well into the polarization instrument, and a compromise solution must be adopted. The construction of a water jacket with four ports to improve the thermal exchange can be recommended, if there are not stringent space requirements.

The bomb base is machined square to permit its clamping with a vise for easy removal of the window plugs. A Teflon plate is fixed to this base to minimize thermal contact between the cell and the polarization instrument. A precision thermistor, Model YSI 44008 with an absolute error of ±0.2 deg, can be used to monitor the cell temperature.

Handling, Loading, and Servicing the High-Pressure Bomb

When filling the bomb with pure ethanol care must be taken to prevent trapping air bubbles. This air, if allowed to stay inside the bomb, will reduce the pumping efficiency and may introduce systematic errors, upon decompression, by shaking the inner cell. In order to avoid them the bomb is placed in the normal horizontal position, filled with alcohol and then tilted and shaken several times until no more air bubbles are seen. The inner cell is then introduced and the top plug secured.

The filling of the inner cell has to be done with the same precautions described for the bomb. This is a critical condition: any residual air within the cell will dissolve with increasing pressure but will form bubbles on decompression thereby giving rise to a large scattering signal.

To achieve this, a technique was developed which requires a certain amount of practice: first, the cell is completely filled using a syringe until a convex meniscus is formed; then the tube that acts as the stopper is also completely filled while partially compressed with two fingers; then cell and stopper are brought together simultaneously while releasing the finger pressure. In this way it is possible to have the sample completely confined in the cell and the stopper, without any air bubbles being trapped.

The cell is then externally rinsed with ethanol and introduced into the bomb. The pressure inlet plug is torqued down and the bomb is ready for use.

No maintenance problems are encountered with this type of bomb; the windows are usually cleaned with a Q-tip moist with ethanol.

Fluorescence Polarization Instrument

A full description of the instrument used has been published by Jameson et al.[19] The only major modification described by the authors was the use of commercially available photon-counting electronics. Our design

---

[19] D. M. Jameson, G. Weber, R. D. Spencer, and G. Mitchell, *Rev. Sci. Instrum.* **49,** 510 (1978).

uses a nonlinear dynode chain with higher voltages in the latter stages of the PMT base (EMI 6256S). Capacity coupling to a fast amplifier discriminator (Ortec Model 9302) is used. This model was selected due to its wide band, high gain amplifier, and integral discriminator capable of counting rates up to 100 MHz. Typically this unit has a rise time of 3 nsec and with the gain used (20×) it has a linear response for input voltages of up to 50 mV. The discriminator output is connected to an Ortec Photon Counter Model 9315. This model features two counters, thus allowing the detection of the parallel and the perpendicular signals simultaneously. Each counter has a range of 10^7 counts. It is also possible to use one of them as a timer, counting input impulses from a 100 kHz internal crystal-controlled clock with a time base switch selectable for increments of 1 msec or 0.1 min. This counter can also be reset and will stop a counting interval in both sections when it reaches the desired time or number of input pulses.

Polarizer Alignment

There are several ways that can be used to align the polarizers. The one that we found more practical involved the use of a solution of glycogen in Millipore-filtered water. The Rayleigh scattering of this solution is used to set the polarizers with transition directions vertical and horizontal with respect to the laboratory reference axis, x, y, z. Let us assume that the direction of propagation of the excitation light (k) is from minus infinity toward plus infinity along the x axis. The sample cuvette is placed at the intersection of the $x, y,$ and z axes. The plane of polarization of the excitation light is said to be vertical (V) when the direction of the electric vector is parallel to the z axis and horizontal (H) when parallel to the y axis. The emission polarizers simultaneously analyze the components of the emitted light, along the y axis, in two planes defined as parallel when the transmission direction of the polarizer is set parallel to the z axis ($\parallel$) and perpendicular ($\perp$) when that direction is set parallel to the x axis.

The first polarizer set is the excitation polarizer. Its vertical position is obtained in the following way: looking down into the cuvette, with the glycogen solution, the intensity of the light scattered upward is minimized by rotating the polarizer in an appropriate polarizer holder. When this is achieved the polarizer is sufficiently close to the vertical position. The polarizer holder was designed to allow exact 90° rotations[20]; thus the horizontal position of the polarizer is already set once the vertical one is fixed by the method described.

To give a reasonable accurate polarizer adjustment the glycogen sample should fulfill the following requirements: no appreciable turbidity

[20] D. M. Jameson, PhD dissertation, University of Illinois at Urbana-Champaign, 1978.

should be present on looking through the solution, and enough scattering should be present to be able to see, with the naked eye, the changes in intensity upon rotation of the polarizer. For this purpose it is also convenient to reduce the environmental lighting to a minimum and to wait until the eye reaches good adaptation to darkness. The excitation wavelength used is 500 nm.

Once the excitation polarizer is set, either the parallel emission polarizer or the perpendicular one can be set by the use of a similar procedure. The photomultipliers are used to determine the directions of preferential transmission of the polarizers. In order to do so, the excitation polarizer is set to its vertical position and the emission polarizer to be positioned is rotated until a minimum number of counts is obtained from the photomultiplier. This means that the transmission direction for the polarizer is close to the position called perpendicular. This is a minimum and not exactly zero because a small number of counts is always given by the light contribution from the parallel emission (the Rayleigh scattering from the glycogen solution is not 100% polarized). This fact prevents a sharp definition of the perpendicular position in a single step. It is necessary to study the polarization measured when the emission polarizer under study is rotated from the previous position to the parallel one. The ratio of the number of counts obtained in this condition must be optimized in order to obtain the highest value of polarization by careful setting of the emission polarizer.

The last step in adjusting the polarizers is achieved by removing the photomultipliers and placing a source of light in front of the emission polarizer just calibrated. In this way a polarized beam is obtained and since its plane of polarization is known, it is very simple to set the other emission polarizer by rotating it until no light transmission is seen by the eye. This position defines the plane of polarization of the second polarizer with respect to the first.

We have outlined the procedure used in the alignment of the polarizers. This method requires several fine adjustments of all the polarizers in order to achieve the state in which the glycogen solution will show a measured polarization of 99.96% and fluorescein in aqueous buffer at 22° will give a polarization of $2 \times 10^{-3} \pm 4\%$ when excited at 350 nm. Other procedures are also used for the testing of the polarization instrument. They are described in the next section.

Fluorescence Polarization Standards

A solution of glycogen in deionized water maintained at room temperature (25°) with excitation at 500 nm can be used to calibrate the polarization instrument, proceeding as previously described. A value of 99.32 ± 0.01% polarization should be obtained with the solution in a standard (10-mm optical path) fluorescence cuvette. This value agrees well with the one reported by Jameson[20] for his own apparatus.

When the high-pressure bomb is employed with glycogen solution inside the round cell normally used in the high-pressure runs, a percentage polarization of 98.22 ± 0.01 is obtained. Considering the fact that the system is optically very different from the one in the previous experiment, the approximate 1% deviation may be neglected.

Fluorescein from J. T. Baker (Kromex) at about 10^{-6} M in 0.01 M Na phosphate buffer may also be studied at room temperature. When excited at 485 nm a polarization value of 0.022 ± 0.002 is obtained. This measurement is done with the standard sample holder[21] and using emission filters (3–69) from Corning. When the sample is excited at 350 nm under the same conditions an essentially zero polarization value is obtained.

Another fluorescein solution, this time 10^{-6} M in glycerol, is measured inside the bomb at different pressures and at room temperature. Upon excitation at 485 nm and at atmospheric pressure a value of 0.453 ± 0.003 is obtained.

Several other compounds have been tested, and some actual comparisons of the same sample have been made using a similar polarization instrument. In all cases very good agreement was obtained, giving us confidence in the polarization determinations obtained while having the samples inside the high-pressure bomb.

High-Pressure Bomb Window Birefringence Correction

The effect of high pressure on matter has been studied in a large number of experiments in many disciplines of science. The information obtained from them is important and in many cases indispensable for our understanding of the properties of matter.

The recent availability of high-pressure apparatus and the techniques developed for making measurements on the sample, e.g., spectroscopic measurements, while the sample is subjected to pressure have been largely responsible for the increased interest in this important field.

Considerable progress has been made in spectroscopic techniques at high pressures since the first measurements of Drickamer. Without pretending to fully cover this particular area of research we can mention

---

[21] E. Fishman and H. G. Drickamer, *Anal. Chem.* **28**, 804 (1956).

important works by Fishman and Drickamer,[21] Drickamer,[22] and Drickamer and Balcham.[23]

In this type of instrumentation the optical properties of the windows used in the high-pressure bomb are among the most important considerations. The windows must be transparent to the electromagnetic radiation of interest while having sufficient strength to withstand the high pressures. Many crystalline substances (i.e., solids whose atoms are arranged in some sort of regular repetitive array) are optically anisotropic.

A typical example of this property is the birefringence of calcium carbonate (calcite). The obvious requirement for high-pressure bomb windows is a material with good transmittance in the visible and ultraviolet regions, low birefringence, and high tensile strength.

It was noted by Campbell et al.[24] that windows used in high-pressure bombs show a pressure-dependent scrambling of polarization apparently as a result of strain-induced anisotropies in the window material (photoelastic effect). This is particularly important here because when studying polarization changes induced in samples by pressure any scrambling of the emitted polarized light produced by the windows should be kept at a minimum.

According to Cantor et al.[25] fused quartz (Ultrasil T-16, supplied by Karl Lambrecht Corp, Chicago) shows very low polarization scrambling between 0 and 2 kbars.

Saphire windows are the most common choice for high-pressure experiments due to their tensile strength and transparency in the UV region. However, these two advantages are insufficient since there is a much higher percentage of polarization scrambling under identical conditions of size and pressure than in fused quartz. Our final choice was F19 Suprasil fused quartz.

Several experiments were performed to check the magnitude of window scrambling employing samples that were not supposed to change the state of polarization of the emitted light between 0 and 2 kbars. For this purpose we selected a fluorophore like fluorescein, which under these conditions should only reflect changes in viscosity of the solvent due to the applied pressure.

The results obtained with fluorescein in glycerol with the system kept

[22] H. G. Drickamer, in "Progress in Very High Pressure Research" (F. P. Bundy, W. R. Hibbard, and H. M. Strong, eds.). Wiley, New York, 1961.
[23] H. G. Drickamer and A. S. Balcham, in "Modern Very High Pressure Techniques" (R. N. Wentorf, ed.). Butterworth, London, 1962.
[24] J. H. Campbell, J. Fisher, and J. Jonas, J. Chem. Phys. 61, 346, (1974).
[25] D. M. Cantor, J. Schroeder, and J. Jonas, Appl. Spectrosc. 29, 393 (1975).

at $-24°$ showed that at 2 kbars the window scrambling resulted in a 30% decrease of the polarization value observed at 1 bar. Moreover, the scrambling is not a linear function of the applied pressure. Therefore, a pressure-dependent window birefringence correction factor must be applied to obtain meaningful results.

The following treatment will show the theoretical approach used in the calculation of a window scrambling factor called α, the means for measuring it, and the changes in the "T" format of the polarization instrument that were prompted by the observations and the theory.

"T" Format Calculations

When the polarization photometer is used in the "T" format, measurements are made by determining the ratio between parallel and perpendicular components excited by vertically polarized light $(S_\parallel^V/S_\perp^V)$ and horizontally polarized light $(S_\parallel^H/S_\perp^H)$. Let

$$r = (S_\parallel^V/S_\perp^V)/(S_\parallel^H/S_\perp^H) \tag{1}$$

The S intensity values refer to the components of the fluorescence after passing through a depolarizing observation window characterized by a scrambling coefficient α. Before passing through this emission window the components of the fluorescence excited by light which is originally vertically polarized, but only incompletely polarized after passing through the excitation window, will be denoted $I_z^V, I_x^V, I_z^H, I_x^H$ in accordance with the scheme of Fig. 10.

The polarization experimentally measured (through the scrambling windows), p', will be given by

$$p' = \frac{r - 1}{r + 1} \tag{2}$$

The values of S and I are related through the following equations:

$$S_\parallel^V = I_z^V(1 - \alpha) + I_x^V\alpha \tag{3}$$
$$S_\perp^V = I_z^V + I_x^V(1 - \alpha) \tag{4}$$
$$S_\parallel^H = I_z^H(1 - \alpha) + I_x^H\alpha \tag{5}$$
$$S_\perp^H = I_z^H + I_x^H(1 - \alpha) \tag{6}$$

In turn $I_z^V \ldots I_x^H$ have been generated by excitation with partially polarized light originating by passage of perfectly polarized light through a window with scrambling coefficient ε. If we designate as $I_\parallel$ and $I_\perp$ the components of the fluorescent light with electric vector, respectively, parallel and perpendicular to the incidence exciting radiation, and if this

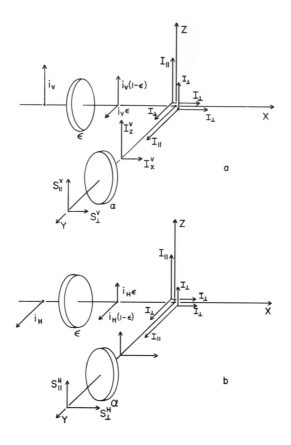

Fɪɢ. 10. Diagrams of the effect of window birefringence upon the intensity of a completely polarized beam of light. For simplicity only 2 windows are drawn. Vectors are not drawn to scale. ε and α are the scrambling coefficients of the excitation and emission windows respectively. (a) Case in which excitation is polarized vertically. (b) Case in which excitation is polarized horizontally. From A. A. Paladini and G. Weber, *Rev. Sci. Instrum.* **52**, 419 (1981). Reprinted by permission of the publisher.

radiation is polarized vertically, that is, in the $\overrightarrow{OZ}$ direction, the following relations can be written:

$$I_z^V = I_{\parallel} \qquad \text{and} \qquad I_x^V = I_y^V = I_{\perp}$$

Similarly if the excitation is polarized horizontally, that is, in the $\overrightarrow{OY}$ direction,

$$I_z^H = I_x^H = I_{\perp} \qquad \text{and} \qquad I_y^H = I_{\parallel}$$

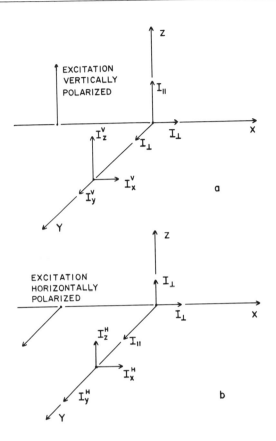

FIG. 11. Limiting cases for excitation polarized vertically (a) and horizontally (b). From A. A. Paladini and G. Weber, *Rev. Sci. Instrum.* **52**, 419 (1981). Reprinted by permission of the publisher.

These relations are shown in Fig. 10a and b.

On vertical excitation through a window with scrambling coefficient ε, as shown in Fig. 10, the values of I_z^V and I_x^V are intermediate between those shown in Fig. 11a and b, namely,

$$I_z^V = (1 - \varepsilon)I_\parallel + \varepsilon I_\perp \tag{7}$$

$$I_x^V = \varepsilon I_\perp + (1 - \varepsilon)I_\perp = I_\perp \tag{8}$$

Similarly, for horizontally polarized excitation passing through the excitation window:

$$I_z^H = I_\perp(1 - \varepsilon) + \varepsilon I_\| \tag{9}$$
$$I_x^H = I_\perp(1 - \varepsilon) + I_\perp \varepsilon = I_\perp \tag{10}$$

If the values of I_z^V, I_x^V, I_z^H, I_x^H given in Eqs. (7) through (10) are introduced into Eqs. (3) through (6) we can obtain by means of Eq. (1) an expression for the observed polarization of the fluorescence as a function of the components $I_\perp$ and $I_\|$ that would be excited by perfectly polarized light and the distorting effects of the excitation and emission windows, described by ε and α, respectively. Proceeding in this way we find

$$S_\|^V = I_\|\varepsilon(1 - \alpha) + I_\perp[1 - \varepsilon(1 - \alpha)] \tag{11}$$
$$S_\perp^V = I_\|(1 - \varepsilon)\alpha + I_\perp[1 - \alpha(1 - \varepsilon)] \tag{12}$$
$$S_\|^H = I_\|(1 - \alpha)\varepsilon + I_\perp[1 - \varepsilon(1 - \alpha)] \tag{13}$$
$$S_\perp^H = I_\|\varepsilon\alpha + I_\perp(1 - \alpha\varepsilon) \tag{14}$$

and therefore

$$r = \frac{S_\|^V S_\perp^H}{S_\perp^V S_\|^H} = \frac{\begin{aligned}I_\|^2(\varepsilon\alpha - x) + I_\perp^2(\alpha - \varepsilon - \alpha\varepsilon - x)\\ + I_\| I_\perp(1 - \varepsilon - \alpha + 2x)\end{aligned}}{\begin{aligned}I_\|^2(\varepsilon\alpha - x) + I_\perp^2(1 - \alpha - \varepsilon + 3\alpha\varepsilon + x)\\ + I_\| I_\perp(\varepsilon + \alpha - 3\alpha\varepsilon)\end{aligned}} \tag{15}$$

where

$$x \equiv \alpha\varepsilon^2 + \alpha^2\varepsilon - \alpha^2\varepsilon^2$$

If we now use the definition of the polarization (p) that would be measured under ideal conditions, that is, without scrambling windows or at atmospheric pressure, namely,

$$\frac{I_\|}{I_\perp} = \frac{1 + p}{1 - p} \tag{16}$$

and Eq. (2), $p' = (r - 1)/(r + 1)$, we can rewrite Eq. (15) as

$$p' = \left[p(1 - 2\alpha - 3\varepsilon + 2x) + \alpha\varepsilon\left(\frac{7p - 1}{2}\right) \right] \Big/$$
$$\left(1 + \frac{\alpha\varepsilon}{2}\left(\frac{1 + 7p^2}{1 - p}\right) - 2xp\left(\frac{1 + p}{1 - p}\right)\right] \tag{17}$$

And since α and ε are small numbers the x term can be assumed to be zero; therefore (17) can be written as

$$p' = \left[p(1 - 2\alpha - 2\varepsilon) + \alpha\varepsilon\left(\frac{7p - 1}{2}\right) \right] \Big/ \left(1 + \frac{\alpha\varepsilon}{2}\left(\frac{7p^2 + 1}{1 - p}\right) \right) \quad (18)$$

Finally, assuming that α and ε are the same and neglecting the terms having the squares of the scrambling coefficients, we can write

$$p' = p(1 - 4\alpha) \quad (19)$$

This equation permits calculation of the true polarization values of samples measured under scrambling conditions induced by pressure, within the approximations already stated and when the "T" format is used.

"L" Format Calculations

When the polarization photometer is used in the "L" format, the measurements are made by determining the ratio between parallel and perpendicular components excited by vertically polarized light ($S_\parallel^V / S_\perp^V$) and calculating the polarization value (p') by using the following equation:

$$p' = \frac{(S_\parallel^V / S_\perp^V) - 1}{(S_\parallel^V / S_\perp^V) + 1} \quad (20)$$

The S values can be replaced in terms of $I_\parallel$ and $I_\perp$ using Eqs. (11) and (12), obtaining

$$p' = \frac{(I_\parallel - I_\perp)(1 - \varepsilon - 2\alpha + 2\varepsilon\alpha)}{(I_\parallel + I_\perp)(1 - \varepsilon)} \quad (21)$$

and working in the same way as in the "T" format deduction we can write p' in terms of p by use of Eq. (16):

$$p' = p\,\frac{(1 - \varepsilon - 2\alpha + 2\varepsilon\alpha)}{1 - \varepsilon p} \quad (22)$$

And assuming $\alpha \equiv \varepsilon$ and small we can simplify (22) to

$$p' = \frac{p(1 - 3\alpha)}{1 - \alpha p} \quad (23)$$

This equation permits calculation of the true polarization values (p) of samples measured under scrambling conditions induced by pressure, within the approximations already stated and when the "L" format is used. The "L" format was used to measure the polarizations, and these were correlated by means of Eq. (23), because the intensities $S_{\parallel}^V$ and $S_{\perp}^V$ are measured through the same emission window in the "L" format. In the "T" format each component is measured through a different emission window and therefore any difference in the scrambling coefficient between these windows will show up in the corrected polarization value as a systematic error hard to correct for. Notice also that the depolarizing effects of the windows in the "T" format have a weight of $4/(3 - p)$ greater that in the "L" format.

If the polarization of a pressure-independent sample like fluorescein dissolved in glycerol is known at atmospheric pressure ($\alpha \simeq 0$) and at any given pressure, then the correction factors for the windows can be calculated using Eq. (19) in the "T" format or Eq. (23) in the "L" format.

Azimuthal Dependence of the Birefringency Effect

The theoretical approach sketched in the last section assumed that when subjected to pressure the windows are uniformly affected on their surface and consequently that there is no azimuthal dependency on this effect. In other words, one assumes the absence of preferential polarization directions along which the depolarizing effects would be reinforced or minimized.

To check for the effects of pressure upon the homogeneity of the windows, the polarization instrument was modified so that a beam of polarized light could pass through the emission windows and be analyzed by means of a rotating polarizer. The azimuthal dependency was studied by analyzing the polarized intensity distribution transmitted for several polarization planes of the first polarizer. As expected, the $\cos^2 \theta$ law (θ being the angle determined between the planes of transmission of the polarizers) was observed, within the experimental error, at atmospheric pressure. Deviations from it were found to be present symmetrically distributed about θ when 2 kbars of pressure were applied to the windows. The intensity distribution was studied for various angles of the plane of polarization of the excitation polarizer, and lack of azimuthal symmetry was found to be less than 10% when 2 kbars of pressure were applied to the windows. However, when a sample is studied under real conditions it is placed between the emission windows and therefore their error contribution to the experimental polarization obtained using the "T" format will be much smaller than 10% (probably less than 5%). The use of the "L" format obviates the need to take this error into consideration.

Viscosity Considerations

Changes in the viscosity of the solvent induced by pressure will result in polarization changes.[26] In order to calculate the scrambling coefficients it is necessary to estimate the polarization changes produced by the increase in viscosity with pressure.

Let us call p_b the polarization of fluorescein measured at 1 bar with solvent viscosity η, p_b' the polarization value that would be measured if no window scrambling were present, p_s the polarization values if only scrambling but no viscosity changes were present, and finally p_{obs} the polarization experimentally observed with all the perturbating factors (viscosity + scrambling) present.

Using Perrin's equation[27]:

$$\left| \frac{1}{p} - \frac{1}{3} \right| = \left| \frac{1}{p_0} - \frac{1}{3} \right| \left| 1 + \frac{RT}{\eta V} \tau \right|$$

where p_0 is the limiting polarization, that is, the polarization in the absence of all extrinsic causes of depolarization, R the gas constant, T the absolute temperature, τ the lifetime of the excited state, η the viscosity of the solution, and V the molar volume of the molecule supposed spherical. This equation permits calculation of the polarization that would be observed at any given pressure if the viscosity of the solvent at this pressure is known, under the additional assumption that p_0, τ, and V are pressure independent. The experimental polarization p' will differ from the calculated value, p, because of the photoelastic birefringence of the windows and under the assumption $\alpha = \varepsilon$, this coefficient can be calculated by Eq. (19) or (23). For the reasons sketched above we chose the "L" over the "T" format and from experiments of the type described the value of α was determined for the range of pressures up to 2 kbars.

Figure 12 shows a plot of α versus pressure. To facilitate further comparison a least-squares fitting of the birefringence coefficient was done with a polynomial of the third degree ($\alpha = b_0 + b_1 X + b_2 X^2 + b_3 X^3$). Best fitting was obtained for the following parameters: $b_0 = -5.32 \times 10^{-4}$, $b_1 = -1.25 \times 10^{-2}$, $b_2 = 3.48 \times 10^{-2}$, $b_3 = -5.26 \times 10^{-3}$, X = pressure in kbar.

[26] F. Perrin, *J. Phys.* **7**, 390 (1926).
[27] F. Perrin, *Ann. Phys.* **12**, 169 (1929).

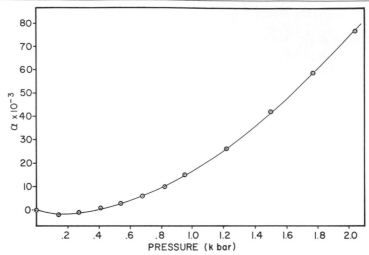

FIG. 12. Scrambling coefficient (α) as a function of the hydrostatic pressure for the "L" format configuration only. From A. A. Paladini and G. Weber, *Rev. Sci. Instrum.* **52,** 419 (1981). Reprinted by permission of the publisher.

Experimental Determinations of Polarization

The fluorescence polarization of 6-propionyl-2-dimethylaminonaphthalene (PRODAN)[28] in isobutanol was measured as a function of temperature and pressure in two independent experiments. Perrin plots, shown in Fig. 13, are in close agreement despite the fact that two different ways of varying the viscosity of the solvent were used.

Figure 14 shows the excitation polarization spectra of PRODAN in glycerol. The corrected and uncorrected data are shown as well as the data obtained at 1 bar. Disagreement is restricted to those wavelengths where there is overlap of strong and weak absorption transitions. Pressure could conceivably have the effect of changing their relative contributions, resulting in intrinsic changes in p_0 with pressure which are unrelated to depolarization by the windows.

Finally, the excitation polarization spectrum of indole in propylene glycol at room temperature was measured at atmospheric pressure and at 2 kbars. No substantial changes were observed under pressure and excellent agreement with the polarization spectrum of indole measured by Valeur and Weber[29] was obtained. In all cases in which we could carry out valid comparisons between corrected and uncorrected values the α fac-

[28] G. Weber and F. Farris, *Biochemistry* **18,** 3075 (1979).
[29] B. Valeur and G. Weber, *Photochem. Photobiol.* **25,** 441 (1977).

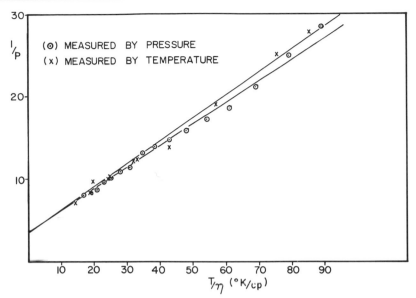

FIG. 13. Perrin plots for PRODAN in isobutanol (Lot DPK, Mallinkrodt, spectral grade) measured by pressure and by temperature. Excitation bandwidth: 0.9 nm. Emission filters Corning 3-74. Pressure data obtained with instrument in "L" format; temperature data obtained in a different photon-counting instrument using "T" format configuration. From A. A. Paladini and G. Weber, *Rev. Sci. Instrum.* **52**, 419 (1981). Reprinted by permission of the publisher.

tors independently determined led to similar agreement as in the cases shown here.

The same correction factors were assumed to apply independently of the wavelengths of excitation and emission, an assumption justified by the results obtained with fluorescein (excitation at 480 nm, emission at 520 nm) and indole (excitation at 280 nm, emission at 320–380 nm). We do not expect further studies of the wavelength dependence to introduce significant changes in the results already obtained.

Final Remarks

The high-pressure bomb as well as the inner sample cell have proved to be both practical and reliable for measurement of spectroscopic changes in biological systems. It is practical because its dimensions, both internal and external, permit its use in several instruments without major changes. The instruments include a spectrophotometer, Beckman Model ACTA MVI, an analog spectrofluorometer, a photon-counting polarization instrument, and a cross-correlation fluorometer. It was also reliable

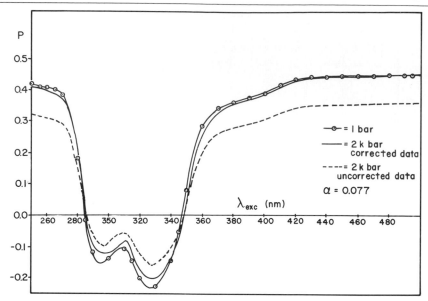

FIG. 14. Fluorescence polarization spectrum of fluorescein in glycerol, temperature 24°, measured at 1 bar, measured at 2 kbars, uncorrected data, and corrected data at 2 kbars. Resolution used was 4 nm from 250 to 310 and 1.7 from 310 to 490 nm. From A. A. Paladini and G. Weber, *Rev. Sci. Instrum.* **52**, 419 (1981). Reprinted by permission of the publisher.

since more than 100 different experiments were performed in less than a year without a single failure due to leaks. This reliability is based on the high degree of precision of the machining of the window plugs and in the fact that loading the samples does not require removal of more than one plug, therefore reducing sealing ring wear. Freedom of contamination from the pressure transmitting fluid and low volume samples are also convenient features of the high-pressure bomb design.

It has been successfully used in measurements of the dissociation of enolase,[30] binding of ethididium bromide to tRNA,[31] and in studies of the effects of pressure on lipid bilayers and natural membranes.[32]

Acknowledgment

The author wishes to thank Professor Gregorio Weber for his guidance during the completion of the work described in this chapter, as well as for critical reading of the manuscript.

[30] A. A. Paladini and G. Weber, *Biochemistry* **20**, 2587 (1981).
[31] P. M. Torgerson, H. G. Drickamer, and G. Weber, *Biochemistry* **19**, 3957 (1980).
[32] L. Parkson, G. Chong, and A. R. Cossins, *Biochemistry* **22**, 409 (1983).

[22] Stopped-Flow Circular Dichroism

By MOLLIE PFLUMM, JEREMY LUCHINS, and SHERMAN BEYCHOK*

Introduction

Stopped-flow circular dichroism (SFCD) is the adaptation of stopped-flow mixing techniques to the measurement of circular dichroism. As a static method for measuring optical activity in absorbing regions of molecules in equilibrium states, CD is widely used in many fields of chemistry. Its applications in biochemistry have been extensively reviewed.[1-5]

Present limitations in the instruments employed to produce rapid mixing, together with those inherent in measurements of CD, combine to restrict SFCD to reactions occurring in the millisecond range or slower. However, within that limitation, the technique is in principle applicable to any reaction sequence involving changes in the CD spectrum down to the millisecond time scale, and may help to elucidate the reaction mechanism.

In this chapter we will discuss the use of SFCD in studying protein conformational changes and, in particular, its application in the investigation of the mechanism of protein folding and assembly. The CD in the far-ultraviolet peptide-absorbing regions (200–240 nm) is, for the most part, due to the secondary structure of the polypeptide chain, so that the formation of such structures in folding reactions can be isolated, independent of other levels of structure. Thus, while time-dependent changes in optical activity of various side chains or prosthetic groups (in the near-ultraviolet and visible regions) may reveal specific aspects of the folding process, far-ultraviolet SFCD is potentially a more powerful tool in that it will almost always yield a directly interpretable result, namely the time-course of secondary structure formation. A multiprobe analysis of a folding reaction, which includes SFCD together with various probes of tertiary structure, should therefore be of decisive help in unraveling the details of protein folding.

* Deceased.

[1] S. Beychok, *Science* **154,** 1288 (1966).

[2] S. Beychok, *in* "Poly-α-Amino Acids" (G. D. Fasman, ed.), p. 293. Dekker, New York, 1967.

[3] J. A. Schellman, *Chem. Rev.* **75,** 323 (1975).

[4] I. Tinoco, Jr. and C. R. Cantor, *Methods Biochem. Anal.* **18,** 81 (1970).

[5] D. Sears and S. Beychok, *in* "Physical Principles and Techniques of Protein Chemistry" (S. J. Leach, ed.), Vol. C. Academic Press, New York, 1973.

Statement of the Protein Folding Problem

The question of how proteins fold to achieve their native three-dimensional structure in accessible time (seconds to minutes) remains one of the great challenges of molecular biology. Assuming that the sequence of amino acids contains the information necessary to dictate the final structure, a direct but complex statement of the folding problem is this: given a sequence, is it possible to predict the three-dimensional structure? The answer, at present, is no, but many important approaches to this question are being taken. An excellent review of the experimental aspects was prepared by Kim and Baldwin[6] and several stimulating papers have been collected together in a recent symposium volume.[7]

It is now widely believed that for proteins of about 100 or more residues, the final folded state is not achieved by a search through all accessible conformational space for the minimum conformational free energy state. Even if only a small number of ϕ, ψ angles is allowed for each residue, the time for such a search appears to be prohibitively long, as Levinthal first noted.[8] This suggests that the folding process is under kinetic control, and points to the need for kinetic methods to elucidate the mechanism. (See Wetlaufer[9-11] and Wetlaufer and Ristow[12] for a general discussion of this problem.)

In discussing protein folding mechanisms, Kim and Baldwin[6] noted that two current models that are useful at this stage are framework and domain folding. The distinction between these is as follows. In the framework model for folding, the hydrogen bonded structure is formed early in the process and the tertiary structure of the molecule is formed by specific interactions of a more or less complete backbone structure. In the domain folding mechanism, one part of the molecule acquires its structure completely while other parts of the molecule may still be folding.

At least in the case of predominantly α-helical proteins, SFCD in the far-UV should be able to make distinctions between these two models. In the last section of this chapter we shall illustrate this in the case of the refolding of hemoglobin (Hb) and myoglobin (Mb) from their unfolded states in 6 M guanidinium chloride (GuHCl), and compare these results with kinetics of refolding from partially unfolded states.

[6] P. Kim and R. L. Baldwin, *Annu. Rev. Biochem.* **51,** 459 (1982).

[7] D. Wetlaufer, "The Protein Folding Problem." Westview Press, Boulder, 1982.

[8] C. Levinthal, *J. Chim. Phys.* **65,** 44 (1968).

[9] D. B. Wetlaufer, *Proc. Natl. Acad. Sci. U.S.A.* **70,** 697 (1973).

[10] D. B. Wetlaufer, *in* "Protein Folding" (R. Jaenicke, ed.). Elsevier, Amsterdam, 1980.

[11] D. B. Wetlaufer, *Adv. Prot. Chem.* **34,** 61 (1981).

[12] D. B. Wetlaufer and S. Ristow, *Annu. Rev. Biochem.* **42,** 135 (1973).

Instrumentation

SFCD in the visible and near-UV was introduced by Bayley and Anson.[13,14] Somewhat later, Nitta et al.[15] reported the application of SFCD for following changes in ellipticity at 270 nm of aromatic amino acid residues.

The first instrument with far-UV capability for measuring polypeptide amide transitions in secondary structure changes was devised by Luchins and Beychok[16] who reported results on acid denaturation of Hb, on refolding of heme-free α chains (α^0) by pH-jump and on refolding of α^0 and β^0 following chain reassociation.

Figures 1 and 2 show, respectively, a denaturation and renaturation reaction, the former with a half-time of 48 msec, the latter, with a half-time of 5.0 sec. These measurements were made on a prototype instrument, using a stabilized 450 W Xe light source and the optics/fluid/electronics subsystems described elsewhere.[16,17]

Our present instrument, substantially advanced over the prototype, is shown schematically in Fig. 3 and is described below.

Functional Description

Optical System. The temperature-stabilized deuterium lamp (LS) is driven by an ultrastable power supply. The resulting low-noise light is focused (LEO) into the high-throughput monochromator (MONO). The monochromatic exit light is collimated (A), optically filtered (F), and then passed through the linear polarizer (P). The monochromatic, linearly polarized beam is refocused (R) and converted into circularly polarized light modulated at 50 kHz (MOD) between left and right circular polarization. The beam traverses the sample in the cuvette (SFC) and impinges on the detector (PMT).

Sample–Fluid System. The drive syringes are filled with reactants via the sample–fluid interface (FI). After valving readjustment, drive actuation can force selected volume ratios of the thermostatted reactant fluids from the drive syringes back through FI. FI valving has been set to direct the reactants through separate channels to the narrow end of FI, where the reactants are mixed homogeneously. From there, the reaction solution is driven into SFC, displacing previously spent reaction fluid which is

[13] P. M. Bayley and M. Anson, *Biopolymers* **13**, 401 (1974).
[14] P. M. Bayley and M. Anson, *Biochem. Biophys. Res. Commun.* **62**, 717 (1975).
[15] K. Nitta, T. Segawa, K. Kuwajima, and S. Sugori, *Biopolymers* **16**, 703 (1977).
[16] J. Luchins and S. Beychok, *Science* **199**, 425 (1978).
[17] J. Luchins, Ph.D. Thesis, Columbia University (1977).

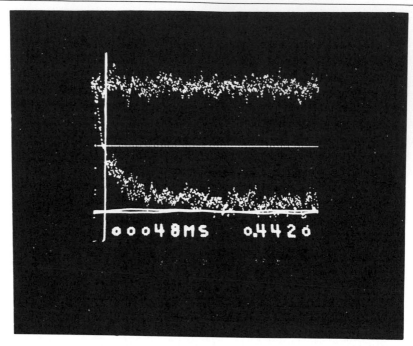

FIG. 1. Acid denaturation of methemoglobin, monitored at 222 nm by SFCD. The protein in water, $3.6 \times 10^{-5} M$ in heme, was reacted by SF mixing with an equal volume of HCl solution, pH 1.3, yielding a reaction solution of final pH 1.6. The upper horizontal trace is the dilution CD value (protein in water, $1.8 \times 10^{-5} M$ in heme; the dilution value being that expected from a one-to-one dilution of reactant protein solution, with no secondary structure change); the middle horizontal line is an electronic screen center marker; and the lower horizontal trace is the final CD value of the denatured methemoglobin. The reaction curve is comprised of two SF runs superimposed, with no averaging; low-pass filter time-constant, t_F, was 4 msec. Approximate half-reaction is indicated by the position of the vertical marker; the corresponding $t_{1/2}$ value is given by the left-hand numeric display. Total time span of photograph, 1.02 sec. For the dilution value trace, $t_F = 4$ msec; final value trace, $t_F = 0.4$ sec; sampling rate for all data, 1 msec per point. From Luchins and Beychok.[16] Copyright 1978 by the AAAS.

swept into the take-up syringe. After a determination, the contents of the take-up syringe are disposed of (S), readying the take-up syringe for the fluid output of a subsequent run.

Electronics System. Given the presence of a sample in SFC with a CD transition at the monochromator wavelength, the beam impinging on PMT exhibits a 50 kHz intensity fluctuation (typically 10^{-3}–10^{-6} of the average intensity). The PMT microcurrent output which follows the beam intensity is converted to voltage and amplified (CV). The voltage output is branched. One branch is stripped of low frequency contributions (HI-PASS) and passed to the synchronous detector referenced to the MOD 50

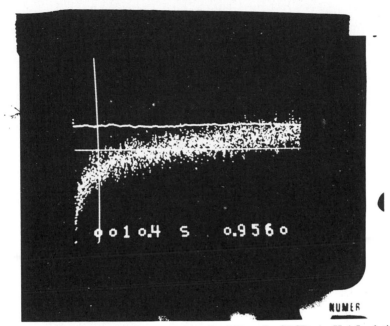

φ o 1 o.4 S . o.9 5 6 o

FIG. 2. Renaturation by pH-jump, monitored at 222 nm by SFCD. A pH 4.5 solution of 4.4×10^{-5} M α^0 monomer had been prepared by mixing five volumes of 5.3×10^{-5} M α^0 monomer at pH 5.7 and 20 mM phosphate with one volume of 1 M KH$_2$PO$_4$. The pH was jumped to 5.8 by SF mixing with an equal volume of 20 mM K$_2$HPO$_4$. The upper horizontal trace is the final (pH 5.8) CD value; the lower horizontal line is an electronic screen center marker. The multiple point at the bottom left is within 5% of the expected dilution CD value (for 2.2×10^{-5} M α^0 monomer at pH 4.5). The data are for three consecutive SF runs superimposed, with no averaging. The t_F for the runs was 40 msec. (The position of the vertical line and the value of the corresponding numeric are arbitrary.) Total time span, 102.4 sec; final CD value obtained with a t_F of 1.25 sec; sampling rate, 100 msec per point. From Luchins and Beychok.[16] Copyright 1978 by the AAAS.

kHz frequency, to follow the amplitude and changes with time in amplitude of the 50 kHz component. To follow the average intensity, the other branch is stripped of high frequency contributions [LO-PASS (a)] and matched to the time constant of the synchronous detector's output [LO-PASS (b)]. The two branches are ratioed to give the CD signal at time t. The output of the divider is acquired upon SF triggering and stored. Data can be edited, smoothed, analyzed, and archived.

Components

Optical. The present LS is a Hanau 200 W deuterium lamp with an ~1 mm^2 plasma and a Suprasil end-window. (We are planning to add a stabi-

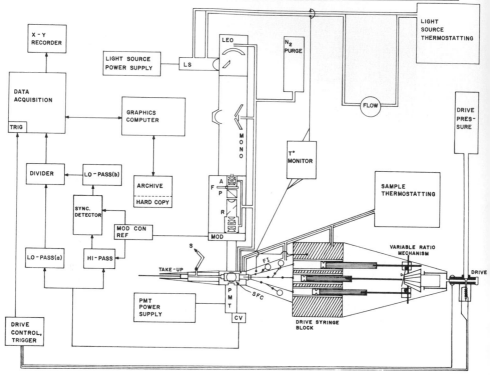

FIG. 3. A, Achromat; CV, post-PMT current to voltage preamplifier; F, optical filter wheel; FLOW, flow monitor/regulator; HI-PASS, LO-PASS, signal processing electronic filters; LEO, light entrance optics; LS, light source; MOD, 50 kHz circular polarizing modulator; MOD CON REF, control unit for MOD, with frequency reference outputs; MONO, high throughput monochromator; P, linear polarizer; PMT, photomultiplier detector; R, imaging system; SFC, stopped-flow cuvette observation block; SYNC. DETECTOR, phase-sensitive amplifier, referenced to MOD frequency.

lized 75 or 150 W Xe source when it has been evaluated.) The light source power supply, which is highly regulated, with highly regulated continuous anode heating, is custom built, as are the light source entrance optics. The source is cooled with a Neslab RTEZ-8 refrigerated circulating bath. LS so configured provides an intense, stable beam with maximum intensity at about 230 nm.

The monochromator is a JY-ISA H.20 holographic grating monochromator (single grating, UV-optimized) custom-modified (slit-mirror distances and mirror angles adjusted in a broad single-point correction) to correct astigmatism. The optical conditioning unit is comprised of (1) a three element CaF_2–silica–CaF_2 achromat compatible with the modified H.20 monochromator, with under $1'$ deviation from 190–450 nm, (2) an optical filter wheel, adjustable external to the light-tight unit container, (3) an ammonium dihydrogen phosphate Rochon linear polarizer with usable transmittance down to ~185 nm, with 4–5° divergence of the extraordinary beam from the undeviated ordinary beam, and (4) a multielement refocusing (and extraordinary beam mask) system with focus set just beyond the windows of SFC. MOD is a Hinds CaF_2 50 kHz piezooptical modulator set for 1/4 wave retardation at the monochromator wavelength, with phase reference output for electronics of detection units.

Fluid Handling. The drive is a custom-modified Berger/Dover Scientific design (fabricated by Commonwealth Technology, Virginia) for rapid mixing of two or three thermostatted solutions with dial-adjustable total volumes and volume ratios (accurate ratios available: 1 : 1, 1 : 2, 1 : 5, 1 : 10; other ratios available: 1 : 20, 1 : 50, 1 : 100) of two of the solutions (the third solution volume is selectable at either zero or the sum of the ratioed volumes). The unit has a 50–200 psi pneumatic drive. There is a closed, monitored sample thermostatting system and a built-in adjustable trigger. Sample solutions come in contact only with Kel-F and Teflon. The SFC observation blocks are custom built with several available pathlengths and designs, all thermostatted. The unit is most suited for straight-through observation (transmittance, CD) with 2, 6, and 10 mm pathlengths, with right-angle observation (fluorescence, fluorescence-detected CD) capabilities built into the 6 and 10 mm designs. All optical windows are of annealed birefringence-free Suprasil A and block bodies are of black Delrin (acetal resin) or black Kel-F.

The entire fluid-handling system is highly compact and allows for ease of optical alignment.

Electronics. Two optically selected, low noise, end-on photomultiplier tubes (PMT) with high UV response, are powered by custom-modified Durrum/Dionex ultrastable PMT power supplies. (One PMT is mounted for straight-through observation, the other, for right-angle observation.) The PMTs receive the optical system's output after the beam has traversed the SFC of the fluid-handling system. The PMT output is immediately passed through an Ithaco 164 current to voltage preamplifier (CV). For CD, the voltage signal is branched, and one branch is passed through a tuned, narrow-bandpass 50 kHz crystal/inductor filter and then into an Ithaco 391A lock-in amplifier (SYNC. DETECTOR), both referenced to the MOD frequency. The selectably low-passed SYNC. DETECTOR output signal (proportional to peak-to-peak magnitude of left minus right circularly polarized transmittance) is fed in as the numerator in a rapid divider circuit. The other branch (for non-CD, this is the only branch) is fed into a custom-built selectable low-pass filter (with adjustable gain and offset) with time constants matching those available on SYNC. DETEC-

TOR output. For CD, this low-passed signal is fed in as the denominator in the divider circuit.

The output is stored in a Nicolet digital oscilloscope, where it is examined for disqualifying experimental artifacts and outputted to hard copy or first to an enhanced Tektronix 4052 Graphics System for averaging, smoothing, analysis, and archiving.

Applications

We will review here a group of experiments dealing with denaturation and refolding of human hemoglobin and myoglobin. The instrument has also been used to study the refolding of unreduced GuHCl-denatured lysozyme.

The denaturation experiments of hemoglobin were carried out in strong acid, as typified by the experiment in Fig. 1, conducted at pH 1.6, 4°. These experiments were mainly performed as calibration type experiments and to examine the potential of the method. The value of the first-order rate constant was seen to be approximately proportional to the first power of hydrogen ion concentration.[16,18] These data, however, are difficult to interpret in terms of (un)folding mechanism, because the final state(s) of the molecule is not known. The same problem is encountered in refolding by pH jump as seen in Fig. 2; here neither initial nor final state is well characterized, and the pH variation of conformation of α^0 is not well understood. A more coherent and interpretable picture emerges from comparison of three other refolding processes: (1) folding induced by chain association, (2) folding induced by heme binding, and (3) folding induced by dilution out of 6 M GuHCl solutions of Hb and Mb.

Folding Induced by Chain Association. Luchins and Beychok[16] studied the refolding of α^0 and β^0 upon recombination at pH 5.7 to form apoHb. This reaction leads to an average increase of 17% in helix content per chain, or 24 residues per chain that are refolded into helical conformation.[19] In terms of molecular ellipticity at 222 nm, there is an increase of 5200 deg cm²/dmol (negative). The SFCD exhibited a single first-order phase, with a half-time ($t_{1/2}$) of 130 sec at 4°.

Folding of α^0 Induced by Binding of Hemin Dicyanide. In this investigation, Leutzinger and Beychok[20] studied the kinetics of heme binding

[18] H. Polet and J. Steinhardt, *Biochemistry* **8,** 857 (1969); W. Allis and J. Steinhardt, *Biochemistry* **8,** 5075 (1969).

[19] C. Lee, M. Pflumm, and S. Beychok, submitted for publication (1985).

[20] I. Leutzinger and S. Beychok, *Proc. Natl. Acad. Sci. U.S.A.* **78,** 780 (1981).

and the ensuing conformational changes by using three stopped-flow techniques: (1) intrinsic fluorescence dependent upon the spatial orientation and distance between the bound heme and the A12(14)α tryptophan, (2) transmittance (absorbance) at or near the Soret band maximum, whose position and intensity depend on the local environment of the heme and the nature of the axial ligands, and (3) far-UV CD, which directly gauges the recovery of secondary structure.

In this reaction, studied at 4°, the α-globin molecule, α^0, is refolded from a state that is 40% helical to α^{heme} (α^h) which is 78% helical, so that between 50 and 55 residues are transferred from nonhelical to helical conformation.[20-23] The process occurs in three discernible steps. In the first, heme binds in a second-order reaction which is diffusion limited. This is accompanied by quenching of fluorescence of Trp A12(α) with an amplitude of about 85%, and an increase in Soret absorbance of about 40% amplitude. In the second step, which is first order, changes occur in fluorescence, absorbance, and 222 nm CD, all probes giving the same value of $t_{1/2} \sim 30$ sec. In this phase, the fluorescence and absorbance changes become complete, and 30–35% of the far-UV molecular ellipticity change occurs (Fig. 4). When the first-order Guggenheim plot of this reaction phase is extrapolated to zero time, it yields the θ_{222} value of α^0. In the third phase, the remaining 60–65% of the total θ_{222} change occurs, in a first-order reaction, with $t_{1/2}$ of 120 sec.

These experiments isolate a level of structure for direct analysis, namely, secondary structure, but in addition examine two features of tertiary structure. These latter are less directly interpretable but they are associated with the conformation of the heme pocket and with fixing of the distance/orientation of heme relative to Trp A12(α). These tertiary elements of structure are recovered while some sizable fraction (60–65%) of the secondary structure is still reforming.

It is interesting to note that the slow phase has essentially the same half-time as is found for the average refolding of the α^0 and β^0 chains when they combine to form apoHb. In the latter case, there is little doubt that the segments being most extensively refolded are at the $\alpha_1\beta_1$ interface involving the G, GH, and H regions at the interface, although the conformation of the heme pocket is also somewhat modified (as reflected in differing heme binding capabilities of α^0 and $\alpha^0\beta^0$). Since the amplitude and relaxation time are the same in the two processes, it is likely that the

[21] Y. K. Yip, M. Waks, and S. Beychok, *J. Biol. Chem.* **247**, 7237 (1972).
[22] M. Waks, Y. K. Kip, and S. Beychok, *J. Biol. Chem.* **248**, 6462 (1973).
[23] Y.-H. Chen, J. T. Yang, and K. H. Chau, *Biochemistry* **13**, 3350 (1974).

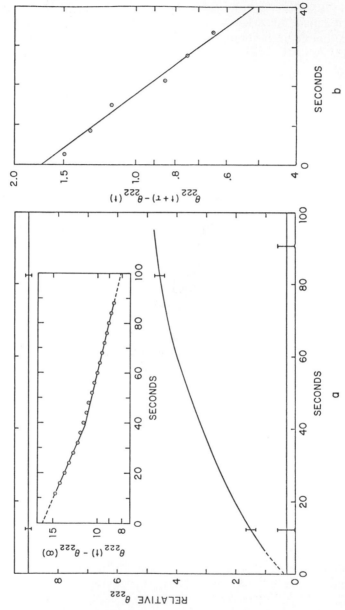

FIG. 4. (a) Far-UV CD recovery for the reaction of α^0 in 20 mM potassium phosphate buffer (pH 5.7) with equimolar hemin dicyanide in 20 mM potassium phosphate buffer (pH 6.7). Final concentration, 2.4 μM; temperature 4°. (Inset) First-order plot of data. (b) Guggenheim plot of the initial phase. From Leutzinger and Beychok.[20]

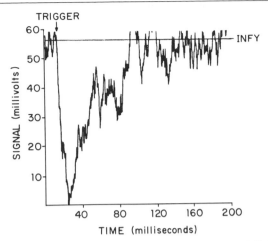

FIG. 5. SFCD of HbCN refolding after 5 fold dilution out of 6 M GuHCl, final pH 7.0, 20°. Monitored at 225 mm, $t_F = 1.25$ msec.

slow phase in the refolding of α^0 following heme binding involves that segment of the molecule. This result points to identification of at least two folding regions in the nascent α^h chains and implies the existence of a refolding intermediate which is conformationally equivalent to the α^0 chain in the apoHb dimer. Evidently, SFCD at 222 nm when heme is added to $\alpha^0\beta^0$ must be done. If a phase with $t_{1/2}$ of 20–30 sec is found, this will be established.

Folding Induced by Dilution of 6 M GuHCl Solutions of Hb and Mb. We have carried out SFCD and SF transmittance (absorbance) (408 or 420 nm) on several samples of HbCN, Mb, and MbCN. The experiments have been done at 20°. The hemoglobin is human hemoglobin and the myoglobin is sperm whale. Figure 5 shows a representative run on HbCN. A 6 M GuHCl solution of the protein is diluted 5-fold with phosphate buffer at pH 7.2. Full scale is 204.8 msec. The data of several runs are averaged to enhance the signal-to-noise ratio.

A typical SF transmittance run at 420 nm is shown in Fig. 6.

The data of Fig. 5 are plotted as a first-order reaction in Fig. 7, covering more than three half-times. A second order plot is curvilinear. Figure 8 shows a plot of the data of Fig. 6. Here, the data are plotted as second order, with equal concentration assumed of heme and protein chains. In this case, a first-order plot was nonlinear in all portions of the plot.

The table summarizes the results. The main points to note are these. The secondary structure of both Hb and Mb returns in a first-order reac-

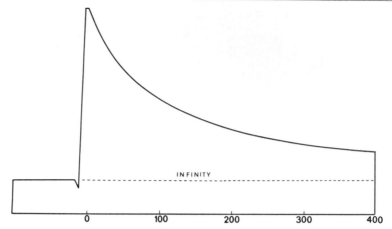

FIG. 6. Transmittance, 420 nm, of MbCN, after 5-fold dilution out of 6 *M* GuHCl, pH 7.0, 20°, t_F = 4 msec. Horizontal scale is in msec.

tion with a half-time of about 30 msec. The data are sufficiently good on averaging that we can specify three (3) half-times in this reaction, but it cannot be shown reliably at present that the return beyond that to the pretrigger value in the remainder of this recovery follows the same reac-

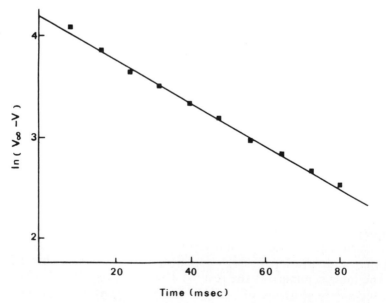

FIG. 7. First-order plot of data of the experiment represented in Fig. 5.

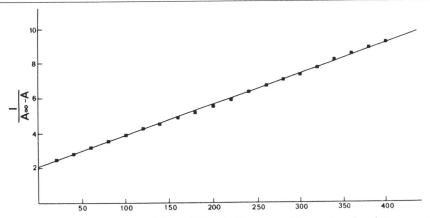

FIG. 8. Second-order plot of data of Fig. 6. The transmittance at each point was converted to absorbance for this plot. Horizontal scale is in msec.

tion, or is of a subsequent process. This is an important point, because in the kinetics of the Soret absorbance recovery, where the signal-to-noise ratio is much more favorable, we definitely see a first-order slower reaction in the last ~15% of the reaction, and we would like to know if this has a counterpart in CD recovery.

The second observation of importance is that the total recovery in Hb is only about 50%, over very long times, whereas it is complete within several seconds in Mb. The failure to obtain more than 50% recovery of native structure in the Hb refolding experiments raises many questions. In this a mixture of fully native and denatured molecules, or a blocked intermediate state? Can this be due to a type III proline isomer[6]? Why does this not happen in Mb? Will this be observed in apoHb? A possible interesting clue may be found in studies on refolding of β^0 when heme is added. In this case, in contrast to Hb, a partially refolded intermediate is formed, which folds slowly to the native form. Since β^h is tetrameric, this may suggest that the tetramer in Hb stabilizes an incorrectly folded form, but this can be tested.

The fact that the Soret absorbance has a fast second order reaction indicates that this is a recombination reaction, and the rate of the reaction suggests that the molecule must first acquire a binding site, although the pocket does not have to be in its completely folded condition, as witness the strong binding by the partially folded central exon product.[24,25] However, a tertiary structure resembling the final state is probably present.

[24] C. S. Craik, S. Buchman, and S. Beychok, *Proc. Natl. Acad. Sci. U.S.A.* **77,** 1384 (1980).
[25] C. S. Craik, S. Buchman, and S. Beychok, *Nature (London)* **291,** 87 (1981).

REFOLDING OF HbCN, Mb, AND MbCN FROM 6 M GuHCl DENATURED STATES

| Protein | SFCD (222 nm) | | | | SF$_{ABS}$ (420 or 408 nm) | | | | | | |
	k_1 (sec^{-1})	$t_{1/2}$ (msec)	Order	Change fast phase (%)	k_1 (sec^{-1} M^{-1})	$t_{1/2}$ (msec)	Order	k_2 (sec^{-1})	$t_{1/2}$ (sec)	Order	Change fast phase (%)
HbCN[a]	2.65×10^4	26	1st	>87.5	1.67×10^5	173	2nd	—	—	1st	>80
MbCN	2.10×10^4	33	1st	>87.5	2.43×10^5	138	2nd	3.4×10^{-1}	2.0	1st	>83
Mb	2.56×10^4	27	1st	>87.5	2.67×10^5	135	2nd	2.0×10^{-1}	3.2	1st	—

[a] In HbCN, total θ_{222} is 54% of native, all occurring in single phase; total recovery of OD$_{420}$ is ~60%.

We can give much more definition to this reaction by measuring the SFCD at the Soret region, which will be considerably more informative than the absorbance, although the signal change here is much smaller than in the far-UV, and these will be technically difficult measurements. Many other questions can be answered (at least partially) by studies of α^h which is monomeric and β^h which is tetrameric.

We wish to recall, here, some important observations which can be culled from several investigations of this system.[26-29] When heme is removed from Mb (which is monomeric), about 15 residues in helical segments are unfolded, and there may be some other conformational changes as well, probably in the heme pocket. When heme is removed from Hb, about the same number of residues are unfolded in each chain and the molecule dissociates to the dimer. However, when heme is removed from α^h or β^h subunits, at least 50 and possibly 55 residues are unfolded. All of this suggests that in Mb, heme removal gives a rather localized change, whereas in the isolated α^h and β^h chains, more than one region of the molecule is involved unless the complementary chains are present to stabilize those segments which do not unfold in apoMb. Together with the studies on the conformation of the isolated central fragment of β^h,[24,25] on the effects of the complementary chains and fragments on this conformation,[25] on the folding steps in α^0 recombination with heme,[20] and on the fragment stability studies of Bucci and co-workers,[29] there is virtually no doubt that the heme pocket as a subdomain of the central exon fragment is an autonomously folded and folding unit, in both equilibrium and kinetic senses. At present, we are vague about exactly which residues constitute this subdomain, and how they are conformationally interrelated with the rest of the molecule, but we know that there are (relatively) stable states of heme-free Mb and Hb in which regions of the molecule have their native conformation while the heme-binding subdomain is unfolded in terms of secondary structure—although tertiary interactions persist.

When these results are considered together, we arrive at a preliminary working model of the folding mechanism of these proteins, in which the secondary structure of the fully unfolded chains is first established in a single phase. The helices so established then come together and fold into a tertiary structure. The latter process involves at least two distinct kinetic phases. It is likely that these reflect the existence of structural subdo-

[26] E. Breslow, S. Beychok, K. Hardeman, and F. R. N. Gurd, *J. Biol. Chem.* **240**, 304 (1965).
[27] K. Javaherian and S. Beychok, *J. Mol. Biol.* **37**, 1 (1968).
[28] G. Collona, C. Balestriera, E. Bisniuto, L. Servillo, and G. Irace, *Biochemistry* **21**, 212 (1982).
[29] D. Franchi, C. Fronticelli, and E. Bucci, *Biochemistry* **21**, 6181 (1982).

mains in folded chains, but the exact relationship remains to be established.

Acknowledgment

Supported in part by Grant PCM-84-04221 from the National Science Foundation and HL 16601 from the National Institutes of Health.

[23] The Measurement of the Kinetics of Lipid Phase Transitions: A Volume-Perturbation Kinetic Calorimeter

By MICHAEL L. JOHNSON, WILLIAM W. VAN OSDOL, and RODNEY L. BILTONEN

Introduction

There is a consensus that a principal type of biological membrane consists of a phospholipid bilayer matrix into which various protein components are immersed. A variety of physical and chemical studies have provided a distinct picture of the time average structure of such membranes. It is also appreciated that the interactions between the phospholipids and the membrane-bound proteins are important in the function and regulation of these proteins. However, most of these ideas are conceived only in terms of a static picture. Little consideration or experimental effort has been put forth to develop an understanding the temporal aspects of such interactions. In fact, there has been little development, either theoretical or experimental, of the description of the cooperative dynamics of the membrane components. This situation exists although it is well known that the phospholipid molecules may diffuse within the plane of a single monolayer of the membrane bilayer or redistribute from one bilayer surface to another, and that proteins can readily diffuse within the membrane.

For some time now, the working hypothesis in our laboratories has been that the lipid matrix of a biological membrane exists in a dynamic equilibrium of ordered and disordered clusters of variable size. It is assumed that the size distribution of these ordered and disordered clusters and their time-dependent fluctuations in size confer special properties upon the membrane which are related to regulation of specific protein functions. Thus, in order to obtain more information about the dynamics of membrane structure a novel volume-perturbation calorimeter was designed to measure the kinetics of lipid phase transitions.

The traditional methods used to determine kinetic parameters associated with systems at presumed equilibrium include perturbation techniques such as temperature and pressure variations or stop-flow methods which perturb the relative concentration of the various components. Generally the resulting temporal change in equilibrium position is monitored using an optical observable. In some biochemical systems such as nucleic acids and proteins these approaches have been successfully employed. However, in phospholipid systems the success has been limited because of the lack of a sensitive natural optical probe and the difficulty in interpreting the existence of multiple relaxation processes. For example, the kinetics of the main transitions of dimyristoyl phosphatidylcholine (DMPC) and dipalmitoyl phosphatidylcholine (DPPC) dispersions have been studied using temperature jump techniques and monitoring changes in turbidity. It was discovered that the relaxation process is complex and at least two short-time relaxations exist, approximately 1 sec and 10 msec.[1] The data were difficult to interpret for two reasons. Either difficulty existed in the interpretation of the observable quantity with regard to its structural basis, or alterations in the membrane system caused by the method used to monitor or induce the temporal changes in the equilibrium position created problems. For example, it is difficult to unambiguously assign the nature of the physical changes in a membrane which generates the observed changes in turbidity. Also, the voltage discharge-induced temperature jump technique produces permanent structural changes in membrane systems. A number of investigators have employed the fluorescence properties of hydrophobic dyes to monitor changes in membrane structure.[2] The difficulty using fluorescence probes is that it is well known that in equilibrium experiments, such as differential scanning calorimetry, the addition of minor contaminants, such as these fluorescence probes, alters the thermodynamic behavior during the phase transition.

Therefore, we have approached the problem with the idea in mind that the data which were to be obtained about the phase-like transitions provide a broad spectrum of kinetic and thermodynamic information that can be compared to equilibrium calorimetric measurements. Furthermore, we wished to obtain information from intrinsic thermodynamic properties of the phospholipid matrix itself rather than, for example, using an added fluorescent probe which might alter the properties of the membrane system. In addition, we decided that it would be essential that the perturbations from equilibrium be small, reversible, and nondestructive. Toward

[1] T. Y. Tsong and M. I. Kanehisa, *Biochemistry* **16,** 2674 (1977).
[2] B. R. Lentz, E. Freire, and R. L. Biltonen, *Biochemistry* **17,** 4475 (1978).

this end we constructed a volume perturbation instrument[3] based on the original design of Clegg *et al.,*[4] as modified by Halvorson.[5] This instrument monitors the induced temperature and pressure changes as indicators of temporal changes in the equilibrium state of the system. This approach was based upon the large volume changes associated with the phase transition of model multilamellar and single lamellar vesicles made from neutral phospholipids. This chapter describes the design and operation of this kinetic calorimeter as it is used to obtain both thermodynamic and dynamic information about the membrane.

The biological relevance of these studies is assumed since we suspect that the dynamic fluctuations of the phospholipid matrix is an important property of the membrane. Because the gel to liquid-crystalline transition of these phospholipid systems is easily detectable we are using it as a model to understand the temporal coupling between lipid fluctuations and protein fluctuations. At this time, however, we will describe only the basic structure and design features of our instrument, its operational characteristics, and report on the type of dynamic and thermodynamic information that can be obtained.

Basis of Our Approach

In our experimental approach, the voltage-dependent extension of a stack of piezoelectric crystals is used to force a sample solution to undergo a small, adiabatic, bidirectional, and periodic volume change. The pressure and temperature changes induced by the volume changes are the driving forces that alter the equilibrium point of the gel to liquid-crystalline transition in a small, periodic fashion. From the comparison of the driving waveform, that is the volume change, with the experimentally observed temperature and pressure waveforms we can deduce the kinetic and thermodynamic properties of the sample solution.

The main phase transiton melting temperatures for multilameller DMPC and DPPC have been shown to have a strong dependence on the hydrostatic pressure.[6,7] The relative excess heat capacity function for DPPC liposomes, as measured by differential scanning calorimetry, at different applied pressures of helium is shown in Fig. 1.[7] Increasing the hydrostatic pressure increases the melting temperature, t_m, for both the low- and high-temperature transitions by approximately 0.024°/atm but

[3] M. L. Johnson, T. C. Winter, and R. L. Biltonen, *Anal. Biochem.* **128**, 1 (1983).

[4] R. M. Clegg and B. W. Maxfield, *Rev. Sci. Instrum.* **47**, 1383 (1976).

[5] H. R. Halvorson, *Biochemistry* **18**, 2480 (1979).

[6] W. Z. Placky, *Biophys. J.* **16**, 138a (1976).

[7] D. B. Mountcastle, R. L. Biltonen, and M. J. Halsey, *Proc. Natl. Acad. Sci. U.S.A.* **75**, 4906 (1978).

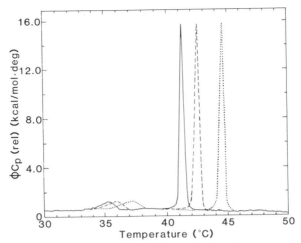

FIG. 1. Relative excess heat capacity of a multilamellar dispersion of DPPC in 0.05 M KCl at a pressure of 1 atm (——) or helium pressures of 68 (– – –) and 136 atm (···). Heat losses through the pressurizing tubing to the high-pressure cell decreased the apparent heat capacity maxima in these experiments; the three baselines were adjusted to be equal at 39°. Redrawn from Mountcastle *et al.*[7]

does not alter the enthalpy change, ΔH, or shape of the transition profile. It should be noted that the phase transition half-width, $\Delta t_{1/2}$, is a sensitive indicator of the degree of cooperativity of the phase transition. Consequently the degree of cooperativity is not affected by pressure, since the shape of the phase transition is unaltered.

An example of the cyclic nature of our instrument can be easily described by reference to Fig. 1. Initially a DPPC solution is equilibrated at 1 atm and 41.8°. These conditions correspond to a position 95% through the main phase transition. The macroscopic volume of the solution is now decreased by an amount sufficient to increase the pressure to 68 atm in approximately 1 msec. The new equilibrium state of the DPPC is described by the dashed line. At this temperature and pressure, the liposome exists in a state which corresponds to about 5% melted. However, at 1 msec following the perturbation, the lipids are still 95% melted. Our instrument subsequently monitors the heat released, i.e., the temperature of the solution, as the DPPC liposomes relax to the new equilibrium position, the 5% position indicated by the dashed line. After sufficient time has elapsed for the system to achieve its new equilibrium position, the macroscopic volume of the system is returned to its original starting volume. The dynamic state of the liposomes would be 5% of the way through the phase transition and the equilibrium position is 95% melted.

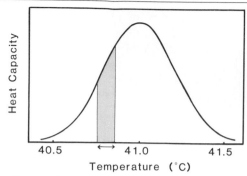

FIG. 2. A simulated excess heat capacity curve for DPPC liposomes showing the portion or the transition traversed by a volume perturbation sufficient to induce a 4 atmosphere pressure change. Note that the temperature is shifted by a factor of 0.024 times the pressure.

At this time, our instrument monitors the absorption of heat as the liposomes relax back to the 95% equilibrium position. In an actual experiment, the situation is more complex than presented so far since the lipid undergoes a microscopic volume change during the phase transition which alters the equilibrium position of the gel to liquid-crystalline transition. The lipids will subsequently decay to a new equilibrium position along a kinetic path involving changes in both heat released and microscopic volume. The possibility for changes in microscopic volume implies that the pressure induced by the macroscopic volume change will not be constant. We therefore alter the macroscopic volume in a known periodic manner and then monitor both temperature and pressure as a function of time.

In normal operation of the instrument, the forced volume change is substantially less than that required to induce a 68 atm pressure change. Typically, the volume change only induces approximately a 4 atm pressure change. This has the desired effect of not forcing the liposomes through the entire phase transition. The situation is shown diagrammatically in Fig. 2. In this simulated example, a solution was equilibrated at 41.0° and 8 atm pressure. After reaching equilibrium the volume was periodically altered to generate a ±2 atm pressure change. This pressure change produces a change in the equilibrium position of the transition equivalent to a temperature change of ±0.05°. Operating at such low volume changes means that our instrument truly makes a small, reversible perturbation from equilibrium in a nondestructive manner. Furthermore, these small perturbations allow the measurement of the kinetics of the phase transition at varying stages within the phase transition. The disadvantage of working with such small volume perturbations is that the resulting time-dependent temperature and pressure changes are smaller. The shaded area of Fig. 2 corresponds to the fraction of the total enthalpy and volume changes which are induced by a 4 atm pressure change.

The phase transition temperatures for multilamellar DMPC and DPPC liposomes have been shown to vary by 0.024°/atm of pressure.[7] Consequently, a 0.1° change in the melting temperature can be accomplished by inducing a pressure change of 4 atm. Such a pressure change can be induced by a fractional volume change of 1.8×10^{-4} in water. Since the half-width of the DMPC transition is only approximately 0.2°, the application of a sufficient volume change to induce a 4 atm pressure change can shift the equilibrium sufficiently to induce as much as 40% of the lipids to shift from their gel to liquid-crystalline state. This implies that a 4 atm increase in pressure on a 0.1 M DMPC solution will raise the temperature by as much as 0.2° since the enthalpy change for the main DMPC transition is approximately 5.0 kcal/mol. This is a maximal estimate since it is estimated from the steepest portion of the transition. It is also a maximal estimate since when this shift in equilibrium position is accomplished under adiabatic conditions, the lipid will release heat and raise the solution temperature, thus partially compensating for the pressure change.

The basic experimental approach outlined here has a number of distinct advantages over previous methods which have been employed to measure the kinetics of lipid phase transitions. First, the perturbations from equilibrium are small and reversible. Second, the method of perturbing the equilibrium, i.e., the volume change, is homogeneous throughout the solution being studied. Third, we are directly measuring thermodynamic state variables of the solution, temperature and pressure, which can be related to the thermodynamic parameters of the solute, e.g., enthalpy, excess heat capacity, and volume changes. Other methods such as changes in turbidity or light scattering are more difficult to interpret in terms of their structural basis. Fourth, this method of monitoring the temporal changes in the equilibrium position does not alter the solute properties. Other techniques, such as the addition of a fluorescent dye, can have large effects on the properties of the solute. Fifth, this apparatus can be used in a "single pulse" mode similar to the temperature jump technique but it has the advantage that it is also bidirectional. Sixth, the apparatus can be used with a periodic forcing function in a manner analogous to phase modulation fluorometry. Seventh, this experimental approach allows the measurement of the kinetics of heat transfer and volume changes over a small portion of the phase transition whereas most previous techniques required the lipids to pass through the entire transition region.

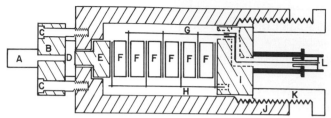

FIG. 3. General diagram of the volume jump cell: (A) pressure transducer, (B) sample compartment support, (C) bolt, (D) sample compartment ring, (E) driving piston, (F) piezo-electric crystal stack, (G) cathode, (H) ground, (I) crystal stack support, (J) main cell body, (K) crystal stack support tightening screw, and (L) high-voltage BNC connector. Redrawn from Johnson et al.[3]

Description of the Apparatus

The volume change is produced by the voltage-dependent extension of a piezoelectric crystal stack (F, in Fig. 3). This extension drives an aluminum piston (E, in Figs. 3 and 4) which forms one wall of the sample compartment where, by the displacement of a 0.001-in. Mylar diaphragm (N, in Fig. 4), the desired volume perturbation is generated. Eighteen piezoelectric disks, Type LTZ-1 (Transducer Products, Goshen, CT), with a fundamental frequency of 500 kHz were stacked with alternating polarity so that they are in parallel electrically but in series mechanically (see Fig. 3). A 0.001-in. brass shim slightly smaller than the crystal diameter and with a protruding tab for electrical contact was placed between each crystal element. The assembled stack was painted with epoxy to hold it together and then coated with high-voltage, arc-resisting, insulating varnish (Red GLPT, GC Electronics, Rockford, IL). The positive and negative polarity tabs were oriented on opposite sides of the stack so that they were easily soldered to the copper rods that formed the electrical connection to the high-voltage BNC connector (L, in Fig. 3). During operation the stack is supported from one side by the crystal stack support, I, and from the other by the driving piston, E. Extension of the stack is proportional to the applied voltage, with desired volume perturbations attainable using moderate voltages (up to 2 kV).

The waveform and frequency of the applied volume pulses are generated by the data acquisition computer (PDP-11/23+). This is implemented by a 12-bit, 400 kHz, continuous performance, digital-to-analog converter (DT3371, Data Translation, Marlboro, MA). The resulting signal is amplified with a Kepco (Flushing, NY) operational power supply (OPS-2000B) with the gain set to 150 to give a driving voltage of approximately 3/4 to 1 kV, which translates to a pressure increment of about 5 atm in our appa-

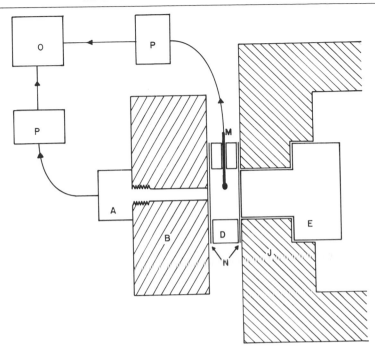

FIG. 4. Expanded view of the sample compartment. Letter codes are the same as in Fig. 3, except (M) temperature transducer, (N) Mylar diaphragm, (O) computer, and (P) signal amplifier. Redrawn from Johnson *et al.*[3]

ratus. We have employed both sine and square wave driving functions for the volume. If a square wave volume perturbation is used and the resulting pressure wave is also a square wave, the data can easily be analyzed in the time domain. A square wave is also advantageous for frequency domain (Fourier) analysis since it contains multiple odd numbered harmonics of the driving frequency. This allows us to collect data relating to multiple frequency perturbations from a single driving frequency experiment. We have successfully employed the amplitude and phase shift of the first, third, fifth, and seventh harmonics which result from a square wave excitation.

The sample compartment (0.8 ml volume) is a right circular cylinder 13 mm in diameter and 6 mm thick (D, in Figs. 3 and 4). The sample is loaded by positioning the apparatus vertically with the tightening screw, K, down, loosening the screw until the driving piston is recessed 1 or 2 mm into the main cell body, stacking in succession the Mylar diaphragm and the sample compartment ring, and then pipetting in the sample until a slight meniscus is formed above the well. Next, a second Mylar diaphragm is stacked on the top surface of the ring, and the sample compartment support, B, is bolted down on top of this diaphragm, taking care to exclude air bubbles during this process. It is required that a low static pressure be applied to the crystals. Any desired static pressure between 1 and 10 atm can be applied to the compartment by turning the tightening screw, K. This static pressure also allows solutions to be used without prior degassing.

Temperature control is maintained by a Neslab Instruments (Portsmouth, NH) RTE-9DD water bath by external circulation through eighteen turns of 3/8-in.-diameter copper tubing silver-soldered directly to the body of the calorimeter and the whole assembly placed in an insulated compartment. The set point of the water bath is externally programmed by the data acquisition computer utilizing a 16-bit digital-to-analog converter. This temperature control has proven to stabilize the calorimeter to better than 0.01°.

For the purposes of monitoring the state changes of the lipid system, two transducers are interfaced with the sample compartment. One, a high-speed thermistor (M, in Fig. 4), was placed in the compartment through a threaded hole drilled in the sample compartment ring (D, in Figs. 3 and 4) and sealed with epoxy. The choice of a thermistor and its location were dictated by our desire to have a fast time response. The primary reason for the slow time response of differential scanning calorimeters is that the temperature sensors, usually thermopiles, have a relatively slow time response and are located in a position physically remote from the aqueous sample. The other transducer (A, in Figs. 3 and 4), directly opposite the driving piston on the other side of the sample compartment, monitors pressure changes through a second Mylar diaphragm.

The signal from the thermistor, a Thermometrics, Inc. (Edison, NJ) FP07DA103N of nominal resistance, 10 kΩ, is processed with two different amplifiers, both built from Analog Devices 2B31L transducer amplifiers (Norwood, MA). The first, which consists of a fixed-resistance Wheatstone bridge and an instrumentation amplifier, is used to produce a signal directly proportional to the temperature of the cell. The second amplifier has a gain of 113 times the first amplifier, and includes a variable resistor in the Wheatstone bridge. By balancing the second bridge at any given temperature and utilizing the high-gain amplifier, a very accurate relative measurement of the time-dependent temperature change can be obtained. This nulling bridge, however, precludes use of the high gain amplifier to determine the absolute temperature of the cell. This bridge is nulled by the data acquisition computer and a 16-bit digital-to-analog converter. The signal from the piezoelectric pressure transducer, a Kulite Semiconductors (Ridgefield, NJ) XTM-1-190-500 with an internal Wheatstone bridge,

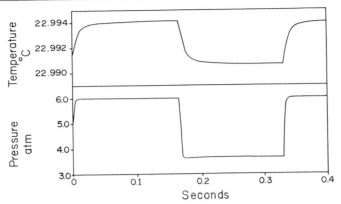

FIG. 5. Response characteristic of water. Redrawn from Johnson *et al.*[3]

is amplified in a straightforward manner analogous to the instrumentation amplifier. Four 1.5 V batteries, coupled to a Zener diode voltage divider are used to excite both transducers at 2.6 V.

Signal-averaging techniques are employed to improve the signal-to-noise ratio, which increases as the square root of the number of observations. For this purpose a 12 channel, simultaneous, direct memory access, analog-to-digital converter (DT3388, Data Translation, Marlboro, MA) is used. The ability to collect simultaneously from multiple inputs, rather than serially, as with conventional analog-to-digital converters, simplifies the interpretation of the phase shifts of the Fourier components we calculate from our data.

Typical Results

For calibration purposes the instrument was first used to study the temperature and pressure response characteristics of water. Assuming that the system is adiabatic and that only pressure–volume work is done on or by the system, one can derive the relationship

$$\left.\frac{\partial T}{\partial P}\right|_s = \frac{TV\alpha}{C_P}$$

where α is the coefficient of thermal expansion, $(1/V)(\partial V/\partial T)$.[5] At 20° and 1 atm $TV\alpha/C_P = 1.6 \times 10^{-3}$°/atm. In Fig. 5 we present the time response of the temperature and pressure in water when a 1000-V square wave is applied to the crystal stack at a frequency of 3 cycles per second. Individual data points are not shown because they all lie within the line width. A value of $\Delta T/\Delta P = 1.5 \times 10^{-3}$°/atm is calculated from the data presented in Fig. 5. This value differs by only 7% from the predicted value. A thermal

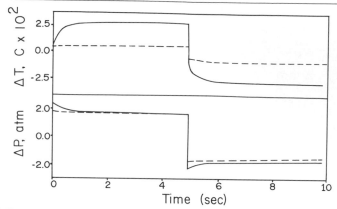

FIG. 6. Response characteristics of DMPC solution. Solid line, $T = 24.15°$; dashed line, $T = 22.85°$. Conditions: 0.2 M DMPC, 15% sucrose, 0.05 M 2-{[2-hydroxy-1,1-bis(hydroxymethyl)ethyl]amino}ethanesulfonic acid, 0.05 M KCl, 0.001 M Na azide, pH 7.4, average pressure, 4 atm. Redrawn from Johnson et al.[3]

response time for the instrument of 3 msec is obtained by a least-squares analysis of the data in Fig. 5, assuming a single exponential decay process. In a similar manner, the response time for the application of the volume change was determined to be 1 msec. Since the sample compartment is in direct contact with the cell body, it is expected that any induced temperature difference will be rapidly dissipated. This dissipation can be observed and its characteristic decay time measured when the apparatus is operated at 0.01 Hz. A least-squares analysis of such an experiment (data not shown) yields an apparent decay time constant for heat transfer from the sample compartment to the environment of approximately 30 sec. The combination of the thermal response time and the thermal dissipation time indicates that this instrument is functional in a time domain which is not readily accessible by other methods which are usually used in the study of membrane dynamics.

The use of the instrument for a more interesting experimental system is shown in Fig. 6, where we present the results obtained with a dispersion of multilamellar vesicles of DMPC; 0.2 M, in 0.05 M 2-{[2-hydroxy-1,1-bis(hydroxymethyl)ethyl]amino}ethanesulfonic acid (No. T-1375, Sigma Chemical Co.) buffer, 0.05 M KCl, 0.001 M Na azide, pH 7.4. The lipid was obtained as a chloroform suspension from Avanti Polar-Lipids, Inc. (Birmingham, AL). The lipid was dried under vacuum, and the buffer was then added to the dried lipid with mixing at 30°. The buffer also contained 15% sucrose to prevent the vesicles from sedimenting in the cell during the course of the experiment.[2] The solid line in Fig. 6 corresponds to a

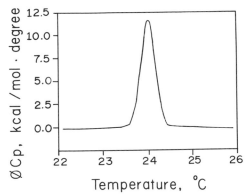

FIG. 7. Excess heat capacity of DMPC vesicles as measured by differential scanning calorimetry. Scan rate was approximately 1°/hr. Conditions as in Fig. 6.

pressure perturbation of 4 atm applied at 24.15°, thereby forcing the system partially through the gel to liquid-crystalline transition, while the dashed line represents a similar perturbation at 22.85°, outside of the phase-transition region. It should be noted that at both of these temperatures a static pressure of 4 atm was applied and that the corresponding melting temperature of the lipid is approximately 24.1° (see Fig. 7). In comparing the two curves, note both the magnitude of the increase in $\Delta T/\Delta P$ (1.5×10^{-2}°/atm at 24.15° vs 3.4×10^{-3}°/atm at 22.85°) and the slower rate of the temperature change in the transition region. The estimated relaxation time for the lipid is approximately 2 sec, which is in reasonable agreement with the results of Yager and Peticolas.[8] It should be noted that the relaxation process is only approximately a single exponential and that the estimated relaxation time is dependent on exactly where in the transition range the experiment is performed. The rapid 1/2 atm decrease in pressure after the application of the pressure pulse applied at 24.15° corresponds to a volume reduction in the lipids as they reorder to decrease the stress on the system induced by the pressure change. Also note that the time response of the thermistor (determined above to be approximately 3 msec) is sufficiently rapid when compared to the response time of the lipids in the phase-transition region, that it can be neglected. The corresponding melting temperature of the lipid is approximately 24.1° with a half-width of approximately 0.2° as measured by differential scanning calorimetry (Fig. 7). The data presented in Fig. 7 were at one atmosphere of pressure. The additional four atmospheres pressure will increase the melting temperature by approximately 0.1°.

[8] P. Yager and W. L. Peticolas, *Biochim. Biophys. Acta* **688,** 775 (1982).

Figure 8 shows data obtained by our instrument which is analogous to the scanning calorimetry data represented in Fig. 7. In Fig. 8, the abscissa is the temperature change induced by a change in pressure and the ordinate is an "apparent" temperature change which is corrected for the variations in ambient pressure. Analysis of the data in Figs. 7 and 8 yields melting temperatures which differ by 0.015° and provide reasonable agreement for the half-width of the transition.

These types of data may be analyzed by a number of methods. First, the data may be analyzed by least-squares fitting to multiple exponential decay rates. Second, the data may be analyzed by least-squares fitting to the numerical solutions of systems of differential equations which describe the instrument and possible molecular mechanisms of the sample being studied. The data can also be analyzed by the use of Fourier methods. It has previously been shown that the frequency dependence of the response for an instrument of this type can be related to the characteristic time constants of the chemical reactions as

$$\frac{\Delta(\text{temperature})}{\Delta(\text{pressure})} = \sum \varepsilon_i \tag{1}$$

$$\varepsilon_i = \frac{\text{amp}_i}{[1 + (\omega\tau_i)^2]^{1/2}} \tag{2}$$

$$\tan[\Psi] = \frac{\sum \varepsilon_i \sin(\Psi_i)}{\sum \varepsilon_i \cos(\Psi_i)} \tag{3}$$

$$\tan[\Psi_i] = \omega\tau_i \tag{4}$$

where amp_i and τ_i are the characteristic maximal amplitudes and the response times for each of i processes, ω is the driving frequency of the pressure change, and Ψ is the phase angle of the resultant temperature change.[4,5,9] We currently feel that the Fourier technique is the best method to analyze the experimental data in a model-independent fashion.

Figure 9 presents a Fourier analysis of the same solution of DMPC as in Fig. 6 at 24.05°. The upper panel is the amplitude of the driven oscillation (dT/dP; °/atm) and the lower panel is the phase angle of the driven oscillation. These data have not been corrected for the dynamic response of the instrument and consequently the amplitude decreases and the phase angle increases as the driving frequency approaches the ~3 msec

[9] G. Weber, *J. Phys. Chem.* **85**, 949 (1981).

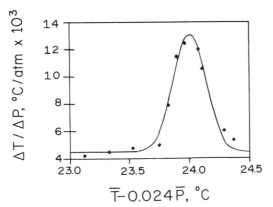

FIG. 8. Amplitude of the pressure-induced temperature change of DMPC vesicles as a function of the effective temperature of the solution. Volume perturbation was at 1.0 Hz. Conditions as in Fig. 6.

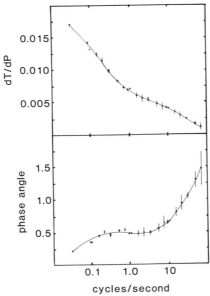

FIG. 9. Amplitude and phase shift of the pressure-induced temperature change of multi-lamellar DMPC vesicles as a function of driving frequency. Effective temperature was 24.05°; other conditions as in Fig. 6.

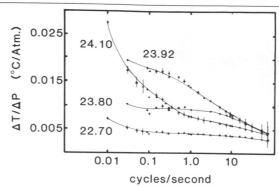

Fig. 10. Amplitude of the pressure-induced temperature change as a function of driving frequency at various temperatures. Conditions as in Fig. 6.

response time of the thermistor. The data also show a very slow component with a characteristic response time of ~8 sec.

The corrections for the dynamic response of the instrument are easy to perform in the frequency domain. The dynamic response correction is accomplished with a series of data from calibration experiments on water, encompassing the temperature and pressure ranges of the experiments involving phospholipid dispersions. For each phospholipid data point, 10 water data points of closest approach (in temperature and pressure) are used in a Cornish–Bowden method to approximate the instrument response characteristics at the temperature and pressure of that phospholipid data point. These can then be used to correct the observed temperature and pressure changes. The analogous corrections are extremely difficult to perform in the time domain.

Figures 10 and 11 show the same type of data as a function of mean temperature and corrected for the dynamic response of the instrument. It should be noted that the volume perturbation is only sufficient to force the lipids 20–30% of the way through the phase transition. Consequently the data at successive temperatures in the phase transition will be characteristic of the kinetics in that limited region of the phase transition. The data at 22.7° are out of the phase transition region and show a low amplitude, due to Joule–Thompson heating, and no phase shift. At the low temperature edge of the phase transition (23.8°) the kinetics are rapid. However, as we traverse through the phase transition the kinetics get slower and slower while the amplitude increases. An analysis of these data at 24.1° indicates that they are a composite of at least four characteristic relaxation times. These are shown in the table along with the corresponding relative amplitudes and confidence intervals. The values of the time constants and amplitudes are very temperature dependent.

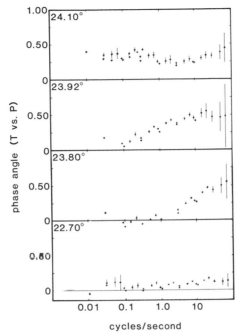

FIG. 11. Phase shifts of the pressure-induced temperature change as a function of driving frequency at various temperatures. Conditions as in Fig. 6.

We conclude that if the gel to liquid-crystalline transition of multilamellar DMPC vesicles were a simple two-state transition the results shown in Figs. 10 and 11 and in the table would be characterized by a single relaxation time. Thus, the transition must be at least a two-step,

APPARENT RELAXATION TIME CONSTANTS OF
DMPC AT 24.1°

τ_i (sec)		Amplitude (%)
11.1	$(6.7, 17.6)^a$	63.2 (45.1, 88.2)
0.69	(0.46, 0.91)	12.1 (10.2, 14.4)
0.22	(0.014, 0.031)	8.2 (5.1, 11.2)
< 0.003^b		16.3 (13.0, 19.6)

^{a} Confidence interval corresponding to ± SEM.
^{b} Faster than instrument response.

three-state process. This general type of mechanism is confirmed by the observation that the apparent melting temperature as measured in our kinetic calorimeter is frequency dependent (data not shown).

Future Developments

We are currently modifying the instrument by the addition of optical components, which will allow the measurement of turbidity, light scattering, and fluorescence. This will enable us to correlate changes in these optical properties with changes in thermodynamic properties. This will also make the study of protein–lipid interactions practical with this instrument.

To date, we have conducted experiments on single-component multilamellar and large unilamellar dispersions of phosphatidylcholines with identical saturated acyl chains. We plan to extend these studies to systems perturbed by the presence of local anesthetics or cholesterol. The effects of varying structure and charge of the lipid head group will also be investigated.

Comments

Light scattering, Raman spectroscopy, and fluorescence changes of marker dyes have been employed in studying phospholipid phase transitions. However, it is difficult to interpret changes in these observables in terms of thermodynamic state variables. The direct monitoring of the temperature and pressure obviates the problem of interpretation of changes in light scattering, fluorescence, etc., in terms of thermodynamic changes in the system. A side benefit of using the solution temperature as a monitor of the time course of the approach to equilibrium is an increase in the signal-to-noise ratio. The signal-to-noise ratio for this instrument is dependent on the number of cycles which are signal averaged. However, a typical value for the temperature uncertainty is $\pm 2 \times 10^{-5\circ}$, which corresponds to a signal-to-noise ratio of more than 2000 in Fig. 6. When similar experiments were performed using light scattering[10] or Raman spectroscopy,[8] the signal-to-noise ratio was notably worse than those obtained with this instrument.

The analysis of the experimental data from this instrument can be performed by several methods. It can measure the temperature and pressure as a function of time after a bidirectional volume change. Data of this type can be least-squares fit to multiple exponential functions to determine the characteristic chemical relaxation times directly. This type of data can

[10] R. M. Clegg, E. L. Elson, and B. W. Maxfield, *Biopolymers* **14**, 883 (1975).

also be used to test directly various molecular mechanisms by least-squares fitting the data to the integral of a system of differential equations which describes the particular molecular mechanism and the response of the instrument. These data also allow the calculation of the amplitude of the temperature and pressure change as a function of the frequency of the volume perturbations (e.g., ΔT vs ω and ΔP vs ω) by the use of Fourier transforms. In addition, these frequency dependencies can be determined directly by the instrument.[5] This flexibility will allow us to measure the frequency and temperature dependence of the enthalpy (ΔH) and heat capacity (ΔC_P) changes for phospholipid phase transitions. A knowledge of these and their frequency dependence should greatly add to our understanding of the molecular interactions involved in membrane structure and function.

Acknowledgments

The authors gratefully acknowledge the contributions of Dr. H. R. Halvorson. Dr. Halvorson supplied us with the detailed plans for his instrument and then spent several days helping us redesign the instrument for our application. This work was supported in part by National Institutes of Health Grants GM28928, AM30302, AM22125, GM07267, and GM26894 and National Science Foundation Grants PCM80-03645 and PCM83-00056.

Author Index

Numbers in parentheses are footnote reference numbers and indicate that an author's work is referred to although the name is not cited in the text.

S

T

Subject Index

L

D-Lactate dehydrogenase, cobalt as probe of metal sites in, magnetic circular dichroism, 280

Lactate dehydrogenase
conformations, reference CD spectra, 228–229, 231, 232
numerical values of CD, 250–251, 253–254
reference CD spectra, 226
secondary structure
comparison of CD estimates and X-ray results, 242–243
Raman amide I vs. X-ray data, 317
from X-ray analysis, 242–246

β-Lactoglobulin
FTIR spectroscopy, second derivative spectra, 302–304
microheterogeneity, 20, 24

β-Lactoglobulin A, 6
in aqueous solution, FTIR spectroscopy, 301–302
asymptotic sedimentation velocity pattern
calculation, 9–10
experimental determination, 13–14
sedimentation at low temperature, boundary formed, 7–8
self-association, 9–10
solvent denaturation, FTIR studies, 309–311

Leghemoglobin, carbonmonoxy, resonance Raman spectroscopy, 369–370

Ligand binding
to metalloporphyrins and hemoproteins, resonance Raman spectroscopy, 363–410
bonding geometry, 363–366
bonding interactions, 363–366
diatomic ligand bonding properties, 363–364
diatomic ligand electronic configurations, 363–364
endogenous ligands, 399–410
exogenous ligands, 366–399
σ and π bonding, 363–366
resonance Raman spectroscopy, 350–412

Ligand-mediated complex formation
between dissimilar proteins, sedimentation patterns, 30–32
involving two ligands, sedimentation patterns, 32–34

Ligand pump, 4

Light scattering. *See also* Small-angle light scattering
to measure polymer formation, 35

Linked reactions, Wyman theory of, 4

Lipid phase transitions
kinetics, 534–551
measurement using volume-perturbation kinetic calorimeter, 536–539
advantages, 539
analysis of data obtained, 550–551
results, 543–550
study, methods employed, 550

Lissamine rhodamine
properties, 480
structure, 479

Lithium, scattering properties, 83

Lithium dodecyl sulfate, mean neutron scattering length density, 104

Low-density lipoprotein, selectively deuterated, small-angle neutron scattering, 106, 107

$(Lys)_n$
circular dichroism spectra, 210–211, 222–225
as model for ideal infinite helix, 233

Lysozyme, 7
conformations, reference CD spectra, 228–229, 231, 232
hydrogen ion titration curve, 413
normalized Raman amide III intensities, 324
numerical values of CD, 250–251, 253–254
percentage helix and extended chain, by FTIR and X-ray, 307
reference CD spectra, 225, 226
secondary structure
comparison of CD estimates and X-ray results, 242–243, 247
Raman amide I vs. X-ray data, 317
Raman amide III vs. X-ray estimates, 325
from X-ray analysis, 242–246